Artificial Intelligence Empowered Smart Energy Systems

Artificial Intelligence Empowered Smart Energy Systems

Edited by

Qiang Yang
Zhejiang University
Hangzhou, China

Gang Huang
Zhejiang University
Hangzhou, China

IEEE Press Series on Power and Energy Systems
Ganesh Kumar Venayagamoorthy, Series Editor

IEEE*Press*

WILEY

Published by John Wiley & Sons, Inc., Hoboken, New Jersey.

Library of Congress Cataloging-in-Publication Data Applied for:

Hardback ISBN 9781394253616

Cover Design: Wiley
Cover Images: © katjen/Shutterstock, © Jackal Pan/Getty Images, © Andriy Onufriyenko/Getty Images

Set in 9.5/12.5pt STIXTwoText by Straive, Chennai, India

Contents

List of Contributors

Jinfan Chen
School of Electrical and Power Engineering
Hohai University
Nanjing, China

and

State Grid Zhejiang Electric Power Co. Ltd.
Luqiao Power Supply Company
Taizhou, China

Peiling Chen
School of Electrical Engineering
Southeast University
Nanjing, China

Ruilong Deng
State Key Laboratory of Industrial Control Technology and College of Control Science and Engineering
Zhejiang University
Hangzhou, China

Ruisheng Diao
ZJU-UIUC Institute
Zhejiang University
Haining, China

Lu Dong
College of Electrical Engineering and New Energy
China Three Gorges University
Xiling District, Yichang, Hubei Province, China

Jiajun Duan
State Grid Smart Grid Research Institute
AI&System Analytics
San Jose, USA

Xueqian Fu
College of Information and Electrical Engineering
China Agricultural University
Beijing, China

Congbin Guo
School of Control and Computer Engineering
North China Electric Power University
Baoding, Hebei, China

Zhengxun Guo
Foshan Graduate School of Innovation
Northeastern University
Guangdong, Foshan, China

and

College of Information Science and Engineering
Northeastern University
Liaoning, Shenyang, China

Tu Lan
ZJU-UIUC Institute
Zhejiang University
Haining, China

Quanlin Leng
School of Electrical and Electronic Information
Xihua University
Chengdu, Sichuan, China

Haifeng Li
Dispatch Center
State Grid Jiangsu Electric Power Company
Nanjing, China

Jincheng Li
Foshan Graduate School of Innovation
Northeastern University
Guangdong, Foshan, China

and

College of Information Science and Engineering
Northeastern University
Liaoning, Shenyang, China

Yulin Li
Faculty of Electric Power Engineering
Kunming University of Science and Technology
Kunming, China

Xinyi Liang
School of Electrical Engineering
Southeast University
College of Automation
Nanjing University of Posts and Telecommunications
Nanjing, China

Zhiwei Lin
School of Electrical and Power Engineering
Hohai University
Nanjing, China

and

School of Engineering
Cardiff University
Cardiff, UK

Chengjun Liu
School of Electrical and Power Engineering
Hohai University
School of Electrical Engineering
Southeast University
Nanjing, China

Liyuan Liu
School of Electrical and Electronic Information
Xihua University
Chengdu, Sichuan, China

Mengxiang Liu
Department of Automatic Control and Systems Engineering
The University of Sheffield
Sheffield, UK

Xuan Liu
School of Control and Computer Engineering
North China Electric Power University
Baoding, Hebei, China

Na Lu
College of Information and Electrical Engineering
China Agricultural University
Beijing, China

Qiaoyu Ma
College of Information and Electrical Engineering
China Agricultural University
Beijing, China

Sha Peng
State Key Laboratory of Industrial Control Technology and College of Control Science and Engineering
Zhejiang University
Hangzhou, China

Jianyu Ren
College of Electrical Engineering and New Energy
China Three Gorges University
Xiling District, Yichang, Hubei Province, China

Yingtao Ru
School of Electrical and Power Engineering
Hohai University
Nanjing, China

and

State Grid Jiangsu Electric Power Co. Ltd.
Liyang Power Supply Company
Changzhou, China

Di Shi
State Grid Smart Grid Research Institute
AI&System Analytics
San Jose, USA

Fangyuan Sun
ZJU-UIUC Institute
Zhejiang University
Haining, China

Jiarong Wang
Faculty of Electric Power Engineering
Kunming University of Science and Technology
Kunming, China

Qianming Wang
School of Control and Computer Engineering
North China Electric Power University
Baoding, Hebei, China

Siqi Wang
State Grid Smart Grid Research Institute
AI&System Analytics
San Jose, USA

Tao Wang
School of Electrical and Electronic Information
Xihua University
Key Laboratory of Fluid and Power Machinery
Xihua University
Chengdu, Sichuan, China

Yishen Wang
State Grid Smart Grid Research Institute
AI&System Analytics
San Jose, USA

Yixiao Wang
College of Electrical Engineering and New Energy
China Three Gorges University
Xiling District, Yichang, Hubei Province, China

Zhiwei Wang
Dispatch Center, State Grid Jiangsu Electric Power Company
Nanjing, China

Hanyan Wu
School of Electrical and Electronic Information
Xihua University
Chengdu, Sichuan, China

Yingjun Wu
School of Electrical and Power Engineering
Hohai University
Nanjing, China

Lei Xi
College of Electrical Engineering and New Energy
China Three Gorges University
Xiling District, Yichang, Hubei Province, China

Chunlei Xu
Dispatch Center
State Grid Jiangsu Electric Power Company
Nanjing, China

Hao Xu
Nanjing Zhongxing New Software Co. Ltd.
College of Automation
Nanjing University of Posts and Telecommunications
Nanjing, China

Bo Yang
Faculty of Electric Power Engineering
Kunming University of Science and Technology
Kunming, China

Feifei Yang
College of Information and Electrical Engineering
China Agricultural University
Beijing, China

Yujian Ye
School of Electrical Engineering
Southeast University
Nanjing, China

Ruixuan Ying
School of Electrical and Electronic Information
Xihua University
Chengdu, Sichuan, China

and

Institute of Multidisciplinary Research for Advanced Materials
Tohoku University
Miyagi, Tohoku Region, Japan

Yongjie Zhai
School of Control and Computer Engineering
North China Electric Power University
Baoding, Hebei, China

Bei Zhang
State Grid Smart Grid Research Institute
AI&System Analytics
San Jose, USA

Chunyu Zhang
College of Information and Electrical Engineering
China Agricultural University
Beijing, China

Xiaoshun Zhang
Foshan Graduate School of Innovation
Northeastern University
Guangdong, Foshan, China

and

College of Information Science and Engineering
Northeastern University
Liaoning, Shenyang, China

Zhenyong Zhang
State Key Laboratory of Public Big Data and College of Computer Science and Technology
Guizhou University
Guiyang, China

Chunyu Zhou
School of Electrical and Electronic Information
Xihua University
Chengdu, Sichuan, China

About the Editors

Prof. Qiang Yang, PhD Qiang Yang (M'03-SM'18) received a PhD degree in Electronic Engineering and Computer Science from Queen Mary, University of London, London, UK, in 2007, and worked in the Department of Electrical and Electronic Engineering at Imperial College London, UK, from 2007 to 2010. He visited the University of British Columbia and the University of Victoria, Canada, as a visiting scholar in 2015 and 2016. He is currently a Full Professor at the College of Electrical Engineering, Zhejiang University, China. His research interests over the years include smart energy systems, large-scale complex network modeling, control and optimization, learning-based optimization and control. He is a member of the National Academy of Artificial Intelligence (NAAI), a Fellow of the British Computer Society (FBCS), the Institution of Engineering and Technology (FIET), and the International Association of Advanced Materials (FIAAM). He is also a Senior Member of IEEE and a Distinguished Member of China Computer Federation (CCF).

In recent years, he has participated in several key research projects supported by a variety of funding bodies, including the National Natural Science Foundation of China, National Key Research and Development Program of China, Chinese National Programs for High Technology Research and Development (863 Program), and National Program on Key Basic Research Project (973 Program). He has published more than 350 technical papers, including more than 200 full journal papers (accumulated impact factor is over 800), and more than 100 international conference papers. He has also applied for over 100 national patents and co-authored 2 books and 10 book chapters. He is also the reviewer of several IEEE/IET journals, e.g., *IEEE Transactions on Power Systems, IEEE Transactions on Smart Grid, IEEE Transactions on Sustainable Energy, IEEE Transactions on Power Delivery, IEEE Transactions on Systems, Man, and Cybernetics, Elsevier Applied Energy, Energy, IET Communications, IET Renewable Power Generation,* and *IET Generation, Transmission & Distribution*. The research interests of the applicant's over the years include smart energy networks, data-driven fault detection and analysis of renewable energy systems, and electric cyber-physical system modeling, control, and optimization.

Professor Yang now leads The Smart Energy Systems Laboratory (SES Lab) of the College of Electrical Engineering at Zhejiang University aiming to deliver a world-leading multidisciplinary research platform to investigate the challenging issues for ensuring the reliable, efficient, and safe operation of multiple energy infrastructures. The research interests of the research team over the years include intelligent systems, smart energy networks, data analytics, cyber-physical system modeling and cyberattack defense, and machine-learning-based control and optimization.

Dr. Gang Huang, PhD Gang Huang (M'18-SM'23) received his PhD degree in Electrical Engineering from Zhejiang University, Hangzhou, Zhejiang, China, in 2018. During 2015–2016, he was with Argonne National Laboratory, Argonne, IL, USA. Presently, he is an Assistant Professor in the College of Electrical Engineering, Zhejiang University, Hangzhou, Zhejiang, China. Prior to this, he served as a senior researcher at Zhejiang Lab, Hangzhou, Zhejiang, China. His primary research interests revolve around artificial intelligence for power and energy systems. He is a Senior Member of IEEE and a Senior Member of China Computer Federation (CCF).

Dr. Huang leads a research group specializing in Power Artificial Intelligence. His research has been supported by the National Natural Science Foundation of China, the Smart Grid-National Science and Technology Major Project, the Ministry of Industry and Information Technology, among others. He has published over 60 papers in leading journals and conferences, holds more than 20 invention patents and three software copyrights, and actively serves the academic community. He is a founding editorial board member of Digital Twins and Applications, an Associate Editor of Protection and Control of Modern Power Systems, a youth editorial board member of Power System Protection and Control and China Electric Power, and a guest editor for special issues of Electric Power Automation Equipment and Electric Power Construction.

Foreword

Artificial intelligence (AI) is revolutionizing industries worldwide, with its transformative impact increasingly evident in the field of smart energy systems. This book, *Artificial Intelligence Empowered Smart Energy Systems*, brings together leading-edge research and innovations at the intersection of AI and energy systems. It provides a comprehensive perspective on the role of AI in enhancing the resilience, security, efficiency, and sustainability of modern energy systems. The authors have crafted an insightful exploration of AI's applications, from cybersecurity and fault diagnosis to renewable energy optimization and virtual power plant coordination. This volume is an indispensable resource for researchers, practitioners, and policymakers striving to harness AI for a sustainable energy future.

Preface

The rapid transformation of energy systems into smarter, more interconnected infrastructures has created unprecedented opportunities and challenges. The integration of artificial intelligence (AI) into smart energy systems has emerged as a pivotal solution for addressing the complexity and dynamic nature of modern power networks. This book, *Artificial Intelligence Empowered Smart Energy Systems*, serves as a compendium of state-of-the-art research and practical solutions across diverse domains of AI and energy systems.

This book delves into the convergence of two pivotal domains: AI and smart energy systems. This comprehensive work, authored by leading experts in their respective fields, not only introduces readers to the fundamental concepts of both AI and smart energy systems but also navigates them through the intricate landscape of their intersection. The book commences with a thorough exploration of AI, elucidating its core principles and techniques without delving into jargon. It then seamlessly transitions into the world of smart energy systems, elucidating their evolution and significance. The heart of the book resides in its exploration of AI's transformative role in reshaping the future of energy landscapes, providing practical insights, and addressing complex questions. Key topics covered include renewable energy forecasting, load monitoring and management, fault diagnosis, resilience-oriented optimization, efficiency-driven control, sustainability-centric approaches, cybersecurity and privacy challenges, and the emergence of virtual power plants and innovative energy markets. Through illustrative examples and real-world applications, this book answers critical questions about the integration of AI in energy systems, ultimately guiding readers toward harnessing AI's potential to shape a sustainable and efficient energy future.

This book represents a cutting-edge fusion of technology and sustainability. Evolving from traditional energy grids, these systems leverage AI to optimize energy generation, distribution, and management. Over the next five years, we anticipate even greater integration of AI into energy infrastructure, enabling more accurate renewable energy forecasting, improved grid resilience, and informed energy consumption choices for individuals. This transformative technology empowers people, industries, and governments to enhance energy efficiency, reduce costs, and meet sustainability goals. It closely aligns with renewable energy, IoT, and climate change mitigation efforts, promising a more efficient and sustainable energy future.

We hope this book fosters collaboration and sparks innovative ideas among researchers and practitioners, ultimately advancing the development of sustainable and intelligent energy systems.

Hangzhou, China
November 30, 2025

Qiang Yang and Gang Huang

Acknowledgments

The completion of this book would not have been possible without the invaluable contributions and steadfast support of many individuals and organizations. We extend our deepest appreciation to the authors whose expertise and dedication were instrumental in shaping the content of this book. Their patience and effort during the revision process were truly commendable.

We also wish to express our gratitude to the reviewers whose insightful feedback and constructive comments greatly enhanced the quality of the chapters. Several authors also took on the dual role of referees, and we deeply value their additional contributions to this endeavor.

This work has been supported by numerous funding agencies and institutions, and we are immensely grateful for their support. We would like to acknowledge the following grants that facilitated the research presented in this book: the National Key Research and Development Program of China under Grant 2023YFB3107603; the National Natural Science Foundation of China under Grants 52177119, 52477087, 62263014, U22B6007, 62293503, 62293500, 62073285, 62373151, 52277108, 52477104, 52007193, and 61703345; Zhejiang Provincial Natural Science Foundation of China under Grants LZ25E070002, LBMHD25E070001, and LR23F030001; the Special Support Plan for Zhejiang Province High-level Talents under Grant 2022R52012; the State Key Laboratory of Alternate Electrical Power System with Renewable Energy Sources under Grant LAPS25015; the Fundamental Research Funds for the Central Universities under Grant 2023JC006; the Youth Talent Program of Sci-Tech Think Tank under Grant XMSB20240710087; and the Guangdong Basic and Applied Basic Research Foundation under Grant 2024A1515030012.

We would like to give special thanks to our families for their unwavering understanding and support throughout this process. Their encouragement has been a source of strength and motivation for us.

Finally, we are indebted to the editorial and production teams, whose professionalism and dedication ensured the successful realization of this book. Their efforts, from concept to publication, were critical to bringing this project to fruition. We hope this book serves as a valuable reference for advancing research and practice in smart energy systems.

Qiang Yang and Gang Huang

1

Machine Learning-Based Applications for Cyberattack and Defense in Smart Energy Systems

Sha Peng[1], Mengxiang Liu[2], Zhenyong Zhang[3], and Ruilong Deng[1]

[1]*State Key Laboratory of Industrial Control Technology and College of Control Science and Engineering, Zhejiang University, Hangzhou, China*
[2]*Department of Automatic Control and Systems Engineering, The University of Sheffield, Sheffield, UK*
[3]*State Key Laboratory of Public Big Data and College of Computer Science and Technology, Guizhou University, Guiyang, China*

1.1 Introduction to Machine Learning

Machine learning (ML) approaches are one of the most extensively utilized artificial intelligence (AI) technologies. They have found increasing applications across diverse domains, including medical diagnosis, image processing, and traffic prediction. With the digital upgrade of energy systems, large amounts of data are available through information and communications technology. Massive amounts of data facilitate the deep applications of ML in energy systems. In this section, an overview of ML approaches is first presented, and then the advantages of ML approaches for energy system cybersecurity are discussed in detail from the aspects of attack, protection, detection, and mitigation.

1.1.1 Overview of Machine Learning Approaches

According to different learning patterns, ML approaches can be divided into four categories, including unsupervised learning, supervised learning, semi-supervised learning, and reinforcement learning. In the following, an essential overview focusing on representative algorithms within each category is given.

1.1.1.1 Unsupervised Learning

Unsupervised learning methods train models using unlabeled datasets. They can automatically discover the properties of data without labels. The seven types of unsupervised learning that have been widely utilized are discussed as follows.

Artificial Intelligence Empowered Smart Energy Systems, First Edition. Edited by Qiang Yang and Gang Huang.

Dimension Reduction and Metric Learning: This technique is adopted to learn a representation function that maps objects into a new feature space while preserving important information. The most frequently used Dimension Reduction and Metric Learning methods are Independent Component Analysis (ICA) and Principal Component Analysis (PCA). Both these techniques rely on matrix decomposition. However, they are computationally inefficient and sensitive to noise and missing data.

Clustering: Clustering aims to discover natural groups in the feature space of the inputs so that objects in the same group have more similar features. Clustering approaches can be classified according to their cluster models. Density-based models perform better for mining non-convex or shape-based clusters by finding areas where data points are concentrated. The task of a centroid-based model is to assign data points to the closest centroid. In distribution-based models, clusters are determined by statistics. Graph-based models transform the inputs into a graph representation. A subset of the graph can be viewed as the prototype of the clusters, in which there is a path from any node to any other node in the subset. Connectivity-based models assume that data points are more relevant to nearby points than the points farther away.

Neural Network (NN): NNs are interconnected networks composed of a collection of adaptive nodes. The topology of NNs simulates the interaction of human nervous systems with real-world objects. Some NNs can be trained with unsupervised learning, such as Autoencoder (AE), Generative Adversarial Network (GAN), and Deep Belief Network (DBN).

Support Vector Machine (SVM): Generally, SVM computes a maximum margin hyperplane that separates the input into two classes. SVM is usually trained in a supervised way. Different from other supervised SVMs, one-class SVM is an unsupervised learning approach that is trained only on normal data.

Rule Induction (RI): RI proceeds by gradually generalizing "if–else–then-" type rules from the input. The training of RI can be assorted into supervised learning and unsupervised learning. Association Rule Mining is a popular unsupervised RI algorithm.

Ensemble Learning: Ensemble learning first develops multiple individual learners and then combines their decisions with an appropriately designed strategy. If every individual learner in the ensemble is trained without labels, this is an unsupervised ensemble. For example, Isolation Forest (iForest) is an unsupervised ensemble learning method based on binary search trees.

Transfer Learning: Transfer learning adopts prior knowledge acquired for one task to tackle those different but similar ones. For example, knowledge learned from identifying bikes could be applied to identify motorcycles. In general, the data for the original model are called source data, while the data for the target task are called target data. The transfer learning methods that do not require labeled target data can be classified as unsupervised learning, such as self-taught clustering and Domain-Adversarial training of NN (DANN).

1.1.1.2 Supervised Learning

Supervised learning methods help learn models using labeled datasets. The nine common types of supervised learning are illustrated as follows.

Regression: Regression is a basic and commonly used ML approach. It produces a probabilistic interpretation value by analyzing relationships among the input variables. The simplest regression model is linear regression.

SVM: SVM deals with problems by adopting a kernel function, which converts an input data space into the desired form. Some notable kernel types used by SVM are Linear Kernel, Polynomial Kernel, and Radial Basis Function (RBF) Kernel.

Decision Tree (DT): DT is a tree-like structure in which each internal node represents a decision on an attribute, while each leaf node corresponds to a classification result. The growth process of DT defines a group of rules. Iterative Dichotomiser 3 (ID3) and Classification and Regression Tree (CART) are commonly used DT algorithms. They separate data with more information gain but use different splitting criteria.

Bayesian: Some ML approaches, such as Naive Bayes and Bayesian networks, leverage the idea of Bayes' theorem. Naive Bayes assumes that each attribute influences the classification results independently. Bayesian network provides a qualitative and quantitative relationship between attributes through a directed acyclic graph and describes the joint probability distribution of attributes through a conditional probability table.

NN: Many NN algorithms employ supervised learning, such as Multilayer Perceptron (MLP), Convolutional Neural Network (CNN), Graph Neural Network (GNN), Recurrent Neural Network (RNN), Long Short-Term Memory (LSTM), and Gate Recurrent Unit (GRU).

RI: Some RI algorithms generate rules through supervised learning. A commonly used supervised RI approach is Non-Nested Generalized Exemplar (NNGE). NNGE classifies a new example based on the distance between the new example and a set of previous exemplars.

Dimension Reduction and Metric Learning: The k-nearest neighbor (kNN) is a well-known supervised Dimension Reduction and Metric Learning algorithm. It assumes that similar data points are close to each other.

Ensemble Learning: An ensemble is based on supervised learning when every individual learner is trained on labeled datasets. AdaBoost, Random Forest (RF), and Gradient Boosting Machine (GBM) are popular supervised ensemble learning approaches.

Transfer Learning: The transfer learning methods that require labeled target data can be classified as supervised learning, such as self-taught learning and multitask learning.

1.1.1.3 Semi-Supervised Learning

Semi-supervised learning is intermediate between supervised learning and unsupervised learning. This approach learns from a small amount of labeled data and a large amount of unlabeled data. So, it would be more applicable to scenarios where it is hard to obtain a reliable supervised model due to the scarcity of labeled data. These scenarios are common in real world as labeling data is expensive and difficult while acquiring sufficient unlabeled data is inexpensive and easier. Under certain assumptions about the underlying distribution of the data, semi-supervised learning can apply unlabeled data to construct a more efficient and robust model. Semi-Supervised SVM (S3VM) is the extension of SVM in semi-supervised learning. The assumption of S3VM is low-density separation. Specifically, S3VM tries to find a hyperplane that separates labeled samples of two classes and cuts through the low-density region of all samples.

1.1.1.4 Reinforcement Learning

Reinforcement learning methods help learn an optimal policy through interacting with the environment. The optimal policy, a mapping from states to optimal actions, aims at maximizing the long-term cumulative reward. Reinforcement learning has been famed as an effective tool for sequential decision-making problems. The learning process of reinforcement learning methods could be explained by concepts of agent, state, action, reward, and environment. Specifically, the agent evaluates the current state, takes an action, and then obtains a reward and the next state from the environment. According to received rewards, the agent learns to take beneficial actions more frequently and gradually filters less beneficial actions. Generally, reinforcement learning algorithms could be classified as model-based methods or model-free methods. Model-based methods use a predictive model of the environment and optimize both the model and the policy. Iterative Linear Quadratic Regulator (ILQR) and Probabilistic Inference for Learning COntrol (PILCO) are popular model-based reinforcement learning methods. On the contrary, in model-free reinforcement learning, the modeling step of environment is skipped, and the optimal action policy is learned directly through trial and error. Q-learning and State-Action-Reward-State-Action (SARSA) are two commonly used model-free reinforcement learning approaches. However, traditional reinforcement learning approaches are confronted with the problem of curse of dimensionality because these approaches use a simple lookup table to represent value function and are applicable only when the state space and action space are small. To overcome this limitation, more advanced deep reinforcement learning algorithms emerge. Deep reinforcement learning uses Deep Neural Network (DNN) to approximate value functions.

1.1.2 Advantages of Machine Learning

ML approaches are increasingly used in medical diagnosis, image processing, stock market trading, and traffic prediction. In smart energy systems, a large amount of data available through the deep integration of control, communication, and computer technologies facilitates the application of ML approaches. Simultaneously, high-performance computing is advancing rapidly, surpassing Moore's Law and providing the necessary hardware foundation for applying ML. The two aspects create favorable conditions for applying ML to smart energy systems. Compared to traditional model-based methods, ML methods offer massive distinct advantages as outlined below:

Modeling: The model-based methods rely heavily on the physical model's accuracy. However, due to the widespread deployment of power electronic devices and intricate system structures, constructing precise energy system models presents significant challenges. Furthermore, in the smart energy system cybersecurity field, various types of cyberattacks also add considerable complexity and unpredictable patterns to the model process. As a result, the application of model-based approaches to energy system cybersecurity is hindered by the intricacies of modeling. Conversely, ML approaches explore and utilize the characteristics of data without relying on an explicit physical model, thereby eliminating the potential adverse impact of an inaccurate physical model.

Computational Complexity: In many instances, the model-based approaches employ a set of nonlinear differential algebraic equations to depict the intricacies of a complex energy system. These equations usually lack closed-form solutions and are typically tackled using iterative numerical techniques. Nonetheless, these iterative methods often entail substantial computational overhead and may not meet the requirement of real-time operation security. Conversely, most ML approaches adopt the idea of offline learning and online decision-making, thus ensuring the real-time functionality of energy systems.

Generalization: The smart energy system is inherently dynamical, with parameters that may undergo frequent changes. The model-based approaches rely on specific parameters for modeling and often exhibit limited generalization capabilities. Similarly, the continual evolution of new types of cyberattacks necessitates approaches with strong generalization abilities in the cybersecurity field. In contrast, data-driven ML approaches typically demonstrate robust generalization capabilities. With the development of ML, transfer learning has

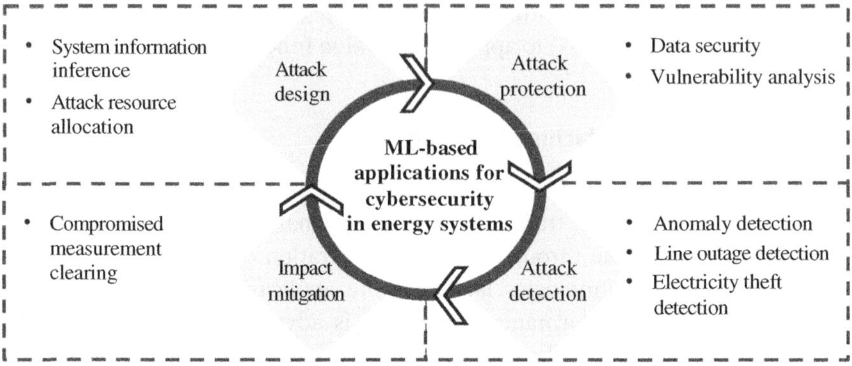

Figure 1.1 The applications of ML in smart energy systems.

emerged as a powerful tool for improving the performance of target learners on target domains by transferring the knowledge acquired from different but related source domains.

Up to now, ML methods have been widely used for cyberattacks and defense in smart energy systems [1, 2]. As shown in Figure 1.1, this chapter summarizes the applications of ML to energy system cybersecurity from the perspectives of attack design and defense countermeasure, respectively. The perspective of attack mainly includes two aspects: system topology inference and attack resource allocation. The perspectives of defense mainly include three aspects: security protection, attack detection, and impact mitigation.

1.2 Machine Learning in Attack Design

A variety of attacks can be launched by cyber intruders. These attacks could be classified as attacks without prior knowledge, including denial of service (DoS), replay attacks, and man-in-the-middle attacks, or attacks with prior knowledge, including False Data Injection Attacks (FDIAs) and line switch attacks. Generally, the latter could severely impact energy systems because it integrates crucial system information for a deliberate attack [3]. Existing ML-based studies on the attacks against energy systems give more attention to attacks with prior knowledge. The attack process can be segmented into three stages. Initially, the attacker formulates an attack model based on the attack objective. Subsequently, the attacker gathers the requisite system information to support the attack model. Given the typically constrained resources available to attackers, the third step involves devising an optimal resource allocation strategy to maximize the attack's impact on the system. Following these steps, the attacker could launch an attack. However, traditional model-based approaches are primarily applicable to the initial step. With excellent data mining and decision-making capabilities, ML approaches make up for the research gap of the latter two important steps and help the attacker infer system information and design a resource allocation policy. The comparison of different ML methods in attack design is shown in Table 1.1.

1.2.1 System Information Inference

The topic of FDIA has attracted considerable attention in the field of energy system cybersecurity in the past few years, as FDIA poses an essential challenge to the reliable operation of energy systems. Existing studies about system information inference have also placed particular emphasis on FDIAs. The attacker tries to launch FDIA with zero knowledge of energy systems, a scenario termed stealthy or blind FDIA. It proves more practical than the works reliant on partial or complete knowledge of the system topology and transmission line admittance because the system information is usually not publicly accessible. These works only require system measurements transmitted over communication networks to infer the system information essential for conducting FDIA.

Unsupervised Learning: The task of the system information inference is to find the system information mixed in the observed measurements, and it could be formulated as the problem of blind source separation. Dimension Reduction and Metric Learning approaches, such as ICA and PCA, are powerful tools for this problem and have been applied to energy system information inference. For instance, [4] constructed a function of loads to represent highly correlated state variables. This work demonstrated that linear ICA, adept at handling non-Gaussian and statistically independent sources, could estimate the Jacobian matrix by multiplying the eigenvectors of the state vector's covariance matrix. In another study [5], PCA was employed to project line measurements onto the first n principal components, subsequently utilizing the resulting PCA matrix as the new topology matrix. However, these studies focused on Direct Current (DC) blind FDIA, and the computational complexity of ICA and PCA remains high. In practice, energy system models operate on Alternating Current (AC) and exhibit real-time changes in system topology and other system parameters. Hence, there's urgent need to explore more advanced ML approaches for the design of complex AC-blinded FDIAs in real-time scenarios.

Moreover, the effectiveness of these works relies on a fixed topology, as historical measurements used for inference correspond to specific topologies. Consequently, they may fail in the presence of Moving Target Defenses (MTDs). Higgins et al. [6] introduced a novel counter-MTD technique based on clustering. Clustering approaches discover natural groups in data. So,

Table 1.1 Comparison of ML methods for attack design.

Issue	Approach		Advantage	Research
System information inference	Unsupervised learning	Dimension reduction and metric learning	It learns latent features, which could be applied to find the system information mixed in the observed measurements	[4, 5]
		Clustering	It discovers data clusters aiding in identifying measurements for diverse topologies and enabling the construction of stronger attacks	[6]
Attack resource allocation	Unsupervised learning	Clustering	It finds data clusters to select attack targets by categorizing system components based on their features	[7]
	Reinforcement learning	Model-free	1. Its sequential decision-making capability can be used to design attack sequences 2. It performs tasks with minimal prior knowledge about the system	[8, 9]
	Supervised learning	Regression	1. Attack impact assessment falls under regression problems, where regression methods demonstrate satisfactory performance 2. Regression methods are easily understood and implemented by system operators	[10]

they could be applied to identify the measurements associated with different topologies. Specifically, this work first identifies measurements relevant to the current topology using a density-based clustering algorithm and then only uses corresponding data to develop the blind attack model.

1.2.2 Attack Resource Allocation

In practice, attackers' resources are often limited. Hence, an effective resource allocation strategy is essential for attack design. Specifically, to maximize the impact of the attack with limited resources, the attacker needs to select appropriate attack targets and accurately assess the impact of the attack. When multiple attacks are executed sequentially, known as sequential attacks, they can yield comparable damages to concurrent attacks or even exacerbate the consequences. For example, sequential line outages are more severe than simultaneous outages. Moreover, sequential attacks can be executed with fewer concurrent resources. Therefore, sequential attacks have lately received significant attention, particularly in designing effective attack sequences.

In recent years, Mixed Integer Linear Programming (MILP) has emerged as the predominant model-based approach for target selection [11], while the multistage hybrid game method has gained popularity for designing attack sequences [12]. Both approaches rely on accurately assessing attack impacts, typically obtained through model-based methods such as power flow calculations. These traditional methods are labor-intensive in modeling and exhibit high time complexity, hindering their real-time applicability. ML approaches offer a promising solution to these limitations and have been applied to address three critical aspects of resource allocation: unsupervised learning for target selection, reinforcement learning for designing attack sequences, and supervised learning primarily for assessing attack impacts.

Unsupervised Learning: Attack targets often exhibit similarities, which can be leveraged through clustering techniques. Clustering categorizes unlabeled data into distinct subsets based on specific features, making it a valuable method for target selection. Paul et al. [7] utilized the k-means to cluster the coordinates of bus locations into groups characterized by proximity. Subsequently, targets with elevated power flow levels were selected from these clustered groups of buses.

Reinforcement Learning: Reinforcement learning has emerged as a promising solution for addressing the problem of designing sequential attacks due to its capacity to tackle sequential decision-making tasks. The state transitions of an energy system can be modeled as a Markov Chain under the condition of having all measurements of power devices. However, attackers typically have access only to partial information. Hence, [8] formulated the optimal strategy of sequential FDIA as a partially observable Markov Decision Process (MDP) and then employed Q-learning with nearest sequence memory to obtain its solution. The nearest sequence memory method helps attackers learn the solution adapted to environmental changes. However, this work overlooks the defenders' protection strategies. Ni and Paul [9] presented an attacker–defender zero-sum game model in an energy system to overcome this limitation and then used Q-learning to solve it. The proposed solution identifies the attacker's optimal attack sequences (critical transmission lines) with the consideration of the defender's actions.

Supervised Learning: Supervised learning demonstrates remarkable effectiveness in uncovering the correlation between inputs and their corresponding labels. Consequently, it can be harnessed to understand the connection between an attack and its resultant impact, aiding attackers with limited information in evaluating the effectiveness of the launched malicious actions. Tan et al. [10] used a linear regression model to assess the effectiveness of an attack. The metric is the remaining time for the system to counteract after being attacked. Based on the predicted impact, an optimal attack consisting of a series of FDIA is constructed.

1.3 Machine Learning in Attack Protection

In order to ensure the cybersecurity of smart energy systems, the defender needs to protect system components from being compromised. The protective measures encompass various strategies such as physical isolation, firewall implementation, data encryption, and redundancy protocols. Up to now, MLs have been widely utilized to deploy defense resources to ensure the energy system's strong anti-intrusion capability. Furthermore, similar to the attack design with limited resources, defense resources are also subject to restricted resources in practical scenarios. Therefore, before deploying protective measures, it is necessary to analyze the energy system's most critical and vulnerable components. ML approaches provide powerful tools for this issue, referred to as vulnerability analysis. The comparison of ML-based protection mechanisms is shown in Table 1.2.

1.3.1 Data Security

Energy system data contains valuable information, such as critical generation and topology parameters as well as demand-side energy usage data. During the wide-area transmission of data, its availability, integrity, and confidentiality can be exposed to serious security threats. ML approaches pave a new way for data transformation. On the one hand, the learning process of ML approaches is more

Table 1.2 Comparison of ML methods for attack protection.

Issue	Approach		Advantage	Research
Data security	Unsupervised learning	NN	It provides a new representation of the original data to protect data from certain cyberattacks, such as sniffing attacks	[13]
Vulnerability assessment	Unsupervised learning	Clustering	It uncovers natural data clusters, facilitating the identification of components requiring protection by categorizing them into disjoint subsets based on their features	[14]
	Reinforcement learning	Model-free	1. Its sequential decision-making capability can be used to pinpoint the most vulnerable components 2. It handles tasks with minimal prior knowledge about the system	[15]

time-efficient than other data protection methods, such as encryption-based and blockchain-based approaches. On the other hand, ML approaches could capture the internal correlation of the data. So, data transformation using ML approaches not only protects the data but also preserves the inherent correlations and utility.

Unsupervised Learning: Some unsupervised learning methods that provide a latent representation of the original data can help ensure the confidentiality of data transferred over energy system communication networks. Keshk et al. [13] proposed a privacy-preserving framework employing a Variational Autoencoder (VAE) to prevent data sniffing.

1.3.2 Vulnerability Analysis

The defender also faces constraints on allocating the limited defensive resources. To address this issue, a critical part of energy system protection is to identify the most vulnerable system components so that the limited resources can be reasonably allocated. This task is equivalent to the issue of designing the attack resource allocation strategy. Specifically, both endeavors tackle the same problem but from the perspectives of the defender and the attacker, respectively. The model-based methods are commonly utilized, and their associated limitations are clearly identified in Section 1.2. With the advantages of handling modeling and computational complexity, ML approaches overcome these limitations and provide insights into vulnerability analysis. The ML methods used for attack resource allocation and vulnerability analysis are also nearly identical as in Section 1.2. Specifically, unsupervised learning is utilized for component classification, while reinforcement learning addresses the sequential attack problem. Notably, there are no applications of supervised learning. One possible reason is that system operators possessing sufficient system knowledge do not require supervised learning assistance to evaluate the impact of attacks.

Unsupervised Learning: Clustering-based approaches excel at identifying similar groups within datasets, making them suitable for partitioning energy system components into distinct subsets based on their internal correlations. The different groups' vulnerability would guide the deployment of defending resources. Anwar et al. [14] leveraged k-means to identify the level of vulnerabilities of each system component in terms of different degrees of FDIA.

Reinforcement Learning: Reinforcement learning is an effective tool for sequential decision-making problems, and therefore it could be used to analyze the sequential attack problem from the defender's perspective. Yan et al. [15] proposed a reinforcement learning-based scheme to identify critical component sequences that led to critical blackouts.

1.4 Machine Learning in Attack Detection

Detection schemes are devised to identify malicious actions, aiming to minimize their impact on energy systems, and trying to prevent attacks at the inception. In recent years, considerable attention has been paid to the application of ML to attack detection. According to different attack impacts, these studies can be divided into three main categories: anomaly detection, line outage detection, and electricity theft detection. Compared with traditional model-based detection methods, ML approaches offer several advantages. First, as the power system increasingly incorporates renewables and power electronics while facing rapidly evolving cyber threats, the complexity of modeling impedes model-based methods. Data-driven ML approaches bypass the tedious building process of the physical system model and thus eliminate the potential impact of an inaccurate model on attack detection. Second, ensuring the secure and stable operation of energy systems necessitates real-time detection, a requirement beyond the capabilities of model-based methods due to their high computational complexity. ML approaches are able to implement real-time detection through offline learning and online decision-making manner. Third, for anomaly detection, maintaining the performance of detection algorithms on incoming data with novel distributions and classes is crucial, as new types of cyberattacks are emerging all the time. This implies that the generalization of detection schemes plays a key role. However, model-based methods have poor generalization ability. Conversely, some ML approaches offer an advanced solution to this problem. Numerous ML methods have been applied to attack detection, as shown in Table 1.3.

1.4.1 Anomaly Detection

Anomaly detection schemes are devised to detect possible outliers in measurements. Anomalies in energy systems can stem from various sources, including equipment malfunctions, system faults, and cyberattacks. For example, FDIA, one of the most severe cyberattacks, is launched through maliciously tampering with system measurements. FDIA can bypass the traditional Bad Data Detector (BDD) by adhering to physical laws, posing a significant risk to energy system security. Numerous research efforts have been devoted to anomaly detection, employing supervised learning, unsupervised learning, reinforcement learning, and semi-supervised learning.

Supervised Learning: Ozay et al. [16] made a first attempt for FDIA detection with ML approaches. This study compared the performance of three supervised learning methods, that are kNN, SVM, and Sparse Logistic Regression. Interpretability is crucial to system operators. Rule-based ML methods are highly expressive. Adhikari et al. [17] proposed a detector based on NNGEs, which creates "if–then–else-" like rules represented by generalized exemplars to classify cyber power contingency, control action, and cyberattacks in real-time. However, generating rules need a thorough knowledge of energy systems and massive manual works. DT has excellent interpretability for system operators with tolerable computation time. Adhikari et al. [18] designed a real-time cyberattack detection system using Hoeffding adaptive trees. The Bayesian method is a useful tool in the uncertain knowledge and probabilistic inference field. Khalid and Peng [19] applied the Bayesian-based approximated filter to each PMU to detect FDIA even in the presence of information loss.

The energy system measurements correlate in a highly nonlinear way. SVM handles nonlinear data efficiently. Deng et al. [20] first predicted measurements using SVM, and then the deviation between the predicted data and the current measurement was fed to an SVM as the training sample to detect FDIA. With the significant increase in computing power and training data, DNN has attracted considerable attention due to its capabilities in extracting nonlinear features. Wang et al. [21] highlighted the influence of the interval width of

Table 1.3 Comparison of ML methods for attack detection.

Issue	Approach		Advantage	Research
Anomaly detection	Supervised learning	Dimension reduction and metric learning	It is simple and can be easily updated online with minimal expenses	[16]
		Regression	Simple and time-efficient	[16]
		RI	Excellent interpretability	[17]
		DT	Good interpretability	Adhikari et al. [18]
		Bayesian	1. Its theoretical foundation offers logical and compelling arguments for detection 2. System operators could incorporate physical knowledge into Bayesian models	[19]
		SVM	It exhibits less susceptibility to overfitting problems and local minima in training	[20]
		NN	It learns intricate nonlinear relationships of measurements, enhancing detection accuracy	[21, 22]
		Ensemble learning	It improves robustness and achieves better performance than individual models	[23]
	Unsupervised learning	Dimension reduction and metric learning	It decomposes the observed measurements into low-rank true measurements and sparse injection data matrices	[24]
		Clustering	It detects anomalies by uncovering natural groups in the feature space	[25]
		NN	It learns intricate nonlinear relationships of measurements, enhancing detection accuracy	[26]

		Ensemble learning	It improves robustness and achieves better performance than individual models	[27]
		Transfer learning	It makes ML models maintain their performance when encountering new attacks	[28]
	Reinforcement Learning	Model-free	1. It detects attacks as quickly as possible with a minimal level of false alarms 2. It performs tasks with minimal prior knowledge about the system	[29]
	Semi-supervised learning	—	Given the labeling challenge in energy systems, it employs a few labeled data and extensive unlabeled data	[30]
Line outage detection	Supervised learning	Regression	Simple, time-saving	[31]
		SVM	It exhibits less susceptibility to overfitting problems and local minima in training	[32]
		Bayesian	1. Its theoretical foundation offers logical and compelling arguments for detections 2. System operators could incorporate physical knowledge into Bayesian models	[32]
		Dimension reduction and metric learning	It is simple and can be easily updated online with minimal expenses	[32]
		NN	It learns intricate nonlinear relationships of measurements, enhancing detection accuracy	[33]

(Continued)

Table 1.3 (Continued)

Issue	Approach		Advantage	Research
Electricity theft detection	Supervised learning	SVM	It exhibits less susceptibility to overfitting problems and local minima in training	[34]
		NN	It uncovers complex user consumption patterns by learning nonlinear relationships between variables	[35, 36]
		Ensemble learning	It improves robustness and achieves better performance than individual models	[37]
		Dimension reduction and metric learning	It is simple and can be easily updated online with minimal expenses	[38]
		Regression	Simple, time-saving	[38]
	Unsupervised learning	RI	Excellent interpretability	[39]
	Semi-supervised learning	—	Given the labeling challenge in energy systems, it employs a few labeled data and extensive unlabeled data	[40]

state variables on the accuracy of attack detection mechanisms and devised a DBN-assisted generic attack detection. The energy system measurements are nonlinearly dependent in a complex topological manner due to interactions and interconnections between system components. Thus, they show an underlying graph structure that is a non-Euclidean space. Based on this, [22] proposed a GNN-based approach for localizing FDIA. This approach first adopts the short-time Fourier transform and a CNN to model the spectral-temporal correlations. Then, the spectral-spatial relationships are modeled with a GNN.

Ensemble learning with multiple classifiers may be able to improve overall learning performance. To distinguish normal, fault, and attack events, [23] employed an ensemble learning-based XGBoost classifier.

Unsupervised Learning: In energy systems, the true measurements matrix typically exhibits low-rank characteristics, as system states change gradually over short periods. Conversely, the injection data matrix tends to be sparse, given that some measurements may be protected, and attackers prefer to exert minimal effort. Because PCA can be utilized to recover each data matrix component, comprising a low-rank component and a sparse component, [24] employed robust PCA to detect FDIA. However, this method is computationally complex and impractical for real-world applications.

Clustering techniques are widely used in detecting system anomalies. A density-based Local Outlier Factor (LOF) analysis was proposed by [25] to detect various types of synchrophasor data quality problems. NNs also provide a solution to separate anomalies. Kundu [26] proposed a global AE-based detection mechanism against FDIA. If the reconstruction error between the AE's input and output is above an estimated threshold, the input measurement can be classified as an anomaly. Ahmed et al. [27] employed an ensemble learning approach, iForest, to defend against FDIA. In this study, PCA is utilized to tackle the dimensionality problem. Energy systems are characterized by their dynamic nature rather than being static. The physical and cyber layers of energy systems experience ongoing changes, resulting in shifts in data distribution and the emergence of new attacks. Transfer learning enables ML to retain its performance on incoming data with new distributions and classes. Zhang and Yan [28] proposed a DANN-based transfer learning framework to train various ML classifiers for detecting unknown attacks.

Reinforcement Learning: Kurt et al. [29] appeared to be the first attempt at online cyberattack detection in energy systems using reinforcement learning. Considering "normal system operation" and "after an attack" as hidden states, this study formulated the problem of how to detect the anomalies as quickly as possible with the minimal level of false alarms, as a partially observable MDP problem. Model-free reinforcement learning techniques excel in navigating uncertain environments, making them suitable for solving this problem.

Semi-Supervised Learning: Supervised learning typically requires extensive labeled data for optimal performance, which can be challenging to obtain in energy systems. Semi-supervised learning offers a solution by leveraging a small portion of labeled data supplemented with a larger amount of unlabeled data. Zhang et al. [30] proposed a semi-supervised deep learning method for FDIA detection. The method combines the idea of AE and GAN. Specifically, the encoder of the AE works as the generator of GAN. Meanwhile, there are two discriminators for labeled and unlabeled data, respectively.

1.4.2 Line Outage Detection

Line outage detection has received limited attention in research, with only supervised learning approaches being explored in this area.

Supervised Learning: Garcia et al. [31] developed a linear regression-based scheme to localize line outages. Ibrahim et al. [32] investigated the performance of three methods, SVM, Naive Bayes, and kNN, in line outage detection. In study cases, Naive Bayes shows the best performance. The above studies only focus on single-line outage detection. To overcome this limitation, [33] adopted the idea of the Bayesian inference and employed an MLP architecture for the real-time identification of a potentially large number of simultaneous line outages.

1.4.3 Electricity Theft Detection

Unlike the above detection schemes based on Remote Terminal Unit (RTU) or Phasor Measurement Unit (PMU) measurements, the electricity theft detection datasets are customers' energy usage. Moreover, electricity theft typically does not lead to insecurity or instability in energy systems; however, it can result in financial losses. Various approaches have been explored to address this issue, including supervised learning, unsupervised learning, and semi-supervised learning.

Supervised Learning: In real-world scenarios, there is often an imbalance between the number of normal users and electricity thieves. SVM is adept at handling unbalanced datasets by assigning different weights to positive and negative examples. Nagi et al. [34] developed an SVM classifier to differentiate between normal users and electricity thieves.

NNs can learn complex nonlinear relationships between dependent and independent variables efficiently. Extensive studies apply NNs to extract user patterns for electricity theft detection. Zheng et al. [35] proposed a CNN model consisting of the wide component and deep CNN component. The wide component, which is a fully connected layer, can learn global features of one-dimensional electricity consumption data, while the deep CNN component can capture the non-periodicity of electricity theft. As RNN can better learn the long-term dependency of data, [36] proposed a detector based on CNN and GRU for electricity theft detection in the distributed generation domain instead of the consumption domain.

Ensemble learning has been applied to electricity theft detection. Punmiya and Choe [37] proposed a theft detector based on gradient-boosting classifiers including XGBoost, CATBoost, and LightGBM methods. Buzau et al. [38] presented a comparative work on supervised learning-based theft detection schemes. This study compared the kNN, linear regression, SVM, and XGBoost, and the results show that XGBoost outperforms the rest of the classifiers.

Unsupervised Learning: Theft samples may be scarce or nonexistent for a particular customer. Consequently, some studies have explored electricity theft detection systems based on unsupervised learning. León et al. [39] presented a predictive model using RI to create a characterization of customer's classes. The final classification model is presented by a rule set. As energy system becomes increasingly complex, generating rules is quite challenging and time-consuming.

Semi-supervised Learning: Though theft examples are rare, semi-supervised learning paves a way to make much use of abnormal patterns. Lu et al. [40] offered an AE-based semi-supervised learning model for irregular consumer detection.

1.5 Machine Learning in Impact Mitigation

In this section, we investigate studies mitigating the impacts of cyberattacks. This area has fewer works and still requires further research. Some studies use game theory or semantic analysis to eliminate the impacts of attacks. In recent years, ML methods have been spotlighted in this field due to their advantage in modeling and generalization. Most existing ML-based studies explore removing attacked measurements through unsupervised learning and supervised learning. The related research is summarized in Table 1.4.

1.5.1 Compromised Measurement Clearing

The aim of compromised measurement clearing is to reduce or eliminate the impacts of malicious data on the energy system operation, which is usually triggered after detecting an anomaly.

Unsupervised Learning: Hao et al. [24] used the robust PCA to recover measurements contaminated by FDIA. Nevertheless, this method is computationally

Table 1.4 Comparison of ML methods for impact mitigation.

Issue	Approach		Advantage	Research
Compromised measurement clearing	Unsupervised learning	Dimension reduction and metric learning	It performs matrix decomposition on the observed measurements matrix to obtain the low-rank true measurements matrix and sparse injection data matrix	[24]
		NN	It can obtain robust nonlinear representations from the input to eliminate biases of attacked measurements	[41]
	Supervised learning	NN	It learns intricate nonlinear relationships of measurements	[42]

complex and the recovery accuracy would be affected by the change of sparsity. Li et al. [41] employed a GAN model to generate real non-tampered measurements based on normal measurements, and then merged the generated measurements and the normal measurements to replace the attacked ones. This method can recover real measurements manipulated by FDIAs while maintaining the integrity of the non-attacked variables. In addition, in order to better defend against time-varying cyberattacks, a novel smooth training technique and an online adaptive window idea are explored to update the parameters of the proposed model.

Supervised Learning: Mestav et al. [42] applied an MLP-based approach to Bayesian state estimation, aiming to eliminate unreliable measurements in real-time. MLP is used to reduce computation complexity in Bayesian estimation. Because Bayesian techniques need prior knowledge about the system, the proposed method is less capable of adapting to a highly dynamic system.

1.6 Future Directions

Drawing from the extensive literature on the application of ML for cyberattacks and defense in smart energy systems, it is clear that ML has gained considerable attention as a valuable tool for guiding system operators. However, this field of research still faces open issues that need further study. In this section, extending from existing literature, possible challenges and potential future research opportunities are summarized.

Attack Resource Optimization: Before launching an attack, attackers need to consider the likelihood of the attack's success and determine the minimum attack resources required. For example, [43] demonstrated that in Cyber-Physical Systems (CPSs), if more than half of the sensing devices are compromised, CPS operators cannot accurately restore system states. In the field of energy system cybersecurity, research on the amount of attack resources needed to achieve attack objectives still needs to be completed. Determining the minimum attack resources involves balancing attack resources, attack effectiveness, and attack stealth. Formulating this problem as an optimization problem for solving can be one research approach.

Cybersecurity Assessment: Cybersecurity assessment refers to estimating the impact of different types of cyberattacks on smart energy systems, ranking these attacks by severity to facilitate the design of defensive measures. The cybersecurity assessment problem is similar to the stability assessment problem, with the distinction that stability assessment involves selecting critical faults based on the current system operating state and conducting N-k fault analyses to assess the impact of various faults on the current state. In contrast, cybersecurity assessment analyzes the impact of different cyberattacks on the current operating state without considering the types of faults caused by the attacks. Significant research attention has been paid to stability assessment problems, while research on cybersecurity assessment still needs to be conducted. Given the diverse range of cyberattacks targeting energy systems, one of the challenges in this problem lies in making assessment schemes applicable and transferable. Due to the underlying correlation among cyberattacks, it may be possible to consider utilizing transfer learning methods to transfer assessment schemes for specific cyberattacks to assessments for new cyberattacks.

Defense Resource Scheduling: Different cyberattacks possess varying characteristics, leading to different impacts on energy system security. Due to the limited defensive resources available to the defending party, it is impractical to implement protective measures against all cyberattacks. Therefore, there is a need to deploy defense resources rationally based on cyberattack severity and corresponding characteristics. Defense resource scheduling involves balancing defense costs and defense effectiveness. This problem is predicated on a clear understanding of the impact of each cyberattack on electric CPS, assessing the tolerance of energy systems to each cyberattack based on their severity, and allocating defense resources accordingly. Given the physical connections or communication relationships among devices in energy systems, when allocating defense resources, it is necessary to consider the concurrent impacts resulting from the integration of defense strategies against different cyberattacks.

Recovery Strategy Planning: Defense against cyberattacks in energy systems consists of four parts: pre-attack protection, attack detection during an attack, mitigation during an attack, and recovery after an attack. However, existing literature primarily focuses on the first three parts. The energy system is closely linked to national security and economic development, and it is essential to swiftly eliminate the impact of attacks and promptly restore safe and stable operations to ensure power availability. Different defense measures need to be taken based on the attack period and the specific electricity usage scenarios, striking a balance between recovery costs and the duration needed for recovery. For example, restoring power in urban commercial districts during daytime blackouts may require a shorter recovery time than nighttime blackouts in school areas.

References

1 Zhang, Z., Liu, M., Sun, M. et al. (2024). Vulnerability of machine learning approaches applied in IoT-based smart grid: a review. *IEEE Internet of Things Journal*. https://doi.org/10.1109/JIOT.2024.3349381.

2 Liu, M., Teng, F., Zhang, Z. et al. (2024). Enhancing cyber-resiliency of DER-based smart grid: a survey. *IEEE Transactions on Smart Grid*. https://doi.org/10.1109/TSG.2024.3373008.

3 Peng, S., Liu, M., Zuo, K., et al. (2023). Stealthy data integrity attacks against grid-tied photovoltaic systems. *2023 IEEE 6th International Conference on Industrial Cyber-Physical Systems (ICPS)*, Wuhan, China. pp. 1–7. https://doi.org/10.1109/ICPS58381.2023.10128033.

4 Esmalifalak, M., Nguyen, H., Zheng, R., and Han, Z. (2011). Stealth false data injection using independent component analysis in smart grid. *2011 IEEE International Conference on Smart Grid Communications (SmartGridComm)*, Brussels, Belgium: IEEE, pp. 244–248. https://doi.org/10.1109/SmartGridComm.2011.6102326.

5 Yu, Z.-H. and Chin, W.-L. (2015). Blind false data injection attack using PCA approximation method in smart grid. *IEEE Transactions on Smart Grid* 6 (3): 1219–1226. https://doi.org/10.1109/TSG.2014.2382714.

6 Higgins, M., Teng, F., and Parisini, T. (2021). Stealthy MTD against unsupervised learning-based blind FDI attacks in power systems. *IEEE Transactions on Information Forensics and Security* 16: 1275–1287. https://doi.org/10.1109/TIFS.2020.3027148.

7 Paul, S., Haq, M.R., Das, A., and Ni, Z. (2019). A comparative study of smart grid security based on unsupervised learning and load ranking. *2019 IEEE International Conference on Electro Information Technology (EIT)*, Brookings, SD, USA: IEEE, pp. 310–315. https://doi.org/10.1109/EIT.2019.8834059.

8 Chen, Y., Huang, S., Liu, F. et al. (2019). Evaluation of reinforcement learning-based false data injection attack to automatic voltage control. *IEEE Transactions on Smart Grid* 10 (2): 2158–2169. https://doi.org/10.1109/TSG.2018.2790704.

9 Ni, Z. and Paul, S. (2019). A multistage game in smart grid security: a reinforcement learning solution. *IEEE Transactions on Neural Networks and Learning Systems* 30 (9): 2684–2695. https://doi.org/10.1109/TNNLS.2018.2885530.

10 Tan, R., Nguyen, H.H., Foo, E.Y. et al. (2017). Modeling and mitigating impact of false data injection attacks on automatic generation control. *IEEE Transactions on Information Forensics and Security* 12 (7): 1609–1624. https://doi.org/10.1109/TIFS.2017.2676721.

11 Mishra, S., Li, X., Pan, T. et al. (2017). Price modification attack and protection scheme in smart grid. *IEEE Transactions on Smart Grid* 8 (4): 1864–1875. https://doi.org/10.1109/TSG.2015.2509945.

12 Wei, L., Sarwat, A.I., Saad, W., and Biswas, S. (2018). Stochastic games for power grid protection against coordinated cyber-physical attacks. *IEEE Transactions on Smart Grid* 9 (2): 684–694. https://doi.org/10.1109/TSG.2016.2561266.

13 Keshk, M., Turnbull, B., Moustafa, N. et al. (2020). A privacy-preserving-framework-based blockchain and deep learning for protecting smart power networks. *IEEE Transactions on Industrial Informatics* 16 (8): 5110–5118. https://doi.org/10.1109/TII.2019.2957140.

14 Anwar, A., Mahmood, A.N., and Tari, Z. (2015). Identification of vulnerable node clusters against false data injection attack in an AMI based smart grid. *Information Systems* 53: 201–212. https://doi.org/10.1016/j.is.2014.12.001.

15 Yan, J., He, H., Zhong, X., and Tang, Y. (2016). Q-learning-based vulnerability analysis of smart grid against sequential topology attacks. *IEEE Transactions on Information Forensics and Security* 12 (1): 200–210. https://doi.org/10.1109/TIFS.2016.2607701.

16 Ozay, M., Esnaola, I., Vural, F.T.Y. et al. (2012). Smarter security in the smart grid. *2012 IEEE Third International Conference on Smart Grid Communications (SmartGridComm)*, IEEE, pp. 312–317. https://doi.org/10.1109/SmartGridComm.2012.6486002.

17 Adhikari, U., Morris, T.H., and Pan, S. (2016). Applying non-nested generalized exemplars classification for cyber-power event and intrusion detection. *IEEE Transactions on Smart Grid* 9 (5): 3928–3941. https://doi.org/10.1109/TSG.2016.2642787.

18 Adhikari, U., Morris, T.H., and Pan, S. (2017). Applying Hoeffding adaptive trees for real-time cyber-power event and intrusion classification. *IEEE Transactions on Smart Grid* 9 (5): 4049–4060.

19 Khalid, H.M. and Peng, J.C.-H. (2016). A bayesian algorithm to enhance the resilience of WAMS applications against cyber attacks. *IEEE Transactions on Smart Grid* 7 (4): 2026–2037. https://doi.org/10.1109/TSG.2016.2544854.

20 Deng, Y., Zhu, K., Wang, R., and Wan, Y. (2019). Real-time detection of false data injection attacks based on load forecasting in smart grid. *2019 IEEE International Conference on Communications, Control, and Computing Technologies for Smart Grids (SmartGridComm)*, IEEE, pp. 1–6. https://doi.org/10.1109/SmartGridComm.2019.8909811.

21 Wang, H., Ruan, J., Ma, Z. et al. (2019). Deep learning aided interval state prediction for improving cyber security in energy internet. *Energy* 174: 1292–1304. https://doi.org/10.1016/j.energy.2019.03.009.

22 Peng, S., Zhang, Z., Deng, R., and Cheng, P. (2023). Localizing false data injection attacks in smart grid: a spectrum-based neural network approach. *IEEE Transactions on Smart Grid* 14 (6): 4827–4838. https://doi.org/10.1109/TSG.2023.3261970.

23 Hu, C., Yan, J., and Wang, C. (2019). Advanced cyber-physical attack classification with extreme gradient boosting for smart transmission grids. *2019 IEEE Power & Energy Society General Meeting (PESGM)*, IEEE, pp. 1–5. https://doi.org/10.1109/PESGM40551.2019.8973679.

24 Hao, J., Piechocki, R.J., Kaleshi, D. et al. (2015). Sparse malicious false data injection attacks and defense mechanisms in smart grids. *IEEE Transactions on Industrial Informatics* 11 (5): 1–12. https://doi.org/10.1109/TII.2015.2475695.

25 Wu, M. and Xie, L. (2016). Online detection of low-quality synchrophasor measurements: a data-driven approach. *IEEE Transactions on Power Systems* 32 (4): 2817–2827. https://doi.org/10.1109/TPWRS.2016.2633462.

26 Kundu, A. (2019). Deep learning techniques for detection of false data injection attacks on electric power grid. PhD Thesis. Texas A&M University.

27 Ahmed, S., Lee, Y., Hyun, S.-H., and Koo, I. (2019). Unsupervised machine learning-based detection of covert data integrity assault in smart grid networks utilizing isolation forest. *IEEE Transactions on Information Forensics and Security* 14 (10): 2765–2777. https://doi.org/10.1109/TIFS.2019.2902822.

28 Zhang, Y. and Yan, J. (2019). Domain-adversarial transfer learning for robust intrusion detection in the smart grid. *2019 IEEE International Conference on Communications, Control, and Computing Technologies for Smart Grids (SmartGridComm)*, IEEE, pp. 1–6. https://doi.org/10.1109/SmartGridComm.2019.8909793.

29 Kurt, M.N., Ogundijo, O., Li, C., and Wang, X. (2018). Online cyber-attack detection in smart grid: a reinforcement learning approach. *IEEE Transactions on Smart Grid* 10 (5): 5174–5185. https://doi.org/10.1109/TSG.2018.2878570.

30 Zhang, Y., Wang, J., and Chen, B. (2020). Detecting false data injection attacks in smart grids: a semi-supervised deep learning approach. *IEEE Transactions on Smart Grid* 12 (1): 623–634. https://doi.org/10.1109/TSG.2020.3010510.

31 Garcia, M., Catanach, T., Vander Wiel, S. et al. (2015). Line outage localization using phasor measurement data in transient state. *IEEE Transactions on Power Systems* 31 (4): 3019–3027. https://doi.org/10.1109/TPWRS.2015.2461461.

32 Ibrahim, A.M., Ezzat, M., and Abdelaziz, A.Y. (2016). Performance comparison of classification methods for line outage detection. *2016 Eighteenth International Middle East Power Systems Conference (MEPCON)*, IEEE, pp. 26–32. https://doi.org/10.1109/MEPCON.2016.7836867.

33 Zhao, Y., Chen, J., and Poor, H.V. (2019). A learning-to-infer method for real-time power grid multi-line outage identification. *IEEE Transactions on Smart Grid* 11 (1): 555–564. https://doi.org/10.1109/TSG.2019.2925405.

34 Nagi, J., Yap, K.S., Tiong, S.K. et al. (2011). Improving SVM-based nontechnical loss detection in power utility using the fuzzy inference system. *IEEE Transactions on Power Delivery* 26 (2): 1284–1285. https://doi.org/10.1109/TPWRD.2010.2055670.

35 Zheng, Z., Yang, Y., Niu, X. et al. (2017). Wide and deep convolutional neural networks for electricity-theft detection to secure smart grids. *IEEE Transactions on Industrial Informatics* 14 (4): 1606–1615. https://doi.org/10.1109/TII.2017.2785963.

36 Ismail, M., Shaaban, M.F., Naidu, M., and Serpedin, E. (2020). Deep learning detection of electricity theft cyber-attacks in renewable distributed generation. *IEEE Transactions on Smart Grid* 11 (4): 3428–3437. https://doi.org/10.1109/TSG.2020.2973681.

37 Punmiya, R. and Choe, S. (2019). Energy theft detection using gradient boosting theft detector with feature engineering-based preprocessing. *IEEE Transactions on Smart Grid* 10 (2): 2326–2329. https://doi.org/10.1109/TSG.2019.2892595.

38 Buzau, M.M., Tejedor-Aguilera, J., Cruz-Romero, P., and Gómez-Expósito, A. (2018). Detection of non-technical losses using smart meter data and supervised learning. *IEEE Transactions on Smart Grid* 10 (3): 2661–2670. https://doi.org/10.1109/TSG.2018.2807925.

39 León, C., Biscarri, F., Monedero, I. et al. (2011). Variability and trend-based generalized rule induction model to NTL detection in power companies. *IEEE Transactions on Power Systems* 26 (4): 1798–1807. https://doi.org/10.1109/TPWRS.2011.2121350.

40 Lu, X., Zhou, Y., Wang, Z. et al. (2019). Knowledge embedded semi-supervised deep learning for detecting non-technical losses in the smart grid. *Energies* 12 (18): 3452. https://doi.org/10.3390/en12183452.

41 Li, Y., Wang, Y., and Hu, S. (2019). Online generative adversary network based measurement recovery in false data injection attacks: a cyber-physical approach. *IEEE Transactions on Industrial Informatics* 16 (3): 2031–2043. https://doi.org/10.1109/TII.2019.2921106.

42 Mestav, K.R., Luengo-Rozas, J., and Tong, L. (2019). Bayesian state estimation for unobservable distribution systems via deep learning. *IEEE Transactions on Power Systems* 34 (6): 4910–4920. https://doi.org/10.1109/TPWRS.2019.2919157.

43 Fawzi, H., Tabuada, P., and Diggavi, S. (2014). Secure estimation and control for cyber-physical systems under adversarial attacks. *IEEE Transactions on Automatic Control* 59 (6): 1454–1467. https://doi.org/10.1109/TAC.2014.2303233.

2

Enhancing Cybersecurity in Power Communication Networks: An Approach to Resilient CPPS Through Channel Expansion and Defense Resource Allocation

Yingjun Wu[1], Yingtao Ru[1,2], Jinfan Chen[1,3], Hao Xu[4,7], Zhiwei Lin[1,5], Chengjun Liu[1,6], and Xinyi Liang[6,7]

[1] *School of Electrical and Power Engineering, Hohai University, Nanjing, China*
[2] *State Grid Jiangsu Electric Power Co. Ltd., Liyang Power Supply Company, Changzhou, China*
[3] *State Grid Zhejiang Electric Power Co. Ltd., Luqiao Power Supply Company, Taizhou, China*
[4] *Nanjing Zhongxing New Software Co. Ltd., Nanjing, China*
[5] *School of Engineering, Cardiff University, Cardiff, UK*
[6] *School of Electrical Engineering, Southeast University, Nanjing, China*
[7] *College of Automation, Nanjing University of Posts and Telecommunications, Nanjing, China*

2.1 Introduction

In this rapidly developing modern society and economy, people's proficiency in computer science and information technology has become increasingly important. The integration of advanced computing, communication, and information technologies into power systems has improved operational efficiency, but it has also increased their exposure to cyberattack risks. This chapter provides a comprehensive perspective on addressing cybersecurity challenges in Power Communication Networks (PCNs), particularly within Cyber-Physical Power Systems (CPPS).

Under this pervasive integration trend, traditional power systems are gradually evolving into CPPS, comprising the PCN and associated power grids. Although this integration enhances power automation, improves power system efficiency, and optimizes user experience, it also introduces numerous latent vulnerabilities. Several real-world incidents highlight the severity of cyberattacks on PCNs. For instance, in 2015, the Ukrainian power grid was damaged by an unknown cyberattack, which manipulated the switchgear of seven substations, resulting in a prolonged, large-scale power outage in the Ivano-Frankivsk region. Similarly, in 2016, the Israel Electricity Authority encountered a cyberattack in which phishing emails delivered malicious code to a power engineer, accidentally introducing a virus into the PCN and requiring the isolation of the infected computer to prevent further damage. In March 2019, a coordinated cyberattack

Artificial Intelligence Empowered Smart Energy Systems, First Edition. Edited by Qiang Yang and Gang Huang.

on Venezuela's PCN occurred in three phases, ultimately leading to an extended outage across 21 regions following the compromise of the Alto Prado substation.

These events underscore the need for an integrative approach that spans network planning, attack-defense strategies, and resource optimization, providing an overall framework to increase the elasticity of PCNs. This chapter aims to contribute to a more secure and robust CPPS by addressing gaps in the existing literature and offering actionable frameworks.

The first section introduces a planning framework for PCNs that balances operational continuity with the ability to resist network attacks. By analyzing the game-theoretic interaction between these objectives, a proposal is made for a deterministic topological programming model that ensures the reliability of information transmission, alongside an uncertainty-based model for anti-cyberattack planning. An enhanced PSO algorithm is then employed to solve these models, with validation performed through both theoretical and practical scenarios.

Subsequently, a systematic classification of common cyberattacks in PCNs is presented, along with an evaluation of corresponding defensive measures. A detailed analysis of attack types—such as Denial-of-Service (DoS), Man-in-the-Middle (MITM), and SAs—clarifies their mechanisms and impacts, while also identifying loopholes within the network architecture. The relationship between defense measures and key security indices—resistibility, recognizability, and recoverability—is highlighted, emphasizing the importance of multilayered defense strategies.

Finally, focusing on the survivability of CPPS communication and information systems, a defensive resource allocation method is developed. Critical power operations are decomposed into atomic services, and corresponding survivability indices are defined. A resource allocation model is then formulated to maximize survivability, which is solved using a modified genetic algorithm and validated through a simulated case study of a CPPS.

2.2 Mechanisms for the Classification and Propagation of Cyberattacks

Based on their prominence in research literature and their real-world frequency in PCNs, this section categorizes cyberattacks. The architecture of a representative PCN is illustrated in Figure 2.1.

The architecture of a standard PCN is composed of three layers: the Backbone Communication Layer (BCL), the Transmission Convergence Layer (TCL), and the Terminal Access Layer (TAL). Each layer serves distinct functions crucial to the safe and efficient operation of the power system.

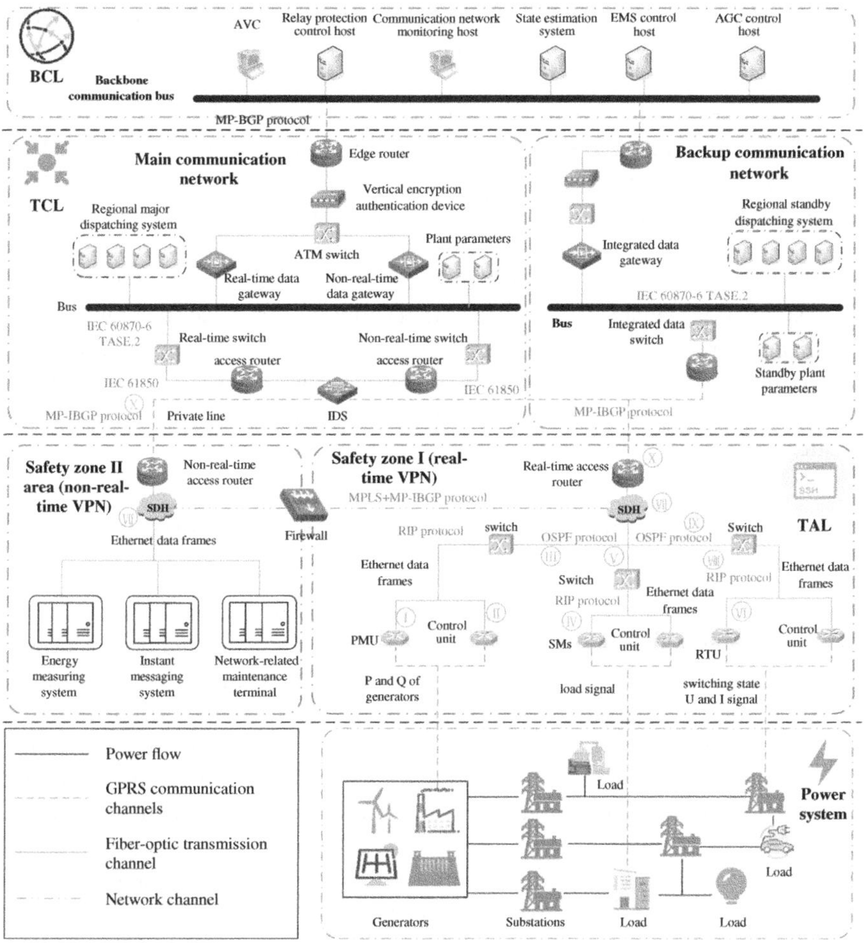

Figure 2.1 The architecture of the PCN. *Source:* Wu et al. [1] / with permission of IEEE.

Serving as the core of the PCN, the BCL is located within the dispatching system and at stations above 220 kV, where communication resources are abundant. The BCL consolidates key central control services for the power dispatching system, including automatic voltage control, relay protection (RP), and load frequency control. Data from the TCL is received and processed here, influencing decisions governing the operation status of the power system.

The TCL, located in another 220 kV station with ample communication resources and typically equipped with a standby communication network, serves as the communication hub for the PCN. It collects and transmits TAL data, using Intrusion Detection Systems (IDS) to guarantee secure data transfer.

Additionally, the TCL issues commands to sustain stable power system operation and accommodates a regional dispatching system to streamline control and dispatch.

The TAL is responsible for gathering power system data and implementing commands issued by the BCL. Segregated into real-time VPN and non-real-time VPN based on security levels, the TAL carries out control services in distinct safety zones. Terminal devices across the power grid collect diverse measurements, while Synchronous Digital Hierarchy (SDH) facilitates real-time, reliable data transmission between safety zones. Employed for safety zone segregation, firewalls work alongside varied transmission protocols that support inter-device communication.

The proliferation of information technology has inevitably led to a growing complexity in the architecture of PCNs. While this complexity enhances grid security and stability control, it simultaneously intensifies network security challenges.

2.2.1 Popularity in Academic Research and Frequency of Actual Cyberattacks

Figure 2.2 illustrates the proportional frequency of cyberattacks, calculated from their representation in academic research and documented occurrences in real-world PCNs over the past decade. Among these, Frequency-Domain Identification Attacks (FDIAs) account for the largest share, representing 47.26% of recorded incidents, followed by DoS attacks at 15.99%. Consequently, attacks such as False Data Injection (FDI), DoS, and Black Hole Attacks (BHAs) have garnered significant scholarly attention in PCN research. In contrast, cyberattacks, including Cross-Site Scripting (XSS), SAs, and Buffer Overflow Attacks (BOAs), have been rarely investigated, each contributing less than 3% of the total occurrences. For analytical clarity, attacks with proportions above 3% are classified as hotspot cyberattacks, whereas those below this threshold are regarded as non-hotspot cyberattacks.

2.2.2 Propagation Mechanism of Hotspot Cyberattacks

This section clarifies the concept of hotspot cyberattacks, systematically categorizes them, and maps their propagation pathways.

2.2.2.1 Denial of Service (DoS) Attack

As one of the most widespread cyberattacks on PCNs, DoS targets terminal communication devices/channels in the TAL, crippling or collapsing these critical components [2, 3]. As depicted in Figure 2.3, attackers flood terminal devices and communication channels with redundant data. This triggers the control

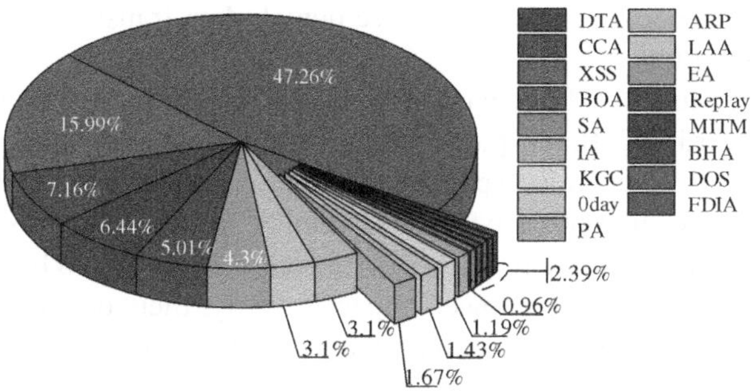

Figure 2.2 Frequency of cyberattacks. *Source:* Wu et al. [1] / with permission of IEEE.

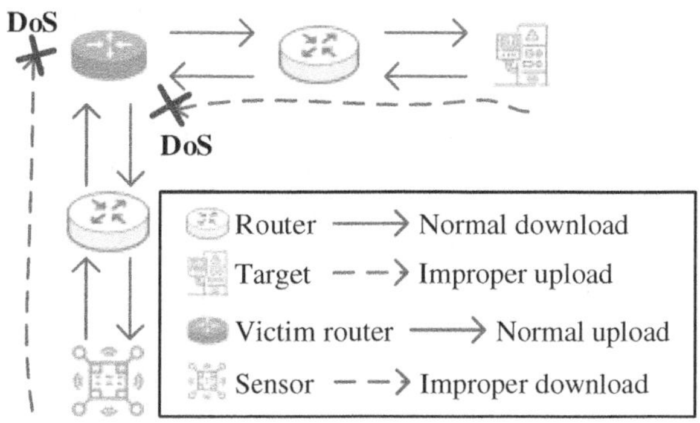

Figure 2.3 The mechanism of DoS attack on the PCN. *Source:* Wu et al. [1] / with permission of IEEE.

center (CC) to produce superfluous action commands, congesting the bandwidth of routers, switches, and servers. Consequently, these impaired devices fail to transmit or receive critical commands [4].

DoS attacks demonstrate intricate polymorphism, encompassing subtypes such as User Datagram Protocol (UDP)/Transmission Control Protocol (TCP) SYN (Synchronize) Flood, Teardrop, Local Area Network Denial (LAND), and Distributed Denial of Service (DDoS) attacks [5].

a) **UDP/TCP SYN Flood Attack:** Among the most prevalent types of DoS attacks, the UDP/TCP SYN Flood Attack primarily exploits the UDP and TCP/IP protocols within communication networks. The attacker floods targeted/random ports of PCN communication devices—such as RTUs, routers, or IAUs—with excessive UDP/TCP SYN messages, frequently using spoofed IP addresses [6, 7]. This induces devices to send response messages to falsified IP addresses. With an extensive volume of redundant request messages, a corresponding deluge of response messages consumes a substantial amount of communication bandwidth. Consequently, the flooded communication device may become unresponsive, causing service denial, data transmission loss, or control system failures.

b) **Teardrop Attack:** Targeting TCP/IP networks, the Teardrop Attack deploys a malicious program ("Teardrop") that exploits fragmentation reassembly flaws to crash victim systems. Attackers inject spoofed overlapping packets into PCN terminal devices/channels, disrupting packet reassembly in target devices. This can trigger kernel panics, system crashes, or forced restarts [8]. While Teardrop Attacks yield comparatively lower impact severity, their defenses prove significantly more straightforward and effective than those for other DoS variants.
c) **Local Area Network Denial (LAND) Attack:** Similar to the Teardrop Attack, the LAND attack involves generating multiple malicious data packets that target specific communication devices. A defining feature of the malicious packets used in LAND attacks is the identity of their source and destination IP addresses. This forces the victim device to dedicate resources to empty connections, thereby trapping it in an infinite loop of self-responses.
d) **Distributed Denial of Service (DDoS) Attack:** Evolving from traditional DoS attacks, the DDoS attack demonstrates enhanced sophistication. Deployed covertly on critical communication devices, multiple hidden DoS programs generate a flood of redundant messages or network requests, aiming to overwhelm and crash them [9, 10]. Unlike traditional DoS attacks, DDoS attacks inflict significantly more damage due to their intelligence and distributed nature.

2.2.2.2 False Data Injection Attacks (FDIAs)

False Data Injection Attacks (FDIAs) directly compromise the integrity and accuracy of power grid state estimation processes. The attack mechanism, as shown in Figure 2.4, involves the deliberate insertion of falsified data into smart measurement devices, including Phasor Measurement Units (PMUs), Remote Terminal Units (RTUs), and various intelligent transformers, within the PCN. The compromised devices then transmit erroneous measurements to the Base Control Center (BCC), resulting in deviations in state estimation and potentially leading to incorrect control decisions across the power grid.

In 2009, Liu et al. [11] introduced the concept of covert FDIA, illustrating how attackers could subvert the grid's integrity by evading conventional Bad Data Detection (BDD) mechanisms reliant on data residue detection. This study highlighted that FDIA could succeed if attackers obtain sufficient information from the PCN [12], inevitably resulting in decreased load, widespread power outages, and substantial economic ramifications [13, 14].

Depending on the level of information available to the attacker, FDIAs are categorized into two types: imperfect FDIA [15] and perfect FDIA [16, 17]. Only minimal data acquisition—for instance, knowledge of the PCN topology

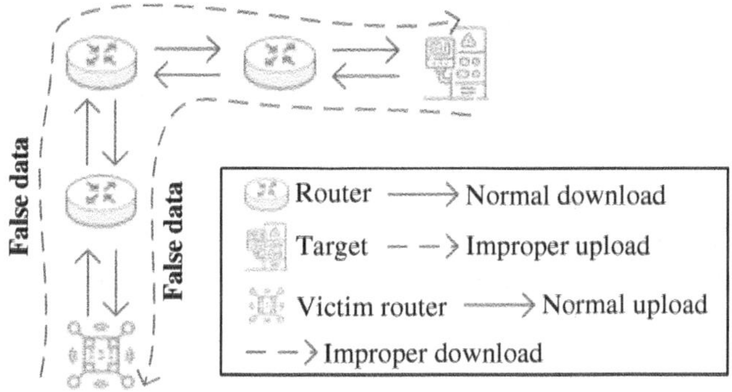

Figure 2.4 The attack mechanism of the FDIA in the PCN. *Source:* Wu et al. [1] / with permission of IEEE.

and line parameters—is needed to execute an Imperfect FDIA [18]. In contrast, a perfect FDIA demands access to full knowledge of the system—such as the bus voltage's phase angle and amplitude—to be executed successfully.

The primary targets of FDIA include disrupting state estimation, tampering with topological data, and manipulating load control instructions within the power system [19]:

i) The CC monitors the power system's operational state by gathering and analyzing data from measuring devices in the Transmission Access Layer (TAL) to ensure power supply reliability and quality [20]. Attackers inject faulty data into measuring devices, such as RTUs, Intelligent Electronic Devices (IEDs), and PMUs, to mislead the state estimator [21].
ii) For accurate topology estimation at the control center (CC), switch statuses and smart meter (SM) data are collected from across the power grid, thereby ensuring that the power system's topology information remains accurately updated [22]. However, an adversary can undermine this process by injecting false data into SMs and fabricating switch statuses, thereby compromising the integrity of the information. This infamous act can result in a discrepancy between the estimated topology and the actual configuration of the power grid, leading to erroneous judgments such as misidentifying open switches as closed. Notably, the success of such attacks depends critically on targeting both switches and SMs with a coordinated stream of faulty data. This necessitates a high level of specialized knowledge and the possession of critical power system information [23, 24].
iii) To disrupt the power market's trading mechanism, an assailant could additionally choose to manipulate the load control system [25]. User data is gathered by the load control system using signal collectors situated within the TAL. To facilitate load control, data analysis, and remote load monitoring, the aggregated data is subsequently processed within the power comprehensive

analysis system [26]. Consequently, by injecting false data into the IEDs within the TAL, the assailant can distort the decision-making of the power comprehensive analysis system. Such erroneous decisions can trigger imbalances in the power market, causing electricity prices to deviate from the optimal range and ultimately incurring substantial economic losses [14].

2.2.2.3 The Black Hole Attack (BHA)

The Botnet-based Hacking Attack (BHA) is a prevalent form of cyberattack that, as shown in Figure 2.5, works by intruding into the PCN's routers, switches, and other transmission devices to grant the attacker control. Devices afflicted with the BHA experience packet loss, resulting in data lacunae [27].

Within wireless sensor networks, the BHA poses a significant threat. The extensive use of diverse wireless sensors within the PCN renders it susceptible to the BHA [28]. Within the PCN, sensor nodes such as PMUs, RTUs, SMs, and PDCs play crucial roles in Data Collection (DC), Transmission, and Processing [29]. The PDC serves as the central hub, whereas the SMs and PMUs function as sensor nodes. Sensor nodes act as source nodes, continuously transmitting data to the PDC [30, 31]. In a BHA scenario, the attacker masquerades as an optimal node fulfilling routing requirements and deceiving sensor nodes into sending data packets to the attacker [32]. Consequently, this leads to a marked deterioration in packet reception rates among other nodes and a corresponding rise in end-to-end latency.

The BHA can be categorized into Home Area Network BHA (HBHA) and Network BHA (NBHA) based on the affected range within the PCN [33]:

a) **Home Area Network BHA (HBHA):** In this scenario, the attacker masquerades as the Phasor Data Concentrator (PDC), leveraging its optimal routing capabilities within the Metering Head-end (MH). The malicious node

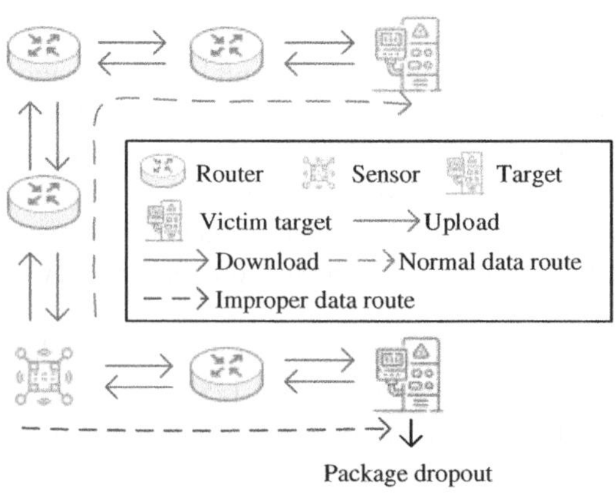

Figure 2.5 The mechanism of the BHA. *Source:* Wu et al. [1] / with permission of IEEE.

proceeds to commandeer SMs, forcing them to transmit data packets to itself, while concurrently injecting falsified information into the MH. Its primary objective is to significantly degrade service at the user level by deliberately inducing packet loss or imposing transmission delays. A crucial clarification is that the attacker exclusively targets the initial transmission channel of the SMs and refrains from any direct assault on the PDC at the CC.

b) **Network BHA (NBHA):** In contrast, attackers can spoof any node within the network, creating a significantly wider range of potential targets than with HBHA. The objective of the NBHA is to diminish overall system utilization by intentionally inducing packet loss or creating transmission delays. Consequently, this attack strategy requires targeting both SMs and PDCs to achieve its intended impact.

2.2.2.4 Address Resolution Protocol (ARP) Spoofing Attack

ARP spoofing is a prevalent cyberattack within PCNs. Figure 2.6 illustrates that this attack exploits vulnerabilities in the TCP/IP protocol's ARP mechanism to fabricate counterfeit gateways or hosts. By doing so, the attacker intercepts data intended for genuine gateways or hosts, subsequently tampering with this data before forwarding it to the intended destination. ARP spoofing attacks constitute a fundamental technique that underlies many DoS and MITM attacks [34].

ARP spoofing attacks typically occur within the Home Area Network (HAN) and the communication infrastructure of the distribution power system.

i) Within the HAN, the attacker poisons the ARP caches of SMs and PDCs by associating their IP addresses with a malicious, attacker-controlled MAC address. This manipulation allows the attacker to impersonate a legitimate

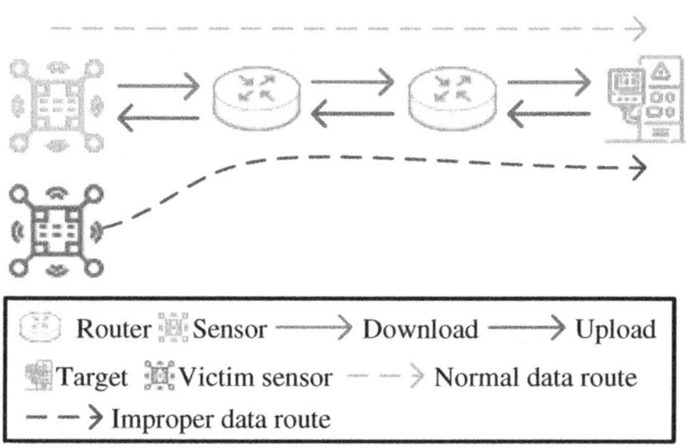

Figure 2.6 The mechanism of the ARP spoofing attack. *Source:* Wu et al. [1] / with permission of IEEE.

MAC address, thereby intercepting and forwarding malicious data to execute the spoofing attack [35].

ii) The lack of authentication and data encryption mechanisms renders the IEC 60870-5-104 protocol within the PCN susceptible to ARP spoofing attacks. As a result, during such an attack, Supervisory Control And Data Acquisition (SCADA) system packets are diverted to forged Communication Nodes (CNs) instead of the intended SMs. After maliciously processing the intercepted data, the attacker returns the tampered packets to the target SM. This can result in disrupted communications or the execution of malicious commands. This ultimately leads to the failure of control and monitoring communication functions [36].

2.2.2.5 Man-In-The-Middle (MITM) Attack

The MITM attack is highly sophisticated [37]. As depicted in Figure 2.7, the attacker employs ARP spoofing [38] to intercept and alter packets from a sender device. Subsequently, maliciously crafted data is sent to the CC [39]. This illustrates an MITM attack where data is first captured and modified before fraudulent information is relayed to the CC. The success of this attack hinges on its stealth; neither the sender nor the CC detects the intermediary, allowing data packets to be intercepted and manipulated with ease [40]. This version is precise, acknowledging that MITM was a key mechanism but not the sole definition of the attack.

The success of an MITM attack in a PCN can lead to catastrophic operational failures. The inherent delays in communication devices and data transmission, as well as the absence of complex authentication measures for monitoring the transmission network within power systems, increase the likelihood of MITM attacks succeeding [22]. MITM attackers typically capitalize on vulnerabilities in both transmission protocols and terminal devices to facilitate their intrusions.

i) Among the communication protocols used in the power grid, IEC61850 stands as the most prevalent in automation. Attackers exploit vulnerabilities within

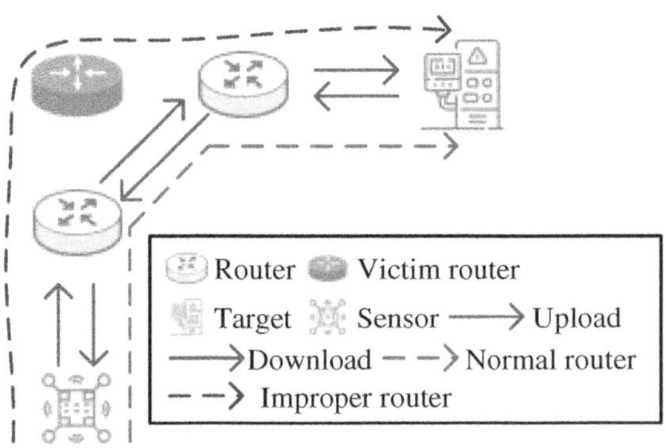

Figure 2.7 The mechanism of the MITM attack. *Source:* Wu et al. [1] / with permission of IEEE.

protocols like Generic Object Oriented Substation Event (GOOSE), Manufacturing Message Specification (MMS), or Sampled Measured Value (SMV) [41], allowing them to manipulate data such as measurements, control signals, and electricity prices. By doing so, attackers disrupt the accurate delivery of packets to target devices.

ii) The attacker compromises terminal communication devices—such as RTUs, PMUs, and SMs—to gain the ability to intercept, modify, or inject fabricated data packets. Subsequently, the malicious data is transmitted to SCADA systems or PDC nodes, leading the CC to make erroneous decisions unbeknownst to them [42, 43].

2.2.2.6 Eavesdropping Attack (EA)

Eavesdropping represents a typical passive attack within PCNs. Although eavesdropping itself is nondisruptive, it is a critical enabler for active attacks. Silently gathering intelligence provides the foundational knowledge required to launch precise MITM, DoS, or FDI attacks [44]. As shown in Figure 2.8, attackers leverage techniques such as traffic mirroring and data replication to gain unauthorized access (UA) to communication streams. Attackers then use cyber analysis software to decipher the intercepted encrypted packets, exposing device locations and communication content [45, 46].

The prevalence of communication vulnerabilities in modern power systems significantly lowers the barrier for executing EAs. This threat is amplified as the grid's communication network evolves toward complex MIMO patterns, which can introduce new attack surfaces if not properly secured [47]. This has resulted in a hybrid communication architecture, where multiple substations and CCs connect to the wide area network via a combination of public channels and private, dedicated lines [48]. This interconnected architecture enables attackers to deploy

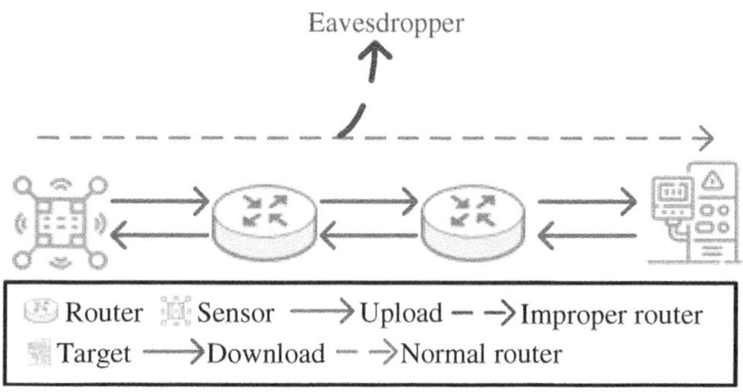

Figure 2.8 The mechanism of Eavesdropping attack. *Source:* Wu et al. [1] / with permission of IEEE.

a concealed MIMO multi-antenna eavesdropper, allowing them to conduct EAs on data packets flowing through public communication lines [49].

The terminal layer of an HAN comprises diverse endpoints, including Smart Home Devices (SHDs), SMs, and Key Distribution Centers (KDCs). The reliance on wireless communication between these terminal devices and the substation creates inherent security risks. Attackers can exploit these vulnerabilities to intercept or manipulate data [50].

2.2.2.7 Replay Attack

Due to imperfect information encryption during information exchange, communication systems are susceptible to replay attacks [51]. In a replay attack, the attacker captures valid data packets from a legitimate session and maliciously retransmits them later to deceive the system. The attacker then replays these previously recorded, valid data transmissions to the target node. Because the attacker replays verbatim copies of previously valid data, the Basic Data Detector (BDD) fails to recognize them as malicious [52]. As shown in Figure 2.9, the attacker's strategy is to replay previously captured data packets—which were initially intended for a prior interval (t–1)—at the current time t with little to no modification.

Replay attacks disrupt the communication network and have a significant impact on the power grid. As a Critical Cyber-Physical System (CCPS), the power system integrates various control systems, but still lacks a perfected security protocol compatible with the entire system, rendering it highly susceptible to attacks [53].

From an attacker's perspective, a replay attack offers a viable vector to inflict direct physical damage on infrastructure [54]. Attackers can transmit malicious data to critical control nodes—such as Energy Management Systems (EMSs), state estimators, and PDCs. This corrupts their data analysis, leading to erroneous outputs. Such manipulation can directly trigger breaker misoperations and

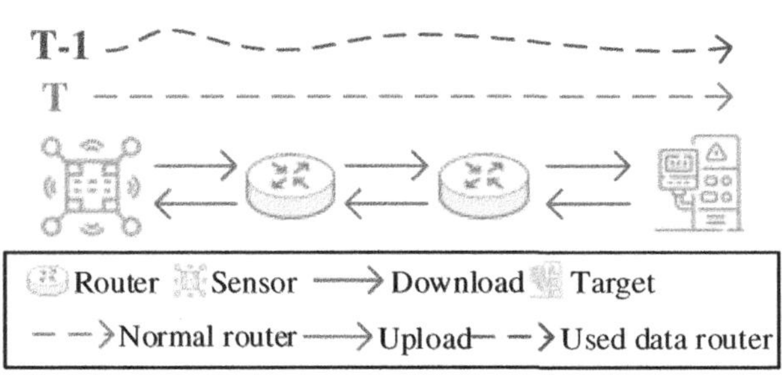

Figure 2.9 The mechanism of replay attack. *Source:* Wu et al. [1] / with permission of IEEE.

unplanned load shedding. If the replay attack also compromises the recovery system, it could result in a prolonged power system outage.

From a system-wide perspective, replay attacks can undermine the integrity of measurement devices in the Wide Area Measurement System (WAMS) and compromise critical communication protocols like MMS and GOOSE within substation automation systems [55, 56].

2.2.2.8 Load Altering Attack (LAA)

Unlike conventional cyberattacks that target CCs or SCADA systems, Load-Altering Attacks (LAAs) directly manipulate electricity demand. As shown in Figure 2.10, this is achieved by compromising end-use appliances and demand-side management systems. By compromising demand response infrastructure, attackers can orchestrate LAAs to manipulate aggregated load, deliberately creating a supply-demand imbalance [57].

Based on their operational strategy, LAAs are categorized as either open-loop or closed-loop. Open-loop LAAs are executed by attackers with limited knowledge of the target system, thereby restricting their impact to a small number of sensitive loads. In contrast, closed-loop LAAs are conducted by attackers possessing a detailed understanding of the system, including node voltages and load variations, often obtained through prolonged monitoring or eavesdropping. This information enables them to dynamically adjust attack strategies in real-time, leading to more severe disruptions [58].

As summarized in Table 2.1, most hotspot cyberattacks are active in nature, a direct consequence of their objective to inflict substantial physical damage on the system. The notable exception is eavesdropping, which remains a passive threat. These active attacks typically compromise the Power Control Network (PCN) by affecting packet integrity and accuracy. Meanwhile, EA is the most common passive cyberattack.

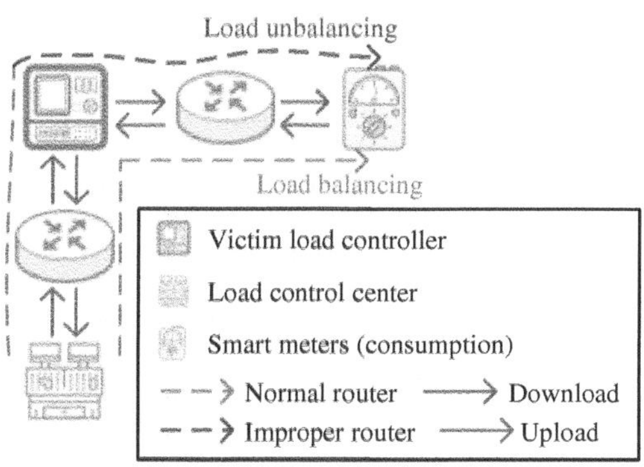

Figure 2.10 The mechanism and path of LAA on PCN. *Source:* Wu et al. [1] / with permission of IEEE.

Table 2.1 The hotspot cyberattack.

Attack form		Mechanism of cyberattack	Impact	Reference or events
DoS attack	UDP/TCP SYN flood attacks	Exploitation of protocol vulnerabilities and forged IP addresses to transmit malicious requests	Exhaustion of communication resources	[59]
	Teardrop attack	Injection of malicious IP packets into the target information node	Operating system paralysis	
	LAND attack	Transmission of spoofed packets with identical source and destination addresses	Malfunction of the target equipment, lacking protective mechanisms	
	DDoS attack	Malfunction of the target equipment, lacking protective mechanisms	Operating system paralysis	
FDIA		Exploitation of protocol vulnerabilities to inject falsified data	Erroneous decision-making by the system	[60]
BH attack		Manipulation of communication devices to drop packets	Significant transmission data loss	[61]
ARP attack		Transmission of pseudo-ARP reply messages to target hosts	Failure of critical data delivery to intended hosts	[62]
MITM attack		Establishment of unauthorized channels between information nodes	Unauthorized data access, tampering, or deletion	[63]
Eavesdropping attack		Use of specialized eavesdropping tools within the same network segment	Extraction of confidential information through packet analysis	[44]
Replay attack		Replay of previously captured data packets to communication devices	System misjudgment and incorrect state assessment	[56]
LAA		Alteration of load data on the demand side	An imbalance between power supply and demand	[64]

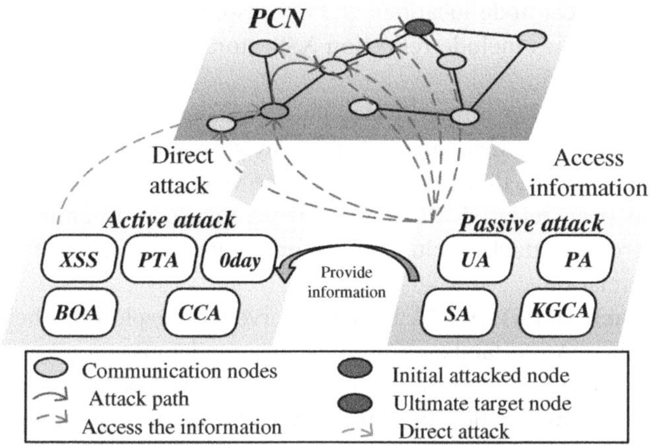

Figure 2.11 Non-hotspot cyberattacks in the PCN. *Source:* Wu et al. [1] / with permission of IEEE.

2.2.3 Propagation Mechanism of Non-Hotspot Cyberattacks

Non-hotspot cyberattacks can be further divided into active and passive categories, as illustrated in Figure 2.11. Passive attacks primarily focus on covertly collecting sensitive information from the PCN through techniques such as eavesdropping and data theft. Although these attacks do not directly disrupt the PCN, the large-scale exposure of critical information they cause poses a severe threat to the entire CPS.

In contrast, active non-hotspot attacks directly target and undermine the PCN through data manipulation and the injection of fabricated commands. Notably, passive attacks often serve as a precursor, providing the necessary intelligence for launching more disruptive active attacks against the targeted system.

2.2.3.1 Active Non-Hotspot Cyberattack

The primary objective of an active non-hotspot cyberattack is to disrupt the PCN's core functions by corrupting data integrity and availability.

a) **Cross-Site Script (CSS or XSS):** In an XSS attack, the perpetrator injects malicious client-side scripts into the PCN's web-based management interface. These scripts then deceive operators into executing unauthorized actions, such as clicking on a harmful link. When executed, the injected malicious script causes the victim's browser to send a crafted request to the web server. The server's response, which includes the malicious payload, is then rendered by the browser, completing the attack. Consequently, the operator's browser executes the malicious link, potentially disclosing sensitive information such as critical node locations or PCN resource configurations. Common XSS vulnerabilities include reflection XSS, storage XSS, and DOM XSS vulnerabilities.

b) **0-day Attack:** A 0-day attacker primarily exploits zero-day vulnerabilities to target devices or communication channels. A zero-day vulnerability is a security flaw in deployed software that is unknown to the vendor and for which no official patch has been released. Current proactive measures to enhance resilience against zero-day attacks include attack graph analysis, penetration testing, and alert correlation.
c) **Path Traversal Attack (PTA):** A PTA breaches servers by exploiting their vulnerabilities to gain UA to databases and execute malicious operations. Attackers can exploit these flaws to infiltrate restricted areas and exfiltrate sensitive assets, including configuration files, user credentials, and system logs. In extreme cases, attackers may execute instructions that result in widespread power system failures, such as generator shutdowns or the manipulation of RP devices, leading to extensive outages.
d) **Buffer Overflow Attacks (BOAs):** BOA exploits buffer overflow vulnerabilities in power system programs. When running, the CC temporarily allocates buffers to store various data types. A buffer overflow occurs when data input surpasses the buffer's designated boundaries, enabling attackers to inject and execute arbitrary code remotely, often without requiring authentication. A BOA can compromise critical power system software and disrupt the stability of grid operations.
e) **Covert Channel Attacks (CCAs):** Attackers exploit weaknesses in physical isolation, including inadequate access control management, a lack of technical protections on storage devices, and delayed detection of library upgrades. Cyber-Physical Attackers (CPAs) can circumvent the physical air-gap protections of a PCN, whether launching from an external or internal vantage point.

2.2.3.2 Passive Non-Hotspot Cyberattack

Passive non-hotspot cyberattacks are inherently designed for reconnaissance rather than immediate disruption, and thus do not directly compromise the security or operational reliability of the PCN. Instead, they primarily target the interception and analysis of power-related information during data transmission or processing, resulting in the leakage of sensitive information. When such passive attacks are combined with active cyberattacks, their impact can be substantially amplified.

a) **Unauthorized Access (UA):** Attackers exploit system authentication vulnerabilities to gain illicit access to databases and retrieve sensitive information.

b) **Password Attack (PA):** Attackers obtain passwords for power system communication equipment through eavesdropping, brute-force cracking, or dictionary attacks. Once inside the system, attackers can issue malicious commands directly to field devices, severely compromising operational integrity.
c) **Sniffer Attack (SA):** SA involves spying on the PCN's state using various sniffers to intercept crucial data. This can be achieved through:
   i) Deploying a packet sniffer on the target device;
   ii) Gaining remote access to a computer within the target network to deploy a packet sniffer;
   iii) Utilizing existing packet sniffers already present on the network.
d) **Key Generation Center Attack (KGCA):** The KGCA targets the Key Generation Center which is responsible for distributing private keys to databases for privacy protection. By stealing or tampering with critical distribution policies, attackers can directly bypass authorization mechanisms and access sensitive data.

As detailed in Table 2.2, while active non-hotspot attacks still constitute the majority, there has been a marked and concerning rise in passive incidents. This rise is attributed to the indirect nature of passive attacks, which have received less research attention. However, passive attacks are crucial for understanding PCN vulnerabilities, as they provide valuable insights into system weaknesses. Therefore, prioritizing research on passive attack mechanisms and defensive measures is imperative.

2.2.4 Targeting Equipment of Cyberattacks in TAL

In the Targeted Attack Layer (TAL), the primary objective of cyberattackers is to compromise the TAL before advancing to the Tactical Control Layer (TCL) and Base Control Layer (BCL). This section examines the interplay between various cyberattack taxonomies and their primary targets—specifically, the critical communication equipment and channels constituting the TAL (Trusted Access Layer). Figure 2.12 depicts this relationship through a three-layer structural diagram, illustrating the interconnected nature of the TAL, the communication network, and the cyberattack layers.

Based on Figure 2.12, the process and mechanisms of cyberattacks targeting the TAL can be analyzed. Table 2.3 summarizes potential vulnerabilities, their corresponding attack types, and the resulting consequences for devices under cyberattack.

Table 2.2 The non-hotspot cyberattacks.

Classification	Cyberattacks	Attack process	References
Active cyberattacks	Cross-site script (CSS or XSS)	Exploits a vulnerability to capture sensitive user credentials or session data, which are then leveraged to manipulate terminal behavior	[65]
	0-day attack	Leverages known but unpatched software or firmware flaws to compromise devices and communication channels	[66]
	Path Traversal Attack (PTA)	Abuses path traversal or weak input/data validation on servers or devices to access arbitrary files or execute unintended operations	[67]
	Buffer Overflow Attack (BOA)	Injects malicious payloads to trigger a buffer overflow on a server, enabling remote code execution and full host compromise	[68, 69]
	Convert Channel Attack (CCA)	Exploits weaknesses in isolation mechanisms between communication zones to bypass segmentation controls and perform malicious actions	[46, 70]
Passive cyberattacks	Unauthorized Access (UA)	Takes advantage of authentication weaknesses to obtain database access and extract sensitive information	[71]
	Password Attack (PA)	Recovers credentials for communication equipment through sniffing/monitoring, enabling disclosure of generation plans or other critical PCN data	[72]
	Sniffer Attack (SA)	Continuously monitors network state via packet capture to intercept and collect essential operational data	[73]
	Key Generation Center Attack (KGCA)	Illegally obtains private keys issued by the KGC, allowing decryption and forgery of ciphertext.	[74]

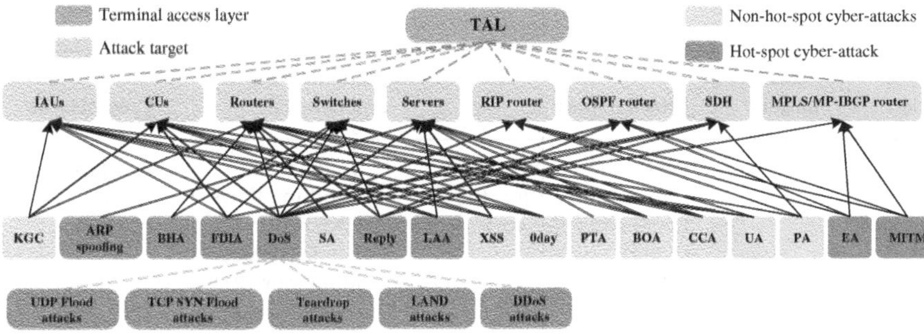

Figure 2.12 Association diagram of the kinds of cyberattacks and victim devices in the TCL. *Source:* Wu et al. [1] / with permission of IEEE.

Table 2.3 Targeting equipment of cyberattacks in TAL.

Classification	Attack target	Attack impact	Vulnerabilities and defects	Cyberattacks
Communication devices	IAUs	IAUs (comprising transmitters, PMUs, and RTUs) fail to collect data (e.g. real-time load, switching state, active/reactive power, and frequency), thereby preventing the data center from accurately assessing the system state.	1. Polling communication protocol flaws 2. Plaintext password storage 3. Cross-site scripting (XSS) 4. OS command execution vulnerability 5. Authentication weakness	A. BOA B. FDIA C. UA D. DoS E. XSS F. 0-day attack G. KGCA H. LAA
	Controller units	The control center (CC) is unable to execute operating commands (e.g. unit dispatch, breaker operation, relay protection).		
	Routers	Raw data collected by IAUs cannot be uploaded to the TCL, and control commands cannot be transmitted to the TAL.	1. Stack overflow 2. Misconfiguration flaws 3. Structured Query Language (SQL) injection 4. Cross-network covert channels 5. Backdoor programs	A. BOA B. PA C. Reply attack D. FDIA E. BHA F. 0-day attack G. CCA H. SA I. KGCA

(Continued)

Table 2.3 (Continued)

Classification	Attack target	Attack impact	Vulnerabilities and defects	Cyberattacks
	Switches	Packets fail to be transmitted within the Local Area Network (LAN).	1. TCP packet handling flaws 2. IPv6 routing header defects 3. Malicious IP option abuse	A. BHA B. DoS C. FDIA
	Servers	Smart grid data cannot be stored or processed in real-time, leading to severe deviations from optimal operational strategies.	1. Cross-network covert channels 2. Stack overflow 3. Path traversal 4. Request forgery 5. SQL injection	A. XSS B. BOA C. ARP spoofing attack D. PTA E. UA
Channels	RIP channels: **I, II, IV, VI**	Original analog signals of the grid cannot be converted into Ethernet frames for transmission.	1. Authentication flaws 2. UDP spoofing	A. PA B. UDP/ TCP SYN flood C. MITM D. SA
	Open Shortest Path First (OSPF) channels: **III, V, VII**	A channel crash prevents normal OSPF protocol transmission, hindering the CC's packet decoding.	1. Local-to-global impact flaws 2. Remote exploitation 3. Self-correction mechanism weakness	A. Reply attack B. MITM C. SA
	SDH: **VII**	Lack of unified transmission of heterogeneous communication services leads to delays or downtime.	1. OAM encryption defects 2. Human-induced vulnerabilities	A. PA B. Reply attack C. FDIA D. DoS
	MPLS+ MP-IBGP channel: **IX**	MPLS + MP-IBGP protocol failure in the SDH network prevents the CC from decoding packets.	1. DoS-related flaws 2. Memory leak	A. Teardrop attack B. MITM C. SA
	MP-IBGP channel: **X**	Packets cannot be transferred using the MP-IBGP protocol to TCL access routers.		

2.3 Power Communication Network Planning Based on Information Transmission Reachability Against Cyberattacks

The deep convergence of computing, communication, and information technologies has ushered in a paradigm shift for power grids. While this integration dramatically enhances operational efficiency, it simultaneously introduces a vastly expanded attack surface, making the grid more vulnerable to cyber-incidents. This chapter investigates the topology planning of PCNs, focusing on the deliberate trade-off between service quality and cyber-threat resilience from the perspective of information delivery.

First, the interplay between efficiency and security is examined, leading to the development of a game-theoretic framework for planning PCN topology. This framework models a strategic interaction where a conventional planner aims to maximize information transmission reachability during normal operations, countered by an adversarial planner who seeks to minimize it under attack scenarios.

Second, to address planning uncertainties, the framework employs a deterministic optimization model for the conventional planner and an uncertainty-aware stochastic model for the adversarial planner.

Third, we address the competitive interaction between the two planners by proposing an enhanced PSO algorithm, designed to solve the complex models arising from this bilevel problem efficiently.

Finally, the proposed framework's effectiveness and practical relevance are validated through comprehensive case studies, encompassing both a benchmark test system and a real-world PCN.

2.3.1 Introduction

Driven by the converging imperatives of energy sustainability, environmental protection, and economic development, smart grids have emerged as a critical focus of research and investment. With Information and Communication Technology (ICT) at their core, smart grids are evolving toward CPPS [75, 76]. Concurrently, the operation of power systems is increasingly reliant on PCNs, rendering them more susceptible to cybersecurity threats. For instance, a cyberattack on the PCN could disrupt regular power system operations [12, 77].

In recent years, considerable research efforts have been dedicated to defending against cyberattacks, yielding valuable insights. Works [78, 79] have analyzed cyberattack behaviors and objectives. Additionally, studies [80, 81] have scrutinized the propagation chain of cyberattack impacts. The study [82] has proposed methods for detecting and identifying cyberattack characteristics.

Other researchers [83] have examined attack mitigation techniques and developed corresponding defensive countermeasures. Furthermore, research [84, 85] has explored cyberattack methods and propagation processes, offering defense strategies. The literature, including the work of Nandi et al. [86], has provided a systematic investigation into the deployment strategies of security defenses, analyzing key decision factors. However, despite advancements in real-time defense measures, achieving comprehensive cybersecurity remains challenging due to the topological vulnerabilities of PCNs [87]. High-profile incidents—from the Stuxnet virus that infiltrated an Iranian nuclear facility (2010), to the cyberattacks causing large-scale blackouts in Ukraine (2015), and those leading to localized outages in Israel (2016)—collectively demonstrate the ability of determined adversaries to circumvent conventional protections.

Hence, there is a pressing need to develop PCN planning methodologies that account for cyberattacks to fortify internal structures. However, such research remains scant. Traditional PCN planning primarily prioritizes economic investment, network reliability, and security requirements [88, 89], without considering the factoring of cyber threats. While a cyberattack-focused public communication network planning method has been proposed [90], it overlooks the interplay between PCNs and the power grid.

This chapter introduces a topology planning methodology for PCNs that incorporates cyberattack scenarios into its core, using game theory as the analytical framework.

2.3.2 Planning Framework for Grid Communication Network Planning Object

The planning framework for the grid communication network is built around three core components: the CC, communication lines, and CNs.

a) **Control Center:** Acting as the central hub, the CC orchestrates grid operations by converting real-time data into intelligent control commands, thereby safeguarding the system's secure and stable performance.
b) **Communication Lines:** These communication lines act as dedicated channels for data transmission, enabling connectivity between the CC and all distributed nodes.
c) **Communication Nodes:** Vital for DC, monitoring, and executing control commands, CNs are strategically positioned. Their locations often coincide with power grid nodes, as their functionality is integral to the power grid's operation.

Considering the interdependence between CNs and power grid nodes, as well as the fixed positioning of the CC, this chapter focuses on optimizing communication lines after the locations of CNs have been determined. This strategy directly

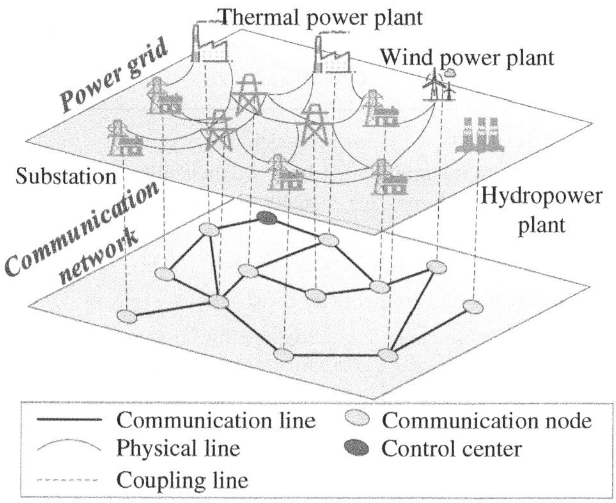

Figure 2.13 Diagram of the relationship between the Power Communication Network and the Power Grid. *Source:* Wu et al. [91] / with permission of IEEE.

supports the central objective of enhancing network efficiency and reliability, with the ultimate aim of ensuring uninterrupted and resilient power grid operation (Figure 2.13).

2.3.2.1 Planning Goals

The planning goals for the PCN revolve around two primary objectives:

a) **Ensuring Operational Data Transmission:** The PCN must efficiently collect, process, and transmit operational data of the power grid. This objective is an indispensable prerequisite for ensuring the secure and stable operation of the power system.
b) **Defense Against Cyberattacks:** Given the escalating cybersecurity threats, the planning process must incorporate robust defense mechanisms to mitigate these risks. A primary objective is to contain the propagation of malicious payloads across the network, thereby isolating breaches and protecting critical assets.

Moreover, the reliability of the PCN is a fundamental pillar for ensuring overall power system stability. To mitigate the risk of network component failures disrupting system functionality, redundant channels should be incorporated into the network design. These redundant channels serve as backups, ensuring uninterrupted data transmission for normal power grid operations even during channel failures.

Table 2.4 The comparison of the two planners.

Key points	Details	
Planner	TP	ACAP
Planning goals	Ensure the ability to transmit information of a regular operation	Improve the ability to defend against cyberattacks
Planning objects	Communication network lines	
Constraints	Economic constraint and connectivity constraint	
Measures	Add lines to form more data transmission paths	Reduce lines that are vulnerable to attack by cyberattacks
Current status	There are various corresponding planning methods	Lack of corresponding planning tools

Since the objectives of operational data transmission and defense against cyberattacks are inherently conflicting, coordination between the two planners—Traditional Planner (TP) and Anti-Cyberattack Planner (ACAP)—is essential. Their planning efforts must be harmonized to achieve a balanced and effective network design that meets operational and security requirements.

Table 2.4 systematically compares the two planners in terms of their objectives, entities, constraints, employed measures, and current states.

2.3.2.2 Planning Framework

The proposed planning framework contrasts two distinct operational scenarios: one under normal conditions and another under cyberattack conditions, with particular focus on their implications for information transmission processes. Figure 2.14 presents a schematic diagram illustrating this comparison.

In a standard PCN information transmission process, a two-way flow occurs over the communication lines: operational data is transmitted upward to the CC, while control commands are sent downward to the relevant nodes. However, under cyberattack scenarios, malicious data—such as tampered information or injected viruses—can propagate along these communication paths, threatening the integrity and reliability of the network.

When planning without consideration of cyberattack defense, the emphasis is on maximizing redundancy in communication lines, offering multiple transmission path options for data. Conversely, a cyber-defense-focused planning strategy prioritizes minimizing the number of communication lines to reduce the network's attack surface and contain the spread of malicious data.

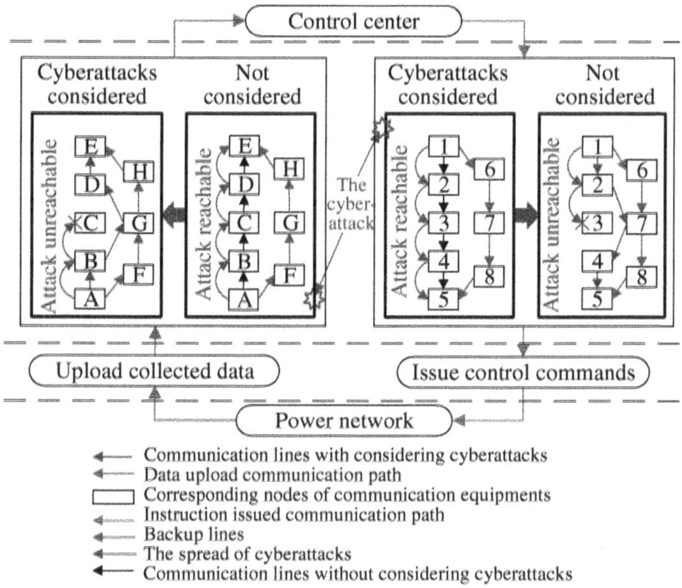

Figure 2.14 A schematic diagram comparing the data transmission process in the Power Communication Network with and without considering a cyberattack. *Source:* Wu et al. [91] / with permission of IEEE.

2.3.3 Topology Planning Model of Power Communication Network

This section systematically constructs topology planning models for both the TP and the ACAP. It begins with the TP's deterministic model, which seeks optimality under predetermined system conditions. In contrast, the ACAP model adopts a probabilistic approach to capture the dynamic nature of cyberattacks, incorporating uncertainties and time-varying probabilities associated with potential threats.

2.3.3.1 Planning Goal of the TP to Ensure Information Transmission of Regular Operation

The primary function of the TP is to safeguard the uninterrupted transmission of routine operational information throughout the power system. This necessitates a comprehensive modeling and analysis of the information transmission mechanisms within the PCN, as systematically outlined in Table 2.4. The analysis places particular emphasis on ensuring operational reliability across diverse system states.

Transmission of Operational Data

1) **Real-Time Performance:** Information transmission within the PCN must satisfy stringent real-time requirements to support power system operations.

Different operational data types may have distinct latency constraints, depending on their criticality [92].

2) **Reliability:** The reliability of the PCN is critical, as it must maintain continuous data flow despite equipment outages or component failures. Here, reliability is defined as the network's ability to sustain information transmission functionality despite the failure of specific communication elements [93].

Given that the existing communication infrastructure already satisfies the real-time performance requirements, the TP primarily focuses on evaluating and optimizing network topology to enhance reliability. The information transmission reliability of the PCN is defined as its capability to sustain uninterrupted data flow amidst multiple, successive component outages.

Planning Objective Function The planning objective function for the TP in a PCN consisting of N CNs is designed to improve information transmission reliability. This is achieved by establishing redundant paths and incorporating backup communication lines and components, ensuring that even in the event of communication line failures, critical information related to the power system's regular operation continues to be transmitted without interruption [94, 95].

The following objective function is established to assess the network's transmission reliability and hierarchically rank the importance of its constituent nodes:

$$\max F(X) = \frac{1}{N}\sum_{n=1}^{N} D_n \cdot Z_n \tag{2.1}$$

Here, $F(X)$ denotes the topology planning objective function for the PCN. X is the binary decision vector indicating the presence or absence of a link between each pair of nodes. $Z_n \in \{0,1\}Z_n \in \{0,1\}$ is a binary indicator for whether node n possesses a backup connection to the next node along its communication path. D_n represents the importance weight assigned to node n.

Its assessed criticality within the network governs the weight D_n assigned to a CN. The importance is evaluated from three perspectives: the node's connectivity (as determined by its associated power node), its information type, and its network reachability. These three dimensions are mathematically formalized as follows [96]:

Node Link Property The significance of a CN is affected by the features of the power node it is linked to. Three key factors are taken into account for assessment:

1) **Voltage Grade of the Power Node:** It mirrors the voltage grade of the power node. Higher voltage grades signify a greater influence;
2) **Power Node Capability:** It stands for the power node's load demand and generation capacity. Higher loads mean greater importance;

3) **Power Node Importance:** It denotes the classification of the power node's importance according to its function.

The computation formula for the significance of a CN according to power node attributes is:

$$D_{n-GNA} = SL_n{}^{*}LS_n{}^{*}LL_n \tag{2.2}$$

where $SL_n \in [0,1]$ represents site level. $LS_n \in [0,1]$ represents load size. $LL_n \in [0,1]$ represents the load level.

Node Transmitted Information Property A CN's importance is determined by the information it transmits, with consideration of both the quantity and significance of information transmitted. The evaluation process goes as follows:

$$D_{n-PGOS} = \sum_{s=1}^{S} \left(PGOS_s{}^{*}Q_S\right) \tag{2.3}$$

$$\sum_{s=1}^{S} PGOS_s = 1 \tag{2.4}$$

where S denotes the total count of routine operations carried out by the grid, $PGOS_s$ stands for the importance of operations in regular operation. Q_s indicates whether information is transmitted through the CN when the routine operation is executed. When information is transmitted through the CN, $Q_s = 1$; otherwise, $Q_s = 0$.

Accessibility of Information Transmission Component malfunctions influence the ability of a CN to be reachable. Nodes with lower reachability have a more significant impact on regular operation and network planning. The assessment procedure proceeds in the following manner:

$$D_{n-FP} = 1 - e^{-f} \tag{2.5}$$

Here, f denotes the average number of failures that occur in the CN within a unit time.

The final calculation formula for D_n is:

$$D_n = D_{n-GNA}{}^{*}D_{n-PGOS}{}^{*}D_{n-FP} \tag{2.6}$$

2.3.3.2 Enhancing Cyberattack Defense Capabilities in ACAP Planning Goals

Strengthening defenses against cyberattacks represents a central objective in the planning framework of Advanced Communication and Automation Platforms (ACAPs) for power systems. This imperative underscores the need to enhance the resilience of modern PCNs. Using Table 2.4 as a reference, a comprehensive analysis of the essential parameters for PCNs is conducted from the perspective of ACAP planning.

The notion of "reachability" is proposed as a key metric to measure the robustness of the network's defense mechanisms against cyberattacks. Reachability refers to the likelihood that a cyberattack can successfully infiltrate the PCN and spread to its intended target node. An attack is deemed "reachable" when it passes through every device on its propagation path; in contrast, "unreachable" attacks cannot extend their effect across the network, resulting in little to no impact on system operations.

Assessing reachability serves as a fundamental basis for evaluating the effectiveness of the PCN's defenses. Building on this basis, a topology planning model customized specifically to ACAPs' requirements is put forward, with reachability used as a key metric to enhance network resilience.

Planning Goal Function The primary goal is to minimize the reachability of potential cyberattacks in the PCN. The ACAP's planning objective function focuses on strengthening defense mechanisms against cyberattacks:

$$\min H(X) = \frac{1}{Y}\sum_{y=1}^{Y} P(S_j|O_x) \tag{2.7}$$

Here, $P(S_j|O_x)$ denotes the probability that the i-th attack successfully invades device S_j when event O_x occurs. O_x stands for the currently happening event labeled x. H represents the objective function for grid communication network topology planning, which is based on the reachability of cyberattacks. Y denotes the quantity of attack modes taken into account. S_j refers to the device targeted during the spread of the cyberattack.

Calculation of Cyberattack Reachability The diverse manifestations and propagation trajectories of cyberattacks entail distinct attack modalities. These modalities encompass the attack form, delineating preparations targeting identified vulnerabilities and the attack path, specifying intrusion points, propagation routes, and target devices. Successful cyberattacks necessitate the exploitation of vulnerabilities across CNs along a selected propagation path. Therefore, attackers adjust their strategies according to their understanding of the PCN's topology; as a result, the ACAP strives to reduce the accessible cyberattack paths that lead to their specified target nodes.

To calculate reachability, constructing a corresponding attack graph is essential. In this graph, nodes represent information devices, and directed edges indicate the information transmission pathways that cyberattacks exploit [97]. The attack graph can be systematically outlined by traversing all CNs and transmission links. Notably, it takes the form of a directed acyclic graph (DAG), where each edge starts from a terminal device (or CC) and ends at another CC (or terminal device). While the physical transmission lines of the PCN might form loops, the properties of information transmission prevent such loops from existing in the attack graph.

The attack graph encapsulates the procedural framework for calculating cyberattack reachability [98].

Calculation of Vulnerability Exploit Success Probability The probability of exploitation success refers to the likelihood that cyberattacks will leverage vulnerabilities present in communication devices. This probability is contingent upon three key attributes: access route (AR), access complexity (AC), and authentication (AU), each characterized by specific levels and corresponding values as delineated in Table 2.5.

The vulnerability exploit success probability is:

$$P(e_i) = AR \cdot AC \cdot AU \tag{2.8}$$

Here, $P(e_i)$ denotes the probability that a cyberattack successfully exploits the vulnerability e_i to infiltrate the current node.

Calculation of the Probability of the Propagation Process The initial step involves determining the prior probability of a cyberattack infiltrating a CN, which indicates the likelihood of successfully compromising the node without being influenced by other factors. Given the prevalent existence of unresolved vulnerabilities within information equipment, cyberattacks exploit various vulnerabilities to breach the current node. Let $Pa(S_j)$ denote the set of nodes that come before S_j on the path where the current node S_j is located, where $S_k \in Pa(S_j)$. If we define

Table 2.5 Values of cyberattack measurement indices.

Indexes	Level	Value
AR	Remote network	High
	Near network	Medium
	Local	Low
AC	Low	High
	Medium	Medium
	High	Low
AU	No	High
	Single	Medium
	Multiple	Low

the set of device nodes in the attack graph as $S = \{S_j | j = 1, 2, \ldots, M\}$, then the probability of successfully intruding into the current node S_j is as follows:

$$P(S_j|Pa(S_j)) = 1 - \prod_{i=1}^{N}(1 - P(e_i)) \tag{2.9}$$

Here, N denotes the number of vulnerabilities that the current node can exploit. $P(S_j|P_a(S_j))$ represents the probability that the current node S_j is successfully invaded on the condition that all previous nodes have been successfully invaded.

Before a cyberattack invades S_j, it must successfully infiltrate all preceding nodes on the path where S_j resides. The prior probability of S_j is therefore the joint probability of the current node and all nodes in its path set, with the calculation formula provided below:

$$P(S_j) = \prod_{k=1}^{K} P(S_k|Pa(S_k)) \tag{2.10}$$

Here, $P(S_j)$ denotes the probability that a cyberattack successfully invades S_j via the path where S_j is situated. K stands for the total number of nodes preceding S_j. Next, the calculation determines the posterior probability, a dynamic evaluation of intrusion outcomes. When a cyberattack has successfully invaded all other CNs, this represents the probability of it successfully infiltrating a specific CN. This assessment considers fluctuations in security conditions, relevant factors, and real-time attack incidents within the PCN. The posterior probability calculation facilitates the dynamic and real-time assessment of cyberattack security risks.

Define $O = \{O_x | x = 1, 2, \ldots, X\}$ as the set of currently observed attack events. O_x denotes that when a cyberattack invades, other nodes are also invaded; the invaded nodes are referred to as evidence points. The formula for the posterior probability of S_j is as follows:

$$P(S_j | O_x) = \frac{P(O_x|S_j) * P(S_j)}{P(O_x)} \tag{2.11}$$

Here, $P(S_j|O_x)$ denotes the probability that S_j is successfully invaded when O_x occurs. $P(O_x|S_j)$ denotes the probability that O_x happens if S_j is successfully invaded. $P(O_x)$ denotes the probability that O_x occurs.

Reachability Computation Process To elucidate the concept and methodology of reachability calculation, consider the example depicted in Figure 2.15, showcasing two PCN diagrams with identical node counts and node importance levels. Figure 2.15a contains 11 communication lines, whereas Figure 2.15b contains 10 communication lines. Consequently, the attack paths may vary between the two PCNs when they are subjected to cyberattacks.

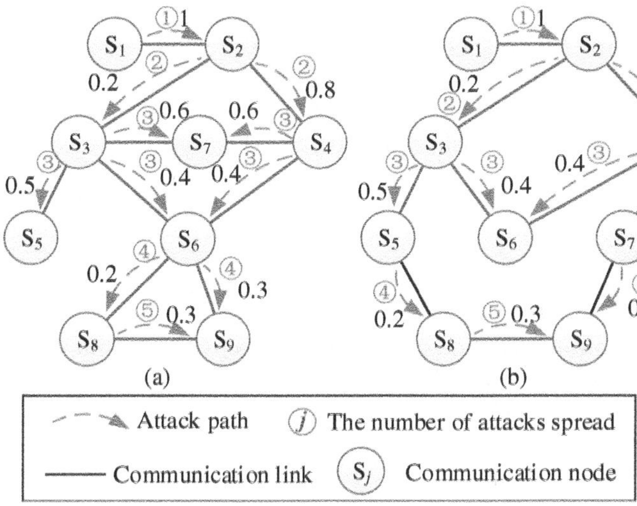

Figure 2.15 Attack diagram corresponding to the Power Communication Network. (a) Power communication network with 11 communication lines; (b) power communication network with 10 communication lines. *Source:* Wu et al. [91] / with permission of IEEE.

In Figure 2.15, it is shown that for the cyberattack, the invading node is S_1 and the target node is S_9. The black lines represent the communication lines within the PCN, and the red line defines the propagation path of the cyberattack. Employing the reachability above calculation method enables the determination of cyberattack reachability.

Let us summarize the steps and findings based on the provided scenario:

1) Prior Probability Calculation:
   - Cyberattack successful invasion probability for S_1: $P(S_1) = 0.3$
   - Cyberattack successful invasion probability for S_2: $P(S_2) = 0.3$
   - Cyberattack successful invasion probability for S_3: $P(S_3) = 0.3 \times 0.2 = 0.6$
   - Cyberattack successful invasion probability for S_1 (given two paths originating from S_1): $P(S_6) = 0.112$
2) Posterior Probability Calculation:
   - When event O_0 occurs (cyberattack successfully invades S_5), the posterior probability of S_6 changes to $P(S_6|O) = 0.592$, and similar adjustments occur for other nodes.
   - When there are no evidence points, the posterior probability of a node being invaded is equal to its prior probability, so $P(S_9|O_1) = P(S_9) = 0.038$.
3) Impact of Topology Change on Reachability:
   - Comparing various network topologies helps verify the impact of changes to the PCN on the reachability of cyberattacks.

- For instance, in Figure 2.15b, reducing the connectivity between S_6 and other nodes decreases the reachability of attacks, while adding connections between S_5, S_7, and S_9 enhances the reliability of information transmission.
- The prior probability of successful invasion of S_9 in Figure 2.15b: $P(S_9) = 0.0447$, indicating reduced reachability compared to Figure 2.15.

These calculations and comparisons elucidate the dynamic nature of cyber-attack invasion probability and the consequential impact of network topology changes on reachability (Table 2.6).

The following constraints are taken into account during the planning process:

1) To attain unified control over the power grid, connectivity constraints are imperative; this necessitates merging the entire PCN into a coherent whole, as outlined by the following formula:

$$G(i) \geq 1, i \in [1, n] \tag{2.12}$$

Here, $G(i)$ denotes the number of paths from node i to the CC. Each node must have at least one line connecting it to the CC, and connectivity is guaranteed by searching for the path between each node and the CC.

2) Construction cost constraints are equally pivotal when it comes to the practical realization of communication line infrastructure, as illustrated below:

$$C(X) = \sum_{j=1}^{M} e_j{}^* c_j \leq C_T \tag{2.13}$$

Here, $C(X)$ denotes the total construction cost function for the communication line. M stands for the total number of connectable lines. $e_j = 0$ or 1 indicates the connection status of line j. c_j represents the construction cost of line j. C_T denotes the total cost of line construction.

Table 2.6 Dynamic change of success probability of nodes being invaded.

Event	Evidence point	Posterior probability $P(S_j \mid O_x)$
O_1	No	0.038
O_2	S_2	0.128
O_3	S_4	0.202
O_4	S_6	0.342
O_5	S_8	0.51

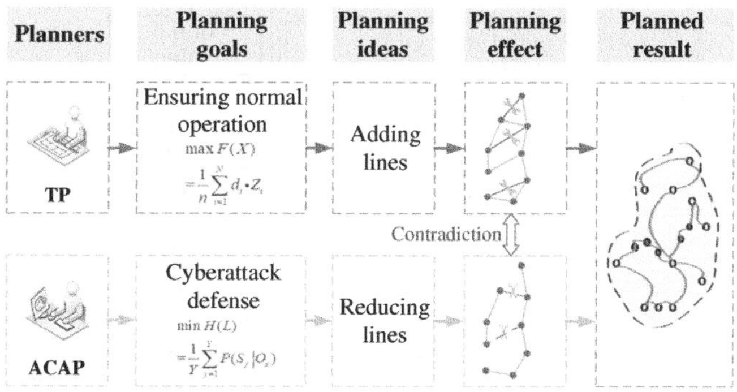

Figure 2.16 Strategic interaction across TP and ACAP. *Source:* Wu et al. [91] / with permission of IEEE.

2.3.4 A Game Theory-Based Planning Method

The objectives pursued by TP and ACAP are inherently incompatible, rendering the procedure of organizing akin to a strategic contest between distinct planners. TP endeavors to enhance the transmission capacity of power system operation services, whereas ACAP focuses on fortifying the PCN's resilience against cyber-attacks. This dialectical relationship portrays a symbiotic yet adversarial dynamic, exemplified in Figure 2.16.

2.3.4.1 Criterion Derived from the Nash Equilibrium

Regarding the interactive game actions depicted in Figure 2.17, the TP initially gathers the necessary operational parameters of the power grid communication network for planning purposes, thereby formulating the fundamental topology eXecutable Traditional Planner (XTP) that optimizes typical operational reliability. Subsequently, the ACAP intervenes to alter the topology, resulting in a revised configuration denoted as eXecutable Anti-Cyberattack Planner (XACAP) with minimized reachability. This iterative process ensues with alternating adjustments made by both TP and ACAP until an equilibrium point is reached [98].

$$\begin{cases} X_{TP}^{*} = \arg\max F\left(X_{TP}, X_{ACAP}^{*}\right) \\ X_{ACAP}^{*} = \arg\max H\left(X_{TP}^{*}, X_{ACAP}\right) \\ X_{TP}^{*} = X_{ACAP}^{*} = X^{*} \end{cases} \tag{2.14}$$

Among the above variables, X_{TP}^{*} denotes the topological decision-making result of the TP at the Nash equilibrium point. Correspondingly, X_{ACAP}^{*} stands for the topological decision-making outcome of entities with the ACAP when the Nash equilibrium state is reached. In contrast, XTP refers to the topological decision

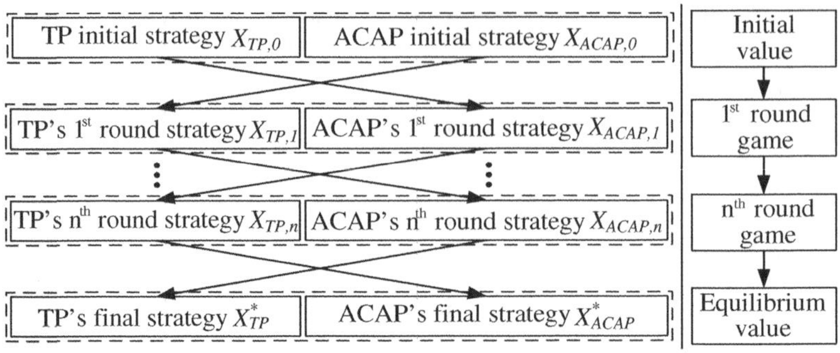

Figure 2.17 The evolutionary game mechanism between the TP and the ACAP. *Source:* Wu et al. [91] / with permission of IEEE.

made by the TP during the iterative process of the game, while XACAP represents the topological decision formulated by the ACAP entities in the same iterative process. Additionally, X^* is defined as the optimal topological decision-making scheme in the entire game system.

2.3.4.2 Model Solution Using Enhanced Particle Swarm-Based Optimization Algorithm

The conventional PSO represents a successive improvement method grounded in colony aptitude techniques. Distinguished by its distinctive search principle, PSO exhibits exceptional convergence speed in practical applications, a trait often absent in alternative algorithms. However, PSO shares common shortcomings with random search algorithms, including susceptibility to local extrema, inadequate convergence precision, and premature convergence.

To circumvent the limitations of traditional PSO, this chapter integrates an enhanced technique outlined in the study [99]. The primary enhancement revolves around generating particles with increased value. The evolutionary dynamics equation governing particle dynamics is as follows:

$$\begin{aligned} V_g &= \omega^* V_g + c_1{}^* rand() * (P_{bestg} - X_g) + c_2{}^* rand() * (G_{best} - X_g) \\ X_g &= X_g + V_g \end{aligned} \tag{2.15}$$

where, $g = 1, 2, \dots, n$, g indicates the iteration count; V_g refers to the speed of the decision X in the g iteration regarding the PCN topology; X_g represents the determination result generated in the g iteration for PCN topological structure; P_{bestg} indicates the topological arrangement outcome X generated in the g iteration; G_{bestg} stands for the swarm extremum of the topological configuration determination X corresponding to the g iteration; $\omega \geq 0$ serves as the inertia coefficient, $c_1, c_2 \geq 0$ act as the learning coefficients, while rand()∈(0,1) denotes the random numerical value within the interval. X_g stands for the planning outcomes of the g iteration, and its elements represent the connection status of the transmission lines. As an illustration, when node i and node j maintain a connection, $x_{ij} = 1$; in the opposite case, $x_{ij} = 0$.

$$X_g = \begin{bmatrix} x_{11} & x_{12} & \cdots & x_{1n} \\ x_{21} & x_{22} & \cdots & x_{2n} \\ \vdots & \vdots & \cdots & \vdots \\ x_{n1} & x_{n2} & \cdots & x_{nn} \end{bmatrix}; x_{ij} = \begin{cases} 1, & connected \\ 0, & else \end{cases}$$

The provided formula generates a total of M particles. To enhance the algorithm's convergence, an additional $M/2$ agents are created, as proposed in [99]. The generation formula is as follows:

$$\begin{aligned} \omega &= \omega_{\max} - \frac{\omega_{\max} - \omega_{\min}}{D_{\max}} * d \\ V_g &= \left(\omega_{\max} - \frac{\omega_{\max} - \omega_{\min}}{d} * D_{\max}\right) * V_g + c_1{}^* rand() \\ &\quad * (P_{bestg} - X_g) + c_2{}^* rand() * (G_{best} - X_g) \\ X_g &= X_g + V_g \end{aligned} \tag{2.16}$$

In the above context, $\omega_{\min}$ and $\omega_{\max}$ correspond to the minimum and maximum weight coefficients, respectively; $D_{\max}$ represents the maximum number of iterative steps; d denotes the step count of the current iteration.

Furthermore, incorporating the model presented in this chapter, the fitness calculation expression for each entity is:

$$L_g = W_1{}^* F(X) + W_2{}^* H(X) \tag{2.17}$$

where, W_1 denotes the weight connected to the end of safeguarding the information transmission efficacy of power grid operation procedures. W_2 stands for the target's weight, which boosts the capacity to fend off cyberattacks.

The procedures for solving the model using the enhanced PSO are presented in Table 2.7.

2.3.5 Simulations

2.3.5.1 System Overview Under Study

The power system employed in this instance is an IEEE 57-node system, presented in Figure 2.18.

Each power node within the IEEE-57 node system corresponds to a communication terminal that is responsible for conveying electrical status data and control signals specific to that node. Such CNs, control hub nodes, and communication links collectively constitute the electrical communication framework. The information nodes associated with power nodes 1 to 57 within the IEEE 57-node network are sequentially assigned numbers from 1 to 57. Additionally, the CC node is designated as number 58.

Table 2.7 Temporal variability of the likelihood of nodes undergoing invasion.

Pseudo code of the planning schemes

```
Setparameter (d_i, c_j, C_T, c_1, c_2, ω_max, ω_min, Q, D_max, W_1, W_2);
Initialize (particle_num);
while (not satisfying the termination condition)
      do
        for g=1 to particle_num;
          for each particle X
                 {
                   L_g=W_1*F(X)+W_2*H(X);
                   if L_g>P_bestg
                         P_bestg=L_g;
                   end
             };
          for each particle X
                 {
                        if L_g>G_bestg
                           G_bestg=L_g;
                            for each particle X
                           {
                               V_g=V_g+c_1*rand()*(P_bestg-X_g)
                                  +c_2*rand()*(G_best-X_g);
                               X_g=X_g+V_g;
                               ω=ω_max-d(ω_max-ω_min)/D_max;
                               V_g=ω*V_g+c_1*rand()*(P_bestg-X_g)
                                  +c_2*rand()*(G_best-X_g);
                               X_g=X_g+V_g;
                           };
                 };
 end
 while (g<=n) && (X_i-X_i-1>Δi )
     do
          {
           g=g+1;
          }
     end
 Output the optimal topology decision X*
```

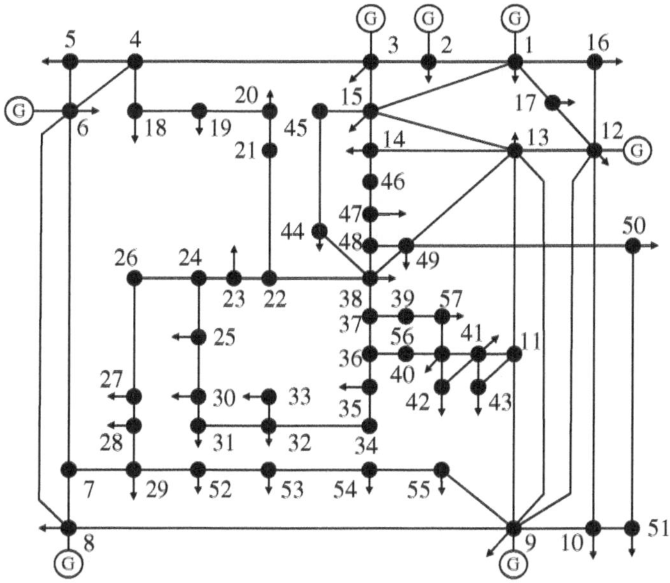

Figure 2.18 IEEE 57-node system. *Source:* Wu et al. [91] / with permission of IEEE.

2.3.5.2 Cyberattacks Factored in During the Planning Phase

Table 2.5 provides concrete values assigned to ease of calculation, as depicted in Table 2.8. The planned topology is subjected to two cyberattacks, each with distinct propagation paths:

Cyberattack 1: Utilizing information node 5 as the attacking node and information node 9 as the target terminal. This attack can propagate through all power info nodes except for the central control terminal.

Cyberattack 2: Employing information node 13 as the intruding node and information node 8 as the target node. This attack can propagate through all information nodes except for the CC.

Based on the defensive condition of individual data system nodes, the likelihood of cyber intrusion into different nodes is determined, following the methodology outlined in the study [100]. This process yields attack diagrams illustrating various network attack scenarios.

2.3.5.3 Planning Results in Different Cases

The utility communication network for the power system associated with the IEEE-57 node is designed based on the network attack diagram, considering three distinct cases:

Table 2.8 The quantities of the cyberattack measurement parameters.

Indexes	Level	Value
AR	Remote network	1
	Near network	0.646
	Local	0.395
AC	Low	0.71
	Medium	0.61
	High	0.35
AU	No	0.704
	Single	0.56
	Multiple	0.45

Case 1: Prioritizing the Assurance of Information Transmission Capability for Power Grid System Operational Services. The objective function's weight is set at $W_1 = 0.8$ and $W_2 = 0.2$, respectively;

Case 2: Balancing both goals equally. The objective function's weights are assigned values of $W_1 = W_2 = 0.5$;

Case 3: Emphasizing the Improvement of Cyberattack Resistance Capacity. These objective function weighting factors are set to $W_1 = 0.2$ and $W_2 = 0.8$, respectively.

The planning outcomes are depicted in Figure 2.19, where blue and red CNs are identified as essential, possessing extensive connections with other nodes. Red CNs are important because of their inherent characteristics, such as linking vital electrical distribution nodes and serving as command centers. Meanwhile, primary telecom nodes are crucial due to their strategic positions within the power communication grid, often located proximal to various essential junctions.

Furthermore, Figure 2.20 illustrates the results of the two objective functions derived from the planning outcomes of the three cases. By adjusting the values of W_1 and W_2, three distinct PCN topologies are obtained (Figure 2.20), each impacting information transmission performance and defense against cyberattacks differently. A decrease in W_1 reduces reliability for Scenario 2 and Scenario 3 by 0.08 and 0.17, respectively, in contrast to Case 1, indicating that data transfer dependability has lower sensitivity to changes in the PCN topology. Conversely, an upturn in W_2 results in a decrease in the accessibility of attacks for Cases 2 and 3 by 0.17 and 0.39, respectively, in comparison to Case 1, highlighting that attack reachability is more responsive to changes in the power communication grid. Accordingly,

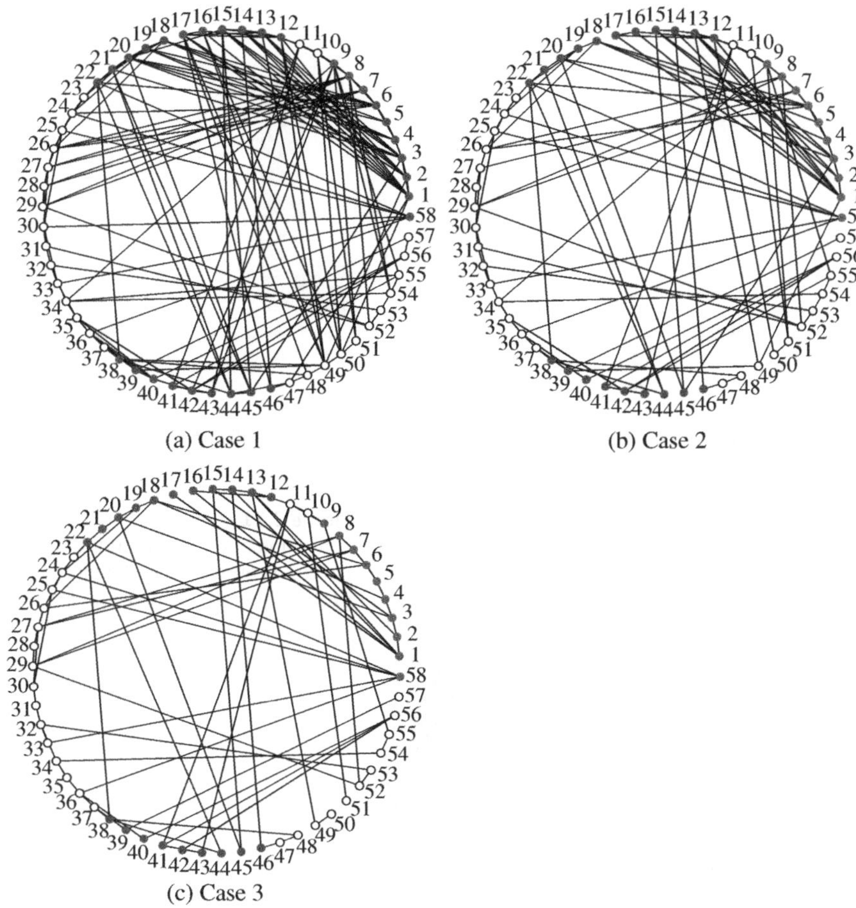

Figure 2.19 Topology planning results. (a) Case 1; (b) Case 2; (c) Case 3. *Source:* Wu et al. [91] / with permission of IEEE.

selecting appropriate figures for W_1 and W_2 is crucial to obtaining a PCN that optimally balances data transfer capability during the regular power system process and defense against cyberattacks.

2.3.5.4 Relative to Planning Methods that Disregard Cyberattacks

For the purpose of assessing the efficacy of the proposed planning approach in this chapter, the communication network topology of the grid derived from the programming strategy presented in [101] served as a reference group, as shown in Figure 2.21.

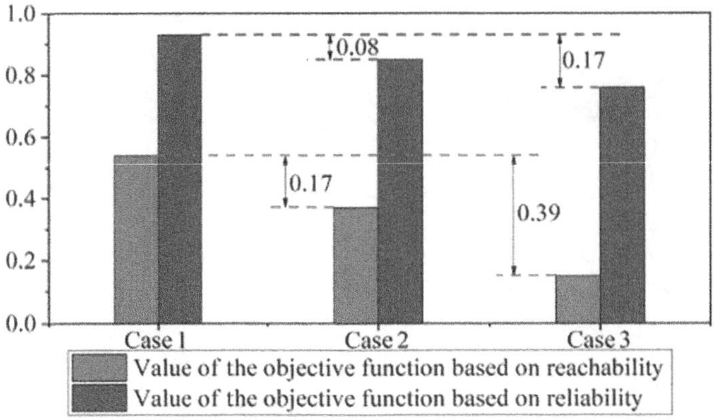

Figure 2.20 Effect comparison of three cases. *Source:* Wu et al. [91] / with permission of IEEE.

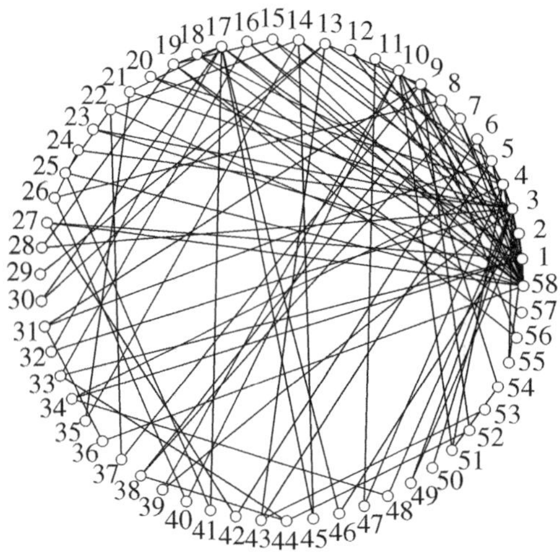

Figure 2.21 Power communication grid planned for a control group. *Source:* Wu et al. [91] / with permission of IEEE.

Given that Case 2 represents a more balanced approach by considering both goals, it is chosen for comparison against the reference group, as depicted in Figure 2.22.

Figure 2.22 reveals that the comparison group exhibits higher accessibility related to cyberattacks, which is attributable to the programming approach proposed by Guo et al. [101], which disregards the influence of cyberattacks. Despite the balanced scheme achieved through the planning method introduced in this chapter, resulting in slightly lower information transmission reliability compared to the control group by 0.14, it significantly outperforms the control group in attack reachability by 0.41. This indicates a considerable enhancement

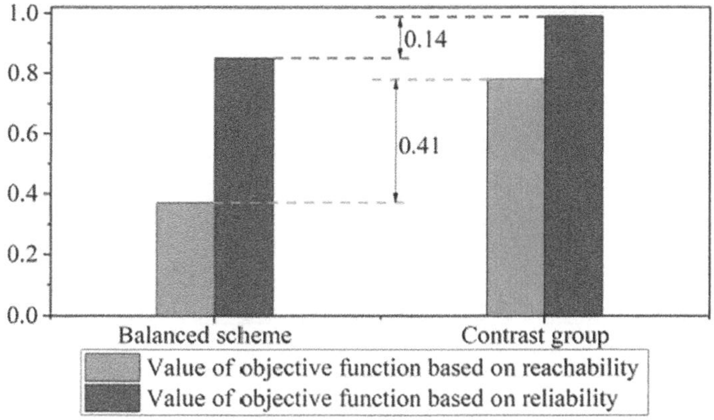

Figure 2.22 Reachability comparison between the balanced strategy and the contrast group. *Source:* Wu et al. [91] / with permission of IEEE.

in the network's resistance to cyberattacks, achieved by sacrificing a minor portion of information transmission reliability. Therefore, the proposed method effectively reduces the accessibility of network attacks and mitigates the risk of cyberattacks to modern power grids.

2.3.5.5 Practical Case Application

The feasibility of the proposed scheme is validated through the planning of a real-world power communication grid, as depicted in Figure 2.23a. Node 37 serves

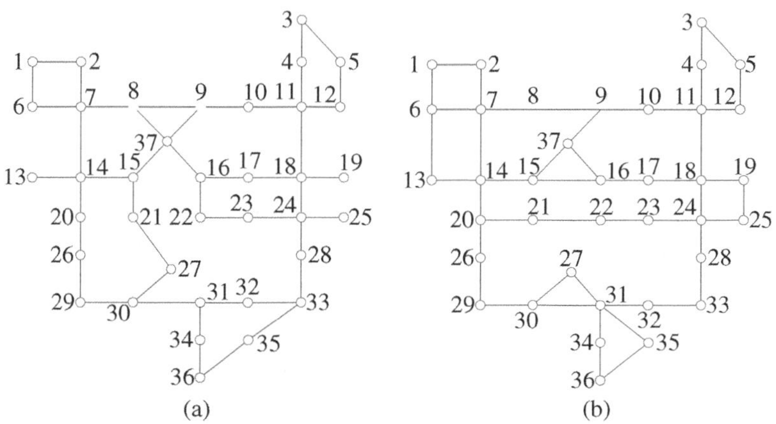

Figure 2.23 Schematic of the city's current Power Communication Network. (a) The existing structure of a city's power communication network; (b) the planning results obtained using the method proposed in this chapter. *Source:* Wu et al. [91] / with permission of IEEE.

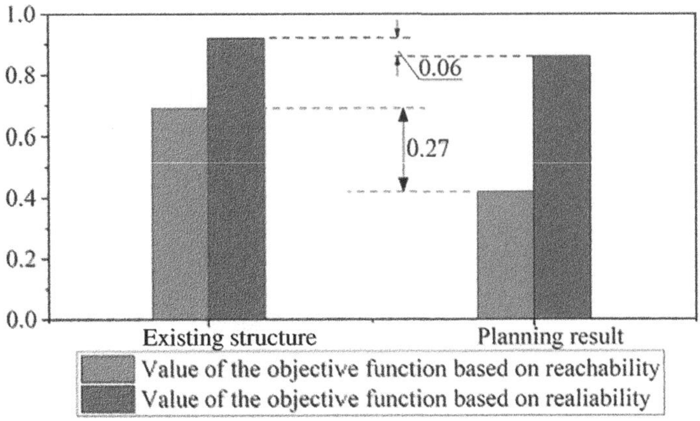

Figure 2.24 Preplanning versus post-planning comparison of the city's Power Communication Network. *Source:* Wu et al. [91] / with permission of IEEE.

as the master control node. The planning result is showcased in Figure 2.23b, employing the method proposed in this chapter.

Comparing the current PCN alongside the planned outcome in Figure 2.24 reveals an enhancement in defense against cyberattacks. Consequently, the method proposed in this chapter demonstrates applicability to real-world systems, yielding favorable outcomes.

Figure 2.23b demonstrates a reduction in five communication lines (8–37, 15–21, 16–22, 21–27, and 33–35) and the addition of seven communication lines (6–13, 15–16, 21–22, 20–21, 19–25, 27–31, and 31–35). Node 37, the control hub node with paramount importance, necessitates increased connectivity (e.g., 6–13, 15–16) to mitigate the impact of network attacks and bolster the reliability of information transfer during power system operation.

Nodes 21, 22, 16, and 17, situated close to the control headquarters, straightforwardly influence cyberattack accessibility. Consequently, their direct connections to the CC are severed, and they reroute their communication through alternative paths. Nodes 33 and 35, while less impactful on the CC, are critical within the system. Diminishing their connectivity and diversifying communication paths mitigate potential disruptions in power system operations if subjected to attacks.

Figure 2.24 presents a comparative analysis of cyberattack accessibility and the dependability of information transmission before and after planning. Post-planning, the PCN significantly reduces the accessibility of cyberattacks while maintaining high dependability in information transmission. Thus, the planned PCN effectively ensures robust information transmission reliability and defense against cyberattacks.

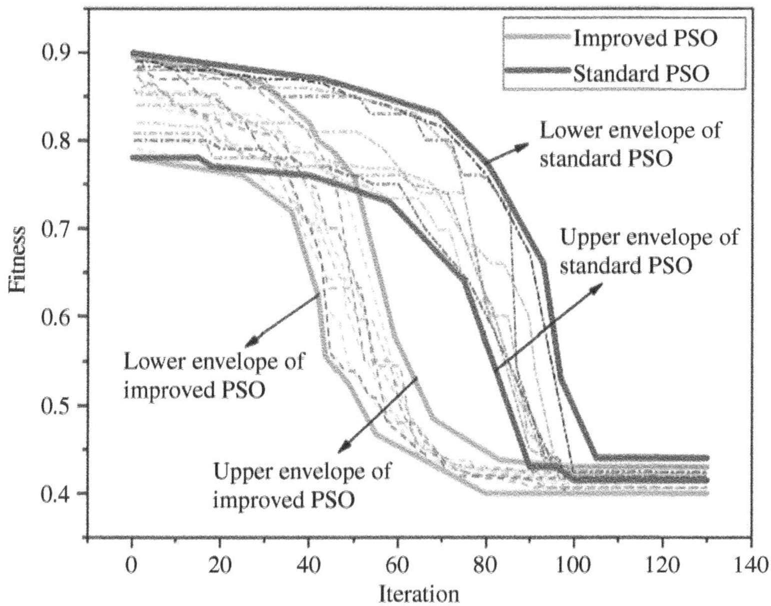

Figure 2.25 Discrepancy in convergence behavior involving standard PSO and improved PSO. *Source:* Wu et al. [91] / with permission of IEEE.

2.3.5.6 PSO-Based Analysis of Planning Outcomes

This section provides a brief comparison of the standard PSO algorithm and its improved version presented in this chapter.

Figure 2.25 illustrates the convergence of the standard PSO and its improved version applied to the practical case. Initially, there is no notable difference in convergence between the two algorithms. However, in the center and later iterative stages, the improved PSO demonstrates superior convergence and stability compared to the standard PSO, enabling easier acquisition of the global best solution.

Figure 2.26 compares the time consumed per iteration of the standard PSO and its improved variant. The modified PSO exhibits an iteration step length with smoother variation, facilitating the attainment of the global optimal solution. Conversely, the gradually increasing iteration step length of the standard PSO makes it prone to prematurely converging to a suboptimal solution.

Figure 2.27 presents the planning outcome of the applied case using the standard PSO. In contrast to Figure 2.23b, it has been observed that the standard PSO is prone to converging to a locally optimal solution, with relevant details presented in communication lines 15–21 and 33–35, highlighted in red dashed lines located in Figure 2.28, which signify suboptimal planning outcomes.

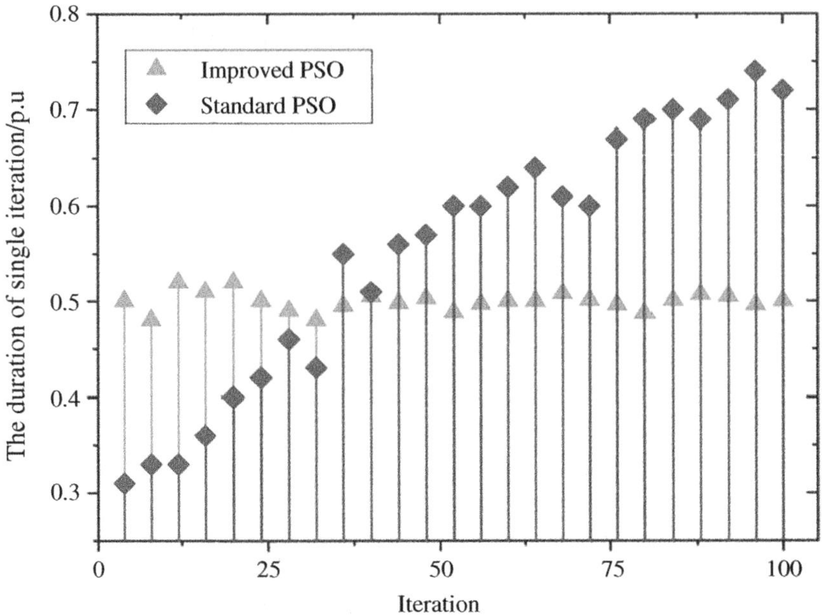

Figure 2.26 Duration per iteration of standard PSO versus improved PSO. *Source:* Wu et al. [91] / with permission of IEEE.

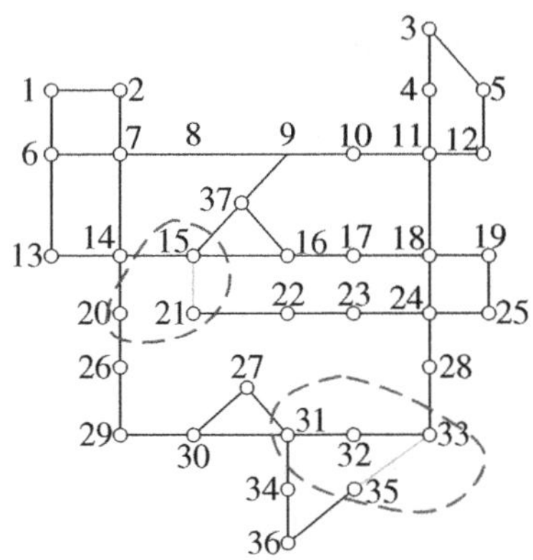

Figure 2.27 Standard PSO-based planning result. *Source:* Wu et al. [91] / with permission of IEEE.

Figure 2.28 illustrates that the modified PSO achieves a decrease of 0.03 in information transmission reliability and a decrease of 0.11 in cyberattack accessibility compared to the standard PSO. Overall, due to the standard PSO's propensity for local convergence, the planning results obtained using the modified PSO exhibit lower attack reachability, indicating greater resilience against cyberattacks.

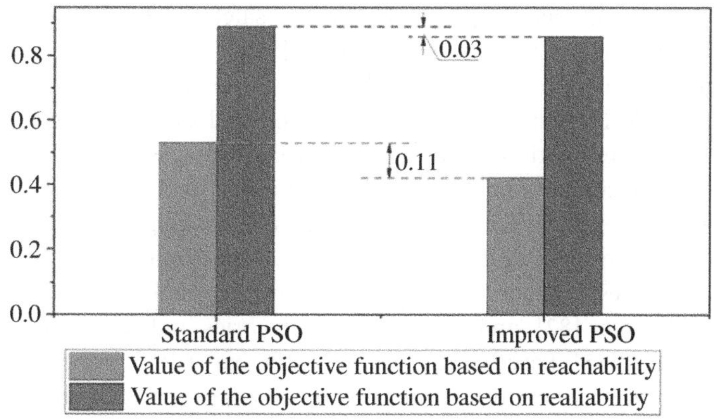

Figure 2.28 Urban power communication network planning comparison: standard PSO algorithm versus improved PSO algorithm. *Source:* Wu et al. [91] / with permission of IEEE.

2.4 Survivability-Oriented Defensive Resource Allocation for Communication and Information Systems Under Cyberattack

This chapter introduces a defensive resource allocation method designed to enhance the survivability of CPPS communication and information systems. First, the communication and information subsystems that support critical power system functions are decomposed into atomic services, enabling fine-grained modeling of their vulnerabilities and dependencies. Survivability evaluation indices, together with their quantitative assessment methods, are then established to measure system robustness under attack scenarios.

Building on these foundations, a model for defensive resource allocation is formulated to maximize survivability while considering the cost-effectiveness of defensive measures. To efficiently solve the proposed optimization problem, a modified genetic algorithm is developed.

Finally, the approach is validated through simulations on a CPPS benchmark constructed from the IEEE 30-bus system. The results demonstrate that the proposed method significantly improves survivability against cyberattacks, providing practical guidance for resilient planning and defense of CPPS communication and information infrastructures.

2.4.1 Introduction

The advent of contemporary advanced communication and information technology has propelled the evolution of the power grid system into the realm of CPPS [102]. While these technologies enhance the safety and operational efficiency of power system operational processes [76], they also increase the power system's susceptibility to cyberattacks due to its increased reliance on communication and information systems. By impairing and even turning off the operational capabilities of these networks, such cyberattacks can pose a significant threat to the secure and stable operation of the electrical grid [78].

Numerous recent studies have focused on the decision-making process for real-time protection approaches against cyber threats. These studies analyze the behavior and objectives of cyberattacks [58, 103], propose methods for their detection and identification [82], and examine the categories and propagation pathways of cyberattacks [85]. Based on the research, the paper [86] proposes a method for deploying security countermeasures using an attack graph to protect organizations. In contrast, the paper [78] utilizes graph-theoretical formulations to address FDI intrusions in intelligent power grids, proposing defense measures for mitigation. However, these defensive methods typically target specific types of cyberattacks, and their effectiveness is not guaranteed. Consequently, despite employing defense strategies, communication and data systems remain vulnerable to security hazards. For instance, the "Stuxnet" worm infamously targeted a nuclear power station in Iran in 2010, leading to its operational shutdown, while a cyber-assault on Ukraine's power grid in 2015 led to a large-scale blackout. Although defense countermeasures were implemented promptly after a cyber-assault disrupted Israel's power grid in 2016, effectively curbing its spread and averting a large-scale blackout, the economic costs were substantial.

Hence, despite existing cyberattack defense measures, cyber-assaults remain that are either unable to be adequately protected against or can be protected against, yet at significant costs. Addressing this challenge necessitates optimizing defense resource allocation and enhancing the intrinsic resilience of CPPS. Several researchers have developed resource allocation approaches to strengthen the resilience of CPPS. The paper [104] explores the survivability of communication and information systems and their role in defending against cyberattacks. Additionally, the studies [78, 104–107] propose survivability defensive methods aimed at enhancing the resilience of communication and information systems within modern intelligent power grids. However, these studies primarily focus on improving communication and information system survivability within CPPS and fail to address how to optimize survivability to the greatest extent from a holistic system perspective. Consequently, an effective defense resource allocation optimization model for enhancing the survivability of communication and information systems within CPPS remains lacking.

This chapter develops a model and approach for defense resource distribution to address the imperative of improving CPPS communication and information system survivability. Initially, critical power business functions of CPPS are disaggregated into atomic services. Subsequently, a set of survivability indices and corresponding calculation methods for these atomic services, along with defense resources, are proposed to enhance survivability. Lastly, a defensive resource distribution model is established to determine the type and quantity of defensive strategies required to improve the survivability of atomic services.

2.4.2 Systematic Evaluation of Survivability for CPPS Communication and Information Systems

The primary role of the communication and information system within the CPPS serves to ensure the reliable execution of power-related operations. In accordance with the regulatory framework established under the guidance of the Economic and Trade Commission of the People's Republic of China and the Electricity Regulatory Commission, these operations are classified into four hierarchical security domains:

- **Area I:** Core operational functions, including the Safety and Stability System (SSS), RP, and Power Dispatch (PD).
- **Area II:** Management-level functions, represented by the Protection Management Information System (PMIS) and the Electricity Market (EM).
- **Area III:** Auxiliary monitoring and environmental information services, such as the Hydrological Information System (HIS) and the Lightning Positioning System (LPS).
- **Area IV:** Supporting administrative services, encompassing Shared Resource Management (SRM) and the Official Document Approval Business (ODAB).

Critical operations within the power grid system, such as those on SSS, RP, and PD in Area I, are deemed indispensable. Hence, in this discourse, survivability refers to the communication and information system's ability to maintain these critical power operations in the face of cyberattacks.

2.4.2.1 Breaking Down Power Businesses into Atomic Services

Directly safeguarding the overall integrity of power operations presents considerable challenges due to their complexity and interdependence. To address this, the concept of atomic services is introduced. Atomic services are defined as minimal, self-contained functional units of power operations, each executable through one or more communication and information components. A key feature of atomic services is that the consequences of cyberattacks on them remain largely independent. This modularization enables critical power operations to be protected by enhancing the viability of individual atomic service units, thereby strengthening the overall resilience of the CPPS.

2.4.2.2 Indexes for Evaluating the Survivability of Atomic Services

Three indices are delineated based on temporal considerations: the Identifiability Index, the Resistibility Index, and the Recoverability Index. They are characterized as follows:

a) **The Identifiability Index:** This measures the CPPS's capability to discern cyberattacks promptly at their onset.
b) **The Resistibility Index:** This gauges CPPS's capacity to uphold critical operations amidst cyberattack propagation.
c) **The Recoverability Index:** This assesses CPPS's aptitude to restore critical operations after a cyberattack.

2.4.2.3 Survivability Evaluation Framework

This section introduces the structure for assessing viability, which is depicted in Figure 2.29. Initially, the crucial operations requiring assurance within the power system are identified according to specific requirements. Subsequently, leveraging atomic services, these critical operations are decomposed into the most minor autonomous units capable of accommodating survivability measures. Finally, the survivability evaluation focuses on scrutinizing atomic services' ability to detect cyberattacks promptly, maintain critical operations during cyberattack propagation, and recover critical operations post-cyberattacks.

2.4.2.4 Calculation Method for Survivability Evaluation Indices

The calculation of survivability indices for atomic services requires explicit consideration of the effects of defensive measures. These effects are evaluated through

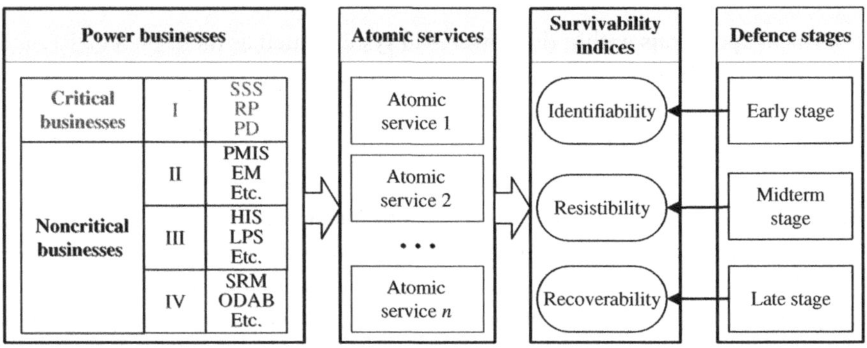

Figure 2.29 The power atomic service-oriented survivability evaluation framework. *Source:* Wu et al. [108] / with permission of IEEE.

simulation following the allocation of defense resources. The presented method is structured into the following steps:

Step 1. Service decomposition and attack library construction.

Critical power operations are decomposed into atomic services according to their execution processes. A cyberattack library is then established, in which attacks are categorized by their threat levels with respect to various elementary services. The library also specifies the likelihood of each attack type targeting specific service categories.

Step 2. Attack sampling.

A random cyber threat is selected from the cyber threat library for evaluation.

Step 3. Simulation of attack consequences.

The influence of the selected adversarial attacks against the system is simulated repeatedly, taking into account the specific characteristics of the attack. During this process, primary data are gathered that are required for calculating survivability indices.

Step 4. Iterative evaluation.

Assuming all cyberattack vectors in the library have been exploited, proceed to Step 5; if not, return to Step 2 for further sampling and simulation.

Step 5. Survivability index computation.

The expected values of the collected simulation data are computed, serving as the basis for calculating survivability evaluation indices for atomic services (Figure 2.30).

This part measures the viability indices. The attack risk index and assault resistance rate index are used to assess resistance. The attack identification rate index and attack identification time index are applied to evaluate detectability, while the attack recovery time index, along with the assault restoration rate index, are used to determine restorability.

Identifiability Index Calculation

A) Index of attack recognition rate

The attack identification rate index for an atomic service is defined as:

$$RI_{Amj} = \sum_{l=1}^{K} RI_{mjl} {}^{*} P_{Dmjl}$$
$$RI_{mjl} = \begin{cases} 1, \text{Attack can be identified} \\ 0, \text{Attack cannot be identified} \end{cases} \tag{2.18}$$

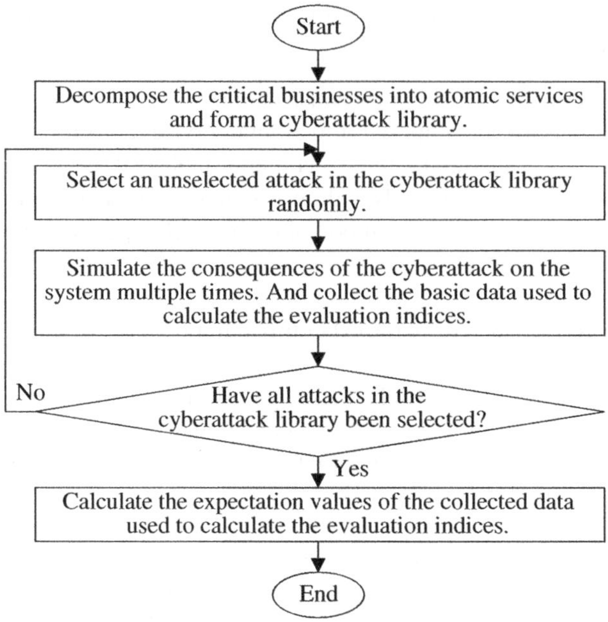

Figure 2.30 Data acquisition process. *Source:* Wu et al. [108] / with permission of IEEE.

The attack identification rate indexes for key business and the whole system are calculated similarly:

$$RI_{Sm} = \sum_{j=1}^{N_m} RI_{Amj}{}^{*} W_{Amj} \tag{2.19}$$

$$RI_{S} = \sum_{m=1}^{M} RI_{Sm}{}^{*} W_{Sm} \tag{2.20}$$

B) Attack identification time index

The attack identification time index for an atomic service is given by:

$$TI_{Amj} = \sum_{i=1}^{K} TI_{mji}{}^{*} P_{Dmji} \tag{2.21}$$

$$TI_{mji} = \begin{cases} 1 & T_{mji} \leq 10\,\text{s} \\ 0.8 & 10\,\text{s} < T_{mji} \leq 1\,\text{min} \\ 0.5 & 1\,\text{min} < T_{mji} \leq 5\,\text{min} \\ 0.2 & 5\,\text{min} < T_{mji} \leq 10\,\text{min} \\ 0 & T_{mji} > 10\,\text{min} \end{cases} \tag{2.22}$$

The attack identification time indexes for the key business and the whole system are computed analogously:

$$TI_{Sm} = \sum_{j=1}^{N_m} TI_{Amj}{}^{*}W_{Amj} \tag{2.23}$$

$$TI_S = \sum_{m=1}^{M} TI_{Sm}{}^{*}W_{Sm} \tag{2.24}$$

Calculation of Resistibility Index

A) Attack hazard metric

The degree of compromise to a basic service unit by an attack is given by:

$$AD_{Amj} = \sum_{i=1}^{K} AD_{mji}{}^{*}P_{Dmji}$$
$$AD_{mji} = \begin{cases} 1, & \text{Atomic service can't work} \\ 0, & \text{Atomic service works properly} \end{cases} \tag{2.25}$$

For the m-th critical business with N_m atomic services, its attack hazard index is:

$$AD_{Sm} = \sum_{j=1}^{N_m} AD_{Amj}{}^{*}W_{Amj} \tag{2.26}$$

Similarly, the attack hazard index for the entire system is calculated:

$$AD_S = \begin{cases} 1 & \exists AD_{Sm} \geq 1 \\ \sum\limits_{m=1}^{M} AD_{Sm}{}^{*}W_{Sm} & else \end{cases} \tag{2.27}$$

B) Attack impedance rate index

For an atomic service, the probability of a successful attack on it is derived from:

$$AS_{Amj} = \sum_{i=1}^{K} AS_{mji}{}^{*}P_{Dmji} \tag{2.28}$$

For the m-th critical business with N_m atomic services, its probability of successful attack is:

$$AS_{Sm} = \sum_{j=1}^{N_m} AS_{Amj}{}^{*}W_{Amj} \tag{2.29}$$

For the entire system, the probability of a successful attack is derived from:

$$AS_S = \begin{cases} 1 & \exists AS_{Sm} \geq 1 \\ \sum\limits_{m=1}^{M} AS_{Sm}{}^{*}W_{Sm} & else \end{cases} \tag{2.30}$$

For the entire system, its attack impedance index is derived from:

$$AP_S = 1 - AS_S \tag{2.31}$$

Recoverability Index Calculation

A) Index of attack recovery time

For a basic service unit, its attack restoration time index is calculated using the formula:

$$RT_{mji} = \begin{cases} T_{Rmji}/T_{0mj} & T_{Rmji} \leq T_{0mj} \\ 1 & T_{Rmji} > T_{0mj} \end{cases} \tag{2.32}$$

$$RT_{Amj} = \sum_{i=1}^{K} RT_{mji}{}^{*}P_{Dmji} \tag{2.33}$$

The attack restoration time indexes for the key business and the whole system are computed similarly:

$$RT_{Sm} = \sum_{j=1}^{N_m} RT_{Amj}{}^{*}W_{Amj} \tag{2.34}$$

$$RT_S = \sum_{m=1}^{M} RT_{Sm}{}^{*}W_{Sm} \tag{2.35}$$

B) Attack recovery rate index

The attack recovery rate index for critical business is calculated as follows:

$$RR_{Sm} = \sum_{i=1}^{K} \left(Q_{Ri}{}^{*}P_{Dmji}\right)/Q_0 \tag{2.36}$$

The attack restoration rate index of the whole system is:

$$RR_S = \sum_{m=1}^{M} RR_{Sm}{}^{*}W_{Sm} \tag{2.37}$$

Method for Fusing Survivability Assessment Indexes The integrated index provides a direct reflection of system survivability. Among all the indices above, smaller values of Attack Detection Score (ADS) and Recovery Time Score (RTS) indicate better system survivability. In comparison, larger values of Resistibility Index (RIS), Identifiability Index (TIS), Attack Recovery Rate (APS), and Attack Recovery Score (RRS) denote enhanced system survivability. The association of unified indexes with survivability is embedded in the composition vector R = [RIS, TIS, APS, RRS, 1-ADS, 1-RTS].

To obtain the fusion index, the calculation formula is as follows:

$$h(R) = a^{*}\frac{1}{E}\|R\|_1 + b\|R\|_{\infty} \tag{2.38}$$

||R||1 and $||R||_\infty$ denote the 1-norm and the infinite norm of the vector R, respectively.

$$||R||_1 = \sum_{e=1}^{E} |R_e|$$
$$||R||_\infty = \max |R_e| \tag{2.39}$$

2.4.2.5 Calculation Method for Survivability Evaluation Indexes

Defensive resources, encompassing both hardware and software measures, are integrated into the CPPS to enhance its resilience. Corresponding to the resilience evaluation indexes, defensive capabilities fall into three types, each aimed at strengthening specific aspects of survivability:

1) **Identifiability Defensive Resources:** These resources bolster the system's identifiability. Typically employed protective measures cover IDSs and deception systems. The associated cost is as follows:

$$C_{ide1}{}^* n_{ide1mj} + C_{ide2}{}^* n_{ide2mj} = C_{idemj} \tag{2.40}$$

2) **Resistibility Defensive Resources:** These resources augment the system's resistance capabilities. Commonly utilized defensive measures encompass traffic filter systems, access regulatory technology, and the installation of camouflage modules. The associated cost is as follows:

$$C_{res1}{}^* n_{res1mj} + C_{res2}{}^* n_{res2mj} + C_{res3}{}^* n_{res3mj} = C_{resmj} \tag{2.41}$$

3) **Recoverability Defensive Resources:** These resources bolster the system's recovery capabilities. Commonly employed defensive measures comprise the establishment of spare modules and data safeguarding and restoration systems. The expense associated with this is as follows:

$$C_{rec1}{}^* n_{rec1mj} + C_{rec2}{}^* n_{rec2mj} = C_{recmj} \tag{2.42}$$

2.4.3 Defensive Resource Allocation Model for Enhancing CPPS Survivability Against Cyber Threats

The enhancement of CPPS survivability against cyber threats necessitates the strategic allocation of defensive resources. By configuring varying quantities and types of defensive measures across atomic services, distinct defense effects can be achieved. Accounting for the constraints inherent in defense resource allocation, diverse schemes for defense resource configuration are formulated. The optimization variable within the framework comprises the quantity and category of protective measures allocated to elementary services. Subsequently, the consequences of cyberattacks are modeled, and the efficacy of survivability protection is assessed through the computation of pertinent indices.

2.4.3.1 Objective Function

The survivability fusion indexes are employed to assess system survivability quantitatively. The target equation is proposed as follows:

$$\max F(L) = h([RI_S, TI_S, AP_S, RR_S, 1 - AD_S, 1 - RT_S]) \tag{2.43}$$

In the above, L denotes the allocation matrix for atomic service protective measures. The columns correspond to different elementary services in quantity, whereas the rows correspond to the types of protective measures in quantity. The elements within the matrix indicate the count of various protective measures assigned to various elementary services.

2.4.3.2 Constraints

The following constraints govern the allocation of resources within the system:

1) Total Defensive Resource Limitation:

$$\sum_{m=1}^{M}\sum_{j=1}^{N_m}(C_{idemj} + C_{resmj} + C_{recmj}) \leq C_0 \tag{2.44}$$

2) Cost Cap Constraint for Each Defensive Resource:

$$\begin{matrix} C_{idemj} \leq C_{ide0} \\ C_{resmj} \leq C_{res0}, \\ C_{recmj} \leq C_{rec0} \end{matrix} \quad \begin{matrix} j = 1, \dots, N_m \\ m = 1, \dots, M \end{matrix} \tag{2.45}$$

2.4.4 Modified Genetic Algorithm for the Proposed Model

This chapter employs an enhanced genetic algorithm, incorporating the following improvements:

1) **Encoding Process:** Decimal encoding is adopted for encoding.
2) **Fitness Function:** The target equation of the proposed framework is not directly utilized as the fitness criterion. Instead, the subsequent fitness criterion is employed:

$$H = \begin{cases} F - F_{\min} + \dfrac{1}{1+e^{-t}}(F_{\max} - F_{\min}) & \dfrac{F - \overline{F}}{\overline{f} - f_{\min}} < k \\ \dfrac{F}{1+e^{t}} & \dfrac{F - \overline{F}}{\overline{F} - F_{\min}} \geq k \end{cases} \tag{2.46}$$

Here, H refers to a revised fitness criterion, with F being the value of the objective function in Eq. (2.43). $F_{\max}$ and $F_{\min}$ represent the peak and trough values of the target function in the absence of considering constraints. $\overline{F}$ denotes the mean of the peak and trough values of the target function.

t denotes an evolutionary algebra, with its figure changing based on the environment.

3) **Operators:** Roulette selection operator, adaptive crossover operator, and adaptive mutation operator are employed.
4) **Chaotic Disturbances:** Chaotic perturbations are embedded into the algorithm. Following selection, recombination, and mutation processes, entities with poor fitness are included in the offspring population, thereby mitigating the risk of the population converging to a local optimal solution.

Below is a detailed account of how the enhanced genetic algorithm is applied to the allocation of defensive resources for ensuring the resilience of communication and information systems.

Improved Genetic Algorithm Flow

Input: Configure the defense effect of different defense measures. Set the maximum iteration count as λ. Assign the significance of core services in the system as W_{Sm}. Assign the importance of elementary services in core services as W_{Amj}. Set the group scale as M_G. Define the encoding method for individuals as C. Determine the termination evolutionary generation as T. Set the crossover rate as P_c. Set the mutation probability as P_v.

Output: Defensive strategy deployment plan.

Initialization: Set the iteration count t to 0 initially. Configure the protective measures regarding elementary services.

Repeat

1) Under the given constraints, randomly assign varying quantities of protective measures to elementary services. During encoding, sort nodes numerically by their identifiers. Represent the defensive resource quantities of atomic services in encoded form to acquire the individual gene C_g, while producing the initial population Q_0.
2) Factor in the defensive effect data derived from the number and type of protective measures deployed on the elementary services. Compute the survivability assessment metrics AP_S, RI_S, TI_S, RR_S, AD_S, and RT_S using the equation in Section II. Calculate the fusion index. Moreover, compute the objective function and derive the fitness of the individual H.
3) Execute crossover, selection, and mutation processes. Remove individual C_n that violates the protection resource cost constraint, and form the offspring population R_t.

(Continued)

4) Perform chaotic disturbance procedures, and randomly reintroduce individuals eliminated for poor fitness to maintain population variability, resulting in population Q_{t+1}.
5) When the termination criterion is fulfilled (i.e., the iteration upper limit is reached), proceed to step (6). Otherwise, return to step (2).
6) Compute the fitness of the final population and acquire the optimum resolution set.

End

2.4.5 Simulations

This section focuses on utilizing the IEEE 30-node system as a case study. A detailed illustration of the system's single-line diagram, alongside its communication and information infrastructure, is provided in Figure 2.31.

2.4.5.1 Introduction of the Studied System

Core Business Elementary Services The SSS comprises a range of essential functions within the power network. These functions begin with DC at power grid nodes, where measurements including frequency, voltage, and electric power are gathered. The collected data are then uploaded (DU) to the substation, where Data Processing (DP) is performed. The processed information is subsequently transmitted to the CC through CN.

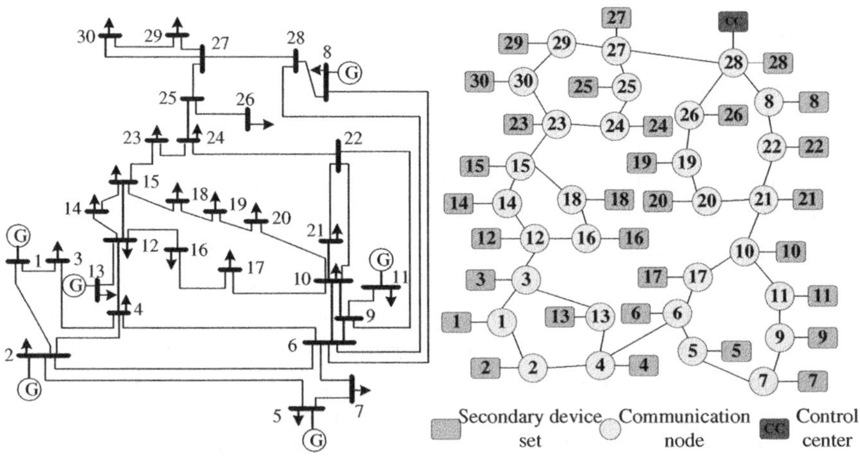

Figure 2.31 IEEE-30 bus system and its communication network schematic. *Source:* Wu et al. [108] / with permission of IEEE.

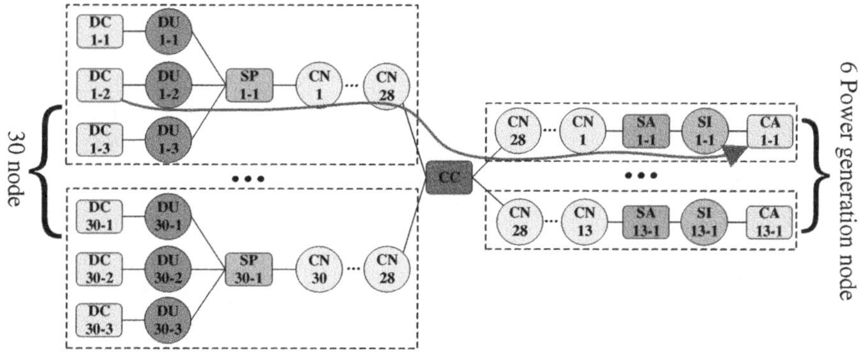

Figure 2.32 Schematic of decomposing SSS into elementary services. *Source:* Wu et al. [108] / with permission of IEEE.

Using the received information, the CC issues operational decisions, which are sent back to the substation through CN. At the substation, the commands are analyzed (SA), issued (SI), and used to control the actions (CA) of electrical components, such as adjusting the power output of a generator.

Accordingly, the SSS can be broken down into a set of elementary services, each representing a minimal, self-contained functional unit within the system. These atomic services are illustrated in Figure 2.32.

The functions of the RP system encompass DC of current, voltage, and phase information, DU, DP, communication via CNs, control decisions at the CC, subsequent CN transmission, Service Analysis (SA), Service Issuance (SI), and Control Actions (CAs) such as regulating the power output of generating units.

In accordance with these operations, the RP can be broken down into a set of elementary services, which are illustrated in Figure 2.33.

The functions of the PD system include DC of frequency, voltage, and power measurements, DU, DP, communication through CNs, decision-making at the CC, subsequent CN transmission, SA, SI, and CAs, such as adjusting power output at power plants and loads.

In accordance with these operations, the PD can be broken down into a set of elementary services, as illustrated in Figure 2.34.

Three primary categories encompass the functions of atomic services: Data Acquisition Atomic Services (DAAS), Data Transfer and Processing Atomic Services (DTPAS), and Control Action Atomic Services (CAAS). The total count of atomic services across all power system operations is recapitulated in Table 2.9.

The CN encompasses all critical services within an identical network topology—notably, the DAAS in the PD and some of the DAAS in the SSS overlap. In conclusion, there are 180 DAAS, 378 DTPAS, and 54 CAAS, resulting in 612 atomic services.

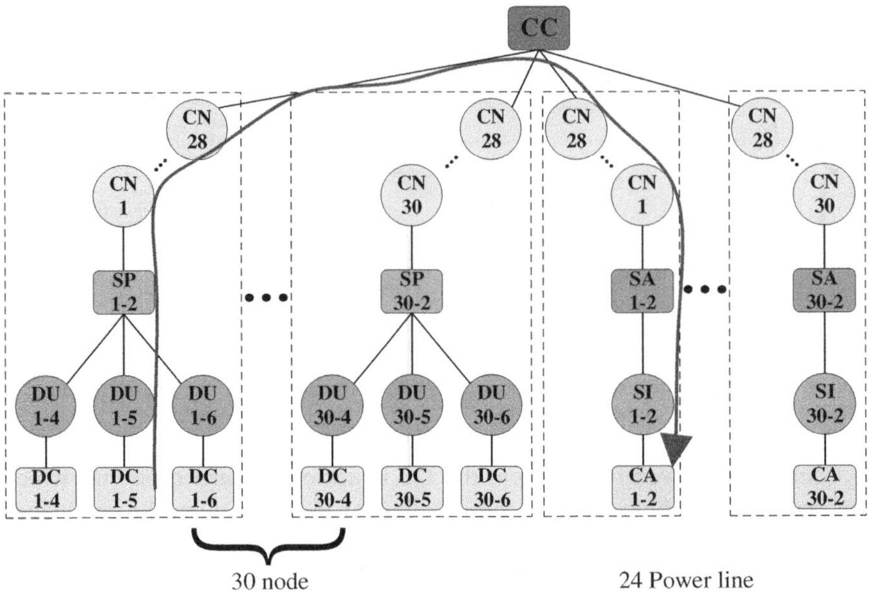

Figure 2.33 Schematic of decomposing RP into elementary services. *Source:* Wu et al. [108] / with permission of IEEE.

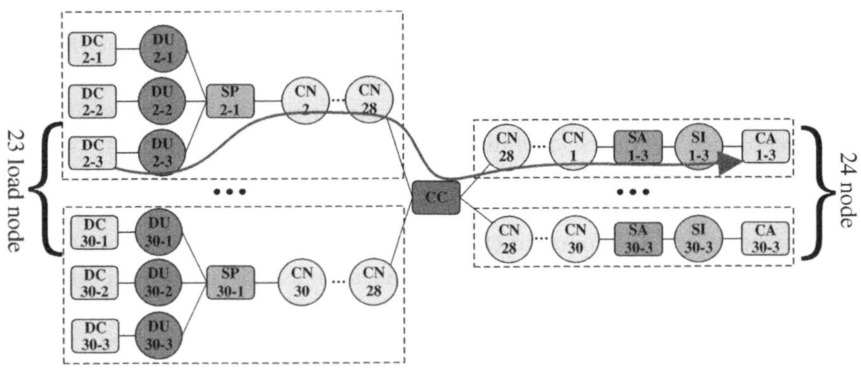

Figure 2.34 Schematic of decomposing PD into elementary services. *Source:* Wu et al. [108] / with permission of IEEE.

Cyber Threat Library In the simulation, three distinct cyber threats denoted as F_1, F_2, and F_3 are introduced. The likelihood of each atomic service being targeted by these cyber threats is outlined in Table 2.10.

Protective Resources Library Seven defensive tactics, which encompass three types of defensive resources, are employed to enhance the resilience of elementary

Table 2.9 Count of atomic services across all power businesses.

Power businesses	Atomic services	Number
SSS	DAAS	90
	DTPAS	162
	CAAS	6
RP	DAAS	90
	DTPAS	198
	CAAS	24
PD	DAAS	69
	DTPAS	170
	CAAS	24

Table 2.10 Likelihood of three atomic services undergoing three cyber threats.

Atomic services	F_1 (%)	F_2 (%)	F_3 (%)
DAAS	50	25	25
DTPAS	25	50	25
CAAS	25	25	50

service modules. Given the challenge of precisely quantifying the cost of these defensive measures, this chapter refrains from specifying their exact values. However, all types of defensive measures have their costs and cost caps outlined in Table 2.11.

Weights of Atomic Services for Various Power Businesses Acknowledging the varying impact of atomic services on the survivability of power businesses, weights assigned to individual atomic services, as presented in Table 2.12, are utilized in the computation of survivability evaluation indices.

2.4.5.2 Simulation Results and Analysis

Impact of Defensive Measures Initially, the influence of protective measures on calculating survivability indices is obtained through simulation. Considering three types of atomic services, each vulnerable to three distinct cyberattacks, a total of

Table 2.11 Composition of defensive resources and their associated costs.

Defensive resource	Resistibility defensive resources			Identifiability defensive resources		Recoverability of defensive resources	
Defensive measure	**Firewall technology**	**Access control technology**	**Deployment of camouflage components**	**Intrusion detection technology**	**Honeypot technology**	**Redundant components**	**Data backup and recovery technology**
Cost	1	1	2	1	2	3	2
Defensive resource cost limit for atomic services	8			6		10	
Defensive resource cost limit for the system				12,000			

Table 2.12 Weights of elementary service modules across various power businesses.

Critical business	SSS	RP	PD
Atomic service	WA1j	WA2j	WA3j
DC	0.003	0.0025	0.003
DU	0.003	0.0025	0.003
SP	0.005	0.0035	0.0035
CN	0.005	0.0045	0.004
SA	0.01	0.0045	0.006
SI	0.01	0.004	0.005
CA	0.01	0.004	0.005

nine attack scenarios are analyzed. The results are presented in Figure 2.35, where the horizontal axis denotes the count of defensive measures deployed. The vertical axis is divided into three categories: (1) the impact of deploying firewall technology, access control, and camouflage components on ADS and APS; (2) the impact of deploying intrusion identification systems and honeypot technology on RIS and TIS; and (3) the influence of incorporating backup components, data backup, and recovery technologies on RTS and RRS.

Defense Resources Allocation Scheme The survivability assessment indices can be computed using the data from Figure 2.35. Subsequently, the defensive resource allocation model is tackled using the modified genetic algorithm. The resulting protection resource allocation plan is laid out in Figure 2.36. Vectors serve to denote the count of protective actions configured, such as $[a, b, c, d, e, f, g]$, indicating the quantities of firewall security mechanisms, access governance technology, camouflage elements, intrusion identification technology, honeypot technology, redundant standby modules, and data backup and restoration technologies as a, b, c, d, e, f, and g, respectively.

Figure 2.36 shows the allocation of numerous defense resources across all CNs, as executing all critical business operations necessitates the involvement of CNs. Additionally, a significant allocation of defense resources is observed for atomic services linked to load or generator nodes. This allocation is attributed to the utilization of data collected from these nodes in both the SSS and the PD. Consequently, a significant portion of defensive resources is allocated toward elementary services related to nodes associated with electrical loads and power generators.

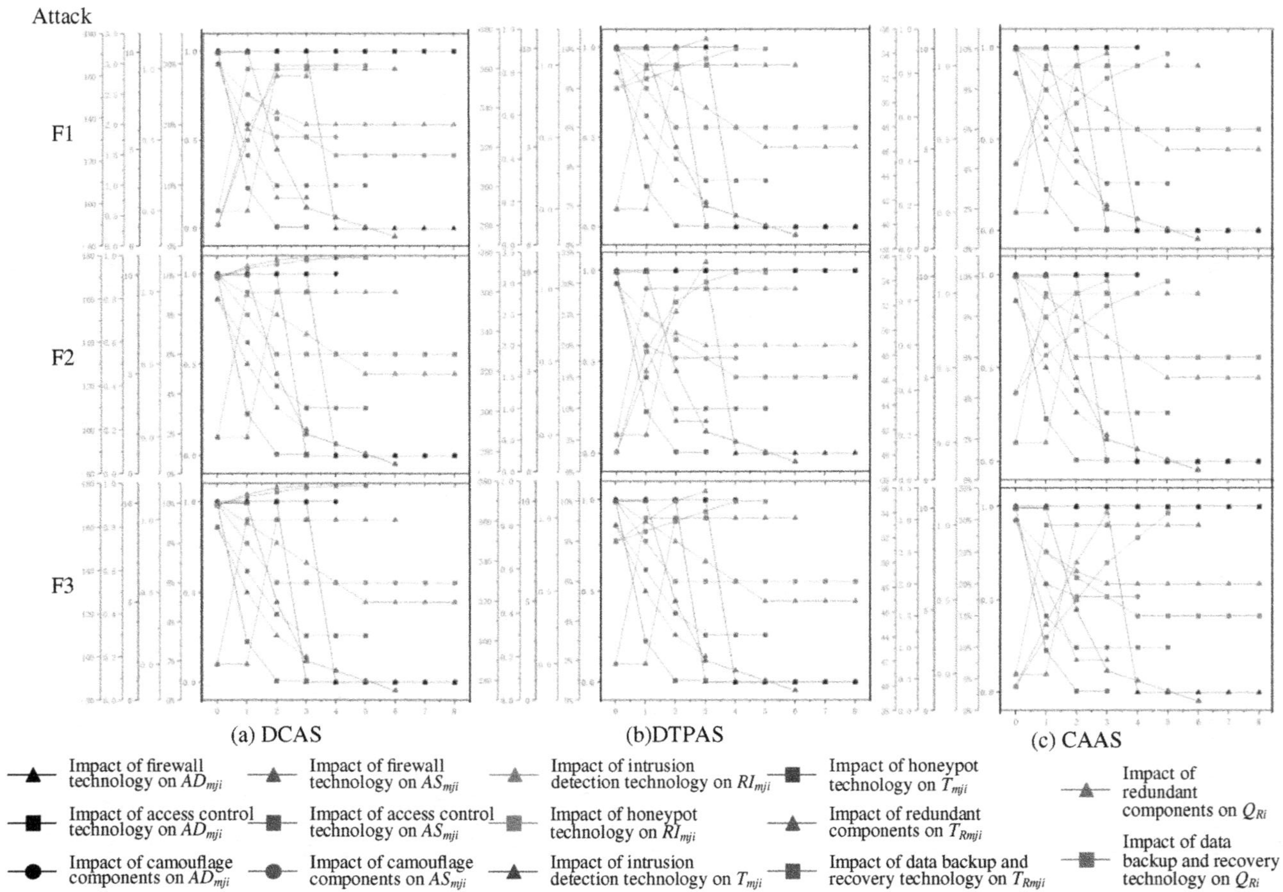

Figure 2.35 The effect of defensive measures. (a) DAAS; (b) DTPAS; (c) CAAS. *Source:* Adapted from Wu et al. [108].

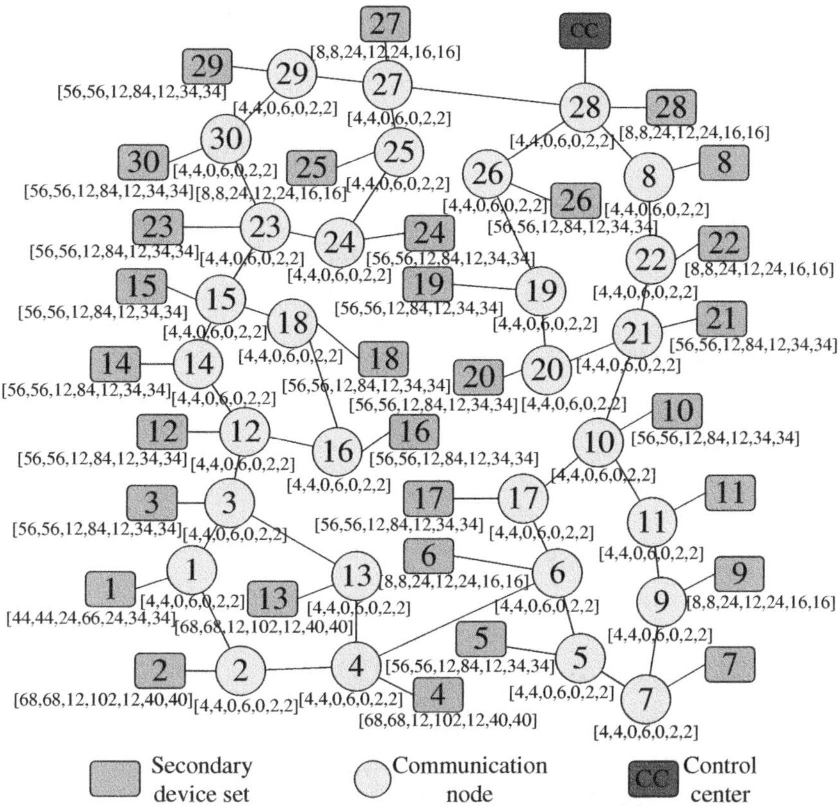

Figure 2.36 Overall allocation of defensive resources. *Source:* Wu et al. [108] / with permission of IEEE.

For a more detailed insight into the arrangement of defense resources on elementary services within the SSS, RP, and PD, refer to Figure 2.37.

Effect of Survivability Improvement A total of 1000 simulations were conducted, in which the system, equipped with defensive resources, was exposed to randomly selected cyberattacks. In parallel, another set of 1000 simulations was performed for the system without any defensive measures. Based on these simulation results, the target function of the suggested model and the integrated viability indices were calculated. The outcomes of this analysis are compiled in Table 2.13.

Table 2.13 demonstrates that following the configuration of defensive resources, the system's survivability experiences a substantial enhancement. This outcome provides empirical evidence supporting the effectiveness of the suggested approach.

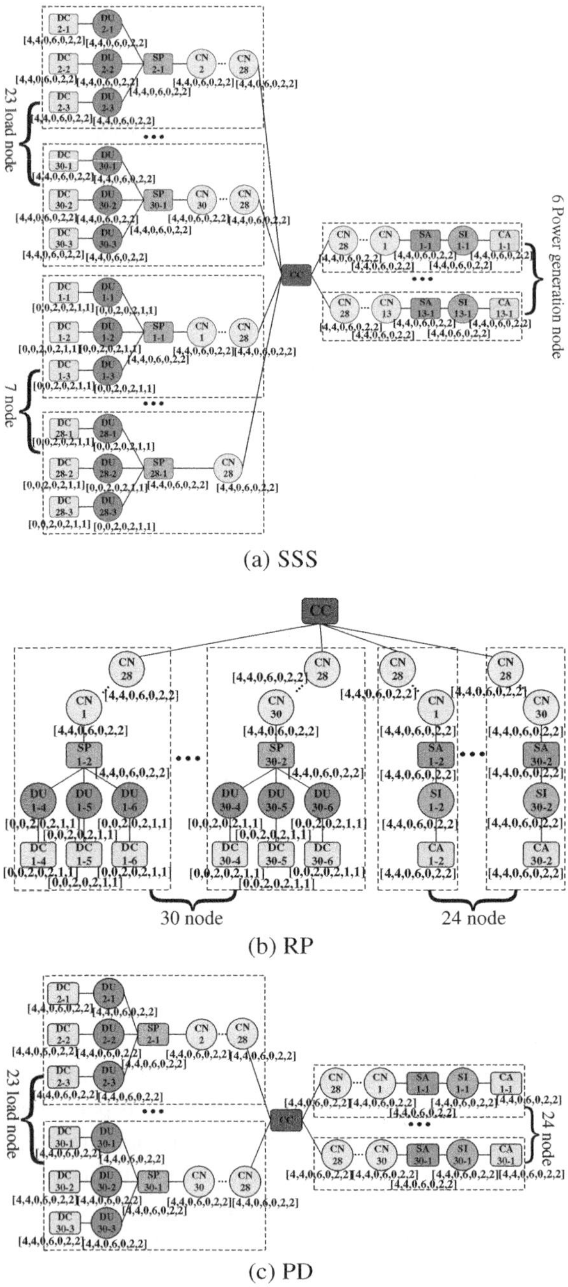

Figure 2.37 The allocation of defensive resources for each business. (a) SSS; (b) RP; (c) PD. *Source:* Wu et al. [108] / with permission of IEEE.

Table 2.13 Viability comparison.

Whether to configure defensive resources	No	Yes
Fusion index	0.5938	0.7413

References

1 Wu, Y., Yingtao, R., Lin, Z. et al. (2023). Research on cyber attacks and defensive measures of power communication network. *IEEE Internet of Things Journal* 10 (9): 7613–7635. https://doi.org/10.1109/JIOT.2022.3181698.

2 Lu, A.-Y. and Yang, G.-H. (2020). Resilient observer-based control for cyber-physical systems with multiple transmission channels under denial-of-service. *IEEE Transactions on Cybernetics* 50 (11): 4796–4807. https://doi.org/10.1109/TCYB.2019.2915942.

3 Peng, C., Li, J., and Fei, M. (2017). Resilient event-triggering H_∞ load frequency control for multi-area power systems with energy-limited DoS attacks. *IEEE Transactions on Power Systems* 32 (5): 4110–4118. https://doi.org/10.1109/TPWRS.2016.2634122.

4 Du, D., Li, X., Li, W. et al. (2019). ADMM-based distributed state estimation of smart grid under data deception and denial of service attacks. *IEEE Transactions on Systems, Man, and Cybernetics: Systems* 49 (8): 1698–1711. https://doi.org/10.1109/TSMC.2019.2896292.

5 Nishanth, N. and Mujeeb, A. (2021). Modeling and detection of flooding-based denial-of-service attack in wireless *Ad Hoc* network using Bayesian inference. *IEEE Systems Journal* 15 (1): 17–26. https://doi.org/10.1109/JSYST.2020.2984797.

6 Hu, B. and Gharavi, H. (2014). Smart grid mesh network security using dynamic key distribution with Merkle tree 4-way handshaking. *IEEE Transactions on Smart Grid* 5 (2): 550–558. https://doi.org/10.1109/TSG.2013.2277963.

7 Kumar, P., Tripathi, M., Nehra, A. et al. (2018). SAFETY: early detection and mitigation of TCP SYN flood utilizing entropy in SDN. *IEEE Transactions on Network and Service Management* 15 (4): 1545–1559. https://doi.org/10.1109/TNSM.2018.2861741.

8 Islam, S.N., Baig, Z., and Zeadally, S. (2019). Physical layer security for the smart grid: vulnerabilities, threats, and countermeasures. *IEEE Transactions on Industrial Informatics* 15 (12): 6522–6530. https://doi.org/10.1109/TII.2019.2931436.

9 Tushir, B., Dalal, Y., Dezfouli, B., and Liu, Y. (2021). A quantitative study of DDoS and E-DDoS attacks on WiFi smart home devices. *IEEE Internet of Things Journal* 8 (8): 6282–6292. https://doi.org/10.1109/JIOT.2020.3026023.

10 Wang, K., Du, M., Maharjan, S., and Sun, Y. (2017). Strategic honeypot game model for distributed denial of service attacks in the smart grid. *IEEE Transactions on Smart Grid* 8 (5): 2474–2482. https://doi.org/10.1109/TSG.2017.2670144.

11 Liu, Y., Ning, P., and Reiter, M.K. (2011). False data injection attacks against state estimation in electric power grids. *ACM Transactions on Information and System Security (TISSEC)* 14 (1): 1–33.

12 Liang, G., Weller, S.R., Zhao, J. et al. (2017a). The 2015 Ukraine blackout: implications for false data injection attacks. *IEEE Transactions on Power Systems* 32 (4): 3317–3318. https://doi.org/10.1109/tpwrs.2016.2631891.

13 Wang, J., Hui, L.C.K., Yiu, S.-M. et al. (2017). A survey on cyber attacks against nonlinear state estimation in power systems of ubiquitous cities. *Pervasive and Mobile Computing* 39: 52–64.

14 Yuan, Y., Li, Z., and Ren, K. (2011). Modeling load redistribution attacks in power systems. *IEEE Transactions on Smart Grid* 2 (2): 382–390. https://doi.org/10.1109/TSG.2011.2123925.

15 Liu, X. and Li, Z. (2014). Local load redistribution attacks in power systems with incomplete network information. *IEEE Transactions on Smart Grid* 5 (4): 1665–1676. https://doi.org/10.1109/TSG.2013.2291661.

16 Che, L., Liu, X., Li, Z., and Wen, Y. (2019). False data injection attacks induced sequential outages in power systems. *IEEE Transactions on Power Systems* 34 (2): 1513–1523. https://doi.org/10.1109/TPWRS.2018.2871345.

17 Hug, G. and Giampapa, J.A. (2012). Vulnerability assessment of AC state estimation with respect to false data injection cyber-attacks. *IEEE Transactions on Smart Grid* 3 (3): 1362–1370. https://doi.org/10.1109/TSG.2012.2195338.

18 Zhao, J., Zhang, G., Dong, Z.Y., and Wong, K.P. (2016). Forecasting-aided imperfect false data injection attacks against power system nonlinear state estimation. *IEEE Transactions on Smart Grid* 7 (1): 6–8. https://doi.org/10.1109/TSG.2015.2490603.

19 Liu, X., Li, Z., Liu, X., and Li, Z. (2016). Masking transmission line outages via false data injection attacks. *IEEE Transactions on Information Forensics and Security* 11 (7): 1592–1602. https://doi.org/10.1109/TIFS.2016.2542061.

20 Zhao, J., Gómez-Expósito, A., Netto, M. et al. (2019). Power system dynamic state estimation: motivations, definitions, methodologies, and future work. *IEEE Transactions on Power Systems* 34 (4): 3188–3198. https://doi.org/10.1109/TPWRS.2019.2894769.

21 Chaojun, G., Jirutitijaroen, P., and Motani, M. (2015). Detecting false data injection attacks in AC state estimation. *IEEE Transactions on Smart Grid* 6 (5): 2476–2483. https://doi.org/10.1109/TSG.2015.2388545.

22 Kim, J. and Tong, L. (2013). On topology attack of a smart grid: undetectable attacks and countermeasures. *IEEE Journal on Selected Areas in Communications* 31 (7): 1294–1305. https://doi.org/10.1109/JSAC.2013.130712.

23 Chakhchoukh, Y. and Ishii, H. (2015). Coordinated cyber-attacks on the measurement function in hybrid state estimation. *IEEE Transactions on Power Systems* 30 (5): 2487–2497. https://doi.org/10.1109/TPWRS.2014.2357182.

24 Hassan, M.U., Rehmani, M.H., and Chen, J. (2020). Differential privacy techniques for cyber physical systems: a survey. *IEEE Communications Surveys & Tutorials* 22 (1): 746–789. https://doi.org/10.1109/COMST.2019.2944748.

25 Tajer, A. (2019). False data injection attacks in electricity markets by limited adversaries: stochastic robustness. *IEEE Transactions on Smart Grid* 10 (1): 128–138. https://doi.org/10.1109/TSG.2017.2733346.

26 Liu, X., Bao, Z., Lu, D., and Li, Z. (2015). Modeling of local false data injection attacks with reduced network information. *IEEE Transactions on Smart Grid* 6 (4): 1686–1696. https://doi.org/10.1109/TSG.2015.2394358.

27 Daoud, L. and Rafla, N. (2019). Analysis of black hole router attack in network-on-chip. *2019 IEEE 62nd International Midwest Symposium on Circuits and Systems (MWSCAS)*, pp. 69–72. https://doi.org/10.1109/MWSCAS.2019.8884979.

28 Liu, Y., Dong, M., Ota, K., and Liu, A. (2016). ActiveTrust: secure and trustable routing in wireless sensor networks. *IEEE Transactions on Information Forensics and Security* 11 (9): 2013–2027. https://doi.org/10.1109/TIFS.2016.2570740.

29 Vineeth, V.V., Radhika, N., and Vanitha, V. (2015). Intruder detection and prevention in a smart grid communication system. *Procedia Technology* 21: 393–399.

30 Ho, Q., Gao, Y., and Le-Ngoc, T. (2013). Challenges and research opportunities in wireless communication networks for smart grid. *IEEE Wireless Communications* 20 (3): 89–95. https://doi.org/10.1109/MWC.2013.6549287.

31 Xu, J. and Zhang, R. (2015). CoMP meets smart grid: a new communication and energy cooperation paradigm. *IEEE Transactions on Vehicular Technology* 64 (6): 2476–2488. https://doi.org/10.1109/TVT.2014.2345415.

32 Ochola, E.O., Mejaele, L.F., Eloff, M.M., and van der Poll, J.A. (2017). Manet reactive routing protocols node mobility variation effect in analysing the impact of black hole attack. *SAIEE Africa Research Journal* 108 (2): 80–92. https://doi.org/10.23919/SAIEE.2017.8531629.

33 Zaidi, S.A.R. and Ghogho, M. (2012). Stochastic geometric analysis of black hole attack on smart grid communication networks. *2012 IEEE Third International Conference on Smart Grid Communications (SmartGridComm)*, pp. 716–721. https://doi.org/10.1109/SmartGridComm.2012.6486071.

34 Hijazi, S. and Obaidat, M.S. (2019). A new detection and prevention system for ARP attacks using static entry. *IEEE Systems Journal* 13 (3): 2732–2738. https://doi.org/10.1109/JSYST.2018.2880229.

35 Sahu, A., Tippanaboyana, H.K., Hefton, L., and Goulart, A. (2017). Detection of Rogue nodes in AMI networks. *2017 19th International Conference on Intelligent System Application to Power Systems (ISAP)*, pp. 1–6. https://doi.org/10.1109/ISAP.2017.8071424.

36 Paukatong, T. (2005). SCADA security: a new concerning issue of an in-house EGAT-SCADA. *2005 IEEE/PES Transmission & Distribution Conference & Exposition: Asia and Pacific*, pp. 1–5. https://doi.org/10.1109/TDC.2005.1547116.

37 Ahmad, F., Kurugollu, F., Adnane, A. et al. (2020). MARINE: man-in-the-middle attack resistant trust model in connected vehicles. *IEEE Internet of Things Journal* 7 (4): 3310–3322. https://doi.org/10.1109/JIOT.2020.2967568.

38 Yang, Y., McLaughlin, K., Littler, T. et al. (2012). Man-in-the-middle attack test-bed investigating cyber-security vulnerabilities in smart grid SCADA systems. *International Conference on Sustainable Power Generation and Supply (SUPERGEN 2012)*, pp. 1–8. https://doi.org/10.1049/cp.2012.1831.

39 Esfahani, A., Mantas, G., Ribeiro, J. et al. (2019). An efficient web authentication mechanism preventing man-in-the-middle attacks in industry 4.0 supply chain. *IEEE Access* 7: 58981–58989. https://doi.org/10.1109/ACCESS.2019.2914454.

40 Alcaraz, C., Lopez, J., and Wolthusen, S. (2017). OCPP protocol: security threats and challenges. *IEEE Transactions on Smart Grid* 8 (5): 2452–2459. https://doi.org/10.1109/TSG.2017.2669647.

41 Zhang, Y., Wang, L., Xiang, Y., and Ten, C.-W. (2015). Power system reliability evaluation with SCADA cybersecurity considerations. *IEEE Transactions on Smart Grid* 6 (4): 1707–1721. https://doi.org/10.1109/TSG.2015.2396994.

42 Kim, J., Tong, L., and Thomas, R.J. (2014). Data framing attack on state estimation. *IEEE Journal on Selected Areas in Communications* 32 (7): 1460–1470. https://doi.org/10.1109/JSAC.2014.2332032.

43 Liu, R., Vellaithurai, C., Biswas, S.S. et al. (2015). Analyzing the cyber-physical impact of cyber events on the power grid. *IEEE Transactions on Smart Grid* 6 (5): 2444–2453. https://doi.org/10.1109/TSG.2015.2432013.

44 Saxena, N. and Grijalva, S. (2017). Dynamic secrets and secret keys based scheme for securing last mile smart grid wireless communication. *IEEE Transactions on Industrial Informatics* 13 (3): 1482–1491. https://doi.org/10.1109/TII.2016.2610950.

45 Cai, H., Zhang, Q., Li, Q., and Qin, J. (2017). Proactive monitoring via jamming for rate maximization over MIMO Rayleigh fading channels. *IEEE*

Communications Letters 21 (9): 2021–2024. https://doi.org/10.1109/LCOMM.2017.2715337.

46 Lu, X., Yang, W., Cai, Y., and Guan, X. (2019). Proactive eavesdropping via covert pilot spoofing attack in multi-antenna systems. *IEEE Access* 7: 151295–151306. https://doi.org/10.1109/ACCESS.2019.2948078.

47 Shafie, A.E., Chihaoui, H., Hamila, R. et al. (2019). Impact of passive and active security attacks on MIMO smart grid communications. *IEEE Systems Journal* 13 (3): 2873–2876. https://doi.org/10.1109/JSYST.2018.2868291.

48 Yang, W., Zheng, Z., Chen, G. et al. (2020). Security analysis of a distributed networked system under eavesdropping attacks. *IEEE Transactions on Circuits and Systems II: Express Briefs* 67 (7): 1254–1258. https://doi.org/10.1109/TCSII.2019.2928558.

49 Mao, Y., He, Y., Zhang, Y. et al. (2020). Secure TDD MIMO networks against training sequence based eavesdropping attack. *IEEE Transactions on Mobile Computing* 19 (12): 2916–2932. https://doi.org/10.1109/TMC.2019.2937081.

50 Tugnait, J.K. (2016). Detection of active eavesdropping attack by spoofing relay in multiple antenna systems. *IEEE Wireless Communications Letters* 5 (5): 460–463. https://doi.org/10.1109/LWC.2016.2585549.

51 Ye, D., Zhang, T.-Y., and Guo, G. (2019). Stochastic coding detection scheme in cyber-physical systems against replay attack. *Information Sciences* 481: 432–444. https://doi.org/10.1016/j.ins.2018.12.091.

52 Deng, R., Zhuang, P., and Liang, H. (2017). CCPA: coordinated cyber-physical attacks and countermeasures in smart grid. *IEEE Transactions on Smart Grid* 8 (5): 2420–2430. https://doi.org/10.1109/TSG.2017.2702125.

53 Trapiello, C., Puig, V., and Rotondo, D. (2021). A zonotopic set-invariance analysis of replay attacks affecting the supervisory layer. *Systems & Control Letters* 157: 105056. https://doi.org/10.1016/j.sysconle.2021.105056.

54 Soltan, S., Yannakakis, M., and Zussman, G. (2019). REACT to cyber attacks on power grids. *IEEE Transactions on Network Science and Engineering* 6 (3): 459–473. https://doi.org/10.1109/TNSE.2018.2837894.

55 Chatterjee, K. and Khaparde, S.A. (2018). Data-driven online detection of replay attacks on wide-area measurement systems. *2018 20th National Power Systems Conference (NPSC)*, pp. 1–6. https://doi.org/10.1109/NPSC.2018.8771807.

56 Ustun, T.S., Farooq, S.M., and Hussain, S.M.S. (2019). A novel approach for mitigation of replay and masquerade attacks in smartgrids using IEC 61850 standard. *IEEE Access* 7: 156044–156053. https://doi.org/10.1109/ACCESS.2019.2948117.

57 Mohsenian-Rad, A. and Leon-Garcia, A. (2011). Distributed internet-based load altering attacks against smart power grids. *IEEE Transactions on Smart Grid* 2 (4): 667–674. https://doi.org/10.1109/TSG.2011.2160297.

58 Amini, S., Pasqualetti, F., and Mohsenian-Rad, H. (2018). Dynamic load altering attacks against power system stability: attack models and protection schemes. *IEEE Transactions on Smart Grid* 9 (4): 2862–2872. https://doi.org/10.1109/TSG.2016.2622686.

59 Hu, Z., Liu, S., Luo, W., and Wu, L. (2020). Resilient distributed fuzzy load frequency regulation for power systems under cross-layer random denial-of-service attacks. *IEEE Transactions on Cybernetics* 1–11. https://doi.org/10.1109/TCYB.2020.3005283.

60 Peng, D.-T., Dong, J., and Peng, Q. (2020). Overloaded branch chains induced by false data injection attack in smart grid. *IEEE Signal Processing Letters* 27: 426–430. https://doi.org/10.1109/LSP.2020.2974097.

61 Rani, P., Kavita, S.V., and Nguyen, G.N. (2020). Mitigation of black hole and gray hole attack using swarm inspired algorithm with artificial neural network. *IEEE Access* 8: 121755–121764. https://doi.org/10.1109/ACCESS.2020.3004692.

62 Shuaib, K., Trabelsi, Z., Abed-Hafez, M. et al. (2015). Resiliency of smart power meters to common security attacks. *Procedia Computer Science* 52: 145–152.

63 Pasetti, M., Ferrari, P., Bellagente, P. et al. (2021). Artificial neural network-based stealth attack on battery energy storage systems. *IEEE Transactions on Smart Grid* 12 (6): 5310–5321. https://doi.org/10.1109/TSG.2021.3102833.

64 Lakshminarayana, S., Adhikari, S., and Maple, C. (2021). Analysis of IoT-based load altering attacks against power grids using the theory of second-order dynamical systems. *IEEE Transactions on Smart Grid* 12 (5): 4415–4425. https://doi.org/10.1109/TSG.2021.3070313.

65 Das, D., Sharma, U., and Bhattacharyya, D.K. (2015). Detection of cross-site scripting attack under multiple scenarios. *The Computer Journal* 58 (4): 808–822. https://doi.org/10.1093/comjnl/bxt133.

66 Sun, X., Dai, J., Liu, P. et al. (2018). Using Bayesian networks for probabilistic identification of zero-day attack paths. *IEEE Transactions on Information Forensics and Security* 13 (10): 2506–2521. https://doi.org/10.1109/TIFS.2018.2821095.

67 Ao, S. and Guo, S. (2011). A enhancement technology about system security based on dynamic information flow tracking. *2011 2nd International Conference on Artificial Intelligence, Management Science and Electronic Commerce (AIMSEC)*, pp. 6108–6111. https://doi.org/10.1109/AIMSEC.2011.6011436.

68 Wang, X., Pan, C., Liu, P., and Zhu, S. (2010). SigFree: a signature-free buffer overflow attack blocker. *IEEE Transactions on Dependable and Secure Computing* 7 (1): 65–79. https://doi.org/10.1109/TDSC.2008.30.

69 Shao, Z., Xue, C., Zhuge, Q. et al. (2006). Security protection and checking for embedded system integration against buffer overflow attacks via hardware/software. *IEEE Transactions on Computers* 55 (4): 443–453. https://doi.org/10.1109/TC.2006.59.

70 Zander, S., Branch, P., and Armitage, G. (2011). Capacity of temperature-based covert channels. *IEEE Communications Letters* 15 (1): 82–84. https://doi.org/10.1109/LCOMM.2010.110310.101334.

71 Khodke, R. and Dhawas, V.N. (2016). Decentralized approach for jamming illegal access and secure data recovery in military networks. *2016 Conference on Advances in Signal Processing (CASP)*, pp. 344–349. https://doi.org/10.1109/CASP.2016.7746193.

72 Alsaleh, M., Mannan, M., and van Oorschot, P.C. (2012). Revisiting defenses against large-scale online password guessing attacks. *IEEE Transactions on Dependable and Secure Computing* 9 (1): 128–141. https://doi.org/10.1109/TDSC.2011.24.

73 Yan, S., Huang, X., Ma, M. et al. (2017). A novel efficient address mutation scheme for IPv6 networks. *IEEE Access* 5: 7724–7736. https://doi.org/10.1109/ACCESS.2016.2636907.

74 Ustun, T.S. and Hussain, S.M.S. (2020). An improved security scheme for IEC 61850 MMS messages in intelligent substation communication networks. *Journal of Modern Power Systems and Clean Energy* 8 (3): 591–595. https://doi.org/10.35833/MPCE.2019.000104.

75 Su, Z., Xu, L., Xin, S. et al. (2017). A future outlook for cyber-physical power system. *2017 IEEE Conference on Energy Internet and Energy System Integration (EI2)*, IEEE. https://doi.org/10.1109/ei2.2017.8245733.

76 Kim, K. and Kumar, P.R. (2012). Cyber–physical systems: a perspective at the centennial. *Proceedings of the IEEE* 100 (Special Centennial Issue), pp. 1287–1308. https://doi.org/10.1109/jproc.2012.2189792.

77 Tang, Y., Chen, Q., Li, M. et al. (2016). Challenge and evolution of cyber attacks in cyber physical power system. *2016 IEEE PES Asia-Pacific Power and Energy Engineering Conference (APPEEC)*, IEEE. https://doi.org/10.1109/appeec.2016.7779616.

78 Ansari, M.H., Vakili, V.T., Bahrak, B., and Tavassoli, P. (2018). Graph theoretical defense mechanisms against false data injection attacks in smart grids. *Journal of Modern Power Systems and Clean Energy* 6 (5): 860–871. https://doi.org/10.1007/s40565-018-0432-2.

79 Amini, S., Pasqualetti, F., and Mohsenian-Rad, H. (2017). Dynamic load altering attacks against power system stability: attack models and protection schemes. *2017 IEEE Power & Energy Society General Meeting*, IEEE. https://doi.org/10.1109/pesgm.2017.8274329.

80 Vellaithurai, C., Srivastava, A., Zonouz, S., and Berthier, R. (2015). CPIndex: cyber-physical vulnerability assessment for power-grid infrastructures. *IEEE Transactions on Smart Grid* 6 (2): 566–575. https://doi.org/10.1109/tsg.2014.2372315.

81 Srivastava, A., Morris, T., Ernster, T. et al. (2013). Modeling cyber-physical vulnerability of the smart grid with incomplete information. *IEEE Transactions on Smart Grid* 4 (1): 235–244. https://doi.org/10.1109/tsg.2012.2232318.

82 Živković, N. and Sarić, A.T. (2018). Detection of false data injection attacks using unscented Kalman filter. *Journal of Modern Power Systems and Clean Energy* 6 (5): 847–859. https://doi.org/10.1007/s40565-018-0413-5.

83 Yu, Z.-H. and Chin, W.-L. (2015). Blind false data injection attack using PCA approximation method in smart grid. *IEEE Transactions on Smart Grid* 6 (3): 1219–1226. https://doi.org/10.1109/tsg.2014.2382714.

84 Letchford, J. and Vorobeychik, Y. (2013). Optimal interdiction of attack plans. In: *AAMAS*, 199–206. Citeseer. https://citeseerx.ist.psu.edu/document?repid=rep1&type=pdf&doi=7360cdcb3c46bbfec1e5c0fd6edca52a0ea12179.

85 Zhao, J., Zhang, G., La Scala, M. et al. (2017). Short-term state forecasting-aided method for detection of smart grid general false data injection attacks. *IEEE Transactions on Smart Grid* 8 (4): 1580–1590. https://doi.org/10.1109/tsg.2015.2492827.

86 Nandi, A.K., Medal, H.R., and Vadlamani, S. (2016). Interdicting attack graphs to protect organizations from cyber attacks: a bi-level defender–attacker model. *Computers & Operations Research* 75 (November): 118–131. https://doi.org/10.1016/j.cor.2016.05.005.

87 Liu, D., Liu, X., Yu, H. et al. (2018). Research on technology application and security threat of internet of things for smart grid. *2018 5th International Conference on Information Science and Control Engineering (ICISCE)*, IEEE. https://doi.org/10.1109/icisce.2018.00110.

88 Huang, X. and Wang, S. (2015). Aggregation points planning in smart grid communication system. *IEEE Communications Letters* 19 (8): 1315–1318. https://doi.org/10.1109/lcomm.2015.2441722.

89 Meng, F., Zhao, H., Yu, B. et al. (2017). An optimized power terminal communication access network planning based on hierarchical genetic algorithm. *2017 IEEE 9th International Conference on Communication Software and Networks (ICCSN)*, IEEE. https://doi.org/10.1109/iccsn.2017.8230150.

90 Jayanthi, M.K. (2017). Strategic planning for information security-DID mechanism to befriend the cyber criminals to assure cyber freedom. *2017 2nd International Conference on Anti-Cyber Crimes (ICACC)*, IEEE. https://doi.org/10.1109/anti-cybercrime.2017.7905280.

91 Wu, Y., Chen, J., Yingtao, R. et al. (2021). Research on power communication network planning based on information transmission reachability against cyber-attacks. *IEEE Systems Journal* 15 (2): 2883–2894. https://doi.org/10.1109/JSYST.2020.3026997.

92 Duan, X., Yang, Y., Tian, D. et al. (2014). A V2X communication system and its performance evaluation test bed. *2014 IEEE 6th International Symposium on Wireless Vehicular Communications (WiVeC 2014)*, IEEE. https://doi.org/10.1109/wivec.2014.6953212.

93 Zhang, R., Zhao, Z., and Chen, X. (2010). An overall reliability and security assessment architecture for electric power communication network in smart grid. *2010 International Conference on Power System Technology*, IEEE. https://doi.org/10.1109/powercon.2010.5666414.

94 Guo, Y. and Xu, M. (2015). Research on reliability evaluation model and path optimization for power communication network. *2015 5th International Conference on Electric Utility Deregulation and Restructuring and Power Technologies (DRPT)*, IEEE. https://doi.org/10.1109/drpt.2015.7432673.

95 Xiaoyan, S. and Hangtian, X. (2016). Analysis on influence factors of the reliability of electric power communication networks. *2016 IEEE International Conference on Electronic Information and Communication Technology (ICEICT)*, IEEE. https://doi.org/10.1109/iceict.2016.7879650.

96 Zhou, M., Rui, L., Qiu, X. et al. (2018). Evaluation of the node importance in power grid communication network and analysis of node risk. *NOMS 2018 - 2018 IEEE/IFIP Network Operations and Management Symposium*, IEEE. https://doi.org/10.1109/noms.2018.8406214.

97 He, X. (2019). Research on intrusion prediction algorithm based on association attack diagram ASAD. *2019 IEEE 11th International Conference on Communication Software and Networks (ICCSN)*, IEEE. https://doi.org/10.1109/iccsn.2019.8905351.

98 Cheng, L. and Yu, T. (2019). Game-theoretic approaches applied to transactions in the open and ever-growing electricity markets from the perspective of power demand response: an overview. *IEEE Access* 7: 25727–25762. https://doi.org/10.1109/access.2019.2900356.

99 Ebrahimi, A., Dehdeleh, V., Boroumandnia, A., and Seydi, V. (2017). Improved particle swarm optimization through orthogonal experimental design. *2017 2nd Conference on Swarm Intelligence and Evolutionary Computation (CSIEC)*, IEEE. https://doi.org/10.1109/csiec.2017.7940168.

100 Fu, R., Huang, X., Xue, Y. et al. (2019). Security assessment for cyber physical distribution power system under intrusion attacks. *IEEE Access* 7: 75615–75628. https://doi.org/10.1109/access.2018.2855752.

101 Guo, J., Han, Y., Guo, C. et al. (2017). Modeling and vulnerability analysis of cyber-physical power systems considering network topology and power flow properties. *Energies* 10 (1): 87. https://doi.org/10.3390/en10010087.

102 Han, Y., Guo, C., Ma, S., and Song, D. (2018). Modeling cascading failures and mitigation strategies in PMU based cyber-physical power systems. *Journal of Modern Power Systems and Clean Energy* 6 (5): 944–957. https://doi.org/10.1007/s40565-018-0407-3.

103 Liang, G., Zhao, J., Luo, F. et al. (2017b). A review of false data injection attacks against modern power systems. *IEEE Transactions on Smart Grid* 8 (4): 1630–1638. https://doi.org/10.1109/TSG.2015.2495133.

104 Garasym, I. (2011). Information security system survivability assessment method based on logical-probabilistic models. *2011 11th International Conference the Experience of Designing and Application of CAD Systems in Microelectronics (CADSM)*, pp. 160–161. IEEE.

105 Abie, H., Bigham, J., Rotondi, D. et al. (2010). Self-healing and secure adaptive messaging middleware for business-critical systems. *International Journal on Advances in Security* 3 (1&2): 34–51.

106 Huang, S., Zhang, H., Wang, J., and Huang, J. (2018). Markov differential game for network defense decision-making method. *IEEE Access* 6: 39621–39634. https://doi.org/10.1109/ACCESS.2018.2848242.

107 Srivastava, A.K., Ernster, T.A., Liu, R., and Krishnan, V.G. (2018). Graph-theoretic algorithms for cyber-physical vulnerability analysis of power grid with incomplete information. *Journal of Modern Power Systems and Clean Energy* 6 (5): 887–899. https://doi.org/10.1007/s40565-018-0448-7.

108 Wu, Y., Hao, X., and Ni, M. (2020). Defensive resource allocation method for improving survivability of communication and information system in CPPS against cyber-attacks. *Journal of Modern Power Systems and Clean Energy* 8 (4): 750–759. https://doi.org/10.35833/MPCE.2019.000148.

3

Multi-Objective Real-Time Control of Operating Conditions Using Deep Reinforcement Learning

Ruisheng Diao[1], Tu Lan[1], Zhiwei Wang[2], Haifeng Li[2], Chunlei Xu[2], Fangyuan Sun[1], Bei Zhang[3], Yishen Wang[3], Siqi Wang[3], Jiajun Duan[3], and Di Shi[3]

[1] *ZJU-UIUC Institute, Zhejiang University, Haining, China*
[2] *Dispatch Center, State Grid Jiangsu Electric Power Company, Nanjing, China*
[3] *State Grid Smart Grid Research Institute, AI&System Analytics, San Jose, USA*

3.1 Introduction

Over the past decades, the electricity sector has undergone significant changes, with ever-increasing penetration of intermittent energy resources, storage devices, and power-electronics-based equipment integrated into both transmission and distribution grids, causing more stochastic and dynamic behavior that affects the secure and economic operation of the power system. Various control measures exist to restore the imbalance of active and reactive power and ensure voltage, frequency, and line flows operate within their normal ranges. Compared to mature controllers that function well at local levels, deriving system-wide optimal controls by coordinating many controllers for real-time operating conditions while complying with all security constraints remains a grand challenge.

Such a control problem is known as security-constrained AC Optimal Power Flow (ACOPF), where objectives can be minimization of total generation costs, total network losses, amount of control actions, or a combination of all. Many prior research efforts were proposed to solve the large-scale non-convex ACOPF problem considering various constraints with mixed-integer variables, including nonlinear programming, quadratic programming, Lagrangian relaxation, interior point method, and Mixed-Integer Linear Programming (MILP). A comprehensive survey is conducted in [1] that summarizes the main challenges and state-of-the-art techniques in solving the real-time OPF problem. In [2, 3], semidefinite programming (SDP) relaxation-based methods are proposed for solving the multi-objective OPF problem. In [4], the authors proposed a quasi-Newton method using second-order information for solving real-time OPF problems.

Artificial Intelligence Empowered Smart Energy Systems, First Edition. Edited by Qiang Yang and Gang Huang.

However, due to the non-convex and complex nature of the problem (NP-Hard), deriving solutions for the ACOPF problem with all security constraints in a real-time environment is very challenging. Thus, relaxation of constraints and simplifications of system models, e.g., direct current (DC)-based OPF, are typically used to obtain feasible and suboptimal solutions. To effectively deal with the abovementioned challenges and derive fast OPF solutions, new approaches are much-needed [5].

The recent success of applying DRL technologies in various control problems, including the game of GO, self-driving cars, and robotics, provides a promising way of deriving instant control actions for reaching feasible and suboptimal ACOPF solutions. Recently, DRL-based algorithms were reported for regulating voltage profiles [6] and transmission line flows [7], load-frequency control [8], and emergency control for enhancing transient voltage behavior [9]. In this chapter, a novel method is proposed to perform multi-objective power flow adjustment in real-time, including regulating regional voltage profiles and reducing transmission losses without overloading lines, using state-of-the-art DRL algorithms. The novelty of the proposed method includes: (1) the operating condition control problem is formulated as a Markov Decision Process (MDP) where reinforcement learning-based algorithms can be applied to derive suboptimal solutions; (2) it provides a general and flexible framework to include various types of control objectives and constraints when training effective AI agents; (3) once properly trained, the AI agent can provide sub-second control actions to regulate bus voltages, line flows, and transmission losses once sensing the current system states, in case of abrupt changes in voltages and line flows; and (4) multi-threading-based training process of DRL agents with periodical model update is developed to ensure long-term control effectiveness and mitigate the overfitting issue.

3.2 Principles of Deep Reinforcement Learning

DRL is a learning problem aiming at maximizing a long-term reward through massive interactions with the environment, as illustrated in Figure 3.1, where s_t is the state observed by the agent at time t, r_t is the reward received by the agent from the environment at time t, and a_t is the action taken by the agent based on its observation and policy. The agent tries to maximize its reward by continuously interacting with the environment and then acquiring the resulting next state together with the corresponding reward feedback.

In such a process, the set of states (S), the set of actions (A), the transition probability from state s to s' with a given action a, denoted as $P(s, a, s')$, the probability of receiving a reward r from action a and state s, denoted as $R(s, a, r)$, and together with a discount factor on the future reward (γ) form an MDP, denoted

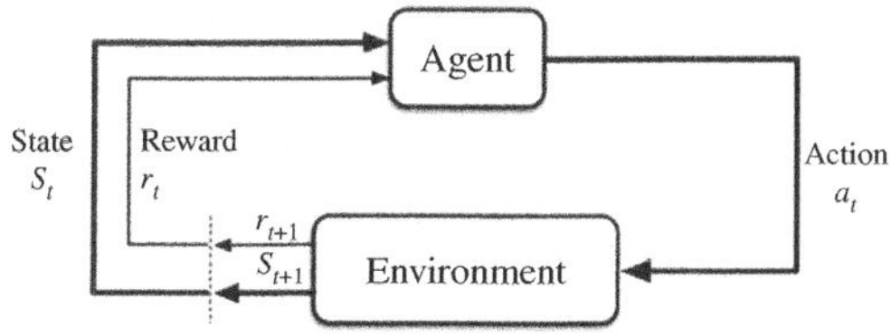

Figure 3.1 The framework of reinforcement learning.

as a tuple$\langle S, A, P, R, \gamma \rangle$. The agent learns to find a policy $\pi(a|s)$, which maps the action from a given state to maximize the long-term reward it receives, as presented in Eq. (3.1). Two other important concepts in RL are state-value function $V^\pi(s)$ and action-value function $Q^\pi(s, a)$, as defined in Eqs. (3.2) and (3.3), respectively. Function $V^\pi(s)$ evaluates how good a state is by calculating the expected reward for an agent starting from such state following a certain policy, whereas $Q^\pi(s,a)$ assesses how good an action is from a certain state by calculating the expected reward for an agent starting from state s with action a and following policy π.

$$G_t = \sum_{t=0}^{\infty} \gamma^t r_t \tag{3.1}$$

$$V^\pi(s) = \mathbb{E}(G_t | s_t = s; \pi) \tag{3.2}$$

$$Q^\pi(s, a) = \mathbb{E}(G_t | s_t = s, a_t = a; \pi) \tag{3.3}$$

In DRL, one usually approximates the policy through a neural network (NN) with parameters θ, denoting the weights and biases in the NN. Therefore, the policy can be expressed as $\pi_\theta(a|s)$, and the agent learns to update θ to achieve the maximum reward.

3.2.1 Deep Q Network (DQN)

One typical example of DRL algorithms is Q-learning, which utilizes a Q-table to map each state and action pair using an action value, $Q(s, a)$, and evaluates the action a taken at state s by considering the future cumulative return R_t. According to the Bellman Equation [10], the cumulative return can be represented as an expected return, shown in Eq. (3.4):

$$Q(s, a) = \mathbb{E}[r_t + \gamma Q(S_{t+1}, A_{t+1}) \mid S_t = s, A_t = a] \tag{3.4}$$

To obtain the optimal action value $Q^*(s, a)$, the Q-learning algorithm looks one step ahead after taking action a at the state s_t, and greedily considers the action a_{t+1} at state s_{t+1} for maximizing the expected target value $r_t + \gamma Q^*(s_{t+1}, a_{t+1})$. Using the Bellman equation, the algorithm can perform online updates to control the Q-value toward the Q-target.

$$Q(s_t, a_t) \leftarrow Q(s_t, a_t) + \eta[r_t + \gamma \max_{a_{t+1} \in A} Q(s_{t+1}, a_{t+1}) - Q(s_t, a_t)] \tag{3.5}$$

where η represents the learning rate. Using a Q-table, both the state and action need to be discrete, thus making it difficult to handle complex problems. To overcome this issue, the DQN method was developed, which uses NNs as a function approximator to estimate the Q-values, $Q(s, a)$, so it can support continuous states in the RL process without discretization of states or building the Q-table. Weights θ of the NN represent the mapping from state to Q-value, and, therefore, a loss function $L_i(\theta)$ is needed to update the weights and corresponding Q-values, using Eq. (3.6) [11]:

$$L_i(\theta_i) = \mathbb{E}_{s,a\sim\rho(\cdot)}\left[(y_i - Q(s, a; \theta_i))^2\right] \tag{3.6}$$

where $y_i = \mathbb{E}_{s'\sim\mathcal{E}}\left[r + \gamma max_{a'} Q(s', a'; \theta_{i-1})|s, a\right]$, and ρ is the probability distribution of the state and action pair (s, a). By differentiating the loss function using Eq. (3.7) and performing stochastic gradient descent, the weights of the agent can be updated [11].

$$\nabla_{\theta_i} L_i(\theta_i) = \mathbb{E}_{s,a\sim\rho(\cdot);s'\sim\mathcal{E}}[(r + \gamma \max_{a'} Q(s', a'; \theta_{i-1}) - Q(s, a; \theta_i))\nabla_{\theta_i} Q(s, a; \theta_i)] \tag{3.7}$$

However, overestimation is a well-known and long-standing problem for all Q-learning-based algorithms. To address this issue, Double DQN (DDQN), which decouples the action selection and action evaluation using two separate NNs, is proposed in [12]. It demonstrates good performance in mitigating the overestimation problem and can obtain better results on ATARI 2600 games than other Q-learning-based methods. In addition, a new model architecture, Dueling DQN, is proposed in [13], which decouples the single-stream DDQN into a state-value stream and an action-advantage stream, and, therefore, the Q-value can be represented as Eq. (3.8).

$$Q(s, a; \theta, \alpha, \beta) = V(s; \theta, \beta) + \left(A(s, a; \theta, \alpha) - \frac{1}{|A|}\sum_{a'} A(s, a'; \theta, \alpha)\right) \tag{3.8}$$

The stand-alone state value stream is updated during every agent learning process. The frequently updated state values and the biased advantage values allow a better approximation of the Q-values, which is the key in value-based methods, allowing more accurate and stable updates for the agent.

3.2.2 Proximal Policy Optimization (PPO)

Like other reinforcement learning algorithms, PPO also originates from a policy gradient, which computes an estimator of the policy gradient and plugs it into a stochastic gradient ascent algorithm [14]. Reference [15] gives a general form of the policy gradient algorithm used to update the parameters θ, as in Eqs. (3.9),(3.10), where η is the learning rate.

$$\nabla_\theta J(\theta) = \nabla_\theta V^\pi(s_0) = \mathbb{E}_\pi[Q^\pi(s,a)\nabla_\theta \ln \pi_\theta(a|s)] \tag{3.9}$$

$$\theta \leftarrow \theta + \eta\nabla_\theta J(\theta) \tag{3.10}$$

However, this plain PG algorithm does not guarantee good convergence [16]. To resolve this issue, the PPO algorithm adopts two advancements for better stability and convergence.

First, an advantage function, defined in Eq. (3.11), is utilized during parameter update instead of the action-value function in Eq. (3.9) to reduce the variance of the estimation [16].

$$Ad^\pi(s,a) = Q^\pi(s,a) - V^\pi(s) \tag{3.11}$$

Second, PPO tries to update the parameters within a certain trust region, and the motivation behind it is illustrated in Figure 3.2. Suppose one starts from point A and tries to climb up to point B; the ascent direction is shown as the arrow. If, unfortunately, a large update step is taken in this direction, one can easily get trapped in the circled valley area, which might be a region hard to escape from. In the real training, the hypersurface is likely much more complex than that in Figure 3.2, and therefore, it is necessary to regulate the update process within a certain trust region. PPO does this by further modifying the $J(\theta)$ in Eqs. (3.9)–(3.13).

$$J(\theta) = \mathbb{E}_{\pi_{\theta_{\mathrm{old}}}}\left[min\left(\frac{\pi_\theta(a|s)}{\pi_{\theta_{\mathrm{old}}}(a|s)}Ad^{\pi_{\theta_{\mathrm{old}}}}, g\left(\varepsilon, Ad^{\pi_{\theta_{\mathrm{old}}}}\right)\right)\right] \tag{3.12}$$

$$g(\varepsilon, Ad) = \begin{cases} (1+\varepsilon)Ad, & Ad \geq 0 \\ (1-\varepsilon)Ad, & Ad < 0 \end{cases} \tag{3.13}$$

where θ_{old} is the present parameter in the NN, and ε regulates the size of the trust region for the update. The minimization operator makes sure that the new policy

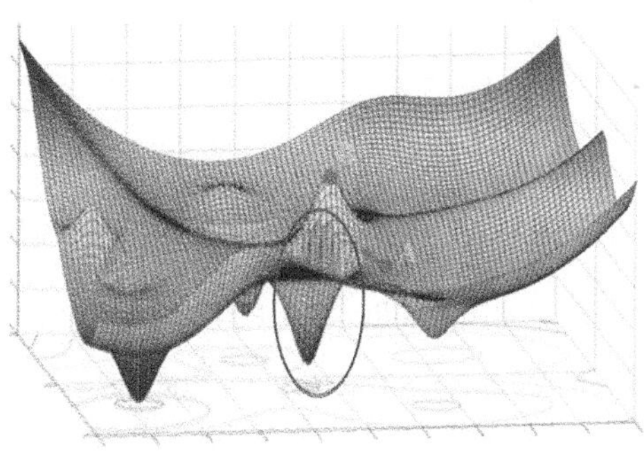

Figure 3.2 Illustration of the motivation behind PPO [16] / MDPI / Public Domain.

does not benefit from going too far away from the old policy and thus regulates the update of the parameter. For example, when Ad is positive, the objective $J(\theta)$ will increase if π_θ $(a|s)$ increases to behave more likely toward the old policy, and the minimum operation determines the ceiling, which is $(1+\varepsilon)Ad$.

Because of these two advancements, the PPO algorithm is usually more stable during an update, which can be very effective when the action space is huge. Considering the fact that a large power grid usually has many controllers for use, the PPO algorithm can be adapted to perform the power flow control for real-time grid operation.

3.2.3 Soft Actor-Critic (SAC)

Similar to standard RL formulation, SAC also employs value function and Q function. However, standard RL aims to maximize the expected return (sum of rewards) $\sum_t E_{(s_t;a_t)\sim\rho_\pi}[R(s_t, a_t)]$ only, while SAC trains a stochastic policy with entropy regularization, which means it maximizes not only the expected return but also the entropy of the policy [17]. Then, the optimal policy π^* is updated using (3.14):

$$\pi^* = \arg \max_\pi \sum_t E_{(s_t;a_t)\sim\rho_\pi}[R(s_t, a_t) + \alpha H(\pi \cdot |s_t))] \tag{3.14}$$

where $H(\pi \cdot | s_t)$ is the entropy of the policy at the state s_t, and the coefficient α controls the tradeoff between exploration and exploitation. Compared to deterministic policy, stochastic policy enables a stronger exploration capability [17], which is especially useful for control problems with relatively small feasible solution regions. In SAC, policy evaluation and improvement are achieved through training the NNs with stochastic gradient descent as well. The value function $V_\psi(s_t)$ and the Q-function $Q_\theta(s_t, a_t)$ are parameterized through NNs with parameters ψ and θ. The soft value function networks are trained to minimize the squared residue error, as shown in (3.15):

$$J_V(\psi) = E_{s_t\sim D}[(V_\psi(s_t) - V_{soft}(s_t)]^2 \tag{3.15}$$

$$V_{soft}(s_t) = E_{a_t\sim\pi}[Q_{soft}(s_t, a_t) - \alpha \ \log \pi(a_t|s_t)] \tag{3.16}$$

Also, the soft Q function is trained by minimizing (3.17):

$$J_\theta(Q) = E_{(s_t,a_t)\sim D}[Q_\theta(s_t, a_t) - \hat{Q}(s_t, a_t)]^2 \tag{3.17}$$

$$\hat{Q}(s_t, a_t) = R(s_t, a_t) + \gamma E_{s_{t+1}\sim p}[V_{\hat{\psi}}(s_{t+1})] \tag{3.18}$$

where $V_{\hat{\psi}}(s_{t+1})$ is the target value network that is updated periodically. Different from the value and Q functions that are directly modeled with expressive NNs, the output of the policy NN follows the Gaussian distribution with mean and covariance [18]. The policy parameters can be learned by minimizing the expected Kullback–Leibler (KL) divergence as (3.19):

$$\begin{aligned} J_\pi(\phi) &= D_{KL}(\pi(\cdot|s_t)\| \left(\exp\left(\frac{1}{\alpha}Q_\theta(s_t,\cdot)\right) - \log Z(s_t)\right) \\ &= E_{s_t\sim D}[E_{a_t\sim\pi_\phi}[\alpha\log(\pi_\phi(a_t|s_t)) - Q_\theta(a_t,s_t)]] \end{aligned} \tag{3.19}$$

3.3 Real-Time Line Flow Control Using PPO

3.3.1 Problem Formulation

The goal is to regulate line flows of targeted lines in real-time through re-dispatching generation without shedding loads. The DRL agent for line flow control can be first trained offline through massive simulations by incorporating the experience of grid operators and then applied online after convergence and verification. A well-trained agent can provide appropriate control action momentarily when facing abnormal conditions. During online training, the agent can further adapt itself to a new, dynamically changing environment. Several important aspects of designing and implementing the PPO-based RL agent, including state, reward, and action, are as follows.

3.3.1.1 State Space

The state space for training the agent can be acquired from Phasor Measurement Unit (PMU) data or EMS snapshots, which include voltage magnitudes at buses of interest, apparent power on transmission lines, and power outputs of generators.

3.3.1.2 Action Space

In general, the RL agent is trained to adjust the real power outputs of generators to regulate line flows. Note that the generator on the slack bus cannot be set since it is used to balance all the power imbalances in the system. Consequently, the unit(s) on the slack bus are excluded from taking control actions. In forming the MDP, active power adjustments to generators are selected as the RL agent's action space $[\Delta P_1, \Delta P_2, \dots, \Delta P_n]$, instead of generator outputs directly from the dispatch $[P_1, P_2, \dots, P_n]$, to correlate states and actions further.

3.3.1.3 Reward Function

The reward function is the feedback received by the RL agent, based on which the agent learns to perform a task well. A well-designed reward function not only guides the agent toward an effective update direction but also speeds up the entire training process. In this work, the reward is designed to be positive when no line-flow limit is violated and negative if any violation. The reward function is described in Eqs. (3.20–3.23) to encourage low generation cost and penalize violations if any:

$$C_{sys} = \sum_{i=1}^{n} C_{gen\_i}(P_{gen\_i}) \tag{3.20}$$

$$D_{overflow} = \sum_{i=1}^{l} \left\{ \begin{array}{l} [min\,(f_i - f_i^{min}, 0)]^2 \\ +[max\,(f_i - f_i^{max}, 0)]^2 \end{array} \right\} \tag{3.21}$$

$$D_{pgen} = \sum_{i=1}^{n} \left\{ \begin{array}{l} \left[min\left(P_{gen_i} - P_{gen_i}^{min}, 0\right)\right]^2 \\ +\left[max\left(P_{gen_i} - P_{gen_i}^{max}, 0\right)\right]^2 \end{array} \right\} \tag{3.22}$$

$$R = \begin{cases} \frac{(-C_{sys}+E1)}{E2}, & \text{if } \textbf{Done} \\ \left(-\frac{D_{overflow}}{E3} - \frac{D_{pgen}}{E4}\right)/E5, & \text{if Not } \textbf{Done} \end{cases} \tag{3.23}$$

where a "**Done**" condition is reached when flows on all lines and outputs of all generators (including ones at the slack bus) are within their corresponding limits. C_{sys} is the total generation cost of the system; C_{gen_i} (.) is the cost function of generator i with the corresponding power generation P_{gen_i} as its input; n is the total number of generators in the system; $D_{overflow}$ is the total line-flow deviation of all lines; f_i, f_i^{min} and f_i^{max} are the line flow, minimum and maximum line limits of line i, respectively; l is the total number of lines in the system; D_{pgen} is the total deviation of power output from all generators; P_{gen_i}, $P^{min}_{gen\_i}$, and $P^{max}_{gen\_i}$ are the power output, the minimum and maximum power capacity from generator i, respectively; R is the reward; $E1$ is added as a bias to make the reward positive when reaching a "Done" condition; $E2, \ldots, E5$ are positive constants to scale the reward to the range of $[-1, 1]$, for speeding up the training process.

3.3.1.4 Training Process of PPO-Based Agents

Two NNs are adopted to approximate the policy and value functions, with parameters denoted as θ and ϕ, respectively. The following variables are further defined: (1) Step_{try} as the maximum number of steps the agent would try to achieve a "**Done**" condition in a single case; (2) $\text{Step}_{\text{batch}}$ as the total number of steps used to update the NNs. The training process is summarized in Table 3.1, where "Iteration" is the RL agent's interaction with the environment; "Episode" is a series of interactions with the environment until "**Done**" or Step_{try} is reached just for a single case; "Epoch" includes all the iterations with the environment until the number of iterations reaches $\text{Step}_{\text{batch}}$ (1 Epoch may contain multiple episodes).

Table 3.1 PPO training process for line flow control.

Training Process on Line Flow Control

1. Initialize θ and ϕ of the networks for policy and value function, set Step_{try} and $\text{Step}_{\text{batch}}$;
2. **for** k = 1, 2, … **do**
3. Initialize cnt_total to be 0
4. **while** cnt_total < $\text{Step}_{\text{batch}}$
5. Initialize cnt_single to be 0; Get a new case
6. **while** Not Done or cnt_single < Step_{try} **do**
7. Run policy $\pi_k = \pi(\theta_k)$, and feed it into the environment to get the next state as well as the reward R_t ***(Iteration)***
8. cnt_single ++1; cnt_total++1 ***(Episode)***
9. **if** cnt_total > = $\text{Step}_{\text{batch}}$
10. **Break**
11. **end if**
12. **end while** ***(Epoch)***
13. Calculate the Monte Carlo reward G_k according to (3.1)
14. Calculate the advantage function according to (3.11) by:
    1) approximating Q value with G_k in step 13
    2) utilize the current output $V(\phi_k)$ from the NN to approximate value function
15. Update the policy NN toward the direction to maximize eq. (3.12)
16. Update the value NN toward the direction to minimize $\mathbb{E}_{\phi_{k\text{old}}}(V(\phi_k) - G_k)^2$
17. **end while**
18. **end for**

3.3.2 Case Studies

The proposed DRL agent for autonomous line flow control is tested on the Illinois 200-bus (200 buses, 38 generators, and 245 lines) system. The detailed configuration of these systems can be found in [19], respectively. The effectiveness of the proposed DRL agent is tested under scenarios with and without contingencies. The environment used for the training process is a high-fidelity power grid simulator to mimic the real power system behavior.

Various operation scenarios are generated by perturbing each load in the Illinois 200-bus system from 70% to 120% of their original values with a fixed power factor. To test the performance of the proposed method, and without loss of generality, all line limits are cut by half for non-transformer branches (except for branches 14–121, 14–149, 26–25, 43–84, 43–132, which are close to the load) to generate branch limit violation issues. Also, constants E1 ~ E5 in Eq. (3.23), are selected as 20,000, 200, 40, 400, and 10, respectively. In addition, 147 branches are selected to be randomly tripped (one at a time) to simulate N-1 contingencies. In this case, the agent tries to adjust the power outputs of 37 generators (except the one on the slack bus) to regulate the flows on all 245 lines. DRL agents with both

Table 3.2 Summary of the testing performance for Illinois 200-bus system.

Scenarios	No contingency		With contingency	
Algorithms	**DDPG**	**PPO**	**DDPG**	**PPO**
Solvable (%)	—	100	—	99.97
1 iteration (%)	—	99.98	—	97.77
Multiple iterations (%)[a]	—	0.02	—	2.2
Avg. cumulative reward[a]	—	0.7449	—	0.7384

a) Unsolvable cases are excluded from the calculation.

Deep Deterministic Policy Gradient (DDPG) and PPO algorithms were attempted to be trained. However, the DDPG agent cannot converge due to the increasing dimension of the action space, which contains 37 continuous control variables. In contrast, the PPO agent can converge well, with the training and testing performances illustrated in Figure 3.3. Also, the detailed statistics on testing performance are summarized in Table 3.2. The results demonstrate that the PPO agent can control all the flows of 245 lines in the entire system by adjusting the power outputs of 37 generators not only under the normal conditions but also under contingencies.

From the test results, the innovative PPO-based agents can respond instantaneously to flow violations under various load conditions, which can be extremely beneficial when an emergency happens. The two advanced improvements in the PPO algorithm ensure more stable performance compared to other algorithms like DDPG. Such stable attributes assure its good performance even on larger power grids with a higher dimension of action space, which is verified through massive numerical experiments on the Illinois 200-bus system.

3.4 Dueling DQN-Based Topology Control for Maximizing Available Transfer Capabilities

Maximizing available transfer capabilities (ATCs) is of critical importance to bulk power systems from both security and economic perspectives, which represents the remaining transfer margin of the transmission network for further energy transactions. Due to environmental and economic concerns, transmission expansion via building new lines for enlarging transfer capabilities is no longer an easy option for many utilities. Compared with re-dispatching generators, shedding electricity demands, and installing FACTS devices, active network

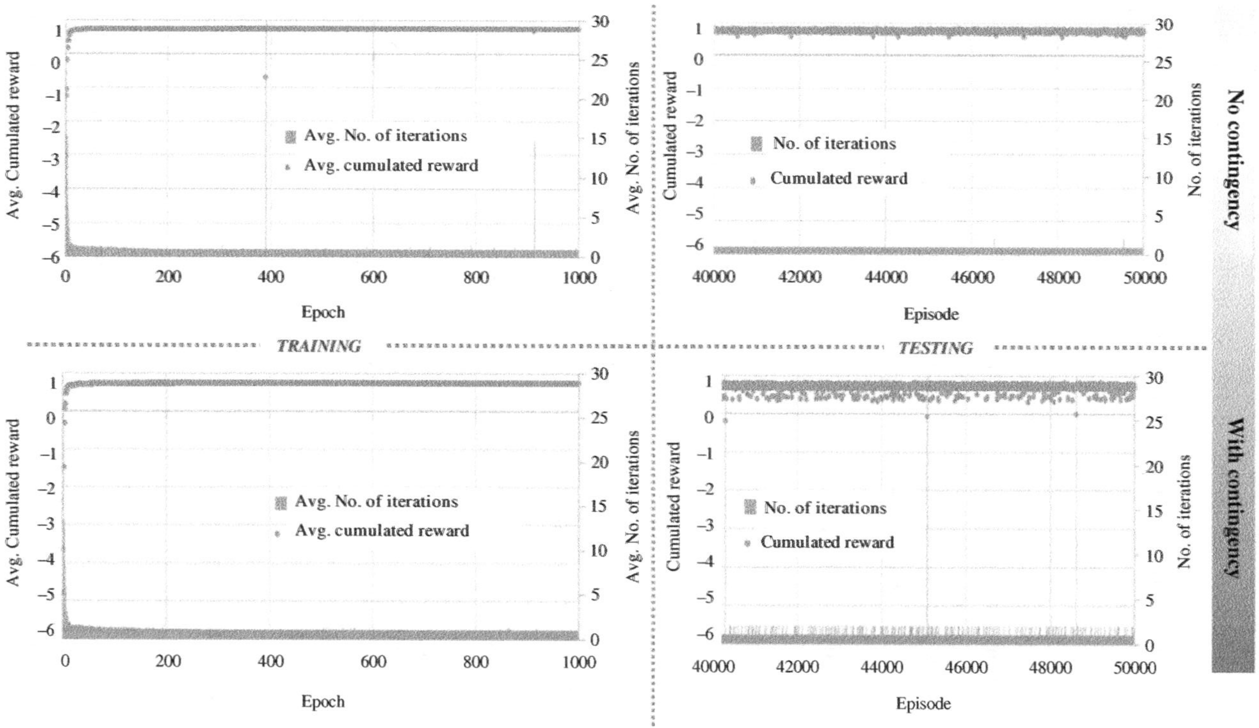

Figure 3.3 Illustration of training and testing process of PPO agents on Illinois 200-bus system.

topology control via transmission line switching or bus splitting for increasing ATCs and mitigating congestions provides a low-cost and effective solution, especially for a deregulated power market or utilities with limited choices. Transmission line switching or bus splitting/rejoining is essentially a multivariate discrete programming problem that is difficult to solve, given the complexity and uncertainties of a bulk power system. Various approaches have been reported to tackle this problem. In [20], an MILP model is proposed with a DC power flow approximation of the power network. In [21], the Transmission Switching (TS) optimization process with DC Optimal Power Flow (DCOPF) is decoupled from a master unit commitment procedure, where the optimal TS schedule is formulated as an MILP problem that is again solved using CPLEX. Reference [22] presents a fast heuristics-based method to speed up the convergence using the aforementioned modeling and solution practice. Similar approaches with variations are also reported in [23, 24], which use a point estimation method for modeling system uncertainties with AC power flow feasibility checking and correction modules.

However, several limitations are observed in existing methods, including: (1) linear approximation in DC power flow without considering all security constraints, which is typically utilized, which affects the solution accuracy for a real power grid. Using full AC power flow with all security constraints for optimization becomes non-convex due to the high nonlinear nature of power grids, which cannot be effectively solved using state-of-the-art techniques without relaxing/sacrificing certain security constraints or solution accuracy. (2) The combination set of lines and bus bars to be switched simultaneously grows exponentially; in addition, sensitivity-based methods are susceptible to changing system operating conditions. Thus, it may take a long time to solve such an optimization process for a large power grid, preventing the solution from being deployed in the real-time environment. To fill these technology gaps, this section discusses a novel method that adopts DDQN-based algorithms with guided exploration and Early Warning (EW) to provide fast and autonomous topology control strategies for maximizing time-series ATCs.

3.4.1 Problem Formulation

Maximizing time-series ATCs via topology control can be modeled as an MDP. In this section, an AC power flow simulator is used to represent the environment and the RL agent state ($s_t^a \in \mathcal{S}$) is a partial observation of the environment state ($s_t^e \in \mathcal{S}$). State s_t^a contains features collected from power flow solvers, including active power outputs and voltage setpoints of generators, loads, line status, line flows, thermal limits, timestamps, etc. The action space $\mathcal{A}$ is formed by including line switching, node splitting/rejoining, and a combination set of both.

An immediate reward r_t at each timestep is defined in Eq. (3.24) to assess the remaining ATCs:

$$r_t = \begin{cases} -1 & \textit{if game over} \\ \frac{1}{N}\sum_{i=1}^{N} max\left(0,\ 1 - \left(\frac{lineflow_i}{thermallimit_i}\right)^2\right) & \textit{otherwise} \end{cases} \tag{3.24}$$

In MDP, a cumulative future return R_t is defined that contains the immediate reward and the discounted future rewards, defined in Eq. (3.25):

$$R_t = r_t + \gamma r_{t+1} + \ldots + \gamma^T r_{t+T} = \sum_{k=0}^{T} \gamma^k r_{t+k} \tag{3.25}$$

where T is the length of the MDP chain, and $\gamma \in [0, 1]$ is the discount factor.

3.4.1.1 Architecture Design

The architecture of training DRL agents for maximizing ATCs is shown in Figure 3.4. First, Imitation Learning (IL) is used to generate a good initial policy for the dueling DQN agent so that exploration and training time can be greatly reduced. Second, a guided exploration method is used to train the agent instead of the traditional Epsilon-greedy exploration. Third, importance sampling is used to increase the mini-batch update efficiency [25]. Moreover, an EW system is designed to increase the system's robustness. Details regarding these techniques are discussed in the following subsections.

3.4.1.2 Dueling DQN Agent

The architecture of dueling DQN is given in Figure 3.5. The original structure is adopted with a batch normalization layer added to the input layer, and the number

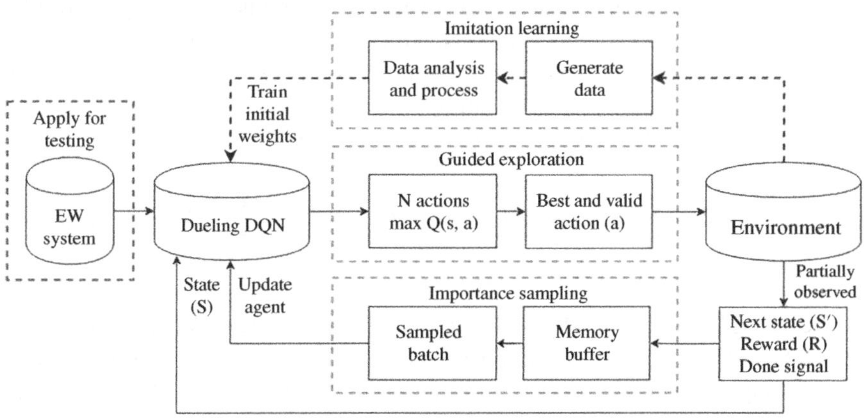

Figure 3.4 Overview of the system architecture.

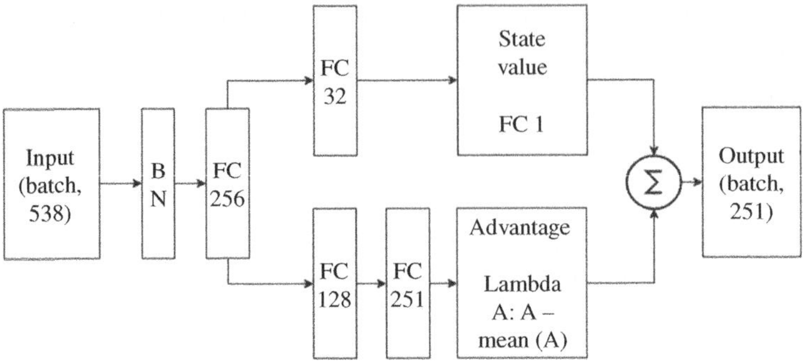

Figure 3.5 Architecture of the Dueling DQN.

of neurons in the hidden layer is modified according to the dimensions of inputs and outputs. The dueling structure decouples the single stream into a state value stream and an advantage stream. The dueling DQN also uses three important techniques in DQN, including (1) an experience replay buffer that allows the agent to be trained off-policy and decouples the strong correlations between the consecutive training data; (2) importance sampling is used to increase the algorithm learning efficiency and final policy quality, by measuring importance of the data using absolute TD-error and giving important data higher priority to be sampled from memory buffer during the training process; and (3) adoption of a DDQN structure, which fixes the q-targets periodically, and then stabilizes the agent updates. The algorithm for training dueling DQN agents is given in Table 3.3.

3.4.1.3 Imitation Learning

IL is essentially a supervised learning method that is used to pre-train DRL agents by providing good initial policies in the form of NN weights. A power grid simulator is used to generate massive data sets, which are then further processed before being used to train the DQN agent. This process allows the RL agent to obtain good $Q(s, a)$ distributions regarding different input states. The loss function used to train the agent is defined as weighted Mean Squared Error (MSE), in Eq. (3.26):

$$J_\theta = \alpha \times \frac{1}{N}\sum_{i=1}^{N}(Q(s,a_i) - \hat{Q}(s,a_i))^2 + \beta \times \frac{1}{|\mathcal{A}| - \mathrm{N}}\sum_{i=N+1}^{|\mathcal{A}|}(Q(s,a_i) - \hat{Q}(s,a_i))^2 \tag{3.26}$$

where α, $\beta \in [0, 1]$, $\alpha + \beta = 1$, $|\mathcal{A}|$ is the size of action space, and vector $\mathbf{Q}(s, a) = [Q(s, a_i), i = 1, \ldots, |\mathcal{A}|]$ is sorted in descending order. The loss function J_θ gives a higher weight to actions, resulting in high scores, which makes the agent more sensitive to score peaks during the training process and, therefore, helps the agent better extract good actions.

Table 3.3 Dueling DQN agent training process.

1: Load pre-trained DQN weights: $\theta = \theta_{imit}$.

2: Initialize Memory Buffer D to capacity N_d.

3: **for** *episode* $\leftarrow 1, M$ **do**

4: Reset the environment, and obtain the state s_0

5: **for** $t \leftarrow 0, T$ **do**

6: Obtain $Q(\cdot|s_t; \theta)$ from the agent and obtain the set of actions with N_g largest Q values.

7: Validate and simulate the actions in the set and choose the valid action a_t with best reward.

8: Execute action a_t in the environment and observe next state s_{t+1}, reward r_t, and d_t.

9: Store the experience $(s_t, a_t, r_t, s_{t+1}, d_t)$ in D, if d_t is True, store multiple times.

10: Sample a minibatch of N_b experience $(s_t, a_t, r_t, s_{t+1}, d_t)$ from D using importance sampling.

11: Calculate q-targets:

$$y_i = \begin{cases} r_i \text{ if } d_t \text{ is True,} \\ r_i + \gamma max_{a'} Q(s_{t+1}, a'; \theta^-) \text{ otherwise} \end{cases}$$

12: Update main network using loss function every N_s step:

$L_i(\theta) = (y_i - Q(s_t, a_t; \theta))^2$

13: Hard copy main network weights θ to the target network θ^-.

14: Set state $s_t = s_{t+1}$.

15: **end for**

16: **end for**

3.4.1.4 Guided Exploration Training

IL provides a good initial policy for snapshots, and then DRL is used to train the agent for long-term planning ability and obtain a globally-concerned policy. For DRL training in this problem, the traditional Epsilon-greedy exploration method is inefficient. First, the action space is large, and the MDP chain is long. Second, the agent is easy to fall into a local optimum. Thus, a guided exploration method is developed, where actions with the N_g highest Q-values are selected at every timestep, the performance of which is simulated and evaluated on the fly. Then, the action with the highest reward is chosen for implementation, and the experience will be stored in the memory. The guided exploration helps the agent to extract the good actions further. With the help of the action simulation function, the training process is more stable, and better experience is stored and used to update the agent. Thus, guided exploration significantly increases training efficiency.

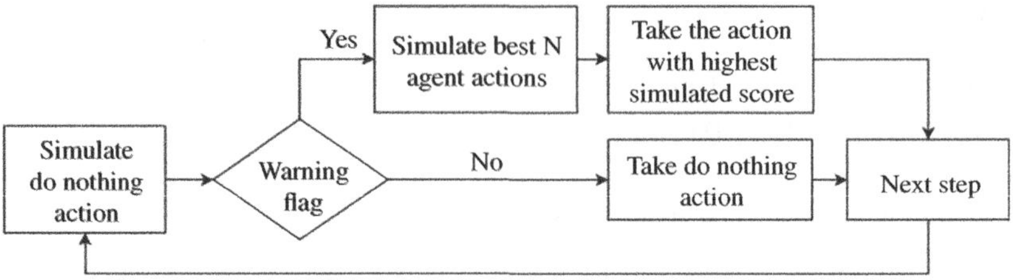

Figure 3.6 Early Warning (EW) workflow.

3.4.1.5 Early Warning

Power systems are highly sensitive to various operating conditions, especially with major topology changes. One bad action may have a long-term adverse effect since the system topology control is successive over a long period of time. The trained DRL agent is not guaranteed to provide good action every time in various complex system states. Thus, an adaptive mechanism, named EW, is developed in this work that can help the agent determine when to apply action and simulate more actions with high $Q(s, a)$ values to increase the error-tolerance and enhance system robustness, with Figure 3.6 illustrating its operation logic.

Initially, at every timestep, it simulates the result of taking no action to the environment, using a Warning Flag (WF) defined in Eq. (3.27).

$$WF = \begin{cases} True & if \ \dfrac{lineflow_i}{thermallimit_i} > \lambda, \ \forall i \in \{1, 2, \dots, 20\} \\ False & otherwise \end{cases} \tag{3.27}$$

If the loading level of a line is higher than a pre-determined threshold λ, a WF is raised. As a result, the N_g top-scored actions are provided by the agent for further simulation. Consequently, the best action with the highest score will be taken. Both guided exploration and the EW mechanism improve the performance and robustness of the proposed RL algorithm.

3.4.2 Case Studies

3.4.2.1 Environment and Framework

The power grid simulator, Python Power Network (Pypownet) [26], is adopted to represent the environment for training RL agents that is able to emulate various operating conditions. The RL agents are trained and tuned through massive interactions with Pypowernet. In this section, the dataset for the IEEE 14-bus model is used to test the performance of the proposed method in autonomous network topology control over long time-series scenarios containing 1000 scenarios with data for 28 continuous days. Each scenario has 8065 timesteps, each representing a 5-minute interval [18]. In this system, there are a total of 156 different

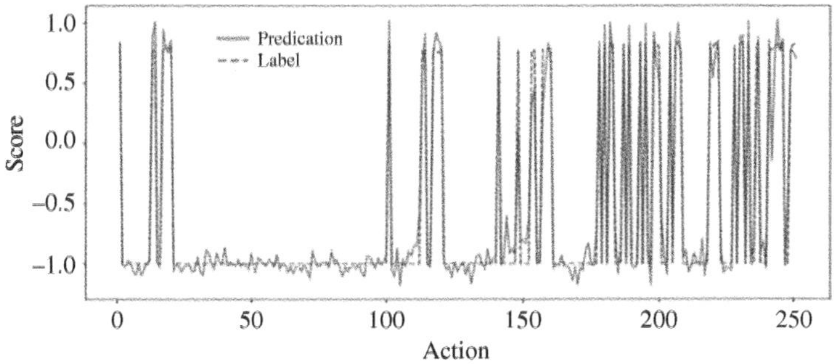

Figure 3.7 Sample DQN prediction after imitation learning (loss function: weighted MSE, optimizer: Adam, learning rate: 1e-3).

node-splitting actions and 20 line-switching actions. Thus, an action space of 3120 is formed by considering null action and all combinations of one node splitting and one line switching without those that can create islands.

3.4.2.2 Effectiveness of Imitation Learning

In the first test, a brute-force method is used to train the RL agent using randomly initialized NN weights and the full action space with a dimension of 3120. As expected, due to the large action space and the long-time sequences, the proposed dueling DQN method did not work well. To solve this problem, the following process is employed to effectively reduce the action space, which includes: (1) 155 node splitting/rejoining actions, (2) 19 line switching actions, and (3) 76 most effective actions with one bus action and one line switching action, and one do-nothing action. In this way, the action space $\mathcal{A}$ is reduced to 251. Then, the IL method introduced above is used to obtain good initial policies. Forty scenarios, each with 1000 timesteps (instead of 8065), are used for IL, yielding a total number of 40,000 sample pairs (state, $Q(s, a)$), which are then separated into a training set (90%) and a validation set (10%). Figure 3.7 shows a sample prediction and label using IL. After training 100 epochs with a batch size of 1, the weighted MSE decreased to around 0.05, indicating NNs can generally catch the peaks and trends and provide relatively effective actions.

3.4.2.3 Improved Performance with Guided Exploration

To shorten the MDP chain and decrease the training difficulty, the 28-day scenarios are divided into single days, each with 288 timesteps. For comparison, the training process of dueling DQN agents with Epsilon-greedy exploration and the proposed guided exploration is depicted in Figure 3.8a and b, respectively.

With Epsilon-greedy exploration, the agent can hardly control the entire 288 timesteps continuously before Episode 7000, without game over, although the

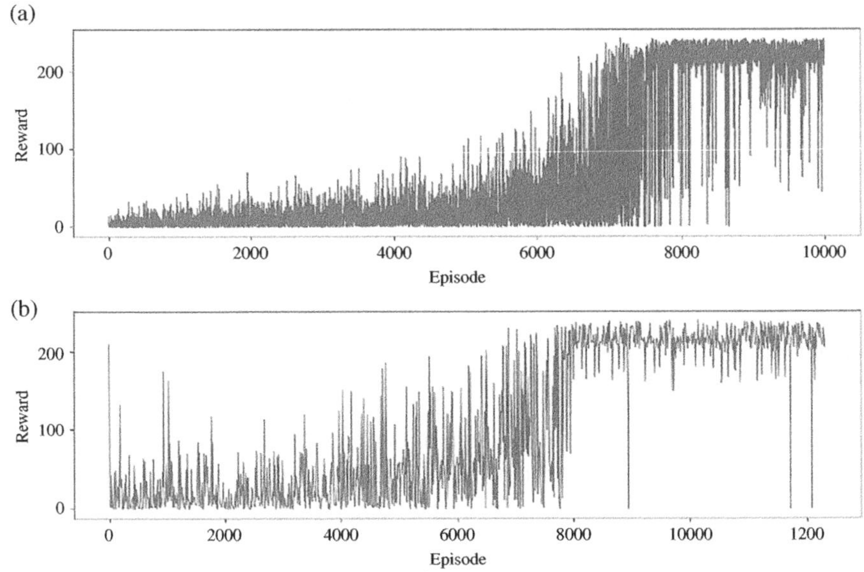

Figure 3.8 (a) Dueling DQN agent training process using epsilon-greedy exploration. (b) Dueling DQN agent training process using guided exploration.

agent's performance keeps improving toward higher reward values. The proposed training process using guided exploration with $N_g = 10$ is shown in Figure 3.8b. The agent can control more steps successfully in the earlier phases of the training process compared to Epsilon-greedy exploration. More importantly, it takes a much shorter time to train an agent with a better policy.

3.4.2.4 Performance Comparison of Different Agents

With the proposed methodology, several case studies are conducted, and their performance is compared in Table 3.4. It is observed that the agent trained only with IL failed in most scenarios. With guided exploration, the agent's performance is greatly improved, where only 7 out of 200 scenarios failed. Using EW (with threshold λ ranging from 0.85 to 0.975), the agent can almost handle all the scenarios well with very few failed, and the scores are much improved. Similarly, 200 long scenarios with 5184 timesteps are tested using DRL agents, where the best score achieved is 82,687.17, using an EW threshold of 0.93. Only 12 scenarios out of 200 experienced bad control performance. Finally, a well-trained agent was submitted to the L2RPN competition with EW $\lambda = 0.885$, which was tested on 10 unseen scenarios, outperformed the other participants, and eventually won the competition. The average decision time for each timestep using the proposed agent is roughly 50 ms.

Table 3.4 Performance comparison of different agents.

Agent	Game over	Mean score all	Mean score w/o dead
Do nothing	91	2471.42	**4534.72**
Only imitation	198	38.21	3820.63
Guided trained	7	4269.63	4424.49
EW $\lambda = 0.85$	**0**	4253.40	4253.40
EW $\lambda = 0.875$	1	4347.56	4369.41
EW $\lambda = 0.90$	**0**	4396.77	4396.77
EW $\lambda = 0.925$	**0**	**4493.27**	**4493.27**
EW $\lambda = 0.95$	**0**	4492.89	4492.89
EW $\lambda = 0.975$	2	4446.12	4491.03

In summary, this section presents a Dueling DQN-based method to maximize time-series ATCs considering various practical constraints. Several innovative techniques are developed, including dueling DQN, and IL for generating good initial policies, reduction of action space via simulation and domain knowledge, guided exploration, and EW for improving DRL agent's stability and robustness. Massive experiments demonstrate that a well-trained AI agent can learn and master optimal topology control problems for a power grid, considering uncertainties.

3.5 Real-Time Multi-Objective Power Flow Control Using Soft Actor-Critic

3.5.1 Problem Formulation

This section discusses a novel method of deriving real-time corrective operational control decisions for power system operating conditions. The control objectives include regulating bus voltages within their secure zones and minimizing transmission line losses while respecting power flow equations and physical constraints, e.g., line ratings and limits of generators. The mathematical formulation of the control problem is given below:

Objective:

$$\text{minimize} \sum_{i,j}^{N} Ploss_{i,j}, (i,j) \in \Omega_L \tag{3.28}$$

Subject to:

$$\sum_{n\in Gi} P_n^g - \sum_{m\in Di} P_m^d - g_i V_i^2 = \sum_{j\in B_i} P_{ij}(y), i \in B \tag{3.29}$$

$$\sum_{n\in Gi} Q_n^g - \sum_{m\in Di} Q_m^d - b_i V_i^2 = \sum_{j\in B_i} Q_{ij}(y), i \in B \tag{3.30}$$

$$P_n^{min} \leq P_n \leq P_n^{max}, n \in G \tag{3.31}$$

$$Q_n^{min} \leq Q_n \leq Q_n^{max}, n \in G \tag{3.32}$$

$$V_i^{min} \leq V_i \leq V_i^{max}, i \in B \tag{3.33}$$

$$\sqrt{P_{ij}^2 + Q_{ij}^2} \leq S_{ij}^{max}, (i,j) \in \Omega_L \tag{3.34}$$

where Eqs. (3.29) and (3.30) represent active and reactive power flow equations, respectively. Equations (3.31) and (3.32) are active and reactive power output constraints of each generator, respectively. Equations (3.33) and (3.34) specify bus voltage secure zones and line flow limits of a power system to be controlled, respectively.

3.5.1.1 Architecture Design

Deriving multi-objective real-time control actions can be formulated as a discrete-time stochastic control process, aka, MDP. Among various DRL techniques, the off-policy SAC method is adopted in this section because of its superior performance in fast convergence and robustness. It maximizes the expected reward by exploring as many control actions as possible, leading to a better chance of finding the optimum. The main flowchart of the proposed methodology is depicted in Figure 3.9, including several key modules:

1) **Module (a):** Figure 3.9a provides the interaction process between the power grid environment (EMS AC power flow solver) and the SAC agent, where the environment receives control action, outputs the corresponding next system states and calculates the reward, and the SAC agent receives states and reward before outputting control actions.
2) **Module (b):** Figure 3.9b shows the offline training process of a SAC agent. Representative system operating snapshots are collected from EMS for preprocessing. System state variables are extracted from those converged snapshots and fed into the SAC agent training submodule, where NNs are used to establish a direct mapping between system states and control actions. The controls are then verified by another AC power flow solution to calculate reward values before updating SAC agent weights to maximize long-term expected reward.
3) **Module (c):** To ensure the long-term effectiveness and robustness of the SAC agent, multiple training processes with different sets of hyperparameters are

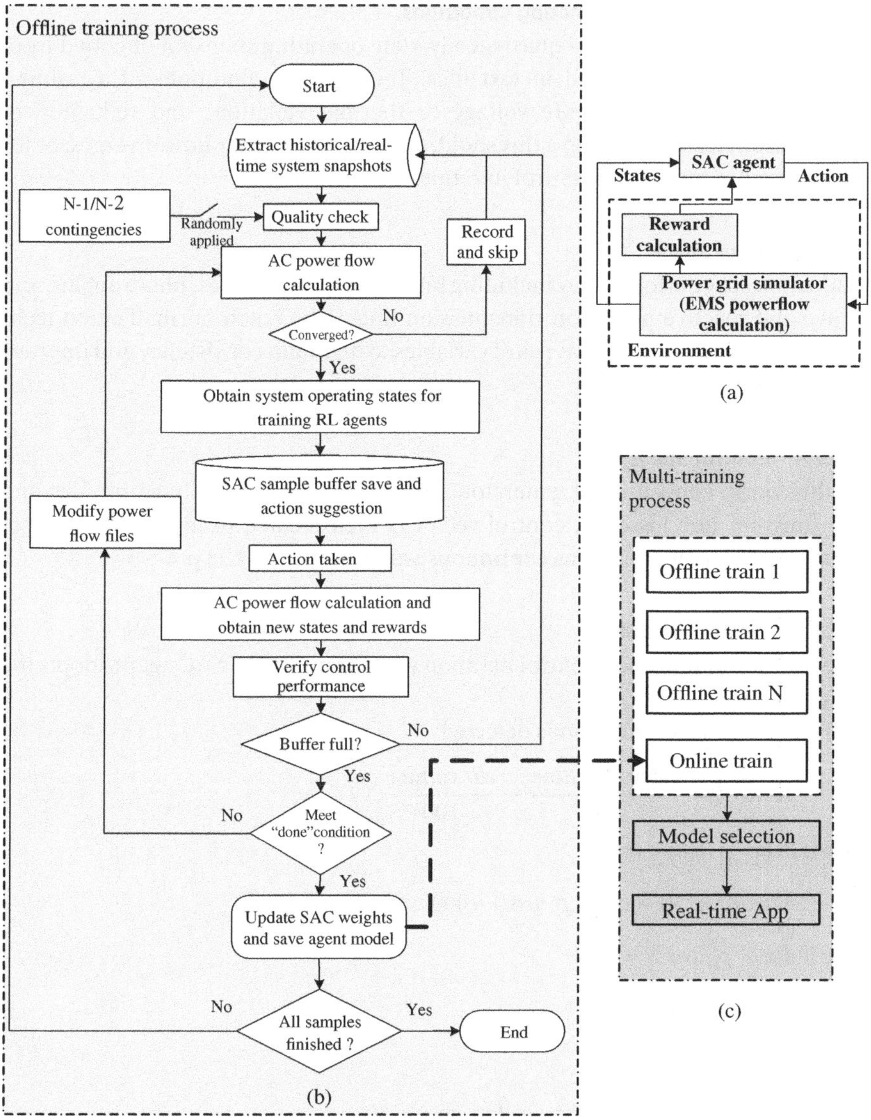

Figure 3.9 Main flowchart and key function modules of the proposed methodology.

launched simultaneously, including several offline training processes and one online training process (initialized by the best offline-trained model), shown in Figure 3.9c. The best-performing model from these processes is selected for application in a real-time environment.

3.5.1.2 Episode and Terminating Conditions

Each episode is defined as a quasi-steady-state operating snapshot obtained from the EMS system and saved in text files. Termination conditions of a training episode can be: (1) no more voltage or thermal violations and reduction of transmission losses reaching a threshold, e.g., 0.5%; (2) power flow diverges; or (3) the maximum number of control iterations is reached.

3.5.1.3 State Space

The action space is formed by including bus voltage magnitudes, phase angles, and active and reactive power on transmission lines. The batch normalization technique is applied to different types of variables to maintain consistency and improve model training efficiency.

3.5.1.4 Control Space

In this work, conventional generators are used to regulate voltage profiles and transmission line losses. A control vector is then created to include voltage set points at each power plant as continuous values, e.g., [0.9,1.1] p.u.

3.5.1.5 Reward Definition

The reward value at each control iteration when training the SAC agent adopts the following logic:

If voltage or flow violation is detected:

$$reward = -\frac{dev\_overflow}{10} - \frac{vio\_voltage}{100}$$

else if $delta\_p\_loss < 0$:

$$reward = 50 - delta\_p\_loss * 1000$$

else if $delta\_p\_loss >= 0.02$

$$reward = -100$$

else:

$$reward = -1 - (p\_loss - p\_loss\_pre) * 50$$

where $dev\_overflow = \sum_i^N (Sline(i) - Sline\_max(i))^2$; N is the total number of lines with thermal violation; $Sline$ is the apparent power of line; $Sline\_max$ is the limit of line apparent power; $vio\_voltage = \sum_j^M (Vm(j) - Vmin) * (Vm(j) - Vmax)$; M is the total number of buses with voltage violation; $delta\_p\_loss = \frac{p\_loss - p\_loss\_pre}{p\_loss\_pre}$; p_loss is the present transmission loss value, and p_loss_pre is the line loss at the base case. The details of training SAC agents are given in Table 3.5.

Table 3.5 Soft actor-critic training process for multi-objective power flow control.

1. Initialize weights of neural networks, θ and ϕ, for policy $\pi(s, a)$ and value function $V(s)$, respectively; initialize weights ψ and $\overline{\psi}$ for the two $Q(s, a)$ functions; initialize replay buffer D; set up training environment, *env*
2. **for:** $k = 1, 2, \ldots$ (k is the counter of episodes for training)
3. **for:** $t = 1, 2, \ldots$ (t stands for control iteration)
4. reset environment $s \leftarrow$ *env.reset*()
5. obtain states and actions $a \sim \pi(\cdot \mid s_t)$
6. apply action a and obtain the next states s_{t+1}, reward value r and termination signal *done*
7. store tuple $<s_t, a_t, r_t, s_{t+1}, done>$ in D
8. $s_t = s_{t+1}$
9. **if** satisfying policy updating conditions, conduct
10. **for** a required number of policy updates, conduct
11. randomly sample from $D, <s_t, a_t, r, s_{t+1}, done>$
12. update Q function, $Q(s, a)$: $\theta_i \leftarrow \theta_i - \lambda_Q \nabla J_Q(\theta_i)$
13. update value function $V(s): \psi \leftarrow \psi - \lambda_V \nabla J_V(\psi)$
14. update policy network $\pi(s, a): \phi \leftarrow \phi - \lambda_\pi \nabla J_\pi(\phi)$
15. update target network $\overline{\psi} \leftarrow \tau\psi + (1 - \tau)\overline{\psi}$
16. update temperature coefficient, α

3.5.2 Case Studies

The proposed SAC-based methodology for multi-objective power flow control was developed and deployed in the control center of SGCC Jiangsu Electric Power Company. To demonstrate its effectiveness, the city-level high-voltage (220 kV+) power network is used, which consists of 45 substations, 5 power plants (with 12 generators), and around 100 transmission lines, serving electricity to the city of Zhangjiagang. Massive historical operating snapshots (full topology node/breaker models for Jiangsu province with ~1500 nodes and ~ 420 generators, at an interval of 5 minutes) were obtained from their EMS system (named D5000 system) where the AC power flow computational module is used as grid simulation environment to train SAC agents. The control objectives are set to minimize transmission losses (at least 0.5% reduction) without violating bus voltages ([0.97–1.07] pu) and line flow limits (100% of MVA rating). Voltage setpoints of the 12 generators in 5 power plants are adjusted by the SAC agent.

The performance of training and testing SAC agents using a time series of actual system snapshots is illustrated in Figure 3.10. From 3 December 2019 to 13 January 2020, 7249 operating snapshots were collected. Two additional copies of the original snapshots were created and randomly shuffled to create a training set (80%) and a test set (20%). For the first ~150 snapshots, the SAC agent struggles to find effective policies (with negative reward values); however, it achieves satisfactory performance thereafter. Several training processes are simultaneously

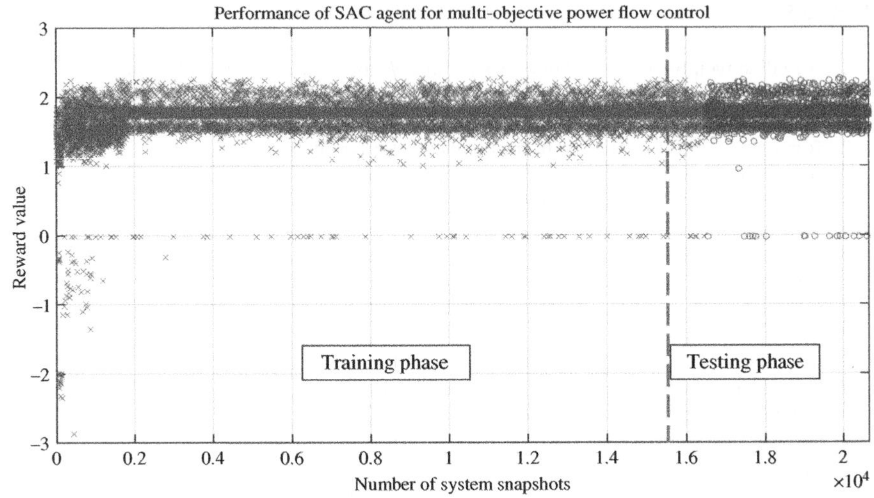

Figure 3.10 SAC agent performance in controlling voltage profiles and line losses.

launched and updated twice a week to ensure control performance. For real-time applications during this period, the developed method provides valid controls for 99.41% of these cases. The average line loss reduction is 3.6412% (compared to the line loss value before control actions). There are 1019 snapshots with voltage violations, in which the SAC agent solves 1014 snapshots completely and effectively mitigates the remaining 5 snapshots.

In conclusion, this section presents a novel methodology for training effective SAC agents with periodic updating for multi-objective power flow control in a real-time operational environment. The detailed design and flowchart of the proposed methodology are provided for reducing transmission losses without violating voltage and line constraints. Numerical simulations conducted on a real power network in a real-time operational environment demonstrate the effectiveness and robustness.

References

1 Mohagheghi, E., Alramlawi, M., Gabash, A., and Li, P. (2018). A survey of real-time optimal power flow. *Energies* 11: 3142. https://doi.org/10.3390/en11113142.

2 Madani, R., Ashraphijuo, M., and Lavaei, J. (2016). Promises of conic relaxation for contingency-constrained optimal power flow problem. *IEEE Transactions on Power Systems* 31: 1297–1307. https://doi.org/10.1109/TPWRS.2015.2411391.

3 Molzahn, D.K., Holzer, J.T., Lesieutre, B.C., and DeMarco, C.L. (2013). Implementation of a large-scale optimal power flow solver based on semidefinite programming. *IEEE Transactions on Power Systems* 28: 3987–3998. https://doi.org/10.1109/TPWRS.2013.2258044.

4 Tang, Y., Dvijotham, K., and Low, S. (2017). Real-time optimal power flow. *IEEE Transactions on Smart Grid* 8: 2963–2973. https://doi.org/10.1109/TSG.2017.2704922.

5 DOE ARPA-E (2022). Grid optimization competition. [Online]. Available: https://gocompetition.energy.gov/

6 Diao, R., Wang, Z., Shi, D. et al. (2019). Autonomous voltage control for grid operation using deep reinforcement learning. *2019 IEEE Power & Energy Society General Meeting (PESGM) 2019 IEEE Power & Energy Society General Meeting (PESGM).*

7 Zhang, B., Lu, X., Diao, R. et al. (2020). Real-time autonomous line flow control using proximal policy optimization. *2020 IEEE Power & Energy Society General Meeting (PESGM).*

8 Yan, Z. and Xu, Y. (2019). Data-driven load frequency control for stochastic power systems: a deep reinforcement learning method with continuous action search. *IEEE Transactions on Power Systems* 34: 1653–1656. https://doi.org/10.1109/TPWRS.2018.2881359.

9 Huang, Q., Huang, R., Hao, W. et al. (2020). Adaptive power system emergency control using deep reinforcement learning. *IEEE Transactions on Smart Grid* 11: 1171–1182. https://doi.org/10.1109/TSG.2019.2933191.

10 Sutton, R.S. and Barto, A.G. (1998). *Introduction to Reinforcement Learning*, vol. 135, 21–22. MIT Press.

11 Mnih, V., Kavukcuoglu, K., Silver, D. et al. (2013). Playing atari with deep reinforcement learning. *arXiv Preprint*. https://doi.org/10.48550/arXiv.1312.5602.

12 van Hasselt, H., Guez, A., and Silver, D. (2016). Deep reinforcement learning with double q-learning. *Proceedings of the AAAI Conference on Artificial Intelligence* 30: https://doi.org/10.1609/aaai.v30i1.10295.

13 Wang, Z., Schaul, T., Hessel, M. et al. (2016). Dueling network architectures for deep reinforcement learning. *Proceedings of the 33rd International Conference on Machine Learning* (pp. 1995–2003). PMLR.

14 Schulman, J., Wolski, F., Dhariwal, P. et al. (2017). Proximal policy optimization algorithms. *arXiv Preprint*.

15 Schulman, J., Moritz, P., Levine, S. et al. (2018). High-dimensional continuous control using generalized advantage estimation. *arXiv Preprint*. https://doi.org/10.48550/arXiv.1506.02438.

16 Genesis (2018). Gradient descent - part 2. [Online]. Available: https://www.fromthegenesis.com/gradient-descent-part-2/.

17 Haarnoja, T., Zhou, A., Abbeel, P. and Levine, S. (2018). Soft actor-critic: off-policy maximum entropy deep reinforcement learning with a stochastic actor. *arXiv Preprint*.

18 RTE France and ChaLearn (2019). L2RPN challenge. [Online]. Available: https://l2rpn.chalearn.org/.

19 Electric Grid Test Case Repository (2016). Illinois 200-bus system: ACTIVSg200. [Online]. Available: http://electricgrids.engr.tamu.edu/electric-grid-test-cases/activsg200/.

20 Fisher, E.B., O'Neill, R.P., and Ferris, M.C. (2008). Optimal transmission switching. *IEEE Transactions on Power Systems* 23: 1346–1355. https://doi.org/10.1109/TPWRS.2008.922256.

21 Khodaei, A. and Shahidehpour, M. (2010). Transmission switching in security-constrained unit commitment. *IEEE Transactions on Power Systems* 25: 1937–1945. https://doi.org/10.1109/TPWRS.2010.2046344.

22 Fuller, J.D., Ramasra, R., and Cha, A. (2012). Fast heuristics for transmission-line switching. *IEEE Transactions on Power Systems* 27: 1377–1386. https://doi.org/10.1109/TPWRS.2012.2186155.

23 Alhazmi, M., Dehghanian, P., Wang, S., and Shinde, B. (2019). Power grid optimal topology control considering correlations of system uncertainties. *IEEE Transactions on Industry Applications* 55: 5594–5604. https://doi.org/10.1109/TIA.2019.2934706.

24 Dehghanian, P., Wang, Y., Gurrala, G. et al. (2015). Flexible implementation of power system corrective topology control. *Electric Power Systems Research* 128: 79–89. https://doi.org/10.1016/j.epsr.2015.07.001.

25 Schaul, T., Quan, J., Antonoglou, I. and Silver, D. (2016). Prioritized experience replay. *arXiv Preprint*. https://doi.org/10.48550/arXiv.1511.05952

26 Lerousseau, M. (2019). A power network simulator with a reinforcement learning-focused usage. [Online]. Available: https://github.com/MarvinLer/pypownet.

4

Smart Generation Control Based on Multi-Agents

Lei Xi, Yixiao Wang, Lu Dong, and Jianyu Ren

College of Electrical Engineering and New Energy, China Three Gorges University, Xiling District, Yichang, Hubei Province, China

4.1 Overview

Since the twenty-first century, the power system has undergone profound changes, with profound changes at all levels of source, network, load and storage. Factors such as rapid access to new energy, integration of power electronic systems, and electric vehicles have changed the composition and characteristics of loads. In addition, the rise of "the integration of internet and big data" and artificial intelligence has also brought a revolutionary impact on the form and regulation of the power system. Automatic Generation Control (AGC) systems are becoming more and more frequent in response to these challenges, and the number of adjustments is high. Entering the era of "grid 2.0" smart grid, the power system is transitioning from "automation" to "intelligence," and the control architecture is also evolving from "centralized optimization" to "decentralized autonomy and centralized coordination." Therefore, AGC following this huge changing trend is further developed into SGC [1] with "smarter," "more optimized," and "more coordinated."

Combined with the practical problems existing in the current AGC and the requirements of the future smart grid for real-time power generation control, the transformation of AGC into SGC must achieve major breakthroughs in the following four core technologies: First, in terms of control structure, SGC should be transformed from the traditional centralized control structure to a decentralized autonomous/centralized coordinated new control system architecture [2]. Second, in terms of control strategy, SGC should be transformed from a traditional AGC linear control strategy to a "multi-agent-" coordinated control strategy with self-learning and self-optimizing capabilities. Third, in terms of control objectives,

Artificial Intelligence Empowered Smart Energy Systems, First Edition. Edited by Qiang Yang and Gang Huang.

the SGC should change from the current AGC only pursuing single-objective optimization of Control Performance Standard (CPS) indicators to "multi-objective" optimization including consideration of real-time energy saving and economic indicators [3]. Fourth, in terms of control adaptability and robustness, the SGC should change from the current AGC to adapt to the weak uncertainty of the grid to adapt to the "strong randomness" environment [4].

4.2 Research on Intelligent Power Generation Control Based on Multi-Agents

4.2.1 Function and Architecture of Multi-Agent System

AGC is the core link in the energy management system. Since its birth, AGC still maintains the centralized control structure of "dispatching center—each power plant/unit" [5]. When the AGC controller tracks a total generation command, the actual power grid dispatcher usually allocates the power command to each unit by the adjustable capacity average distribution method [6]. However, the traditional centralized control method needs to collect the operation data of all units in real time, and the amount of information transmitted is large. When the scale of the unit increases, the convergence time of the centralized control method will be greatly improved, which makes it difficult to meet the real-time control requirements of AGC, and the optimization effect is not good.

In order to solve the problems of massive data, communication bottlenecks, and coordination and interaction, the development trend of smart grid EMS is "decentralized autonomy," and the traditional centralized EMS is gradually transforming into a decentralized, integrated, flexible and open-distributed EMS, and a hybrid structure of distributed/centralized [7]. As the core component of EMS, AGC must also adapt to this changing trend and carry out structural and strategic changes. One of the basic problems of decentralized control system is to achieve consistency among all agents. In a multi-agent network, all agents are said to be consistent when they reach a consensus on the value of the variable. They care about their neighbors through information exchange [8]. Compared with the traditional AGC power allocation optimization algorithm, the core of the consistency algorithm lies in decentralized autonomy. Each agent only needs to obtain the information of the local and neighboring agents in real time, and the transmission information is small, the optimization time is short, and the ideal convergence value can be obtained.

The multi-agent system [9] realizes the interaction between agents, making each agent more in line with the real system in the autonomous decision-making control process, and can naturally realize task decomposition and parallel computation, but its interactive mathematical process is more complex. Figure 4.1 shows

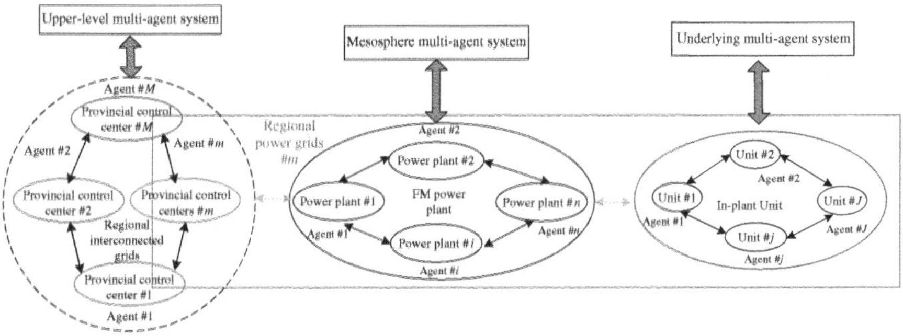

Figure 4.1 The framework of hierarchical multi-agent system for Smart Generation Control (SGC).

the hierarchical multi-agent system framework for intelligent power generation control. Specifically, the power generation control process of each regional power grid is mainly divided into three processes, namely:

1) the multi-agent interaction and decision-making process between the provincial control center, using the short-time cooperative control between the regional power grid to improve the control performance indicators of the entire interconnected system, in which each regional power grid will make a decision to determine its total power instruction after the interaction, in order to release to the next layer of FM power plant;
2) the multi-agent interaction and decision-making process between the FM power plants in the region, using the cooperative control between the power plants to achieve the optimal power command allocation of the power plants with different characteristics, in order to meet the economy and rapidness of the balanced power disturbance;
3) the multi-agent interaction and decision-making process between different units in the bottom power plant, using the cooperative control between units to achieve the optimal power command distribution of the unit under different working conditions, in order to meet the safety and economy of the power plant command response.

4.2.2 Virtual Power Generation Tribe Control Based on Multi-Agent Theory

The concept of Virtual Generation Tribe (VGT) is a new form of intelligent power generation control based on multi-agent consistent coordination theory proposed by the author. The VGT-based AGC decentralized control framework proposed in this chapter mainly consists of two levels: (1) the regional power grid is divided into several territorial power grids (VGT) according to geographical distribution, and the communication network is added between each tribe and each unit; (2) in each AGC control cycle, each agent only communicates with neighboring agents and learns the total power instructions calculated by the dispatching center (VGT leader or tribal leader) at its level. After executing certain protocols, each unit can get its own power generation instructions.

The basic concepts related to virtual power generation tribe can be defined as follows: (1) Virtual power generation tribe: In fact, it is to add a new generation scheduling and control layer between the original medium regulation AGC and the Power Plant Controller (PPC). It is a decentralized and autonomous form, matching with the "territorial" power grid. The generator set group composed of large power plant PLC, active distribution network AGC, micro-network AGC, and load regulation system in the territory is called VGT. Among them the tribe is mainly divided by the high-voltage liaison line in the regional power grid; leader-follower mode is adopted for communication and cooperation among VGTS, and the leader is responsible for power disturbance balance; the leader is the dispatching center of the regional power grid, which is mainly responsible for the cooperative operation among various tribes; the follower is the dispatching end of the ordinary tribe, which is mainly responsible for interaction and coordination with the leader. (2) Leader: there is only one leader in each tribe. The leader is the dispatching center of all kinds of generator sets in the tribe, and is responsible for communication and cooperation with the provincial dispatching end (superior) and the parents of each family in the tribe. (3) Family: a generator set group with similar power generation regulation characteristics in the tribe, such as thermal power set group and gas set group. Among them, the model power generation control unit with a certain dispatch leadership ability in the family is selected as the parent. Generally speaking, a power plant can be used as a family, in which the power plant control center is the parent. (4) Family members: an independent power generation control unit, mainly for communication and cooperation with parents. Generally speaking, a unit in a power plant is a family member (Figure 4.2).

4.2.2.1 First-Order Multi-Agent Consistency Algorithm

(1) Multi-agent consistency graph theory

A graph G refers to a binary set (V_G, E_G) where: non-empty finite set $V_G = \{v_1, v_2, \ldots, v_n\}$ is the set of vertices. The set E_G of unordered or ordered element pairs $e_k = (v_i, v_j)$ in vertex set V_G is the edge set. If e_k is an unordered pair of elements, the graph G is called an undirected graph. If e_k is an ordered pair of elements, then the graph G is called a directed graph. Obviously, undirected graphs can be thought of as special directed graphs. A graph that contains no self-rings and multiple edges is called a simple graph.

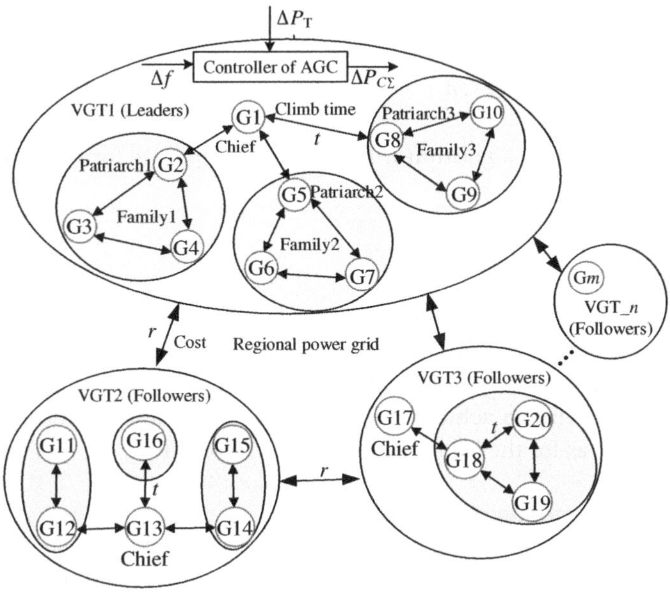

Figure 4.2 Schematic representation of the VGT information topology.

Let $G = (V, E)$ be the graph with n vertices, and matrix $\boldsymbol{A}$ be the adjacency matrix of G, where a_{ij} is the ith row, the jth column element in $\boldsymbol{A}$. For undirected simple graphs with no weight, $\boldsymbol{A}$ is symmetric and the diagonal element a_{ij} is 0. Also, the Laplacian matrix $\boldsymbol{L} = [l_{ij}]$, where:

$$l_{ij} = \begin{cases} -a_{ij}, & j \neq i \\ \sum_{j=1, j\neq i}^{n} a_{ij}, & i = j \end{cases} \tag{4.1}$$

where, l_{ij} is an element of $\boldsymbol{L}$. The communication topology considered in this section is a simple graph with no weight and no direction.

(2) Discrete-time first-order consistency algorithm

Let $x_i(k)$ be a certain state of the agent i at k time, such as voltage, frequency, power generation cost and other physical characteristics. It is generally assumed that a multi-agent system reaches a consistent state if and only if there is $x_i(k) = x_j(k)$ for any i and j. Considering that the communication and transmission data in the actual system usually have a certain time interval, the discrete-time first-order consistency algorithm is used for research. For the leaderless discrete-time consistency algorithm [10], it can be described as:

$$x_i[k+1] = \sum_{j=1}^{n} d_{ij} x_j[k], \quad i = 1, 2, \cdots, n \tag{4.2}$$

where, d_{ij} is the element in the row random matrix $\boldsymbol{D}$, defined as follows:

$$d_{ij} = |l_{ij}| \Big/ \sum_{j=1}^{n} |l_{ij}|, \quad i = 1, 2, \cdots, n \tag{4.3}$$

For a discrete time consistency algorithm with a leader, the updated formula for its leader is:

$$x_i[k+1] = \sum_{j=1}^{n} d_{ij} x_j[k] + \varepsilon g(x[k]) \tag{4.4}$$

In the formula: ε is a positive real number, which reflects the convergence performance of the consistency algorithm, called the convergence coefficient; $g(x[k])$ is called the input deviation of the consistency network, which provides a reference direction for other agents to achieve consistency. The update rules for the other agents are the same as for the leaderless consensus algorithm.

4.2.2.2 Consistent AGC Power Allocation Algorithm

(1) AGC power dynamic distribution model

On the basis of the proposed AGC framework based on virtual generation tribe, the objectives of regulation cost and climb time are considered in the process of dynamic power allocation of AGC, its mathematical model is described as follows:

$$\begin{cases} \min f_1 = \sum_{i=1}^{n} \sum_{m=1}^{M_i} C_{im} \Delta P_{im} \\ \min f_2 = \min \max\limits_{\substack{m=1,\ldots,M_i \\ i=1,\ldots,n}} \left(\Delta P_{im} / \Delta P_{im}^{\text{rate}} \right) \\ s.t. \Delta P = \sum_{i=1}^{n} \sum_{m=1}^{M_i} \Delta P_{im} \\ \Delta P \cdot \Delta P_{im} > 0 \\ \Delta P_{im}^{\min} \le \Delta P_{im} \le \Delta P_{im}^{\max} \\ i = 1, 2, \cdots, n; m = 1, 2, \cdots M_i \end{cases} \tag{4.5}$$

where, ΔP is the total power instruction calculated by PI controller; ΔP_{im} represents the AGC generation power instruction of the m unit of VGT$_i$; C_{im} is the adjustment cost factor of the m unit of VGT$_i$; $\Delta P_{im}{}^{\text{rate}}$ represents the adjustment rate of the m unit of VGT$_i$; $\Delta P_{im}{}^{\max}$ and $\Delta P_{im}{}^{\min}$ represent the power maximum and minimum standby capacity of the m unit of VGT$_i$, respectively; n is the number of VGT; M_i is the unit number of VGT$_i$.

Therefore, when regional power grid loads are disturbed, in order to make VGT with lower average regulation cost bear more power disturbance, this paper selects regulation cost as the consistent state variable among VGT, and adopts the "leader-follower" power distribution algorithm. In order to make the unit with a faster adjustment rate bear more power, this paper selects the power climb time as the consistent state variable between the units, and adopts the "leader-family" mode power distribution algorithm.

(2) The consistency of the adjustment cost of the virtual power generation tribe

After selecting the adjustment cost as the consistency state variable between VGT, the adjustment cost of the *ith* VGT can be defined as follows:

$$r_i = C_i \cdot |\Delta P_i| \tag{4.6}$$

where, C_i represents the average adjustment cost coefficient of the *i*th VGT; ΔP_i represents the AGC generation power instruction of the *i*th VGT. According to the formula, the adjustment cost of the VGT_i is updated consistently as follows:

$$r_i[k+1] = \sum_{j=1}^{n} d_{ij}[k] r_j[k] \tag{4.7}$$

At the same time, in order to ensure power balance, the leader's adjustment costs should be updated as follows:

$$r_i[k+1] = \begin{cases} \sum_{j=1}^{n} d_{ij}[k] r_j[k] + \mu \Delta P_{\text{error}}, & \text{if } \Delta P > 0 \\ \sum_{j=1}^{n} d_{ij}[k] r_j[k] - \mu \Delta P_{\text{error}}, & \text{if } \Delta P < 0 \end{cases} \tag{4.8}$$

where, μ denotes the power error adjustment factor, and $\mu > 0$; ΔP_{error} is the deviation between the total power instruction and the total power of all VGTs, which is defined as follows:

$$\Delta P_{\text{error}} = \Delta P - \sum_{i=1}^{n} \Delta P_i \tag{4.9}$$

When the total power instruction, if $\Delta P_{\text{error}} > 0$, then the adjustment cost r_i needs to increase. If $\Delta P_{\text{error}} < 0$, then the adjustment cost r_i needs to decrease. When $\Delta P < 0$, the adjustment cost increases and decreases in the opposite trend.

When the adjustment cost consistency algorithm is used among VGT, the power of one VGT may exceed its total maximum power. At the same time, the smaller the maximum adjustment cost $r_i^{\max}$ of the VGT, the faster the power limit is reached. When the limit value is reached, the VGT should be withdrawn from the network information topology, and the neighboring VGT should be able to receive the information status and modify the corresponding row random matrix elements. When the power limit is reached, the power of the VGT and its adjustment cost time are as follows:

$$\Delta P_i = \begin{cases} \Delta P_i^{\max}, & \text{if } \Delta P_i > \Delta P_i^{\max} \\ \Delta P_i^{\min}, & \text{if } \Delta P_i < \Delta P_i^{\min} \end{cases} \tag{4.10}$$

$$r_i = r_i^{\max} = \begin{cases} C_i \cdot \left|\Delta P_i^{\max}\right|, & \text{if } \Delta P_i > \Delta P_i^{\max} \\ C_i \cdot \left|\Delta P_i^{\min}\right|, & \text{if } \Delta P_i < \Delta P_i^{\min} \end{cases} \tag{4.11}$$

where, $\Delta P_i^{\max}$ and $\Delta P_i^{\min}$ represent the power maximum and minimum reserve capacity of the ith VGT, respectively.

At the same time, the connection weights with the ith VGT all become zero, that is:

$$a_{ij} = 0, j = 1, 2, \cdots, n \tag{4.12}$$

Therefore, the VGT with the largest adjustment cost $r_i^{\max}$ is selected as the leader in this paper to ensure that the VGT leader is not replaced.

(3) The power climb time consistency of the unit

After selecting the power climb time as the consistency state variable of the unit, the power climb time of the mth unit of VGT_i can be defined as follows:

$$t_{im} = \Delta P_{im} / \Delta P_{im}^{\text{rate}} \tag{4.13}$$

where, ΔP_{im} represents the power generation command of the mth unit of VGT_i. $\Delta P_{im}^{\text{rate}}$ can be calculated as follows:

$$\Delta P_{im}^{\text{rate}} = \begin{cases} P_{im}^{\text{rate}+}, & \text{if } \Delta P_i > 0 \\ P_{im}^{\text{rate}-}, & \text{if } \Delta P_i < 0 \end{cases} \tag{4.14}$$

where, $P_{im}^{\text{rate}+}$ represents the rise adjustment rate limit of the mth AGC unit of VGT_i; $P_{im}^{\text{rate}-}$ represents the descending regulating rate limit of the mth AGC unit of VGT_i.

The power climb time consistency of the mth AGC unit of VGT_i is updated as follows:

$$t_{im}[k+1] = \sum_{w=1}^{M_i} d_{wm}[k] t_{iw}[k] \tag{4.15}$$

At the same time, in order to ensure the power balance of VGT, the leader's climb time should be updated as follows:

$$t_{im}[k+1] = \begin{cases} \sum_{w=1}^{M_i} d_{wm}[k] t_{iw}[k] + \mu_i \Delta P_{\text{error}-i}, & \text{if } \Delta P_i > 0 \\ \sum_{w=1}^{M_i} d_{wm}[k] t_{iw}[k] - \mu_i \Delta P_{\text{error}-i}, & \text{if } \Delta P_i < 0 \end{cases} \tag{4.16}$$

where, μ_i represents the power error adjustment factor of VGT_i, and $\mu > 0$; $\Delta P_{\text{error-}i}$ is the deviation of the total power instruction of VGT_i and the total power of all its units, which is defined as follows:

$$\Delta P_{\text{error}-i} = \Delta P_i - \sum_{m=1}^{M_i} \Delta P_{im} \tag{4.17}$$

When the total VGT_i power command $\Delta P_i > 0$, if $\Delta P_{\text{error-}i} > 0$, then the climb time t_{im} needs to be increased, if $\Delta P_{\text{error-}i} < 0$, then the climb time needs to be reduced. When $\Delta P_i < 0$, the trend of increase and decrease of climb time is opposite.

Similar to the cost consistency algorithm, when the climb time consistency algorithm is used between units, it is possible to cause the power of some units to exceed their maximum power. At the same time, the smaller the maximum power climb time $t_{im}^{\max}$ of the unit, the faster the power limit is reached. When the power limit is reached, the power of the unit and its climb time are as follows:

$$\Delta P_{im} = \begin{cases} \Delta P_{im}^{\max}, & \text{if } \Delta P_{im} > \Delta P_{im}^{\max} \\ \Delta P_{im}^{\min}, & \text{if } \Delta P_{im} < \Delta P_{im}^{\min} \end{cases} \tag{4.18}$$

$$t_{im} = t_{im}^{\max} = \begin{cases} \Delta P_{im}^{\max} / P_{im}^{\text{rate}+}, & \text{if } \Delta P_{im} > \Delta P_{im}^{\max} \\ \Delta P_{im}^{\min} / P_{im}^{\text{rate}-}, & \text{if } \Delta P_{im} < \Delta P_{im}^{\min} \end{cases} \tag{4.19}$$

where, $\Delta P_{im}^{\max}$ and $\Delta P_{im}^{\min}$ represent the power maximum and minimum reserve capacity of the mth unit of VGT_i, respectively.

The connection weights with unit m all become zero, that is:

$$a_{mw} = 0, w = 1, 2, \cdots, M_i \tag{4.20}$$

Therefore, in order to ensure that the leader is not replaced, we select the unit with the largest power climb time $t_{im}^{\max}$ as the leader. At the same time, in order to ensure that some units exit the network information topology and the new network topology can still maintain strong connectivity, we form the same unit $t_{im}^{\max}$ as a family, and the leader has information exchange with each family.

(4) Power error adjustment factor analysis

It can be seen that the values of the power error regulating factors μ and μ_i directly affect the convergence performance of the consistency algorithm. Generally speaking, if the regulator value is too small, the VGT leader's or leader's consistency variable update amplitude is smaller, and the convergence speed is slower. On the contrary, when the value is too large, the consistency variable update amplitude is too large, which may directly lead to the algorithm not converging. Therefore, choosing the appropriate regulator is very important to the convergence speed and stability of the consistency algorithm.

For a regional power grid containing n VGTs, when the VGT leader consistency variable increases $\mu \Delta P_{\text{error}}$, the average consistency variable of each agent increases $\mu\Delta P_{\text{error}}/n$. The total power disturbance increment of the regional power grid is:

$$\Delta P_{\text{in}} = \sum_{i=1}^{n} \frac{\mu \Delta P_{\text{error}}}{nC_i} \tag{4.21}$$

In the regulation cost consistency algorithm, take the power deviation $|\Delta P_{\text{error}}| < \varepsilon$ as the convergence condition, then the sufficient condition of the algorithm convergence is $|\Delta P_{\text{error}} - \Delta P_{im}| < \varepsilon$, that is:

$$\begin{cases} \dfrac{n(\Delta P_{\text{error}} - \varepsilon)}{\sum_{i=1}^{n} \frac{\Delta P_{\text{error}}}{C_i}} < \mu < \dfrac{n(\Delta P_{\text{error}} + \varepsilon)}{\sum_{i=1}^{n} \frac{\Delta P_{\text{error}}}{C_i}}, & \text{if } \Delta P_{\text{error}} > 0 \\ \dfrac{n(\Delta P_{\text{error}} + \varepsilon)}{\sum_{i=1}^{n} \frac{\Delta P_{\text{error}}}{C_i}} < \mu < \dfrac{n(\Delta P_{\text{error}} - \varepsilon)}{\sum_{i=1}^{n} \frac{\Delta P_{\text{error}}}{C_i}}, & \text{if } \Delta P_{\text{error}} < 0 \end{cases} \tag{4.22}$$

where, ε is the maximum allowable power error of the regional power grid, and $\varepsilon > 0$.

Similarly, in the regulation cost consistency algorithm, take the power deviation $|\Delta P_{\text{error-}i}| < \varepsilon_i$ as the convergence condition, then:

$$\begin{cases} \dfrac{M_i(\Delta P_{\text{error-}i} - \varepsilon_i)}{\Delta P_{\text{error-}i} \sum_{m=1}^{M_i} \Delta P_{im}^{\text{rate}}} < \mu_i < \dfrac{M_i(\Delta P_{\text{error-}i} + \varepsilon_i)}{\Delta P_{\text{error-}i} \sum_{m=1}^{M_i} \Delta P_{im}^{\text{rate}}}, & \text{if } \Delta P_{\text{error-}i} \cdot \Delta P_i > 0 \\ \dfrac{M_i(\Delta P_{\text{error-}i} + \varepsilon_i)}{\Delta P_{\text{error-}i} \sum_{m=1}^{M_i} \Delta P_{im}^{\text{rate}}} < \mu_i < \dfrac{M_i(\Delta P_{\text{error-}i} - \varepsilon_i)}{\Delta P_{\text{error-}i} \sum_{m=1}^{M_i} \Delta P_{im}^{\text{rate}}}, & \text{if } \Delta P_{\text{error-}i} \cdot \Delta P_i < 0 \end{cases} \tag{4.23}$$

where, ε is the maximum allowable power error of VGT_i, and $\varepsilon_i > 0$.

4.2.2.3 Robust Consistency Algorithm in Nonideal Communication Networks

(1) Communication noise and delay

As shown in Figure 4.3, when this situation is taken into account, the consistency variable obtained by agent i from agent j can be described as:

$$y_{i,j}[k] = x_j[k - \tau_{i,j}(k)] + \omega_{i,j}(k) \tag{4.24}$$

where, x_j is the state of agent j; $\tau_{i,j}(k)$ and $\omega_{i,j}(k)$ are the communication delay and noise from agent j to agent i, respectively, and have $\tau_{i,i}(k) = 0$ and $\omega_{i,i}(k) = 0$.

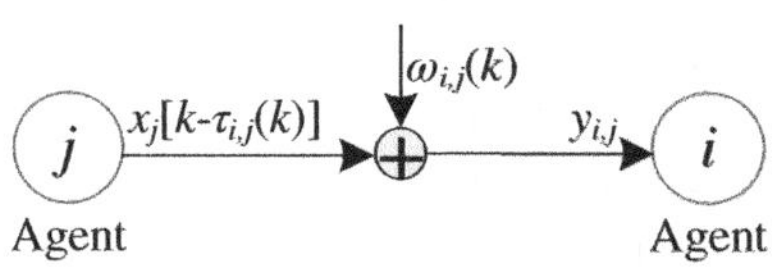

Figure 4.3 A schematic representation of the time delay and noise in the transmission of conformance variables.

Figure 4.4 Two possible variable topology scenarios.

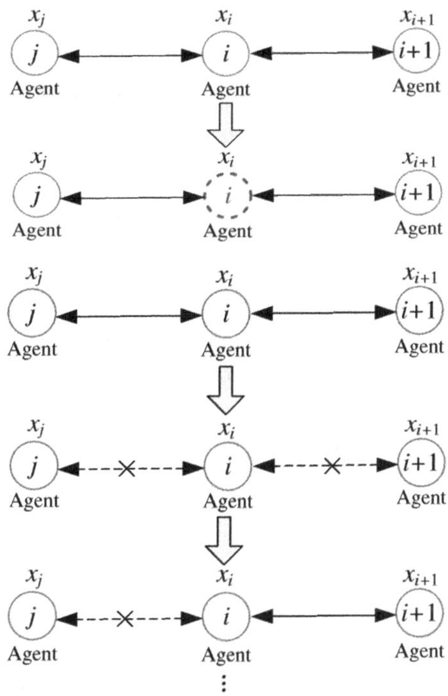

(2) Variable topology

As shown in Figure 4.4, due to the entry and exit of communication agents at any time, coupled with occasional communication failures, the interaction network between multiple agents is often topologically variable. The existence of the variable topology also makes the adjacency matrix $\boldsymbol{A}$ of the communication network dynamically change, which will have a great influence on the convergence of the simple first-order consistency algorithm. Supposing that the adjacency matrix $\boldsymbol{A}$ is a relatively simple (0,1) matrix, its elements will vary as follows:

(1) When a group exits the communication network, the corresponding elements of the adjacency matrix can be updated as follows:

$$a_{ij} = 0, j = 1, 2, \cdots, n, \quad \text{if } \Delta P_i < \Delta P_i^{\min} \text{ or } \Delta P_i > \Delta P_i^{\max} \tag{4.25}$$

(2) When a communication failure occurs between agents, the corresponding elements of the adjacency matrix can be updated as follows:

$$a_{ij} = \begin{cases} 1, & \text{with probability } p_{ij} \\ 0, & \text{with probability } 1 - p_{ij} \end{cases} \tag{4.26}$$

where, $0 \leq p_{ij} \leq 1$ represents the probability of successful communication between agents i and j. $j = 1, \ldots, n$ and $j \neq i$.

(3) Consistency gain function

In order to eliminate the influence of communication delay and noise on the algorithm at the same time, this paper uses the consistent gain function to update the state information of the agent, as follows:

$$x_i[k+1] = x_i[k] - c(k)\sum_{j=1}^{n} l_{ij}[k](x_j[k - \tau_{i,j}(k)] + \omega_{i,j}(k)) \tag{4.27}$$

where, $l_{ij}[k]$ is the element value of (i, j) of the Laplacian matrix; $c(k)$ stands for the consistent gain function.

In order to ensure the convergence of the algorithm, the consistent gain function $c(k)$ must simultaneously satisfy the following two conditions:

$$\sum_{k=0}^{\infty} c(k) = +\infty \tag{4.28}$$

This formula can ensure the convergence of the agent consistency variable.

$$\sum_{k=0}^{\infty} c^2(k) < +\infty \tag{4.29}$$

This formula can improve the convergence robustness of the consistency algorithm in the case of communication noise and delay.

Similar to the consistency algorithm under an ideal communication network, the adjustment cost consistency algorithm and climb time consistency algorithm considering a nonideal network can be described as follows:

$$r_i[k+1] = \begin{cases} R_i[k], & \text{for all followers} \\ R_i[k] + \Delta P_{\text{leader}}, & \text{for leader} \end{cases} \tag{4.30}$$

$$t_{iw}[k+1] = \begin{cases} T_{iw}[k], & \text{for all family member} \\ T_{iw}[k] + \Delta P_{\text{chief-}i}, & \text{for chief} \end{cases} \tag{4.31}$$

where, R_i represents the consistency update item of the adjustment cost; P_{leader} represents the leader's input increment; T_{iw} represents the consistent update item of climb time; $P_{\text{chief-}i}$ represents the input increment of the clan leader. These variable values can be calculated separately as follows:

$$R_i[k] = r_i[k] - c^r(k)\sum_{j=1}^{n} l_{ij}[k]\left(r_j\left[k - \tau^r_{i,j}(k)\right] + \omega^r_{i,j}(k)\right) \tag{4.32}$$

$$\Delta P_{\text{leader}} = \begin{cases} \mu\Delta P_{\text{error}}, & \text{if } \Delta P_{C\Sigma} > 0 \\ -\mu\Delta P_{\text{error}}, & \text{if } \Delta P_{C\Sigma} < 0 \end{cases} \tag{4.33}$$

$$T_{iw}[k] = t_{iw}[k] - c^t(k)\sum_{v=1}^{m_i} l^{t(i)}_{wv}[k]\left(t_{iv}\left[k - \tau^{t(i)}_{wv}(k)\right] + \omega^{t(i)}_{wv}(k)\right) \tag{4.34}$$

$$\Delta P_{\text{chief-}i} = \begin{cases} \mu_i \Delta P_{\text{error-}i}, & \text{if } \Delta P_i > 0 \\ -\mu_i \Delta P_{\text{error-}i}, & \text{if } \Delta P_i < 0 \end{cases} \tag{4.35}$$

where, r and t represent the state values of adjustment cost and climbing time. respectively. Superscript (i) represents the parameters of the unit in the ith tribe; $\boldsymbol{L}^{(i)} = [l_{wv}{}^{(i)}] \in R^{mi \times mi}$ represents the Laplacian matrix of the communication network in the i VGT.

(4) Virtual and real consistency variables

In order to adapt to the variable topology of the communication network, the virtual and real consistency variables are proposed here. The former is mainly used as the interaction information between agents and the input information of the consistency interaction, which is not constrained by the power of the unit. The latter is mainly used to calculate the actual scheduling power of the unit to meet its operation constraints. Therefore, the conversion relationship between them can be described as follows:

$$r_i^{\text{ac}} = \begin{cases} r_i, & \text{if } r_i < r_i^{\max} \\ r_i^{\max}, & \text{if } r_i \geq r_i^{\max} \end{cases} \tag{4.36}$$

where, r^{ac} and t^{ac} represent the real adjustment cost and the state value of climb time, respectively.

Based on the real consistency variable, the generation power instructions of the tribe and the unit can be calculated as follows:

$$\Delta P_i = \begin{cases} r_i^{\text{ac}}/C_i, & \text{if } \Delta P_{C\Sigma} > 0 \\ -r_i^{\text{ac}}/C_i, & \text{if } \Delta P_{C\Sigma} < 0 \end{cases} \tag{4.37}$$

$$\Delta P_{iw} = \begin{cases} t_{iw}^{\text{ac}} \cdot UR_{iw}, & \text{if } \Delta P_i > 0 \\ -t_{iw}^{\text{ac}} \cdot DR_{iw}, & \text{if } \Delta P_i < 0 \end{cases} \tag{4.38}$$

(5) The application and solving process of robust consistency algorithm

In general, the flow of robust consistent power dynamic distribution algorithms in nonideal communication networks can be seen in Figure 4.5.

4.3 Intelligent Power Generation Control for Islands and Microgrids

With the development and utilization of small-scale, low-carbon, and nontraditional renewable energy sources and their integration into low-voltage distribution networks in the form of distributed power sources, traditional power grids are transforming from passive power distribution networks with unidirectional power transmission to active power distribution networks with bidirectional power

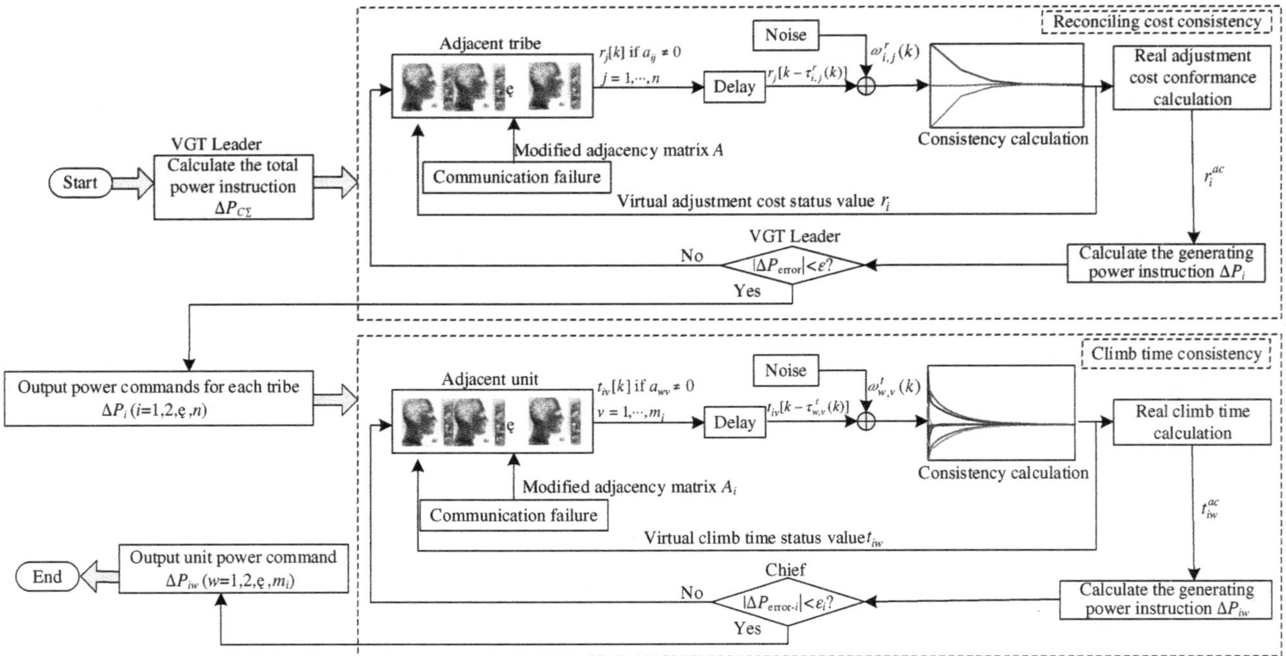

Figure 4.5 Flow chart of power dynamic distribution for robust consistency algorithm.

transmission. Flexible and intelligent control systems and advanced distribution technology are the basis for safe and efficient use of renewable distributed energy clean power, and the construction of intelligent distribution network is imperative.

Generally speaking, the increasing of distributed energy and the integration of active load will lead to more and more randomness and uncertainty of active power and load disturbance in intelligent distribution network. Therefore, the Wolf pack hunting strategy is proposed, and the AGC unit carries out frequent information exchange under the designed framework to quickly calculate the optimal power allocation instructions, so as to realize the optimal cooperative control of each region and the power autonomy of the island grid, and improve the utilization rate of renewable energy and reduce the power generation cost.

Wolf hunting strategies are used for collaborative decentralized area optimization. The traditional AGC used in Islanding Smart Distribution Network (ISDN) mainly includes: (1) obtaining the total power command through the controller (usually PI controller) according to the local frequency deviation; (2) the total power instructions are distributed to each unit in a fixed proportion of adjustable capacity according to the real-time control situation. However, an AGC based on a Wolf pack hunting strategy consists of: (1) obtaining total power commands via the MAS-SG framework based on local frequency bias; And (2) the leader tracks the total power command, with each unit communicating with its neighbors, all of whom are dynamically assigned through the MAS-CC framework to obtain their respective power commands. The application of Wolf pack hunting strategy is not limited by centralized calculation and power instruction allocation of a single centralized controller. In fact, if some agents fail, others are able to continue exchanging information and achieve a new consistency. Since there is usually more than one communication channel between agents, AGC performance can remain optimal even if one channel fails.

4.3.1 AGC Cooperative Control Based on Equal Incremental Rate Consistency Algorithm

4.3.1.1 Intelligent Distribution Network Decentralized Autonomy Framework

The Economic Dispatch Problem (EDP) of power system is actually the problem of how to minimize the operating cost under the premise of ensuring the safety and stability of the system. However, when the load disturbance occurs in the distribution network or the power disturbance occurs in the wind, light, and other power sources, the AGC controller quickly distributes the total power to each FM unit, and then the generating power of the whole island intelligent distribution network unit often deviates from the optimal economic operation point. Therefore, a consistency algorithm based on equal incremental rate is proposed to ensure that the output of each unit can still operate optimally and

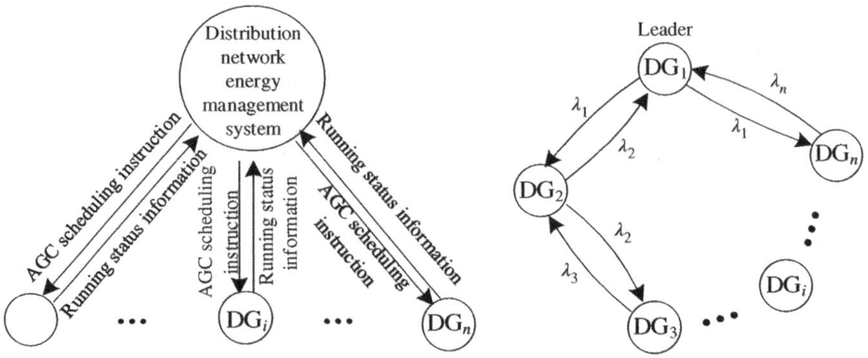

Figure 4.6 AGC Power Distribution Framework for Smart Distribution Network. (a) Traditional centralized control framework; (b) Conformance cooperative control framework.

economically after AGC regulation, and effectively coordinate the cooperation between secondary frequency modulation power distribution and tertiary frequency modulation power distribution.

As shown in Figure 4.6, the traditional AGC control process of an isolated island distribution network includes: (1) Collecting the local frequency deviation, tracking and obtaining an AGC total power command through a certain controller(using PI control); (2) Collecting the real-time operating status of all AGC units, and allocating power instructions to each AGC unit according to certain strategies or algorithms. In contrast, under the cooperative consistency control framework, the process of AGC control includes: (1) One of the AGC units (i.e., the leader) is responsible for tracking the total AGC power instructions; (2) Each unit exchanges information with neighboring units, and in each AGC control cycle, all units get their own AGC power adjustment instructions after reaching an agreement on a certain state quantity or target through the consistency algorithm.

From the point of view of controller reliability, co-conformance control is not subject to centralized calculation and instruction assignment of a single centralized controller, the residual agents can still process and interact with information and reach an agreement. From the perspective of data transmission reliability, the co-consistency control agents usually contain more than one communication channel. When some communication lines fail, each agent has complementary information, and the control performance of AGC can still keep the current optimal condition.

4.3.1.2 AGC Power Allocation Model of Smart Distribution Network

The cost of generating electricity commonly used in economic dispatch can be expressed approximately as a quadratic function:

$$C_i(P_{Gi}) = a_i P_{Gi}^2 + b_i P_{Gi} + c_i \tag{4.39}$$

where, P_{Gi} represents the output of the ith unit; C_i represents the generation cost of the ith unit; a_i, b_i, c_i represent the coefficients of the generation cost of the ith unit, respectively.

Therefore, after the AGC power distribution, the generation cost of each unit will change as follows:

$$C_i(P_{Gi,\mathrm{actual}}) = C_i(P_{Gi,\mathrm{plan}} + \Delta P_{\mathrm{G}i}) = \alpha_i \Delta P_{\mathrm{G}i}^2 + \beta_i \Delta P_{\mathrm{G}i} + \gamma_i \tag{4.40}$$

where, $P_{Gi,\mathrm{actual}}$ is the actual generating power of ith unit; $P_{Gi,\mathrm{plan}}$ is the planned generation power of ith unit; P_{Gi} is the AGC regulating power of ith unit; α_i, β_i, and γ_i are respectively the dynamic coefficients of the power generation cost of the ith unit after considering the power disturbance, where $\alpha_{i=} a_i$, $\beta_i = 2a_i P_{Gi,\mathrm{plan}} + b_i$, $\gamma_i = a_i P_{Gi,\mathrm{plan}}{}^2 + b_i P_{Gi,\mathrm{plan}} + c_i$.

For a system containing n AGC units, the regulation objectives of AGC can be described as:

$$\begin{cases} \min\ C_{\mathrm{total}} = \sum_{i=1}^{n} \left(\alpha_i \Delta P_{\mathrm{G}i}^2 + \beta_i \Delta P_{\mathrm{G}i} + \gamma_i\right) \\ s.t. \Delta P_{\Sigma} - \sum_{i=1}^{n} \Delta P_{\mathrm{G}i} = 0 \\ \Delta P_{\mathrm{G}i}^{\min} \le \Delta P_{\mathrm{G}i} \le \Delta P_{\mathrm{G}i}^{\max} \end{cases} \tag{4.41}$$

where, C_{total} represents the actual total cost of power generation, and is the objective function of AGC power distribution; ΔP_{Σ} represents the total power instruction tracked by AGC; $\Delta P_{\mathrm{G}i}^{\min}$ and $\Delta P_{\mathrm{G}i}^{\max}$ represent the minimum and maximum adjustable capacities of ith unit, respectively.

According to the equal incremental rate criterion, when the generation cost of each unit is equal to the partial differential derivative of its AGC-regulated power, C_{total} can reach the minimum value, that is:

$$\frac{dC_1(P_{\mathrm{G1,actual}})}{d\Delta P_{\mathrm{G1}}} = \frac{dC_2(P_{\mathrm{G2,actual}})}{d\Delta P_{\mathrm{G2}}} = \cdots = \frac{dC_n(P_{\mathrm{G}n\mathrm{,actual}})}{d\Delta P_{\mathrm{G}n}} = \lambda \tag{4.42}$$

where, λ is the slight increase rate of power generation cost. Therefore, λ is selected as the consistency variable of multi-agent network, and it is calculated as follows:

$$\lambda_i = 2\alpha_i \Delta P_{\mathrm{G}i} + \beta_i \tag{4.43}$$

where, λ_i is a slight increase in the rate of power generation cost of ith unit.

4.3.1.3 Consistency Algorithm of Equal Incremental Rate

(1) Algorithm description

The equal incremental consistency algorithm can be described as:

$$\lambda_i[k+1] = \sum_{j=1}^{n} d_{ij} \lambda_j[k], \quad i = 1, 2, \cdots, n \tag{4.44}$$

In order to satisfy the power balance constraint, the leader's consistency update process introduces the power deviation quantity as follows:

$$\lambda_i[k+1] = \sum_{j=1}^{n} d_{ij}\lambda_j[k] + \varepsilon \Delta P_{\text{error}} \tag{4.45}$$

where, ΔP_{error} is the difference between the total power instruction of AGC and the total regulated power of all units, that is:

$$\Delta P_{\text{error}} = \Delta P_{\Sigma} - \sum_{i=1}^{n} \Delta P_{\text{G}i} \tag{4.46}$$

Considering the constraints of unit output, the consistency update formula can be expressed as:

$$\begin{cases} \lambda_i = \lambda_{i,\text{lower}}, \ \Delta P_{\text{G}i} < \Delta P_{\text{G}i,\text{min}} \\ \lambda_i[k+1] = \sum_{j=1}^{n} d_{ij}\lambda_j[k], \ \Delta P_{\text{G}i,\text{min}} \le \Delta P_{\text{G}i} \le \Delta P_{\text{G}i,\text{max}} \\ \lambda_i = \lambda_{i,\text{upper}}, \ \Delta P_{\text{G}i} > \Delta P_{\text{G}i,\text{max}} \end{cases} \tag{4.47}$$

where, $\lambda_{i,\text{lower}}$ and $\lambda_{i,\text{upper}}$ are the minimum and maximum values of the consistency variables of agent i, respectively.

(2) Virtual consistency variable

When the power limit of a certain unit is exceeded, the consistency variable of the unit is taken as its limit value, and it will no longer participate in the updating of other consistency variables. The jumping of the update rule means that the topological matrix $\boldsymbol{D}$ dimension and its element d_{ij} will change frequently. As shown in Figure 4.7, when the adjustment instruction of unit G5 exceeds its adjustable

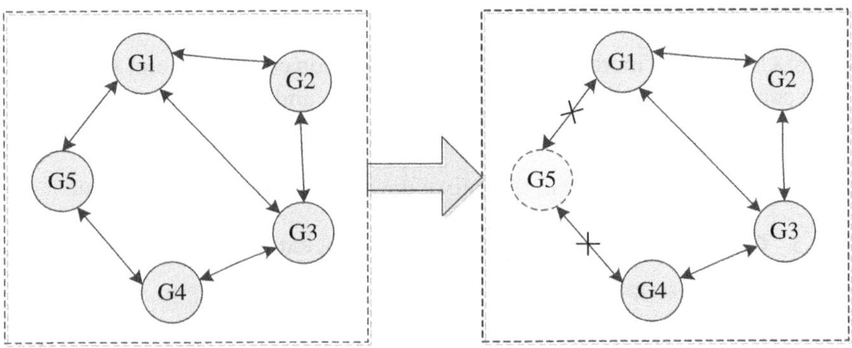

Figure 4.7 Schematic diagram of communication topology change when power limit is exceeded.

capacity, the consistency update process will be withdrawn, and the communication topology represented by the solid line changes. If G5 happens to be the leader group, the leader of each group needs to be re-selected. In addition, considering the "plug and play" demand of intelligent distribution networks, how to take reasonable measures to solve the problem of time-varying topology is particularly important.

In view of the above problems, the concept of a virtual consistency variable is proposed. Compared with the real consistency variable considering the power constraint of the unit, the iterative update of the virtual consistency variable does not consider the power constraint of the unit. After obtaining the virtual consistency variable $\lambda_{i,\text{virtual}}$ at each step, the real consistency variable λ can be calculated λ_i as follows:

$$\lambda_i = \begin{cases} \lambda_{i,\text{lower}}, & \lambda_{i,\text{virtual}} < \lambda_{i,\text{lower}} \\ \lambda_{i,\text{virtual}}, & \lambda_{i,\text{lower}} \leq \lambda_{i,\text{virtual}} \leq \lambda_{i,\text{upper}} \\ \lambda_{i,\text{upper}}, & \lambda_{i,\text{virtual}} > \lambda_{i,\text{upper}} \end{cases} \tag{4.48}$$

As shown in Figure 4.8, since the virtual consistency variable does not take into account the power constraint of the unit, the interactive network topology of the virtual consistency variable will not change. The real consistency variable of the unit can be obtained by modifying the updated value of the virtual consistency variable according to the power constraint of the unit.

The experimental results show that the introduction of a virtual consistency variable can effectively solve the problem of frequent change of real consistency variable updating topology caused by power constraints, and greatly reduce the computational complexity of the algorithm. In view of the time-varying topology caused by "plug and play," we can design the communication topology of all the

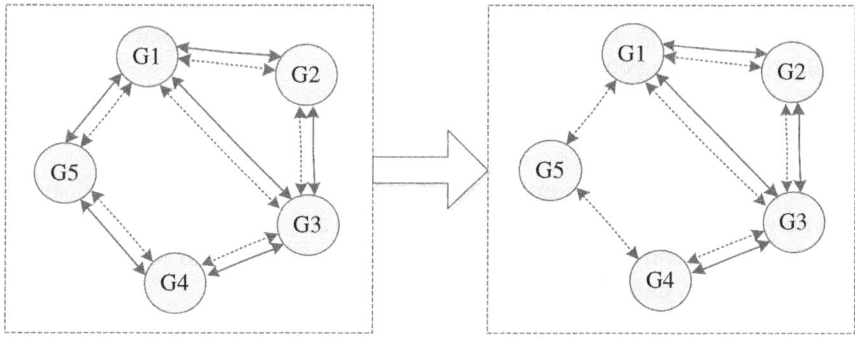

Figure 4.8 The virtual consistency variable is introduced into the schematic diagram of the communication topology change when the power limit is exceeded.

units that can be put in advance, and make the units be put in the communication "virtual connection" state through the virtual consistency variable.

Once the units are put into use, the real consistency variable can be modified according to the constraints of the units, without any topology modification.

(3) ε analysis of convergence coefficient

The value of convergence coefficient ε affects the convergence speed of the consistency algorithm. If the value of the convergence coefficient ε is too small, the incremental amplitude of the system introduced by the leader is small and the convergence is slow. If the value of the convergence coefficient ε is too large, the incremental amplitude of the system introduced by the leader in a single iteration step is too large, which may lead to the algorithm which cannot converge. Therefore, the convergence coefficient must be reasonably selected.

Through simulation research, the author found that when the power deviation ΔP_{error} drops to 10% of the maximum value, the time consumed until convergence reaches more than 90% of the total convergence time. It can be seen that with the reduction of power deviation ΔP_{error}, the speed of consistent convergence is greatly reduced. In order to improve the convergence speed reasonably under the condition of ensuring the convergence of the algorithm, a method of changing the convergence coefficient ε is proposed. Different from the previous constant convergence coefficient ε, when ΔP_{error} is less than a certain value (such as 10% of the maximum power deviation), the leader's update amplitude $\varepsilon\Delta P_{\text{error}}$ is taken as a fixed value. The experimental results show that this improvement can greatly improve the convergence speed of the algorithm. It is worth noting that the value of $\varepsilon\Delta P_{\text{error}}$ should meet certain requirements to ensure the convergence of the algorithm.

For a system containing n agents, each time the leader consistency variable increases $\varepsilon\Delta P_{\text{error}}/n$, the average consistency variable of each agent increases. The total power increment of the AGC unit is $\sum(\varepsilon\Delta P_{\text{error}}/2n\alpha_l)$. Take the power deviation $\Delta P_{\text{error}} < \Delta P_{\text{error}}^{\max}$ as the termination condition, then the sufficient condition for algorithm convergence is $\varepsilon\Delta P \leq \Delta P_{\text{error}}^{\max}$.

$$|\varepsilon\Delta P_{\text{error}}| \leq \frac{\Delta P_{\text{error}}^{\max}}{\sum_{i=1}^{n} {}^{1}/_{2n\alpha_i}} \tag{4.49}$$

where, $\Delta P_{\text{error}}^{\max}$ is the maximum allowable power error of isolated island distribution network, and $\Delta P_{\text{error}}^{\max} > 0$. Take the value of $\varepsilon\Delta P_{\text{error}}$ as the right term of the right-hand side term of the above inequality, which can take into account the convergence stability and convergence speed of the algorithm.

References

1 Zhang, X.S., Yu, T., Yang, B., and Li, L. (2016). Virtual generation tribe based robust collaborative consensus algorithm for dynamic generation command dispatch optimization of smart grid. *Energy* 101: 34–51. https://doi.org/10.1016/j.energy.2016.02.009.

2 Li, J.W., Yao, J.G., Yu, T., and Zhang, X.S. (2021). Distributed deep reinforcement learning for integrated generation-control and power-dispatch of interconnected power grid with various renewable units. *IET Renewable Power Generation* 16: 1316–1335.

3 Zhang, Y., Qin, H., Yan, H.M., Shi, S.S. and Fang, C. (2019) Predictive control strategy of energy storage system in distribution system with wind power. *2019 IEEE Sustainable Power and Energy Conference (iSPEC)*, pp. 2609–2613. https://doi.org/10.1109/ispec48194.2019.8975378.

4 Li, J.W., Yu, T., and Zhang, X.S. (2022). Coordinated load frequency control of multi-area integrated energy system using multi-agent deep reinforcement learning. *Applied Energy* 306.

5 Xi, L., Chen, J.F., Huang, Y.H. et al. (2018). Smart generation control based on deep reinforcement learning with the ability of action self-optimization. *SCIENTIA SINICA Informationis* 48 (10): 1430–1449.

6 Xi, L., Li, H.K., Zhu, J.Z. et al. (2022). A novel automatic generation control method based on the large-scale electric vehicles and wind power integration into the grid. *IEEE Transactions on Neural Networks and Learning Systems.* https://doi.org/10.1109/tnnls.2022.3194247.

7 Yu, T., Xi, L., Yang, B. et al. (2016). Multiagent stochastic dynamic game for smart generation control. *Journal of Energy Engineering* 142: 4015012.1–4015012.1. https://doi.org/10.1061/(asce)ey.1943-7897.0000275.

8 Yin, L.F., Yu, T., and Zhou, L. (2018). Design of a novel smart generation controller based on deep Q learning for large-scale interconnected power system. *Journal of Energy Engineering* 144: https://doi.org/10.1061/(asce)ey.1943-7897.0000519.

9 Zhang, X.S., Yu, T., Yang, B. et al. (2015). Approximate ideal multi-objective solution Q(λ) learning for optimal carbon-energy combined-flow in multi-energy power systems. *Energy Conversion and Management* 106: 543–556. https://doi.org/10.1016/j.enconman.2015.09.049.

10 Zhang, X.S., Yu, T., Pan, Z.N. et al. (2018). Lifelong learning for complementary generation control of interconnected power grids with high-penetration renewables and EVs. *IEEE Transactions on Power Systems* 33: 4097–4110.

5

Power System Fault Diagnosis Method Under Disaster Weather Based on Random Self-Regulating Algorithm

Tao Wang[1,2], *Liyuan Liu*[1], *Ruixuan Ying*[1,3], *Chunyu Zhou*[1], *Hanyan Wu*[1], *and Quanlin Leng*[1]

[1] *School of Electrical and Electronic Information, Xihua University, Chengdu, Sichuan, China*
[2] *Key Laboratory of Fluid and Power Machinery, Xihua University, Chengdu, Sichuan, China*
[3] *Institute of Multidisciplinary Research for Advanced Materials, Tohoku University, Miyagi, Tohoku Region, Japan*

5.1 Introduction

With the rapid development of electric power systems, the difficulty of ensuring their economic, safe, and stable operation increases. Affected by natural disasters, human attacks, system errors, and other factors, a power grid is hard to avoid failures. Consequently, the fault diagnosis of power grids has always played an important role in the rapid analysis of fault events and fast restoration of power supply. Consequently, a variety of fault diagnosis methods of power grids based on smart technology have been proposed, such as expert systems, neural networks, Petri nets, Bayesian networks, spiking neural P systems, evidence theory, and optimization technology.

Fault diagnosis methods based on the optimization technology are characterized by strict mathematical logic, easy implementation, and quick response. Therefore, the researchers have developed great interest in them, which have been widely used in real power systems. For example, in [1], the fault component estimation was modeled as a 0–1 integer–integer programming problem. Accordingly, an analytical model for fault diagnosis based on the protection action principle was built and a Genetic Algorithm (GA) was employed to solve it. Reference [2] proposed an analytical model for fault diagnosis of power systems considering the failures of Protective Relays (PRs) and Circuit Breakers (CBs), which was solved by the forbidden algorithm Tabu Search (TS). Reference [3] redefined the desired state function of PRs and CBs, and constructed an improved analytical model for fault diagnosis of power grids. Meanwhile, a diagnosis method based

Artificial Intelligence Empowered Smart Energy Systems, First Edition. Edited by Qiang Yang and Gang Huang.

on the Genetic Algorithm-Tabu Search (GATS) for power systems was proposed for solving the optimization objective function. And then, Reference [4] assigned different contribution factors to different types of PRs and CBs in the model of [3], and a fault diagnosis method based on a binary coding brainstorming optimization was proposed to solve the analytical model. Reference [5] completely preserved the coupling relationship between faulty components and protection devices. Besides, a fully analytical model was proposed and an improved discrete particle swarm algorithm was employed to calculate it.

Furthermore, Reference [6] proposed a new analytical model for the fault diagnosis of power systems based on existing analytical model-based methods, in which the redundancy and temporal sequence of alarm information were considered. Moreover, it was solved by GATS. Reference [7] introduced the chance-constrained programming model to diagnose faults in power systems, which was solved by a genetic algorithm based on Monte Carlo simulation. Besides, in Reference [8], the expected states of various types of protection devices were modeled as a whole, based on which a new fault diagnosis method based on topological description was proposed and the analytical model was solved by a genetic algorithm. In Reference [9], a fault feature matching index of external environment was introduced to the existing fault diagnosis analysis model of power systems. Besides, a simulated annealing genetic algorithm was employed to solve the improved analysis model.

The aforementioned works show that both fault diagnosis analysis models and optimization algorithms are key components of fault diagnosis methods based on optimization technology. Although the models in References [1–4] have promoted the diagnosis technique, they cannot effectively diagnose faulty components when the alarm information is seriously distorted. This is because all the above methods take the alarm messages as the actual action of protection devices. Consequently, the models in References [5–9] distinguish the actual states of protection devices from the alarm information and set different empirical weights in the objective functions to improve the diagnostic capability of analytical models. Unfortunately, the corresponding models are effective only in some specific scenarios. Therefore, up to now, it is still difficult to accurately diagnose faults in complex fault scenarios by using the optimization technology, lacking high adaptability.

Meanwhile, the weather factors in external environment under disaster weather are closely related to power system faults. However, the most of aforementioned works diagnose faults only based on the fault information obtained from the internal power system. In addition, with the increasing dimension of variables and complexity of analytical models, the difficulty of solving diagnosis problems has also increased, resulting in a high demand for optimization algorithms.

To effectively deal with the above issues, this paper proposes a novel fault diagnosis method under disaster weather based on a random self-regulating algorithm. It includes two core parts, namely an improved analytic model considering disaster weather and a random self-regulating algorithm for solving it. In this paper, faults can be diagnosed accurately when the operations of protection devices are failed or incorrect, as well as when the fault alarm information is missing or false. The main contributions of this paper are described as follows.

1) Existing works mainly use empirical weights to avoid misdiagnosis or missed diagnosis caused by abnormalities and alarm distortion of protection devices. To address the negative impact of empirical weights on diagnostic accuracy, this paper proposes a self-regulated expectation trust factor, a self-regulated alarm trust factor, and a self-regulated weather trust factor to improve the fault diagnosis analysis model. The first two trust factors are calculated based on the fault alarm data and self-test information of protection devices at the fault time so as to avoid misdiagnosis and missed diagnosis caused by the abnormal operation of protection devices and distorted alarm information. Moreover, based on the fault alarm information and meteorological data of protection devices, the self-regulated meteorological trust factor is calculated to improve the reliability of meteorological information, which is also critical to improving diagnostic accuracy.
2) To promote global optimization ability of the high-dimensional diagnosis analytic model and overcome the insufficient generalization ability caused by fixed setting of algorithm parameters, a random self-regulating algorithm with a bionic self-regulating function is proposed. The function has two characteristic operators, namely the increment operator of guiding probability and the iterative mutation operator. The increment operator of guiding probability can effectively balance the algorithm convergence rate and stability so as to increase its global optimization ability. The iterative mutation probability and hypercube density probability in the second operator can not only prevent the algorithm from excessive mutation during optimization process but also maintain the population richness well, which greatly increases the generalization ability of the algorithm.

5.2 Analytic Model for Fault Diagnosis

The analytic model-based method is to construct an objective function that can embody the logical relationship between operating states of protection devices and suspicious faulty components. Then, the fault diagnosis problem is formulated as a mathematical optimization problem. In the aforementioned work, a weighted sum method was used to calculate objective functions. However, the same trust factors were assigned to different types of protection devices. Consequently, the corresponding methods had difficulties in the identification of abnormal fault alarm information. Besides, the fault feature matching indices of external environment were gradually considered in the analytic model. However, due to the lack of trust factors for the objective function, it was difficult to accurately diagnose actual states of protection devices.

Therefore, to improve the universality of trust factors as well as the diagnostic ability of analytic models for power systems, this paper comprehensively considers the regulation law of trust factors and the influence of weather factors on fault diagnosis results. Accordingly, an objective function is proposed as follows.

$$\min E(B) = E_{ex}(B) + E_{al}(B) + E_{we}(B) \tag{5.1}$$

where $B = [D\ R\ C]$ is the independent variable of the objective function, which is also called the fault hypothesis. Specifically, $D = [d_1, d_2, \dots, d_{nd}]$ is a set of suspicious faulty components, where d_i $(1 \leq i \leq n_d)$ represents the state of the i-th suspicious faulty component; $d_i = 0$ and $d_i = 1$ represent that the component d_i is normal and faulty, respectively. $R = [r_1, r_2, \dots, r_{n_r}]$ is a set of protection relays associated with D, where r_i $(1 \leq i \leq n_r)$ represents the state of the i-th PR; $r_i = 1$ and $r_i = 0$ represent that the PR r_i has acted and does not work, respectively. $C = [c_1, c_2, \dots, c_{n_c}]$ is a set of CBs associated with D, where c_i $(1 \leq i \leq n_c)$ represents the i-th CB; $c_i = 1$ and $c_i = 0$ represent that the CB c_i has tripped and does not work, respectively. Moreover, $E_{ex}(B)$ represents the abnormal states of protection devices, $E_{al}(B)$ expresses the distortion of alarm information and $E_{we}(B)$ indicates how the fault probability of a line is with its actual state.

The objective function $E(B)$ describes the uncertainty about the operation of protection devices and the corresponding alarm information. The value of $E(B)$ reflects the confidence level of the fault hypothesis. The smaller the function value of $E(B)$, the higher its credibility.

The objective function is composed of three functions, namely $E_{ex}(B)$, $E_{al}(B)$ and $E_{we}(B)$, which are described in detail as follows.

The $E_{ex}(B)$ reflects the failed and incorrect operations protection devices, including PRs and CBs. It is shown as follows.

$$E_{ex}(B) = \sum \delta_{rex_i} \left| r_i(B) - r_i^{ex}(B) \right| + \sum \delta_{cex_j} \left| c_j(B) - c_j^{ex}(B) \right| \tag{5.2}$$

where $r_i(B)$ and $c_j(B)$ represent the actual states of PRs and CBs, respectively; $r_i^{ex}(B)$ and $c_j^{ex}(B)$ represent the expected states of PRs and CBs, respectively; δ_{rex_i} and δ_{cex_j} represent the self-regulating expectation trust factors of PRs and CBs. The smaller the function value of $E_{ex}(B)$, the smaller the number of corresponding protection devices.

The $E_{al}(B)$ reflects the missing and improper alarms of protection devices, which is expressed as follows.

$$E_{al}(B) = \sum \gamma_{ral_i} \left| r_i(B) - r_i^{al} \right| + \sum \gamma_{cal_j} \left| c_j(B) - c_j^{al} \right| \tag{5.3}$$

where r_i^{al} and c_j^{al} represent the alarm states of protection devices, namely the fault alarm information; γ_{ral_i} and γ_{cal_j} represent the self-regulating warning trust factors of PRs and CBs, respectively. The smaller the function value of $E_{al}(B)$, the smaller the number of corresponding protection devices.

The $E_{we}(B)$ reflects how the fault probability of a line with its actual state, which uses weather data to judge the faults caused by disastrous weather. It is shown as follows.

$$E_{we}(B) = \sum T_{we_k} \left| d_k(B) - D_k^p \right| \tag{5.4}$$

where $d_k(B)$, D_k^p and T_{we_k} represent actual states, fault probabilities and self-regulating weather trust factors of lines, respectively. The smaller the function value of $E_{we}(B)$, the more accurate the fault probabilities of lines.

5.2.1 Three Types of Self-Regulating Trust Factors

The uncertainty in the alarm information of protection devices may increase due to the impact of disaster weather. Consequently, the diagnostic accuracy of traditional fault diagnosis analytic models will decrease because they use the same trust factor for different types of protection devices. Therefore, the corresponding diagnostic methods cannot effectively identify the abnormal action signals of protection devices as well as the distortion states of fault alarm information.

To address above issue, this paper proposes three types of self-regulating trust factors, namely self-regulating expectation trust factor δ_{ex}, self-regulating warning trust factor γ_{al}, and self-regulating weather trust factor T_{we}. The δ_{ex} is used to weaken negative effects caused by the failed and incorrect operations of protection devices, γ_{al} is employed to deal with the missing and improper reports of alarm information, and T_{we} is used to handle adverse effects of the mismatch between fault probabilities of lines and their actual states.

Specifically, if there are different kinds of uncertain factors, then they will be distinguished via the self-test information of protection devices. Afterward, in our analytic model, δ_{ex} is related to the abnormal states of protection devices and γ_{al} is related to the distortion of fault alarm information. Consequently, the analytic model can categorically identify these factors by regulating δ_{ex} and γ_{al}. Moreover, T_{we} will vary based on weather and alarm information because it is composed of the fault probability of lines and the alarm information of protection devices. Thus, T_{we} can strengthen the correlation between different kinds of information.

The specific descriptions of the three types of self-regulating trust factors are described in Section 5.2.1.1.

5.2.1.1 Self-Regulating Expectation Trust Factor and Self-Regulating Warning Trust Factor

In practical project applications, a protection device usually has two small probability events [5], which are: (1) a protection device misses to report its alarm information while it acts incorrectly; (2) a protection device misreports its alarm

Table 5.1 The combination and evaluation of expected, actual and alarm states of PRs.

$D = [d_1, d_2, \ldots, d_{nd}]$	d_i $(1 \leq i \leq n_d)$	$d_i = 0$	Result
0	0	0	Accurate
0	0	1	The alarm of $d_i = 1$ is missing.
0	1	0	Small probability
0	1	1	The d_i operates incorrectly.
1	0	0	The $R = [r_1, r_2, \ldots, r_{n_r}]$ fails to operate.
1	0	1	Small probability
1	1	0	The alarm of D is false.
1	1	1	Accurate

information while it fails to act. To reduce the complexity of diagnostic problems, both the two situations are not considered in this paper.

And then, how to obtain the self-regulating expectation trust factor and the self-regulating warning trust factor is described as follows.

First of all, take a PR as an example. The combination and evaluation of its expected states, real states and alarm states are obtained, as shown in Table 5.1. Suppose r^{ex}, r, and r^{al} represent the expected, actual, and alarm states of a PR.

When a PR operates abnormally or its alarm information is distorted, its uncertainty index a_r is established based on its expected and alarm states, as follows.

$$\alpha_r = \overline{r^{ex}} r^{al} \oplus r^{ex} \overline{r^{al}} \tag{5.5}$$

where the "$\oplus$" represents the logical "*or*" operation of variables, and the logical "*not*" operation of a variable is represented by the overline. If $\alpha_r = 1$, then the PR operates abnormally or its alarm information is distorted. If $\alpha_r = 0$, then the PR operates reliably or its alarm information is accurate.

In the same way, when the fault alarm information of a CB is unreliable, its uncertainty index a_c is established based on its expected and alarm states, as follows.

$$\alpha_c = \overline{c^{ex}} c^{al} \oplus c^{ex} \overline{c^{al}} \tag{5.6}$$

Similarly, if $\alpha_c = 1$, then the CB trips abnormally or its alarm information is distorted. If $\alpha_c = 0$, then the CB trips reliably or its alarm information is accurate.

Second, with the development of online detection function of protection devices, the self-test technology has gradually been applied to power systems [10], which can detect whether the alarm information of protection devices is distorted or not. When the alarm state of a protection device is abnormal, it will send out a self-test alarm. Suppose that when fault alarm data is missed or reported incorrectly with self-test information, the information may not be wrong [11]. Set the self-test alarm index of a PR as s_r. If the PR sends out a self-test alarm, then $s_r = 1$; otherwise, $s_r = 0$. Similarly, set the self-test alarm index of a CB as s_c. If the CB sends out a self-test alarm, then $s_c = 1$; otherwise, $s_c = 0$.

Finally, when the fault alarm information of protection devices is abnormal, δ_{ex} and γ_{al} need to be adjusted. Specifically, if the state of a protection device is abnormal, then the self-regulating expected trust factor needs to be decreased. If the alarm information is distorted, then the self-regulating alarm trust factor needs to be lowered. And then, a_r and s_r are used to establish the self-regulating expectation trust factor δ_{rex} and the self-regulating warning trust factor γ_{ral} of PRs, as follows.

$$\delta_{rex} = 1 - \mu\alpha_r(1 - s_r) \tag{5.7}$$

$$\gamma_{ral} = 1 - \alpha_r s_r \tag{5.8}$$

where μ represents the expected adjustment coefficient and it is a real number in [0, 1]. Usually, μ is set as 0.4 via experimental test results.

In the same way, the self-regulating expectation trust factor δ_{cex} and the self-regulating warning trust factor γ_{cal} of CBs are obtained via Eqs. (5.9) and (5.10).

$$\delta_{cex} = 1 - \mu\alpha_c(1 - s_c) \tag{5.9}$$

$$\gamma_{cal} = 1 - \alpha_c s_c \tag{5.10}$$

5.2.1.2 Self-Regulating Weather Trust Factor

First of all, the fault risk for each line should be calculated according to the risk analysis method based on gray fuzzy theory in [12]. Specific calculation rules are described as follows.

a) The fault risk of lines is divided into four levels according to the external environment, which are high risk, comparative high risk, comparative low risk, and low risk.
b) Triangular membership function is used to express the fuzzy relationship between various disaster weathers and the fault risk of lines. Accordingly, a gray fuzzy discrimination matrix $\underset{\otimes}{\tilde{\mathbf{R}}}$ and a weight matrix $\underset{\otimes}{\tilde{\mathbf{W}}}$ of different weather factors are established.
c) Gray fuzzy theory is used to comprehensively evaluate the fault risk of lines (i.e., $\underset{\otimes}{\tilde{\mathbf{B}}}$), which is calculated via Eq. (5.11).

$$\underset{\otimes}{\tilde{\mathbf{B}}} = \underset{\otimes}{\tilde{\mathbf{W}}} \cdot \underset{\otimes}{\tilde{\mathbf{R}}} \tag{5.11}$$

Table 5.2 Fault probability corresponding to different fault risk of lines.

Fault risk	Fault probability value
High	0.875
Comparative high	0.625
Comparative low	0.375
Low	0.125

Second, the fault probability range of lines is evenly divided into four intervals, and each one corresponds to a fault risk level of lines. The midpoint of each interval is taken as the fault probability (i.e., D_k^p) of its corresponding fault risk, as shown in Table 5.2.

Since this paper focuses on the fault diagnosis under disaster weather, the fault probability of lines is considered in the analytic model if and only if the fault risk levels of lines are high and comparatively high. When $D_k^p < 0.625$, we uniformly set $D_k^p = 0$.

It is worth noting that power system faults caused by disaster weather have a certain degree of contingency. Therefore, a line with a high-risk level may not fail. Consequently, a line fault judgment index t_{cp} is introduced to reduce the impact of such a situation on diagnostic results as follows.

$$t_{cp} = \sum_{\oplus, r_{is}^{alarm} \in L(d_k)} r_{is}^{alarm} \otimes c_{js}^{alarm} + \sum_{\oplus, r_{ir}^{alarm} \in L(d_k)} r_{ir}^{alarm} \otimes c_{jr}^{alarm} \tag{5.12}$$

where "⊗" represents the logical multiplication operation; $\sum_{\oplus, r_{is}^{alarm} \in L(d_k)} r_{is}^{alarm}$ and $\sum_{\oplus, r_{ir}^{alarm} \in L(d_k)} r_{ir}^{alarm}$ represent the results of logical "*or*" operations of alarm information for main, primary backup, and secondary backup PRs at sending and receiving ends of lines, respectively; $L(d_k)$ represents a set of PRs for line d_k; c_{js}^{alarm} and c_{jr}^{alarm} represent the alarm information of CBs at sending and receiving ends of lines, respectively. If $t_{cp} \geq 1$, then we set $t_{cp} = 1$; otherwise, $t_{cp} = 0$.

Finally, the self-regulating weather trust factor T_{we_k} is calculated via Eq. (5.13), which is used to adjust the impact of weather on the objective function of fault diagnosis.

$$T_{we_k} = 2t_{cp}D_k^p \tag{5.13}$$

5.2.2 Expected States of Protection Devices

The expected states of different kinds of protection devices are described as follows.

5.2.2.1 Expected States of Main Protective Relays

Suppose r_m is the main PR of component d_i. If d_i fails, then r_m should act. Therefore, the expected state r_m^{ex} of r_m is described as:

$$r_m^{ex} = d_i \tag{5.14}$$

5.2.2.2 Expected States of Primary Backup Protective Relays

Suppose r_p is the primary backup PR of component d_i. If d_i fails as well as its main PR r_m also does not operate, then r_p should act. Therefore, the expected state r_p^{ex} of r_p is described as:

$$r_p^{ex} = d_i\overline{r_m} \tag{5.15}$$

5.2.2.3 Expected States of Secondary Backup Protective Relays

Suppose r_s is the secondary backup PR of component d_i. If d_i fails as well as both its main and primary backup PRs do not operate, then r_s should act. Moreover, when $d_x \in D(r_s)$ fails, if CBs on associated paths from r_s to d_x do not trip, then r_s should act, where $D(r_s)$ represents a set of power system components within the protective range of r_s except for d_i. Therefore, the expected state r_s^{ex} of r_s is described as:

$$r_s^{ex} = d_i\overline{r_m r_p} \oplus \sum_{d_x \in D(r_s)} \left(d_x \otimes \prod_{c_p \in p(r_s, d_x)} \overline{c_p} \right) \tag{5.16}$$

5.2.2.4 Expected States of Breaker Failure Protective Relays

Suppose r_f is the breaker failure PR of CB c_j. If c_j fails to trip when a fault occurs but the tripping signal has been sent to c_j, then r_f should act. Therefore, the expected state r_f^{ex} of r_f is as follows.

$$r_f^{ex} = \overline{c_j} \otimes \sum_{\oplus, r_i \in R(c_j)} r_i \tag{5.17}$$

where $N(c_j)$ represents a set of PRs that can make c_j to trip.

5.2.2.5 Expected States of Circuit Breakers

Expected state c_j^{ex} of CB c_j is described as follows. If any PR in $R(c_j)$ operates, then c_j should trip.

$$c_j^{ex} = \sum_{r_i \in R(c_j)} r_i \tag{5.18}$$

5.3 Random Self-Regulating Algorithm

The random self-regulating algorithm is composed of a solution matrix **B** and a control matrix **P**. It is controlled by a bionic self-regulating function, as shown in Figure 5.1.

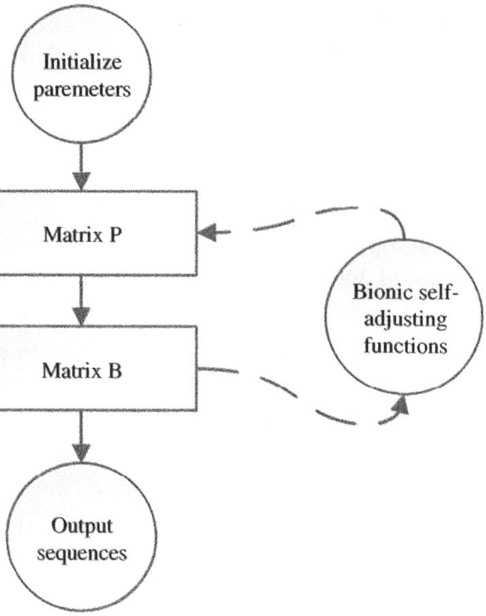

Figure 5.1 Schematic diagram of random self-regulating algorithm.

The row vectors of matrix **P** are regarded as guide indices so as to guide changes in the organism. It is composed of guiding probabilities $p_{i,j}$ $(0 \leq p_{i,j} \leq 1;\ 1 \leq i \leq H;\ 1 \leq j \leq m)$, where H and m represent the number of rows and columns of matrices **P** and **B**, respectively. It is as follows.

$$\mathbf{P} = \begin{bmatrix} p_{11} & p_{12} & \cdots & p_{1m} \\ p_{21} & p_{22} & \cdots & p_{2m} \\ \vdots & \vdots & \ddots & \vdots \\ p_{H1} & p_{H2} & \cdots & p_{Hm} \end{bmatrix} \tag{5.19}$$

The bionic self-regulating function regards the matrix **B** as a population and its row vector as an organism. It is composed of the bit $b_{i,j}(0 \leq b_{i,j} \leq 1;\ 1 \leq i \leq H;\ 1 \leq j \leq m)$, as shown in Eq. (5.20). The matrices **B** and **P** form a bijection, $\mathbf{B} = f_1(\mathbf{P})$ and $f_1(\cdot) := \{(p_{i,j} \geq 0.5)b_{i,j} = 1\} \wedge \{(p_{i,j} < 0.5)b_{i,j} = 0\}$.

$$\mathbf{B} = \begin{bmatrix} b_{11} & b_{12} & \cdots & b_{1m} \\ b_{21} & b_{22} & \cdots & b_{2m} \\ \vdots & \vdots & \ddots & \vdots \\ b_{H1} & b_{H2} & \cdots & b_{Hm} \end{bmatrix} \tag{5.20}$$

5.3.1 Random Self-Regulating Algorithm

Input parameters of the random self-regulating algorithm include H, m, p_j^a and T_{maxgen}, where p_j^a represents the selection probability of each element of the matrix **B**, T_{maxgen} represents the maximum number of iterations of the algorithm.

In this paper, p_j^a is set as 0.5. Besides, several important variables are involved in the algorithm, such as $\mathbf{B}_i$, $f_{\mathbf{B}_i}$, $\mathbf{B}_{best}$, $\mathbf{B}_{bad}$, $\mathbf{G}_{bad}$, and Δ, where $\mathbf{B}_i$ ($\mathbf{B}_i = (b_{i,1}, b_{i,2}, \dots b_{i,m})$) represents the i-th solution vector in $\mathbf{B}$, $f_{\mathbf{B}_i}$ represents the function value of $\mathbf{B}_i$, $\mathbf{B}_{best}$ represents contemporary optimal solution, $\mathbf{B}_{bad}$ represents the contemporary worst solution and $\mathbf{G}_{bad}$ represents the global worst solution; Δ represents the increment operator of each element in $\mathbf{P}$, i.e., the increment operator of guiding probability. The output of a random self-regulating algorithm is $\mathbf{G}_{best}$, which represent the global optimal solution. The steps of the random self-regulating algorithm are described as follows, whose flowchart is shown in Figure 5.2.

Step (1): Initialize parameters H, m, p_j^a, and T_{maxgen}.
Step (2): Set $T = 1$.
Step (3): Calculate the objective function based on the initialized solution matrix **B**.

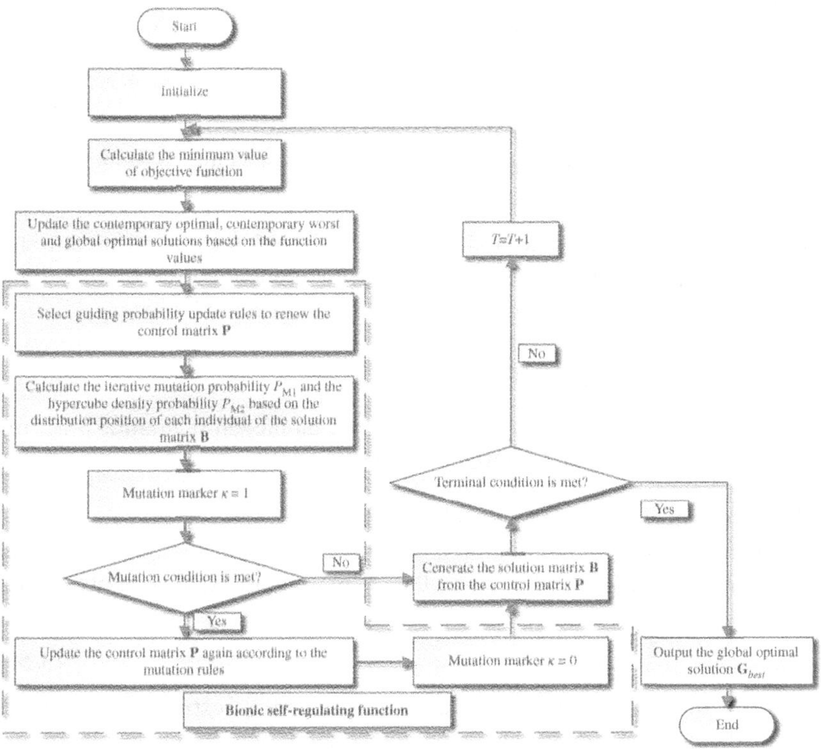

Figure 5.2 Flowchart of random self-regulating algorithm.

Step (4): Update the contemporary optimal solution $\mathbf{B}_{best}$, the contemporary worst solution $\mathbf{B}_{bad}$ and the global optimal solution $\mathbf{G}_{best}$ based on the objective function values.

Step (5): Select guiding probability update rules and update the control matrix **P**. For $i = 1, \ldots, H, j = 1, \ldots, m$, do steps (5.1) and (5.2).

Step (5.1): If $rand \le p^a_{i,j}$, then do Eq. (5.21), where $rand \in [0, 1]$ is a uniformly distributed random number and $k_1 \neq k_2 \neq i$.

$$b_{i,j} = \begin{cases} B_{k1,j}(\text{if } f_{B_{k1}} > f_{B_{k2}}) \\ B_{k2,j}(\text{if } f_{B_{k1}} \le f_{B_{k2}}) \end{cases} \tag{5.21}$$

If $rand > p^a_{i,j}$, then do the process as follows.

$$b_{i,j} = \mathbf{G}_{best,j} \tag{5.22}$$

Step (5.2): Update $p_{i,j}$ by Eq. (5.23).

$$p_{i,j} = \begin{cases} p_{i,j} + \Delta \quad (\text{if}\,(b_{i,j} == 1)) \\ p_{i,j} - \Delta \quad (\text{if}\,(b_{i,j} == 0)) \end{cases} \tag{5.23}$$

Step (6): Calculate the iterative mutation probability P_{M1} and the hypercube density probability P_{M2} based on the distribution position of each individual of solution matrix **B**.

Step (7): Update the control matrix **P** again according to the mutation rules.

Step (8): Generate the solution matrix **B** from the control matrix **P**.

Step (9): If the terminal condition is met, then output the global optimal solution $\mathbf{G}_{best}$; otherwise, go to step (2).

After performing the above steps, the random self-regulating algorithm outputs a binary code sequence. If a code is 1, then the component corresponding to its serial number is a faulty one; otherwise, the component is not faulty.

5.3.2 Bionic Self-Regulating Function

Both global convergence capability and speed of the bionic self-regulating function are improved based on the self-regulation of Δ and iterative mutation.

5.3.2.1 Increment Operator of Guiding Probability

The increment operator of guiding probability (i.e., Δ) expressed by Eq. (5.24) is used to calculate the change of $p_{i,j}$ as well as to reasonably regulate convergence speed of the algorithm.

$$\Delta = k_a(p_{\text{lim}\,it} - p_{ij}) \tag{5.24}$$

First, the difference between the guiding probability extreme value p_{limit} ($p_{\text{limit}} = 1 \vee 0$) and p_{ij} is used to calculate the increment of p_{ij} in the next iteration, and then the adjustment coefficient k_a is used to avoid the range of difference from being too large.

If a constant is used as k_a, then it cannot be associated with the variation of the function value. In this case, k_a can be calculated by Eq. (5.25) based on the objective function values of $\mathbf{B}_i$, $\mathbf{B}_{best}$, $\mathbf{B}_{bad}$, $\mathbf{G}_{best}$, and $\mathbf{G}_{bad}$ as well as the selected update rule that is given in step (5.1).

$$k_a = \begin{cases} \dfrac{f_{\mathbf{B}_i} - f_{\mathbf{B}_{best}}}{f_{\mathbf{B}_{bad}} - f_{\mathbf{B}_{best}}} & \left(\text{if } rand \leq p_{i,j}^a\right) \\[2ex] \dfrac{f_{\mathbf{B}_i} - f_{\mathbf{G}_{best}}}{f_{\mathbf{G}_{bad}} - f_{\mathbf{G}_{best}}} & \left(\text{if } rand > p_{i,j}^a\right) \end{cases} \tag{5.25}$$

where $f_{\mathbf{G}_{best}}$, $f_{\mathbf{G}_{bad}}$, $f_{\mathbf{B}_{best}}$, and $f_{\mathbf{B}_{bad}}$ represent the function values of $\mathbf{G}_{best}$, $\mathbf{G}_{bad}$, $\mathbf{B}_{best}$ and $\mathbf{B}_{bad}$, respectively.

In the early algorithm operation stage, optimization results may change a lot under different iterations, and the chaotic values of $f_{\mathbf{G}_{best}}$, $f_{\mathbf{B}_{best}}$, and $f_{\mathbf{B}_{bad}}$ are high. As the number of iterations increases, the parameters $f_{\mathbf{G}_{best}}$ and $f_{\mathbf{B}_{best}}$ gradually tend to become a constant, and the convergence curve of $f_{\mathbf{B}_{bad}}$ has a small fluctuation range, as shown in Figure 5.3.

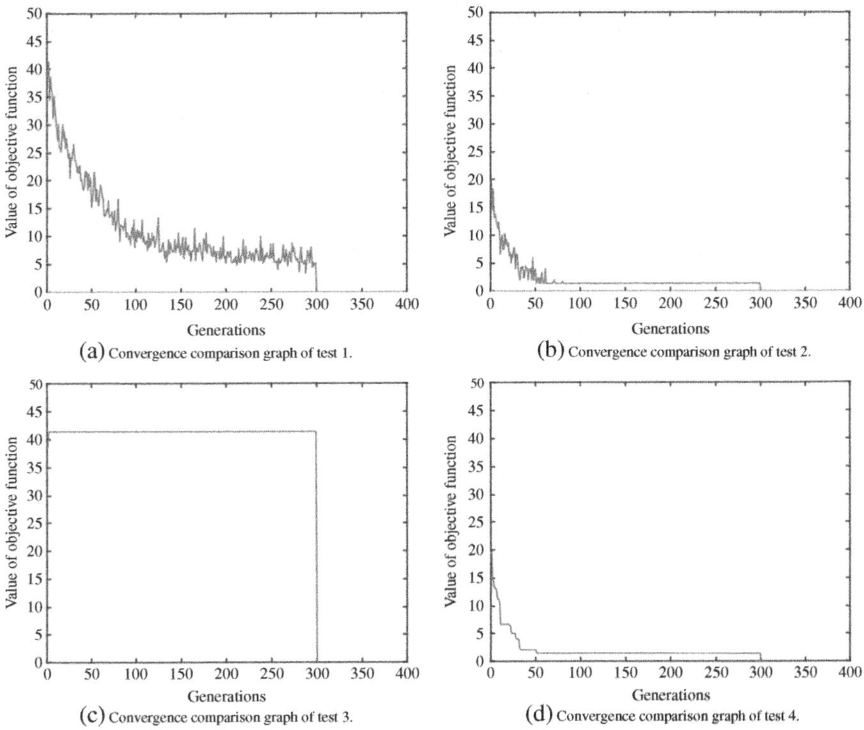

Figure 5.3 Convergence comparison graph.

At the later stage of the iteration, k_a can be decomposed into a linear function with respect to $f_{\mathbf{B}_i}$ $\left(k_a = \frac{f_{\mathbf{B}_i}}{f_{\mathbf{B}_{bad}}(or f_{\mathbf{G}_{bad}}) - f_{\mathbf{B}_{best}}(or f_{\mathbf{G}_{best}})} - \frac{f_{\mathbf{B}_{best}}(or f_{\mathbf{G}_{best}})}{f_{\mathbf{B}_{bad}}(or f_{\mathbf{G}_{bad}}) - f_{\mathbf{B}_{best}}(or f_{\mathbf{G}_{best}})}\right)$, where parameters $f_{\mathbf{G}_{bad}}$, $f_{\mathbf{G}_{best}}$, and $f_{\mathbf{B}_{best}}$ are approximate constants. This process shows that the change of k_a is only closely related to the iteration and is not limited to the choice of objective function. In this way, the subjective influence of setting k_a by expertise is avoided.

5.3.2.2 Iterative Mutation Operator

The iterative mutation operator can promote the algorithm to have the ability of cumulative mutation probability in the optimization process, which can prevent the algorithm from falling into local optima and improve its global convergence ability.

The iterative mutation process is controlled by two probability conditions, defined as follows.

$$(number_{rand} < P_{\mathrm{M1}}) \wedge (number_{rand} > P_{\mathrm{M2}}) \tag{5.26}$$

where the iterative mutation probability P_{M1} is controlled by the number of iterations and the hypercube density probability P_{M2} is controlled by the hypercube density of population. The P_{M2} is used to maintain population diversity and prevent it from being overdense.

When the condition of iterative mutation is satisfied, the matrices **P** and $\mathbf{B}_{best}$ will mutate. The pseudo-code of mutation process is shown in Figure 5.4, where *bestindex* is the index of contemporary optimal solution, and *randi* ([0, 1]) is a number generated randomly from 0 or 1.

```
if rand<0.42
 for i=1,2,...,H do
  for j=1,2,...,m do
   if i ≠ bestindex
    p_ij = rand
   else
    for z=1,2,...,ceil(T_gen / T_maxgen)•m do
     B_best,j = randi([0,1])
    end for
   end if
  end for
 end for
end if
```

Figure 5.4 Pseudo-code of iterative mutation.

Iterative Mutation Probability. The iterative mutation P_{M1} is used to avoid the algorithm drooping into local optimal solutions early and prevent the loss of global optimal solution caused by the excessive mutation in the later stage.

At the early stage of the iteration, the lower threshold for P_{M1} is set. At the same time, the value of P_{M1} is increased to enhance the possibility of the mutation. Accordingly, the possibility of drooping out local optimal solutions is improved.

The maximum mutation probability in the later stage is calculated by Eq. (5.27). It can be seen that the mutation probability at the later stage is low.

$$P_{M1} < \sum_{i=160}^{200} \frac{1}{i} \approx 0.23 \tag{5.27}$$

The pseudo-code of P_{M1} is shown in Figure 5.5.

The mutation probability P_{M1} is generated by iterative accumulation by Eq. (5.28).

$$P_{M1} = \kappa \cdot P_{M1} \tag{5.28}$$

where κ represents mutation flag, which can avoid excessive accumulation of mutation probability at the later stage that may lead to missing the global optimum. If a mutation occurs, κ is set to 0; otherwise, κ is set to 1.

Hypercube Density Probability of Population. The hypercube density probability (i.e., P_{M2}) reflects the distribution of population individuals in the population hypercube, which means the population diversity. The hypercube population density is obtained based on the volume of the hypercube and the number of individuals, and then it is converted to the final value of P_{M2} by the log and sigmoid functions. The pseudo-cod of P_{M2} is shown in Figure 5.6,

Figure 5.5 Pseudo-code of P_{M1}.

```
P_M1 = P_M1•κ
P_M1 = P_M1 + 1/gen
if T_gen < 0.8•T_maxgen do
  if P_M1 > 0.25 do
    P_M1 = 0.75
  end if
else
  P_M1 = P_M1
end if
if mutation do
  κ=0
end if
κ=1
```

Follow the codes in Fig. 5, and then

$$q = 100 \cdot rand \cdot \left|\frac{\log(density1)}{\log(density0)}\right|$$

$$P_{M2} = \frac{1}{1+\exp(-q/50)}$$

Figure 5.6 The calculation code of P_{M2}.

where *density* represents hypercube population density, *density0* represents the initial one and *density1* represents the current one; the log function is used to solve the high-dimensional volume expansion problem; $100 \cdot rand$ is used to map q to interval (0, 100) and expand the range of random probability; the *sigmoid* function is used to map $q/50$ to (0.5, 1).

The calculation equations of *density* are shown as follows and its pseudo-code is shown in Figure 5.7. First, the population hypercube volume is calculated according to Eqs. (5.29)–(5.32). And then, the population abundance is calculated by Eq. (5.33). Finally, the distribution of population individuals is obtained in the population hypercube.

$$center = \sum_{i=1}^{H} \frac{1}{H \cdot p_i} \tag{5.29}$$

$$norm = \|p_i - center\|_2 \tag{5.30}$$

$$digonal = 2 \cdot max(norm) \tag{5.31}$$

$$volume = length^\wedge m \tag{5.32}$$

$$density = \frac{H}{volume} \tag{5.33}$$

where *center* is the geometric center vector, *norm* is 2-Norm of the control matrix row vector $\mathbf{P}_i$ $(i \in 1, 2, \ldots, H)$, *digonal* is the longest diagonal of the population hypercube and *volume* is the population hypercube volume.

5.3.3 Fault Diagnosis Process

Fault diagnosis steps of the proposed method are described as follows.

Step (1): Determine power outage areas by the network topology analysis approach.

Step (2): Establish the fault hypothesis (i.e., the set B) for each power outage area with the suspicious faulty components in set D, the related PRs in set R and the related CBs in set C.

Step (3): According to the D, R, and C, construct the fault diagnosis objective function (i.e., Eq. (5.1)) and the expected states of protection devices (i.e., r_{ex} and c_{ex}).

Step (4): According to the r_{ex} and c_{ex}, alarm information and self-test information, calculate the self-regulating expected trust factor δ_{ex} and the self-regulating

Figure 5.7 Pseudo-code of density.

```
for j=1,2,···,m do
  for i=1,2,···,H do
    center(j)=center(j)+P(i,j)/H
  end for
end for
center = center(:,1)
for i=1,2,···,H do
  for j=1,2,···,m do
    norm(i)=norm(i)+(P(i,j) − center(j))^2
  end for
end for
norm = norm(:,1)
for i=1,2,···,H do
  norm(i) = norm(i)^0.5
end for
[digonal,maxindex] = max(norm)
digonal = 2•digonal
length = digonal
volume = length^m
density = H/volume
q = 100•rand•|log(density1)/log(density0)|
P_M2 = 1/(1+exp(−q/50))
```

alarm trust factor γ_{al} by Eqs. (5.5)–(5.10). And then, calculate the self-regulating weather trust factor by Eqs. (5.12)–(5.13) based on the fault probability of lines and the alarm states of protection devices.

Step (5): Employ the random self-regulating algorithm to solve fault diagnosis objective function so as to obtain the global optimal solution $\mathbf{G}_{best}$.

Step (6): End the calculation and output diagnostic results.

5.4 Experiment and Analysis

A typical power system previously discussed in [1, 5] is employed to verify the effectiveness and superiority of the proposed method, as shown in Figure 5.8, which includes 28 components, 40 CBs, and 124 PRs.

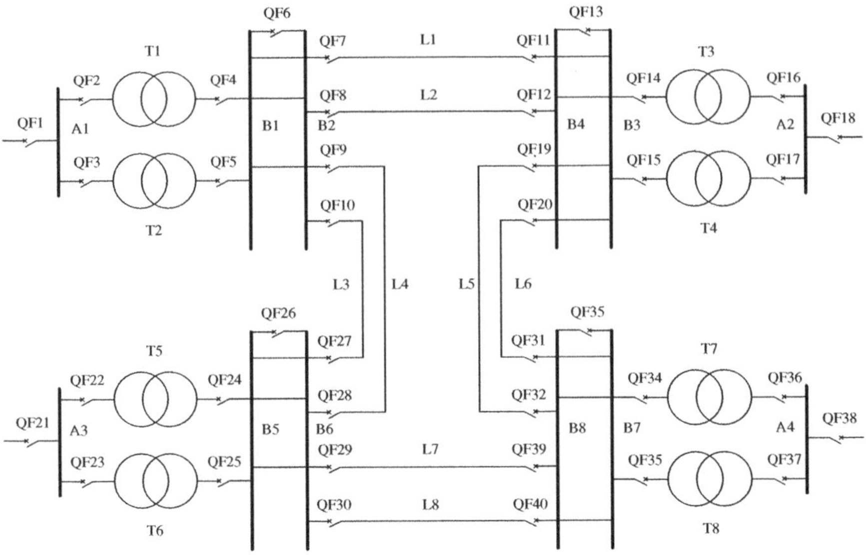

Figure 5.8 A typical power system.

The symbols in Figure 5.8 are described as follows. A1–A4 and B1–B8 represent buses, L1–L8 represent transmission lines, T1–T8 represent transformers, QF1–QF40 represent CBs, A1m–A4m, and B1m–B8m represent the PRs of buses; L*x*S*y* and L*x*R*y* (y = m, p, or s) represent the PRs at the sending and ending ends of transmission line x, where m, p, and s represent the main, first backup and second backup PRs, respectively; QF1f–QF40f represent the PRs of breaker failure protection.

Besides, the computer used for computational work was an Intel(R) core(TM) i7-4720HQ CPU @ 2.60GHZ. The proposed method was carried out by using the software MATLAB.

5.4.1 Case Study

Four typical fault cases are simulated and their comparison results between the proposed method and the ones in [5, 9] are shown in Table 5.3 in appendix, which involves the scenarios with failed or incorrect operations of protection devices as well as the improper or missing of their alarm information. We can see that, for cases 1–2, all three methods can accurately diagnose faulty components. Meanwhile, the failed operations and missing alarm information of protection devices can also be acquired. For case 3, the method in [5] yields multiple diagnosis results, while the results acquired by our method and the one in [9] are accurate and unique. For case 4, the fault scenario is very complicated and

Table 5.3 The diagnostic results.

No.	Alarm information	Self-test information	Fault risk of lines	Results		
				Reference	Reference	The proposed method
1	B1m, L2Rs, L3Rs, QF5, QF7, QF9, QF12, QF27	—	L2 and L3 are comparative high risk	The faulty component is B1; QF6 fails to operate.	The faulty component is B1; QF6 fails to operate.	The faulty component is B1; QF6 fails to operate.
2	A3m, B6m, QF21, QF22, QF23, QF24, QF25, QF27, QF28, QF29, QF30	QF26f	—	The faulty components are A3 and B6; QF26 fails to operate; the alarm of QF26f is missing.	The faulty components are A3 and B6; QF26 fails to operate; the alarm of QF26f is missing.	The faulty components are A3 and B6: QF26 fails to operate; the alarm of QF26f is missing.
3	T6p, L3Sm, L3Rm, QF10, QF25, QF26, QF29	T6p, QF27f	L3 is high risk	The faulty component is L3; QF27 fails to operate; the alarm of QF27f is missing; the alarm of T6p is false.	① The faulty component is L3; QF27 fails to operate; The alarm of QF27f is missing; the alarm of T6p is false. ② The faulty components are L3 and T6; T6m and QF27 fail to operate; The alarms of QF27f and QF23 are missing.	The faulty component is L3; QF27 fails to operate; the alarm of QF27f is missing; the alarm of T6p is false.
4	T2m, T2p, B6m, QF5f, L1Rm, L4Sm, QF3, QF6, QF7, QF9, QF24, QF26, QF30	L1Rm, L4Rm, QF28	L1 and L4 are high risk	The faulty components are B6, T2, L1 and L4; T2p operates incorrectly:QF5 fails to operate; the alarms of L1Sm, L4Rm, QF11 and QF28 are missing.	The faulty components are B6, T2 and L4; QF5 fails to operate; the alarms of L4Rm and QF28 are missing: the alarms of T2p and L1Rm are false.	The faulty components are B6, T2 and L4; T2p operates incorrectly; QF5 fails to operate, the alarms of L4Rm and QF28 are missing: the alarm of L1Rm is false.

only the proposed method can still acquire accurate results. The method in [9] misdiagnoses faulty components. Moreover, although the method in [5] correctly diagnoses faulty components, the alarm information of protection devices is misdiagnosed. Consequently, the proposed method is effective and superior with high fault tolerance.

To improve the intelligibility, case 4 is taken as an example to describe the proposed method in detail.

The fault scenario is described as follows. In thunderstorm weather, there are thunder and lightning orange warnings, downpours, and strong breezes. Both the line L4 and transformer T2 occur the phase A earthing fault. Bus B6 occurs short-circuit fault between phases A and B. Action information of PRs and CBs are shown as follows. The main PR T2m of transformer T2 operates and the CB QF3 is tripped off, whereas CB QF5 fails to operate. The PR of breaker failure protection (i.e., QF5f) operates to trip off CBs QF6 and QF7. The first backup PR T2p operates incorrectly. The main PRs L4Sm and L4Rm of line L4 operate, and CBs QF9 and QF28 are tripped off. The main PR B6m of bus B6 operates, and CBs QF24, QF26, and QF30 are tripped off. The alarm information is affected by lightning during power transmission, and certain alarm information is distorted. The distorted alarm information includes: the fault alarm information of main PR L1Rm is false; the fault alarm information of main PR L4Rm and CB QF28 are missing.

Protection devices related to received alarm information include T2m, T2p, QF5f, L1Rm, L4Sm, B6m, QF3, QF6, QF7, QF9, QF24, QF26, and QF30, while the ones related to received self-test alarm information include L1Rm, L4Rm, and QF28. Based on the received alarm information, power outage areas are determined with the network topology method. Moreover, according to the received weather data, the grey fuzzy comprehensive judgment method is used to determine the fault risks of lines L1 and L4, which are all at high risk. Thus, the fault probability of L1 and L4 is calculated as 0.875.

The specific form of the fault hypothesis is: $B = [d_1, \dots, d_5, r_1, \dots, r_{23}, c_1, \dots, c_{10}]$. Suppose the suspicious fault components in power outage areas are B1, B6, T2, L1, and L4, which correspond in turn to $D = \{d_1, d_2, d_3, d_4, d_5\}$. The related PRs include T2m, T2p, T2s, B1m, B6m, L1Sm, L1Rm, L1Sp, L1Rp, L1Ss, L1Rs, L4Sm, L4Rm, L4Sp, L4Rp, L4Ss, L4Rs, QF5f, QF7f, QF9f, QF24f, QF28f, and QF30f, which correspond in turn to $R = \{r_1, r_2, \dots, r_{23}\}$. The related CBs include QF3, QF5, QF6, QF7, QF9, QF11, QF24, QF26, QF28, and QF30, which correspond in turn to $C = \{c_1, c_2, \dots, c_{10}\}$.

According to the received alarm information, the states of PRs and CB are marked as $R^{al} = \left(r_1^{al}, r_2^{al}, \dots, r_{23}^{al}\right) = (1\ 1\ 0\ 0\ 1\ 0\ 1\ 0\ 0\ 0\ 0\ 1\ 0\ 0\ 0\ 0\ 0\ 1\ 0\ 0\ 0\ 0\ 0)$ and $C^{al} = \left(c_1^{al}, c_2^{al}, c_3^{al}, \dots, c_{10}^{al}\right) = (1\ 0\ 1\ 1\ 1\ 0\ 1\ 1\ 0\ 1)$, respectively. According to the received self-test information, the self-test alarm indexes of PRs and CBs are marked as $S_r = (s_{r_1}, s_{r_2}, \dots, s_{r_{23}}) = (0\,0\,0\,0\,0\,0\,1\,0\,0\,0\,0\,0\,1\,0\,0\,0\,0\,0\,0\,0\,0\,0\,0)$ and $S_c = (s_{c_1}, s_{c_2}, \dots, s_{c_{10}}) = (0\,0\,0\,0\,0\,0\,0\,0\,1\,0)$, respectively.

According to the above variables associated with fault scenarios, the object function is established by Eqs. (5.1)–(5.4). Then, the optimization problem is solved by the random self-regulating algorithm. The parameters are set as follows. The numbers of population size and maximum iteration are set as 50 and 300, respectively; the number of coding digits of the fault hypothesis is 38; the first five digits represent the actual states of suspicious faulty components; the 6th to 28th digits represent the actual states of PRs; the 29th to 38th digits represent the actual states of CBs. The global optimal solution found is (0 1 1 0 1 1 1 0 0 1 0 0 0 0 0 0 1 1 0 0 0 0 1 0 0 0 0 0 1 0 1 1 1 0 1 1 1 1). The corresponding fault diagnosis results are: bus B6, transformer T2, and line L4 are faulty components; PRs T2m, T2p, B6m, L4Sm, L4Rm, and QF5f operate; CBs QF3, QF6, QF7, QF9, QF24, QF26, QF28, and QF30 trip off. According to the fault diagnosis results, we can identify that PR T2p operates incorrectly, CB QF5 fails to operate, the alarms of L4Rm and QF28 are missing and the alarm of L1Rm is improper.

5.4.2 Accuracy Test

In this subsection, the convergence speeds of the four methods analyzed above are tested to further verify the effectiveness and superiority of the proposed Random Self-regulation Algorithm (RSA). In the test, the *iteration* is set as 800. The comparison results are shown in Figure 5.9.

We can see that the RSA can get the best performance for cases 1, 3, and 4. For case 2, the convergence speed of the RSA is only a little bit slower than that of the Binary Particle Swarm Optimization (BPSO). At the same time, both Simulated Annealing Genetic Algorithm (SAGA) and Optimization Spiking Neural P System (OSNPS) converge slowly in four cases. For cases 2 and 3, the convergence speed gap between the BPSO and RSA is small because the dimensions of diagnosis problems are low. However, since the dimensions of diagnosis problems in cases 1 and 4 are high, the solving difficulty increases. Consequently, the BPSO falls into local convergence in case 1 and it converges more slowly than the RSA in case 4. It is worth noting that the dimension of diagnosis problem in case 4 is 38 and the RSA is still good enough for solving the fault diagnosis model.

At the same time, all four cases are tested 1000 times and solved with the above four algorithms. The corresponding results are shown in Table 5.4. It shows that the RSA has the highest accuracy rate for every case. For case 2, all four algorithms achieve 100% accuracy because of its simple failure scenario. Since cases 1 and 3 are more complicated than case 2, their solving difficulty rises and the diagnostic accuracy of each algorithm decreases. The solution difficulty of case 4 is the highest because of the large number of faulty components and the multiple

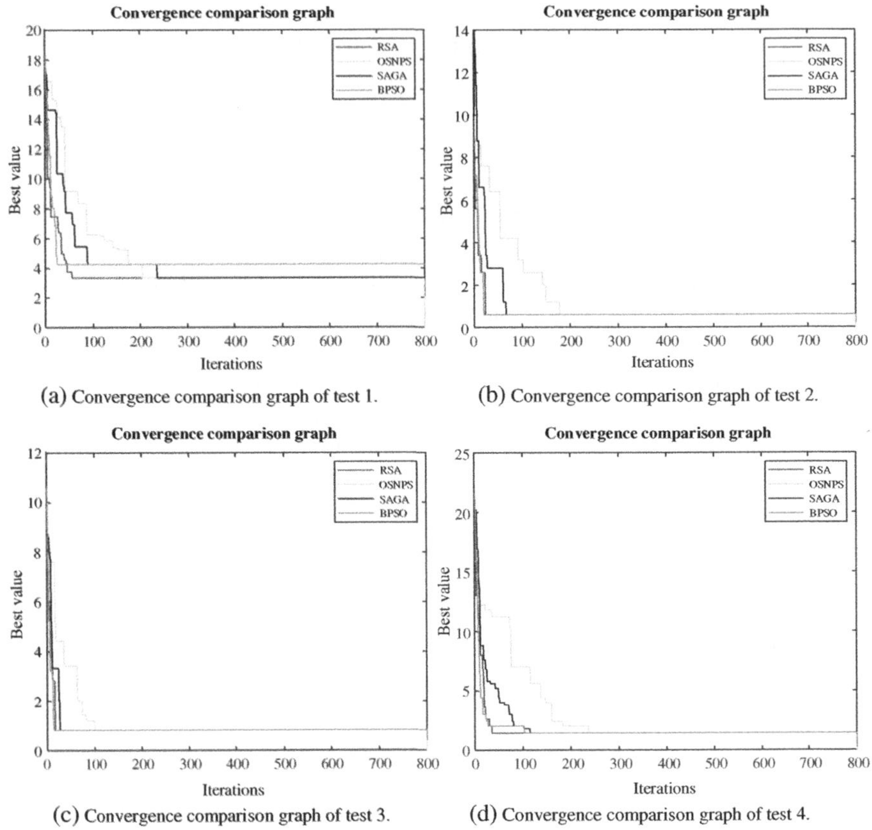

(a) Convergence comparison graph of test 1.

(b) Convergence comparison graph of test 2.

(c) Convergence comparison graph of test 3.

(d) Convergence comparison graph of test 4.

Figure 5.9 Convergence comparison graph.

distortions of alarm messages. Compared with cases 1–3, all the diagnostic accuracies of the BPSO, OSNPS, and SAGA decrease significantly in case 4, while the one proposed RSA is still high. Therefore, considering convergence speed and diagnostic accuracy comprehensively, we can see that the RSA is appropriate and effective for solving the analytical model of grid fault diagnosis and the overall performance of fault diagnosis method based on RSA is good.

To solve diagnostic problems of power systems caused by abnormal protection devices and alarm information distortion under disaster weather, this paper proposes a novel fault diagnosis method based on a random self-regulating algorithm. To comprehensively and efficiently use the fault alarm information, self-test information of protection devices and weather data, three types of self-regulating trust factors are designed to avoid the subjectivity of empirical weights and improve the fault tolerance and diagnostic accuracy of analytic models. Moreover,

Table 5.4 Example of a lengthy table which is set to full-text width.

No.	BPSO [13]	OSNPS [14]	SAGA [15]	RSA
1	100%	96.8%	99.7%	100%
2	100%	100%	100%	100%
3	90.4%	98.4%	99.5%	99.5%
4	90.9%	88.3%	88%	99.1%

a bionic self-regulating function based on the solution and control matrices is proposed to establish the random self-regulating algorithm. The algorithm has two capabilities, namely the self-regulation of the increment operator of guiding probability and the iterative mutation of the control matrix, based on which the optimization and global convergence speeds are improved. Finally, the random self-regulating algorithm is used to solve the fault diagnosis problem of power systems, and the case studies verify its feasibility and effectiveness.

In the future, the evidence theory [16] will be considered to effectively deal with uncertain information due to its powerful ability. Besides, abundant multi-source fault data including remote measurements and temporal information will also be considered in diagnosis analytic models of power systems. To reduce the building difficulty of models, multi-object functions for fault diagnosis will be established. Accordingly, multi-objective optimization algorithms will be designed to solve them [17, 18].

References

1 Wen, F.S. and Han, Z.X. (1995). Fault section estimation in power systems using a genetic algorithm. *Electric Power Systems Research* 34 (3): 165–172.

2 Guo, W., Wen, F., Ledwich, G. et al. (2010). An analytic model for fault diagnosis in power systems considering malfunctions of protective relays and circuit breakers. *IEEE Transactions on Power Delivery* 25 (3): 1393–1401.

3 Lin, X., Ke, S., Li, Z. et al. (2010). A Fault diagnosis method of power systems based on improved objective function and genetic algorithm-tabu search. *IEEE Transactions on Power Delivery* 25 (3): 1268–1274.

4 Xiong, G.J., Shi, D.Y., Zhang, J. et al. (2018). A binary coded brain storm optimization for fault section diagnosis of power systems. *Electric Power Systems Research* 163: 441–451.

5 Liu, D.B., Gu, X., and Li, H. (2011). A complete analytic model for fault diagnosis of power systems. *Proceedings of the Csee* 31 (34): 85–92.

6 Zhang, Y., Chung, C.Y., Wen, F. et al. (2016). An analytic model for fault diagnosis in power systems utilizing redundancy and temporal information of alarm messages. *IEEE Transactions on Power Systems* 31 (6): 4877–4886.

7 Song, H., Dong, M., Han, R. et al. (2018). Stochastic programming-based fault diagnosis in power systems under imperfect and incomplete information. *Energies* 11 (10): 2565.

8 Xu, B., Yin, X.G., Wu, D.L. et al. (2019). An analytic method for power system fault diagnosis employing topology description. *Energies* 12 (9): 1770.

9 Shang, H.Y., Wang, C.Y., Chen, M.H. et al. (2018). Feature matching method for power system fault diagnosis under disastrous weather conditions. *Power Construction* 39 (10): 37–43.

10 Yuan, H., Qu, G., Zhuang, W.J. et al. (2014). Exploring the content of secondary equipment condition monitoring in power networks. *Automation of Electric Power Systems* 38 (12): 100–106.

11 Chen, J.C., Zhang, Y.J., Huang, G.Q. et al. (2021). An optimization model for fault diagnosis in power grids taking into account the confidence of protection and circuit breaker alarm information. *Power System Protection and Control* 49 (4): 28–36.

12 Chen, X.T., Wang, T., Ying, R.X. et al. (2021). A fault diagnosis method considering meteorological factors for transmission networks based on P systems. *Entropy* 23 (8): 1008.

13 He, Z., Chiang, H.D., Li, C. et al. (2009). Fault-section estimation in power systems based on improved optimization model and binary particle swarm optimization. *IEEE Power and Energy Society General Meeting* 1–8.

14 Wang, T., Zeng, S.K., Zhang, G.X. et al. (2015). Fault section estimation of power systems with optimization spiking neural P systems. *IRomanian Journal of Information Science and Technology* 18 (3): 240–255.

15 Shi, Q.Y., He, W., and Yang, F. (2012). Research on fault magnetic field diagnosis method of grounding grid based on simulated annealing genetic algorithm. *Power System Protection and Control* 40 (20): 5.

16 Deng, Y. (2020). Uncertainty measure in evidence theory. *Science China Information Sciences* 63 (11): 210201.

17 Zhang, X.Y., Zhou, K.F., Pan, H.B. et al. (2018). A decision variable clustering-based evolutionary algorithm for large-scale many-objective optimization. *IEEE Transactions on Evolutionary Computation* 22 (1): 97–112.

18 Zhang, X.Y., Zhou, K.F., Pan, H.B. et al. (2020). A network reduction-based multi-objective evolutionary algorithm for community detection in large-scale complex networks. *IEEE Transactions on Cybernetics* 50 (2): 703–716.

6

Statistical Machine Learning Model for Production Simulation of Power Systems with a High Proportion of Photovoltaics

Xueqian Fu, Feifei Yang, Qiaoyu Ma, Na Lu, and Chunyu Zhang

College of Information and Electrical Engineering, China Agricultural University, Beijing, China

6.1 Introduction

Simulating random photovoltaic (PV) electricity production is crucial for electricity planning, operational scheduling, and long-term market transactions, especially in systems where PV accounts for a significant portion of the power generation. Our approach integrates style transfer and bidirectional long short-term memory (Bi-LSTM) to construct a Generative Adversarial Network (GAN), capturing uncertainty and temporal dynamics in PV generation for enhanced realism and accuracy. With the increasing integration of PV power generation, the operational conditions of the power system have become a complex and variable issue. Considering the difficulties such as power optimization faced by complex PV scenarios, the currently popular adaptive neighborhood clustering (CAN) method [1] is used to extract typical scenarios of PV output.

The continuous rise in global energy consumption, increasing at an annual rate of approximately 2%, has highlighted the adverse effects of relying on conventional energy sources like petrochemicals and coal. These traditional energy sources contribute significantly to the greenhouse effect, posing serious threats to Earth's future climate conditions [2]. Hence, the urgent need to transition toward renewable energy sources and prioritize sustainable development has become imperative for humanity [3]. PV energy stands out for its eco-friendly characteristics, low maintenance requirements, and widespread availability of solar energy [4]. However, even minor fluctuations in PV energy within short timeframes can disrupt the reliability, stability, and efficiency of power grids [5]. Thus, refined predictive modeling for PV power systems is essential for effective grid management. Moreover, precise power load forecasting, coupled with optimized dispatch strategies, aids in the efficient allocation of power generation

Artificial Intelligence Empowered Smart Energy Systems, First Edition. Edited by Qiang Yang and Gang Huang.

and distribution, mitigating uncertainties in PV power generation and facilitating demand response [6]. This approach helps maintain a balance between power supply and demand, minimizing energy wastage due to surplus generation. Therefore, precise forecasting of ultrashort-term PV power generation and power load is imperative for the routine functioning and safety of power systems.

6.2 Methodology

6.2.1 Long Time Scale

6.2.1.1 Bidirectional LSTM Style-Based Generative Adversarial Networks

The proposed GAN consists of a generator (G) and discriminator (D), engaged in an adversarial process during training, as depicted in (6.1). In this game, the generator's objective is to maximize V(D, G), while the discriminator's goal is to minimize it. Through continuous adversarial training, the generator strives to generate realistic data samples, and the discriminator endeavors to differentiate between real ones and generated ones. This iterative process fosters improvement in the generator's ability to produce authentic samples.

$$\min_G \max_D \mathrm{V}(D, G) = \mathrm{E}_{x \sim P_{ori}(x)}[\log D(P_{ori}(x))] + \mathrm{E}_{x \sim P_{gen}(x)}[\log(1 - D(P_{gen}(x)))] \tag{6.1}$$

where, $P_{ori}(x)$ is real data distribution, $P_{gen}(x)$ is generated data distribution.

In Figure 6.1, the generator within the GAN is depicted, consisting of two distinct networks: Mapping Network and Synthesis Network. Initially, the Mapping Network is responsible for mapping real PV scenarios into a latent space vector, capturing the fundamental features inherent in the input PV scenario. Following this, the Synthesis Network employs style transfer techniques, utilizing the latent vector, to modulate the data flow within each block. This process facilitates style blending, enhancing the diversity of the generated PV scenarios.

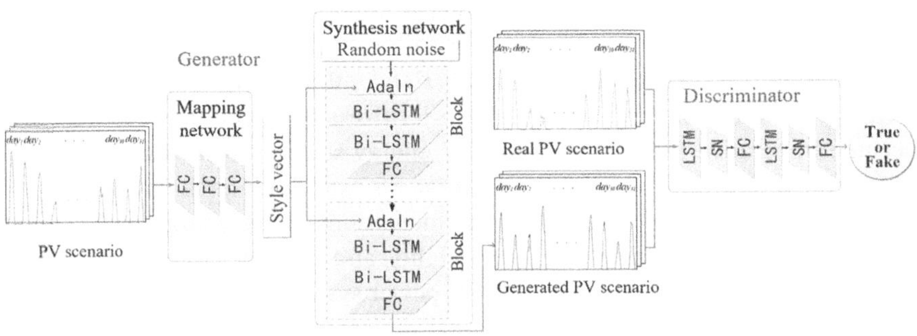

Figure 6.1 Proposed BL-style GAN.

By incorporating styles and content from various PV scenarios, the generator ensures diversity in the output scenarios.

To achieve style transfer, we adopt the Adaptive Instance Normalization (AdaIN) method [7], as depicted in Eq. (6.2). This method enables the adjustment of the statistical characteristics of the input PV scenario to match those of a reference PV scenario, effectively transferring the style of the reference scenario to the generated scenario. This mechanism significantly contributes to enhancing the diversity and realism of the generated PV scenarios, contributing significantly to the overall effectiveness of the proposed GAN framework.

$$\text{AdaIn}(x, y) = \sigma(y)\left(\frac{x - \mu(x)}{\sigma(x)}\right) + \mu(y) \tag{6.2}$$

where, x and y represent generated scenario and encoded reference scenario, while σ and μ denote the mean and standard deviation.

In this neural network architecture, the Mapping Network comprises several Fully Connected (FC) layers aimed at disentangling the features of the input PV scenario. Conversely, the Synthesis Network is composed of multiple blocks, each incorporating the AdaIN operation to modulate latent variables onto the block's input data. Subsequently, two stacked Bi-LSTM layers are employed within each block to capture the inherent temporal characteristics of PV scenarios.

Given the typical time-series nature of PV scenario data, conventional Convolutional Neural Networks (CNNs) often struggle to effectively capture temporal features due to their limitations in handling time dependencies. Therefore, we have chosen to utilize a Bi-LSTM network to extract the inherent temporality present in the modulated data. Unlike LSTM, which processes data flow in a singular direction, Bi-LSTM incorporates information from both preceding and subsequent time steps when analyzing sequence data. This bidirectional approach, illustrated in Eqs. (6.3) and (6.4), allows for a more comprehensive capture of contextual information within the sequence data of PV scenarios, as highlighted by recent studies [8].

$$h_t^{forward} = LSTM^{forward}\left(x_t, h_{t-1}^{forward}\right) \tag{6.3}$$

$$h_t^{backward} = LSTM^{backward}\left(x_t, h_{t+1}^{backward}\right) \tag{6.4}$$

where, $LSTM^{forward}$ is forward LSTM, and $LSTM^{backward}$ is backward LSTM. x_t is the input of the PV scenario sequence at time t, and $h_{t-1}^{forward}$ $h_{t+1}^{backward}$ are the hidden layer states of LSTM at t. The final output is obtained by combining the hidden states from both directions.

$$h_t = \left[h_t^{forward}, h_t^{backward}\right] \tag{6.5}$$

The discriminator is equipped with LSTM units, with spectral normalization applied to the LSTM kernel. This strategy is crucial for reinforcing model stability, mitigating mode collapse, and enhancing the quality of generated samples, as evidenced by recent studies [9]. As depicted in Eq. (6.6), spectral normalization functions by regulating gradient flow, achieved through constraining the Spectral Norm (SN) of the weight matrix. This methodology is instrumental in fortifying the stability of deep learning models.

$$\mathrm{SN}(W) = \frac{W}{\sigma_{\max}(W)} \tag{6.6}$$

where, W is the weight matrix of the LSTM, and its maximum singular value is denoted as $\sigma_{\max}(W)$

The LSTM computes the output for each memory cell, and the final discrimination result is determined through a voting mechanism. Eq. (6.7) illustrates how the loss function for the GAN can be derived from the discriminator's outputs.

$$LossD = \mathrm{CE}(L_{\mathrm{ori}}, 1) + \mathrm{CE}(L_{\mathrm{gen}}, 0) \tag{6.7}$$

where L_{ori} is the identification result for real scenario data, which corresponds to a truth label of 1, and L_{gen} is the identification result for generated weather data, which corresponds to a truth label of 0. Cross-entropy (CE) is employed as the loss metric to calculate the discriminator's loss. In addition, to better ensure the probabilistic features of generated PV scenarios, we have designed a loss function based on mean and variance.

$$LossS = |\ \sigma(X) - \sigma(\widehat{X})\ | + |\ \mu(X) - \mu(\widehat{X})\ | \tag{6.8}$$

The *LossS* is employed to enforce consistency in the mean and variance between generated PV scenarios and real PV scenarios.

6.2.1.2 Clustering with Adaptive Neighbors

As PV power generation becomes increasingly integrated, the operational conditions of the power system have become more complex and variable. This necessitates careful management and adaptation to ensure efficient and reliable electricity supply. Considering the difficulties such as power optimization faced by complex PV scenarios, this section introduces the use of an unsupervised method—clustering. Here, the currently popular CAN method [1] is used to extract typical scenarios of PV output. Please note that the following provides an application idea. The overall technical roadmap is shown in Figure 6.2.

Theorem 6.1 ***(Laplacian Matrix) [10]*** The Laplacian matrix $L = D - W$ is composed of the degree matrix D, which is diagonal, and the adjacency matrix W. The Laplacian matrix has some nice properties:

1) The matrix L is symmetric;
2) All eigenvalues of the Laplacian matrix are real numbers;

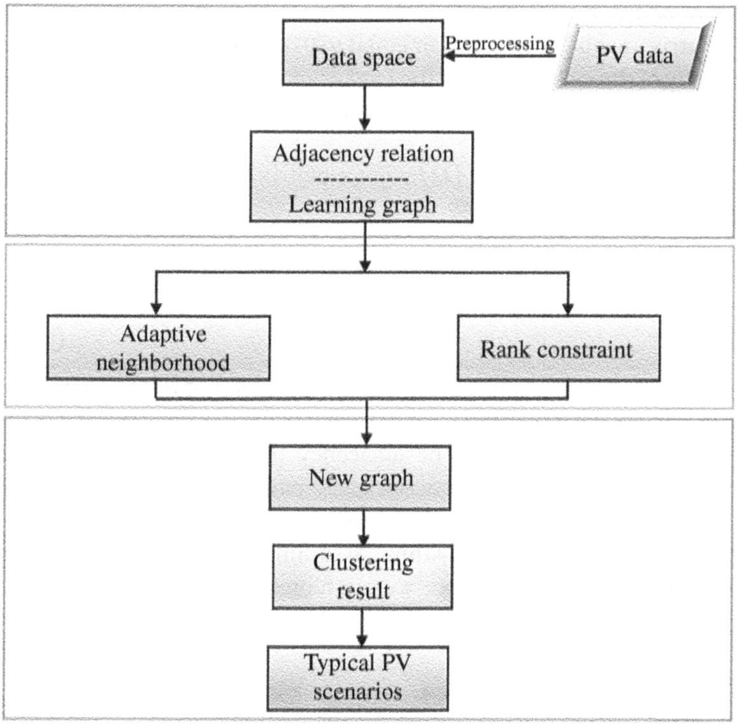

Figure 6.2 Technology roadmap.

3) For any vector f, we have the following form:

$$f^T Lf = \frac{1}{2}\sum_{i,j=1}^{n} w_{ij}(f_i - f_j)^2 \tag{6.9}$$

It is easy to obtain the following form using the definition of the Laplacian matrix:

$$\begin{aligned} f^T Lf &= f^T Df - f^T Wf \\ &= \sum_{i=1}^{n} d_i f_i^2 - \sum_{i,j=1}^{n} w_{ij} f_i f_j \\ &= \frac{1}{2}\left(\sum_{i=1}^{n} d_i f_i^2 - 2\sum_{i,j=1}^{n} w_{ij} f_i f_j + \sum_{j=1}^{n} d_j f_j^2\right) \\ &= \frac{1}{2}\sum_{i,j=1}^{n} w_{ij}(f_i - f_j)^2 \end{aligned} \tag{6.10}$$

4) The Laplacian matrix is a positive semi-definite matrix, so the n real eigenvalues of this matrix are all greater than or equal to 0, i.e., $0 = \lambda_1 \leq \lambda_2 \leq \cdots \leq \lambda_n$;
5) The multiplicity of the eigenvalue 0 of the Laplacian matrix is equal to the number of connected components in the graph represented by the similar matrix.

Theorem 6.2 ***(Ky Fan's Theorem) [11]*** For a matrix $F \in R^{n\times c}$, the following minimization problem holds:

$$\min_F \frac{1}{2}\sum_{i=1}^{n}\sum_{j=1}^{n}\|f_i - f_j\|_2^2 a_{ij} = Tr(F^T LF)$$
$$s.t. \quad F^T F = I \tag{6.11}$$

where, the ith row of $F \in R^{n\times c}$ represents the function value $f_i \in R^{c\times 1}$ assigned to each data point, and $Tr(\cdot)$ denotes the trace of the matrix.

The CAN algorithm [1] considers the adjacency probability between data, effectively exploring the similarity between data pairs in complex manifold datasets. By learning the similarity graph, it assigns samples to clusters. Below is the specific model of the CAN algorithm. Assuming a dataset comprising n samples, each characterized by d features, we can represent it as a matrix $X = \{x_1, x_2, \ldots, x_n\} \in R^{n\times d}$.

$$\min_A \sum_{i=1}^{n}\sum_{j=1}^{n}\left(\|x_i - x_j\|_2^2 a_{ij} + \gamma a_{ij}^2\right)$$
$$s.t. \quad \forall i, a_i^T \mathbf{1} = 1, 0 \leq a_{ij} \leq 1, rank(L) = n - c \tag{6.12}$$

where, $L = D - (A + A^T)/2$ is the Laplacian matrix of A, and $D \in R^{n\times n}$ is a diagonal matrix with the ith diagonal element being $\sum_{j=1}^{n}(a_{ij} + a_{ji})/2$. By balancing the objective function between these two parts, we can obtain an optimal solution $A = (a_{ij})$. Furthermore, this matrix is a non-negative matrix, and based on the similarity matrix, neighborhoods can be assigned to each data point, thus corresponding to graph G. By leveraging property 5 of the Laplacian matrix, introducing the rank constraint of the Laplacian matrix ensures that the neighborhood assignment of A is ideal.

Assuming $\sigma_i(L)$ is the ith smallest eigenvalue of the Laplacian matrix L, according to property 4 of the Laplacian matrix, we have $\sigma_i(L) \geq 0$. Introducing a sufficiently large parameter value λ, based on Ky Fan's theorem, Eq. (6.12) can be transformed into the following form:

$$\min_{A,F} \sum_{i=1}^{n}\sum_{j=1}^{n}\left(\|x_i - x_j\|_2^2 a_{ij} + \gamma a_{ij}^2\right) + 2\lambda Tr(F^T LF)$$
$$s.t. \quad \forall i, a_i^T \mathbf{1} = 1, 0 \leq a_{ij} \leq 1, F \in R^{n\times c}, F^T F = I \tag{6.13}$$

Iterate to solve the above clustering model:

1) When the similarity matrix A remains fixed, Eq. (6.13) transforms into the following form:

$$\min_{F\in R^{n\times c}, F^TF=I} Tr(F^TLF) \tag{6.14}$$

The ideal solution F in this objective function consists of the eigenvectors associated with the c lowest eigenvalues of the matrix L.

2) When the similarity matrix F remains fixed, Eq. (6.13) transforms into the following form:

$$\min_{A} \sum_{i=1}^{n}\sum_{j=1}^{n}\left(\left\|x_i - x_j\right\|_2^2 a_{ij} + \gamma a_{ij}^2 + \lambda\left\|f_i - f_j\right\|_2^2 a_{ij}\right)$$
$$s.t.\quad \forall i, a_i^T\mathbf{1} = 1, 0 \le a_{ij} \le 1 \tag{6.15}$$

From this form, it can be seen that the model is independent for different i. Therefore, for each i, we can solve the following form:

$$\min_{a_i} \sum_{j=1}^{n}\left(\left\|x_i - x_j\right\|_2^2 a_{ij} + \gamma a_{ij}^2 + \lambda\left\|f_i - f_j\right\|_2^2 a_{ij}\right)$$
$$s.t.\quad a_i^T\mathbf{1} = 1, 0 \le a_i \le 1 \tag{6.16}$$

Transform Eq. (6.16) into the following form:

$$\min_{a_i^T\mathbf{1}=1, 0\le a_{ij}\le 1} \left\|a_i + \frac{1}{2\gamma} r_i\right\|_2^2 \tag{6.17}$$

where, $d_{ij}^x = \left\|x_i - x_j\right\|_2^2$, $d_{ij}^f = \left\|f_i - f_j\right\|_2^2$, and $r_i \in R^{n\times 1}$ are vectors with the ith element being $r_{ij} = d_{ij}^x + \lambda d_{ij}^f$.

Equation (6.13) can be solved using Algorithm 6.1.

Algorithm 6.1 solves Eq. 6.13

Input: Data matrix $X \in R^{n\times d}$, Number of clusters c, Trade-off parameters γ and λ;
Output: Similar matrix $A \in R^{n\times n}$ containing the number of connected components accurately;

1: Initialize the similarity matrix $A \in R^{n\times n}$;
2: **while** not convergence **do**
3: Update F according to Eq. 6.14;
4: Update A according to Eq. 6.17;
5: **end while**

6.2.2 Short Time Scale

6.2.2.1 Decomposition Strategy and Attention-Based Long Short-Term Memory Network

The proposed model in this section refers to the author's relevant paper [12]. The goal of the model is to anticipate the value of the target sequence for the subsequent τ time step, as outlined in Eq. (6.18). In our framework, Y symbolizes the prediction target sequence, and $Y = (y_1, y_2, \dots, y_T)^T \in R^T$ denotes this sequence within a specific time window T. The decomposition algorithm generates derived variables, denoted as X, which can be articulated as $X = (x^1, x^2, \dots, x^n)^T = (x_1, x_2, \dots, x_T) \in R^{n\times T}$, where n represents the total count of derived variables. Here, $x^k = \left(x_1^k, x_2^k, \dots, x_T^k\right)^T \in R^T$ signifies the initial k derived variable, while $x_t = \left(x_t^1, x_t^2, \dots, x_t^n\right)^T \in R^n$ encapsulates the numerical outputs of all derived variables at the specified time, t.

$$\hat{y}_{T+\tau} = F(x_1, x_2, \dots, x_T) \tag{6.18}$$

where, $F(\cdot)$ stands for the nonlinear mapping function that we endeavor to acquire in order to forecast $\hat{y}_{T+\tau}$.

To begin, we utilize Ensemble Empirical Mode Decomposition (EEMD) to break down the target sequence into simpler subsequences, each providing unique insights into the sequence [13]. This process yields derived variables. Subsequently, the historical values of these derived variables serve as input for the prediction model. We employ a recurrent neural network (RNN) enhanced with a dual-stage attention mechanism to forecast future values in the time series. Within the encoder, the input attention mechanism dynamically selects pertinent derived sequences, automatically extracting relevant variables and assigning varying attention weights based on their impact on prediction. In the decoder, the temporal attention mechanism chooses pertinent encoder hidden states throughout the entire duration of the sequence. The LSTM model is combined with the dual-stage attention mechanism and trained concurrently through standard backpropagation. This approach, termed EEMD-DARNN (dual-stage attention-based recurrent neural network), adeptly selects pertinent input features and captures temporal relationships over extended periods amongst sequential data. The overarching framework of this method is depicted in Figure 6.3.

EEMD, a technique for time-frequency localization analysis, offers adaptability and comprehensiveness. It breaks down intricate signals into a limited set of Intrinsic Mode Functions (IMFs) and residue, considering the inherent time scale attributes of the data. This approach alleviates mode aliasing issues encountered in EMD. In EEMD, each IMF is determined as the average of multiple sets of IMF components obtained through EMD. The process involves adding white Gaussian

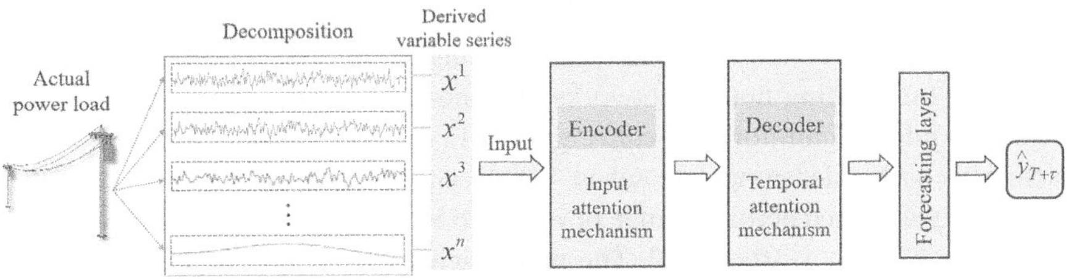

Figure 6.3 The overarching framework of this method.

interference in the initial signal, conducting EMD, and subsequently averaging the EMD components.

$$x_{mq}(t) = x_m(t) + n_q(t) \tag{6.19}$$

$$x_{mq}(t) = \sum_{n=1}^{N-1} y_{mqnEMD}(t) + r_{mqnEMD}(t) \tag{6.20}$$

$$x_m(t) = \sum_{n=1}^{N-1} y_{mnEEMD}(t) + r_{mnEEMD}(t) \tag{6.21}$$

where, $r_{mnEEMD}(t) = \frac{1}{Q}\sum_{q=1}^{Q} r_{mqnEMD}(t), y_{mnEEMD}(t) = \frac{1}{Q}\sum_{q=1}^{Q} y_{mqnEMD}(t), r_{mqnEMD}(t),$ $y_{mqnEMD}(t)$ and $q = 1, 2, \ldots, Q$ signify residue and the ultimate IMF acquired through EEMD, residue and n individual IMFs extracted from the decomposition of $x_{mq}(t)$ via EMD, and the quantity of occasions of introducing white Gaussian noise, respectively.

The attention mechanism functions as a means of resource allocation, assigning varying weights to input features to ensure that crucial information remains prominent despite changes in step size. Much like the selective visual attention observed in humans, its primary objective is to sift through vast amounts of data and spotlight the most relevant information for the given task at hand. The dual-stage attention mechanism operates within an encoder-decoder framework. The impact of input parameters on forecasting power load at each time step is actively evaluated by the input attention mechanism. It can effectively represent the influence of external inputs in the encoder layer. This module ensures that the total attention weights of external inputs sum up to 1, facilitating the selection of vital external inputs during feature extraction across different time points. Meanwhile, the decoder layer incorporates a mechanism for temporal attention, which captures prolonged temporal dependencies within the encoder, assigning weights to encoder hidden states throughout the entire time window.

Model parameters. The model undergoes a parameter tuning phase to unlock its full potential and enhance its performance. Adapting parameters to suit the dataset's characteristics involves employing grid search during experimentation, wherein parameters are individually selected within predetermined ranges. The model comprises three parameter types, outlined as follows: (1) Hyperparameters essential for training neural networks are fine-tuned by varying batch size and learning rate to monitor loss convergence. Eventually, optimal performance is achieved with a learning rate set at 0.0001 and a batch size of 64. (2) A grid search is performed across a range of values to identify the best T (time window), including T in {3, 5, 10, 15, 25}. The model achieves its highest performance on the validation set with $T = 10$, leading to its selection as the optimal size. (3) The hidden sizes (m and p) are fine-tuned through grid search over a predefined set of values, $m = p$ in {16, 32, 64, 128, 256}. Ultimately, $m = p = 128$ is chosen as it delivers superior performance. For consistency in the experiments, the comparison models' relevant parameters are configured to match those of EEMD-DARNN. Furthermore, the precision of each model is assessed for power load forecasting over multiple forecast horizons. During training, we utilize Adam optimizer in conjunction with the minibatch Stochastic Gradient Descent (SGD). All models undergo ten iterations, and we report their mean accuracy for comparative analysis. The backpropagation approach is employed to adjust parameters with a Mean Square Error (MSE) objective function (Eq. (6.22)). The input data undergo normalization using z-score method (Eq. (6.23)), and final outcomes are derived through inverse normalization.

$$MSE = \frac{1}{N}\sum_{i=1}^{N}\left(\hat{y}_t^i - y_t^i\right)^2 \tag{6.22}$$

$$z = \frac{x - \mu}{\sigma} \tag{6.23}$$

Evaluation criteria. We utilize four evaluation metrics to assess the efficacy of our proposed approach in forecasting power load [14]:

$$R^2 = \frac{\sum_{i=1}^{N}\left(\hat{y}_t^i - \bar{y}\right)^2}{\sum_{i=1}^{N}\left(y_t^i - \bar{y}\right)^2} \tag{6.24}$$

$$RMSE = \sqrt{\frac{1}{N}\sum_{i=1}^{N}\left(\hat{y}_t^i - y_t^i\right)^2} \tag{6.25}$$

$$MAPE = \frac{1}{N}\sum_{i=1}^{N}\left|\frac{\hat{y}_t^i - y_t^i}{y_t^i}\right| \times 100\% \tag{6.26}$$

$$MAE = \frac{1}{N}\sum_{i=1}^{N}\left|\hat{y}_t^i - y_t^i\right| \tag{6.27}$$

N, $\bar{y}$, y_t, and $\hat{y}_t$ denote the overall sample count, the mean of observed values, and the observed and predicted values at time t, respectively.

6.2.2.2 Forecasting with Dynamic Mask

Initially, we posit that the PV power output at time t is denoted as v_t. The objective is to utilize the sequence of PV power generation $V = v_0, v_1, v_2, \ldots, v_t$ over the period $T = 0, 1, 2, \ldots t$ to predict the subsequent PV power generation v_{t+1} at time $t+1$. It is important to note that during this forecasting process, the model does not incorporate any additional input data.

The concept of masking in our model draws inspiration from the architecture of a conventional denoising autoencoder. Within our framework, the Masked module delineates the input sequences and subsequently obscures sections of lesser importance, as determined by the resulting values. The overall structure of the model is illustrated in Figure 6.4.

The core of our Masked-Block encompasses a pooling layer in conjunction with an MLP. The procedural essence of masking, as encapsulated in Eq. (6.28), unfolds through two pivotal stages: 1) Ascertainment of the edge importance vector corresponding to sequence positions; 2) Hierarchical sorting of the edge importance vector followed by its application in descending order of significance. For an input sequence $Z \in \mathbb{R}^{L \times C}$, a Global Average Pooling (GAP) layer alongside an MLP is initially applied to map and derive vectors emblematic of the significance of sequence positional features:

$$Q = \text{MLP}(GAP(Z)), Q \in \mathbb{R}^{L \times 1} \tag{6.28}$$

Subsequently, the sorted feature importance vector is amalgamated with the original input's positional data to attribute a distinct feature importance value

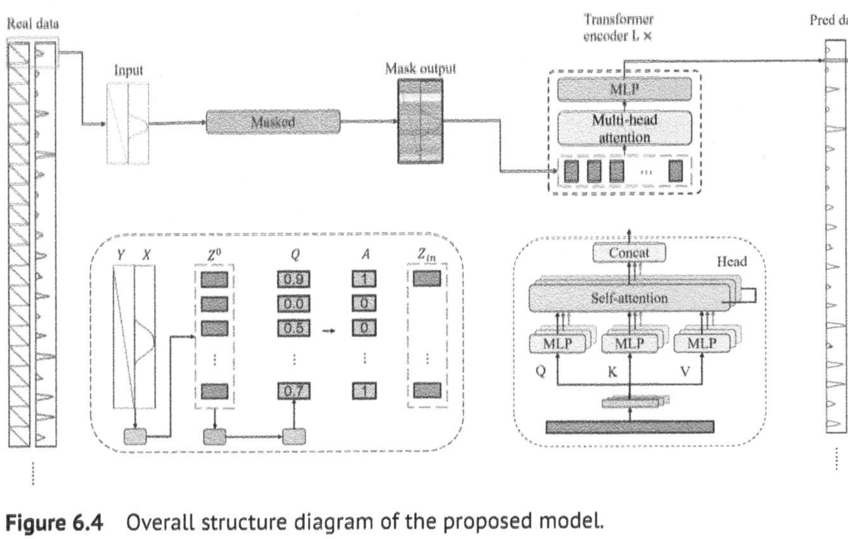

Figure 6.4 Overall structure diagram of the proposed model.

to each temporal step. Data masking then proceeds in accordance with a predetermined proportion α, modeled akin to a regression dilemma linked to model gradients:

$$A^k = M \cdot Q^{k^T}, M \in \mathbb{R}^{(L-\alpha)\times 1}, m_i = 1 \tag{6.29}$$

In this framework, Q epitomizes the vector of eigen importance, whereas A signifies the vector of positions. This intricate process ensures that the informational tag Q along with the representational tag A^k undergo slow and elaborate updates.

Finally, the feature extraction facet of our envisaged PV forecasting model is realized through an encoder, structured around the Transformer architecture. This encoder transduces the historical PV input sequence $x = x_1, x_2, \dots, x_t$ along with the temporal sequence $y = y_1, y_2, \dots, y_t$ into a continuous representation of the feature sequence $z = z_1, z_2, \dots, z_t$ (Eq. (6.30)), subsequently mapping these to forecasted PV outcomes v_{t+1} via an MLP. The MLP linearly amalgamates historical PV and temporal input data to synthesize an encoded fusion between features. The nonlinear activation function between layers, facilitated by a ReLU function, acts as a nonlinear feature filter encoded through MLP projection. The historical PV input and temporal series, postlinear full connection and nonlinear projection, are transformed into the feature representation z (Eq. (6.30)).

$$z = f(x, y) = \text{ReLU}(\text{MLP}(\text{concat}(x, y))) \tag{6.30}$$

The Transformer structure, composed of self-attentive layers, delves deep into the internal features of vector z, amalgamating PV and temporal attributes. Each self-attention layer comprises a multi-head self-attention mechanism and an MLP layer, enabling the model to perform complex mining and utilization of the interrelationships present in the data. This mechanism is instrumental in enhancing the models predictive accuracy for future PV power generation.

$$v_{t+1} = Transformer(z^{0'}) = \sum i = 1^L MSA(z^{0'}) \tag{6.31}$$

To mitigate the risk of information loss during the vector z fusion process, the model integrates residual connections, thus preserving the integrity of the information throughout the model. Moreover, it leverages previously generated output sequences as supplementary data for the generation of subsequent sequences, enhancing the model's contextual awareness and forecasting accuracy.

$$MSA(Z^l) = Z^l + MLP\left(MultiHead\left(Z^l\right)\right) \tag{6.32}$$

$$MultiHead - Attention(Q, K, V) = Concat\left(MH_1, MH_2, \dots, MH_n\right) W^{MSA} \tag{6.33}$$

$$MH_i = Att\left(Q_{MH_i} W_i^Q, K_{MH_i} W_i^K, V_{MHi} Wi^V\right) \tag{6.34}$$

The matrices W_i^Q, W_i^K, and W_i^V, along with W^{MSA}, serve as projections within the attention mechanism, enabling the model to allocate suitable weights to various parts of the input data, contingent upon their pertinence to the current task. This weighted mechanism ensures that the model pays attention to the most pertinent information when predicting future PV power generation.

$$Att(Q, K, V) = softmax\left(\frac{QK^T}{\sqrt{d_k}}\right)V \tag{6.35}$$

The MLP component of the PV-Transformer is enhanced with GELU (Gaussian Error Linear Units) activation functions, providing nonlinear capabilities to the model's architecture. This ensures that each sequence in the model is precisely aligned with its corresponding time step, thereby capturing the temporal dynamics effectively.

$$z_p = \text{MLP}(\text{GAP}(Z)) \tag{6.36}$$

The final output z_p represents the PV power generation forecast at a specific time point, encapsulating the culmination of complex data processing and feature extraction through the PV-Transformer architecture. This intricate framework, combining masked data processing and advanced Transformer mechanisms, sets a new precedent in forecasting accuracy for PV power generation, enabling a deeper understanding of the underlying dynamics and facilitating more reliable renewable energy management strategies.

6.3 Case Studies

6.3.1 Long Time Scale

6.3.1.1 Year-Round Photovoltaic Scenario Generation

Training GAN networks for monthly PV scenarios requires a substantial amount of data. Depending solely on historical data derived from actual statistics may not be adequate. Therefore, this study initially collects a significant volume of PV data obtained from weather data and PV models. Subsequently, the model undergoes training using the acquired PV data.

For experimentation in this study, weather data includes temperature, direct radiation, and diffuse radiation sourced from multiple meteorological stations in Guangzhou, Guangdong Province, China. The data, obtained from MERRA-2, spans from 1975 to 2021, covering all January months across 47 years, and is sampled hourly.

Furthermore, a sliding window technique is utilized to introduce certain temporal variations, acting as a basic form of data augmentation. This involves shifting the data distribution forward and backward in time by one day and two days within the time series. Consequently, a total of 1316 samples are generated, forming the PV scenario dataset used for model training.

For comparison, we employ 3 popular time-series data generation and deep learning models for new energy scenario generation: RGAN (Recurrent GAN) [15], StyleGAN2 [16], and DCGAN (Deep Convolutional GAN) [17]. In the subsequent comparative analysis, each model generates 1316 samples of January weather data, which are then thoroughly compared with real data for analysis purposes.

Figure 6.5 presents the t-SNE computation results for both the proposed GAN and benchmark models to assess the diversity of the generated PV scenarios. The proposed GAN exhibits exceptional results in terms of diversity, with the distribution range of the generated scenarios closely aligning with that of the original

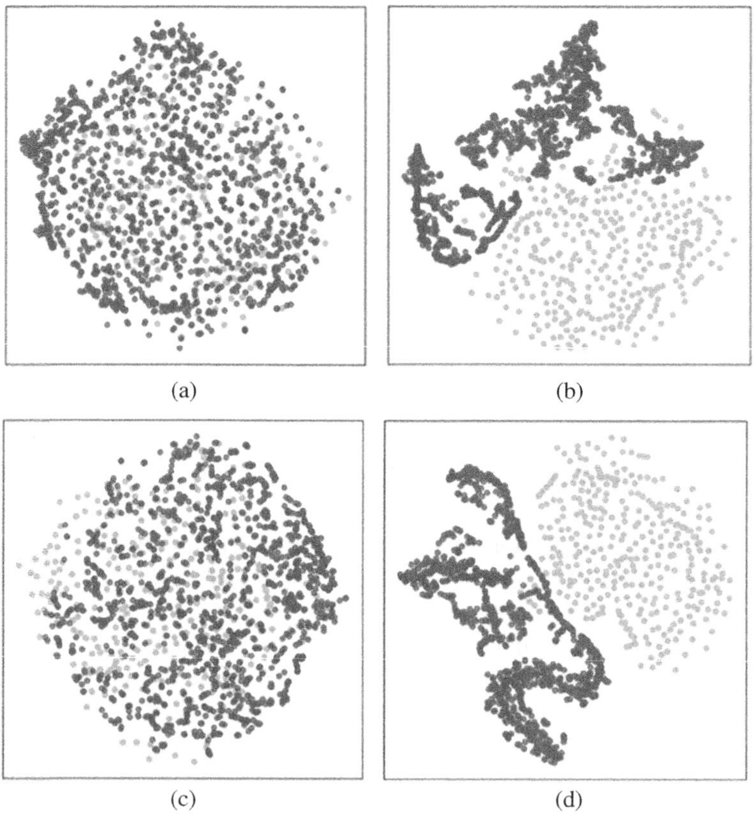

Figure 6.5 Result of monthly PV scenario from proposed GAN (a), TimeGAN (b), StyleGAN2 (c), DCGAN (d) using t-SNE.

scenarios. TimeGAN follows closely, with the distribution range of generated scenarios slightly narrower than that of the original scenarios. In contrast, the distribution ranges of scenarios generated by StyleGAN2 and DCGAN deviate from those of the original scenarios.

Both the proposed GAN and TimeGAN are grounded in LSTM for sequential generation, harnessing the recurrent structure to enhance the capture of temporal dependencies and sequential information within the data. This architecture enables the models to generate more diverse samples by effectively simulating various patterns and variations present in the input data. Additionally, the proposed GAN employs a style transfer strategy, further augmenting the diversity of generated PV scenarios compared to TimeGAN.

While StyleGAN2 and DCGAN demonstrate some diversity in generating scenarios, they encounter challenges due to the inherent limitations of CNN in effectively handling time-series data. These models encounter difficulties in grasping the distribution feature of actual PV scenarios, resulting in generated scenarios that often deviate significantly from the distribution of authentic scenarios.

Table 6.1 provides discriminative scores for the proposed GAN alongside benchmark models to assess the temporal fidelity of generated PV scenarios. The proposed GAN model demonstrates the lowest discriminative score for generated scenarios, indicating that the binary classification model based on GRU struggles to distinguish between original and generated PV scenario data. This suggests that GAN, leveraging Bi-LSTM, produces scenarios with temporal features exhibiting minimal deviation from real scenarios, showcasing similar patterns in temporal characteristics.

Among the benchmark models, TimeGAN performs the best in terms of temporality. This can be attributed to TimeGAN's incorporation of unsupervised adversarial losses on real and synthetic sequences, along with the introduction of stepwise supervised losses using original data. This approach ensures better preservation of the temporal features of PV scenarios.

StyleGAN2 and DCGAN, utilizing CNNs, struggle with capturing crucial dependencies within time-series data. As a result, they often fail to accurately reproduce the temporal characteristics in the generated scenarios.

Compared to other methods, the proposed GAN demonstrates superiority, showcasing the effectiveness of integrating style transfer for diversity and utilizing Bi-LSTM for temporal fidelity in generated scenarios. Leveraging the

Table 6.1 Results for discrimitive score.

Model	Proposed	TimeGAN	StyleGAN2	DCGAN
Score	0.3068	0.4375	0.4716	0.4621

diversity inherent in the time-series data of the PV scenario dataset, this approach holds promise for extension to various other time-series data types, including wind power scenarios.

6.3.1.2 Typical Scenarios Extraction

Experimental settings. In this section, to showcase the efficacy and supremacy of the clustering technique in extracting typical PV output scenarios, we'll ll employ additional synonymous or closely related terms for both rephrasing and embellishment, of PV data from five PV plants in the Yulara Solar System (2023) are used, spanning November 2020 to October 2021.

The original time resolution is 5 minutes, missing data is imputed, and the data is transformed into hourly resolution PV output data. The output processes are divided into daily time windows, resulting in 365 sets of data with each input data dimension being 120. The dataset is normalized using Min-Max normalization to scale all data points in all dimensions to the range [0,10].

In real-world scenarios, it's common that we can't predetermine the number of clusters and must instead depend on suitable techniques to ascertain the appropriate cluster count. [18]. In this article, we rely on interpretable variance for assessment, which is delineated as the proportion of variance among groups compared to the overall variance in the dataset. Initially, the K-means method is used for pre-clustering, running 2 to 100 clusters. To avoid the randomness of initialization, the clustering process is run 100 times to obtain stable clustering results. When the number of clusters is set to 13, the point shows the fastest growth rate, and the interpretable variance reaches 85% for the first time. Therefore, the number of clusters for this scenario is determined to be 13. This method allows for selecting a smaller number of clusters to represent most of the overall information in the data, while also saving computational costs to some extent.

Model validation. To validate the comprehensiveness and typicality of clustering methods in extracting typical PV output scenarios, the yearly PV output data from 5 PV power stations are used as original input to the clustering model. This results in 13 clusters of PV output scenarios for each of the 5 PV power stations. The clustering outcomes are depicted in Figures 6.6–6.10, with the horizontal axis denoting the output time and the vertical axis indicating power values. The lines illustrate the centroid trajectories for each classification. The distribution of each scenario obtained by the method in this article is shown in Table 6.2.

Among the 13 clusters obtained, the fluctuation characteristics of PV output samples in each cluster are significantly different, usually changing based on the sunlight and temperature conditions throughout the day. The PV power generation scenarios are divided into 3 levels: Level I: Periods with little to no sunlight, typically from 00:00 to 06:00 and 18:00 to 23:00; Level II: Periods with increasing (decreasing) sunlight and temperature, generally from 06:00 to

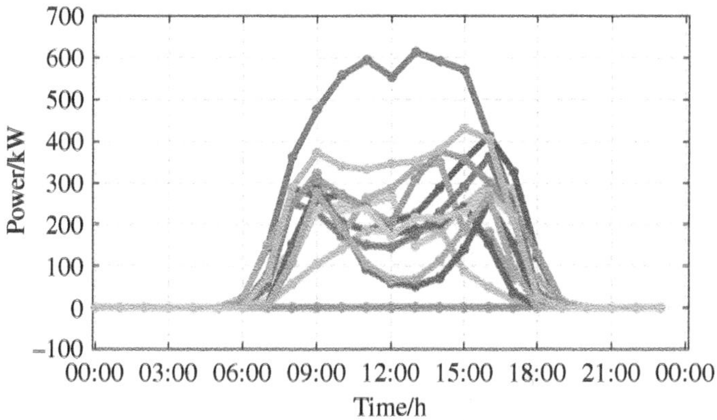

Figure 6.6 Typical scenarios of the 1st PV station.

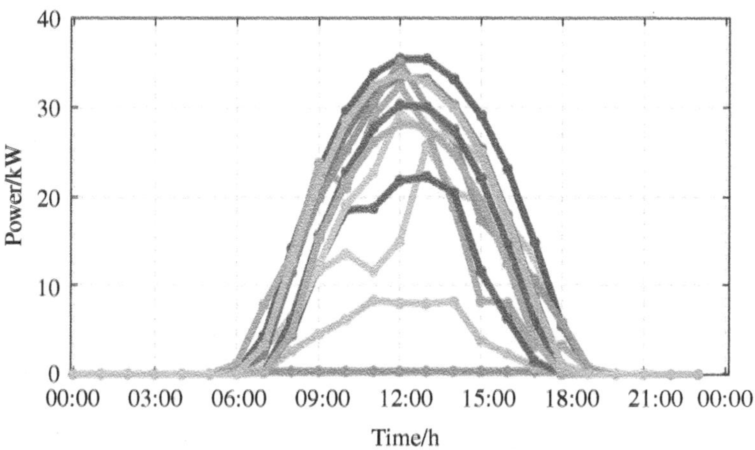

Figure 6.7 Typical scenarios of the 2nd PV station.

10:00 and 16:00 to 18:00. The PV power output varies noticeably but is not at its maximum; Level III: Periods with abundant sunlight, typically from 10:00 to 16:00, when the PV power output is high and can supply more power to the load.

In the clustering model of this paper, when selecting cluster centers, the mean value of each class sample is used as the reference. The sample from each cluster that is closest to the mean value of that class is chosen as the representative scenario for that cluster. This approach preserves the authenticity of the samples and the fluctuation of the original output, making subsequent optimization scheduling research more practically meaningful.

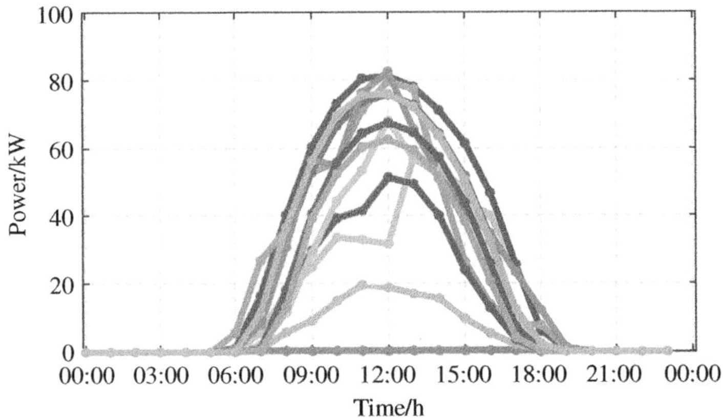

Figure 6.8 Typical scenarios of the 3rd PV station.

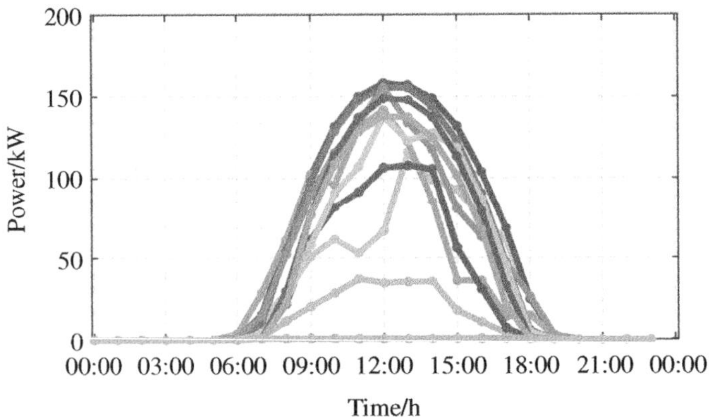

Figure 6.9 Typical scenarios of the 4th PV station.

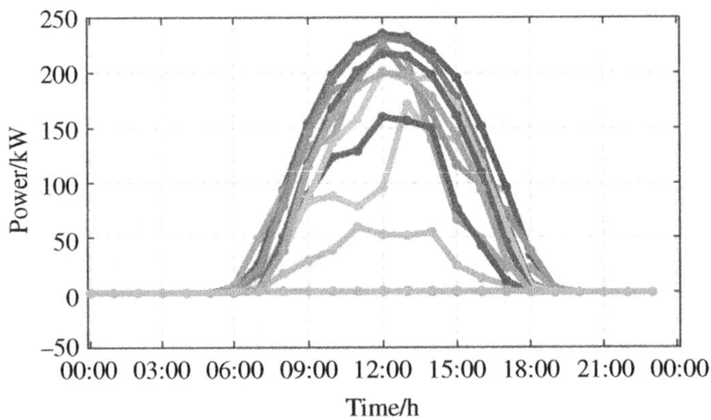

Figure 10 Typical scenarios of the 5th PV station.

Table 6.2 The proportion of each typical scenario.

Scenarios	Probability	Scenarios	Probability	Scenarios	Probability
1	0.1288	6	0.0384	11	0.2658
2	0.0164	7	0.0603	12	0.0055
3	0.0137	8	0.1342	13	0.1918
4	0.1014	9	0.0082		
5	0.0274	10	0.0082		

6.3.2 Short Time Scale

6.3.2.1 Power Load Forecast

The results in this section refer to the author's relevant paper [12]. This research utilized electricity load data from agricultural activities in Guangzhou, Guangdong Province, China. This dataset yields 96 records every day, totaling 35,136 records. It constitutes a univariate time series. For the entire year of 2020, data was collected every 15 minutes. Following a 6:2:2 ratio, it was partitioned into training, validation, and test sets.

This section contrasts the outcomes of five models across various forecast horizons. To provide a more intuitive demonstration of the proposed approach, observed and predicted values of this model are compared when $\tau = 1$, as shown in Figure 6.11 (only 2000 sample points are displayed for clarity of comparison). This visualization effectively portrays ultrashort-term power load forecasting.

Four deep-learning models are employed for comparison. LSTM [19] and gated recurrent unit (GRU) [20], extensions of RNN, mitigate the challenge of gradient vanishing and grasp extended dependencies within sequential data over time

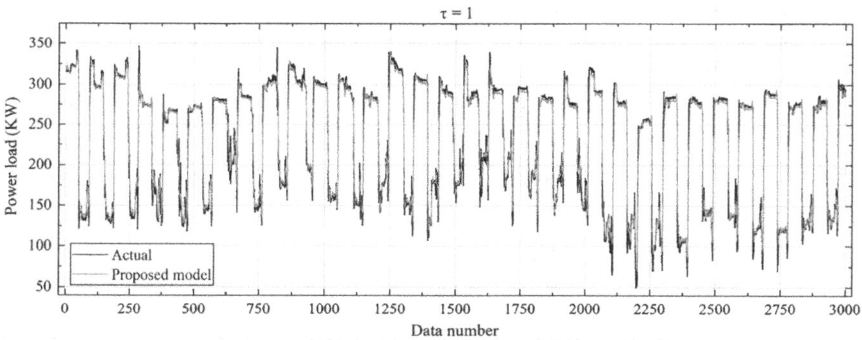

Figure 6.11 Comparison of actual values and predicted results.

through their encoder-decoder frameworks. LSTM incorporates memory, input, forget, and output gates to extract meaningful features and enhance expressive capabilities, while GRU offers a more streamlined configuration with enhanced training velocity and effectiveness. ResNet [21] features a deep network architecture with residual modules, mitigating network degradation and enabling faster convergence compared to traditional networks. Fully Convolutional Networks (FCNs) [22] revolutionize CNNs by replacing full connection layers with convolutional layers, allowing for input of variable size, increased efficiency, and better preservation of spatial information. All models utilize historical values of the target data as input variables, with parameters aligned with those of EEMD-DARNN for consistency in the experiment. Table 6.3 presents a comparison of forecast accuracy among the five methods.

EEMD-DARNN surpasses every baseline method across all forecast horizons. Specifically, in the shortest forecasting horizon (15 minutes), in contrast to ResNet (the least effective approach), EEMD-DARNN demonstrates an increase in R^2 by 0.065, and reductions in MAE, RMSE, and MAPE by 6.687, 9.788, and 0.030, respectively. Conversely, when in contrast with LSTM (the most effective approach), EEMD-DARNN shows increases in R^2 by 0.030, and decreases in MAE, RMSE, and MAPE by 1.651, 5.205, and 0.014, respectively. At the longest prediction horizon (1 hour), in contrast to ResNet (the worst approach), EEMD-DARNN exhibits increases in R^2 by 0.060, and reductions in MAE, RMSE, and MAPE by 6.385, 7.335, and 0.033, respectively. Conversely, in contrast with LSTM (the best approach), EEMD-DARNN demonstrates increases in R^2 by 0.016, and decreases in MAE, RMSE, and MAPE by 1.398, 2.283, and 0.003, respectively. However, it's observed that the model's accuracy diminishes as the forecast horizon extends, particularly in regions with 'inflection points', indicating that the method is more

Table 6.3 Accuracy comparison among the five models.

Horizon	Model	LSTM	GRU	Resnet	FCN	Proposed
	metrics					
15 min	R^2	0.940	0.940	0.905	0.930	0.970
	RMSE	17.503	17.505	22.086	18.963	12.298
	MAPE	0.054	0.054	0.070	0.056	0.040
	MAE	8.986	8.988	14.022	9.853	7.335
1 h	R^2	0.925	0.923	0.881	0.905	0.941
	RMSE	19.588	19.849	24.640	21.992	17.305
	MAPE	0.068	0.067	0.098	0.095	0.065
	MAE	12.745	12.272	17.732	15.592	11.347

Table 6.4 PV prediction evaluation index of different models (Test Set of Belgian).

	RMSE	R^2	MAE	COR
CNN	31.88	0.9983	15.97	0.9983
TCN	28.98	0.9986	13.57	0.9993
RNN	302.96	0.8478	156.55	0.9211
LSTM	30.09	0.9985	16.21	0.9992
GRU	23.02	0.9991	9.54	0.9996
N-BEATS	24.22	0.999	8.47	0.9995
MIXER	30.09	0.9985	16.21	0.9993
Transformer	24.73	0.999	14.48	0.9995
ASMNET	21.42	0.9992	8.76	0.9996

effective for shorter-term load forecasting. Enhancing forecast accuracy in such 'inflection points' regions constitutes a key area for further research.

6.3.2.2 Photovoltaic Power Generation Forecast

The data used in this study was sourced from the publicly available Belgian PV power generation data provided by Elia Group. It includes all PV electricity records from 2012 to the present. Additionally, the company has also supplied forecast results for some models with industrial potential.

In the experiment, the proposed model is benchmarked against established deep sequential models, which are, respectively, CNN [23], TCN [24], RNN [25], LSTM [23], GRU [26], Transformer [27], MIXER [28], and N-BEATS [29]. To prove the validity of our proposed model. RNN is distinguished by their architecture where nodes are interconnected in a sequence that progresses through time steps. LSTM and GRU belong to the same architecture, and on top of this, a "gate" structure is added to better aggregate past inputs. Convolution can aggregate neighbor data. In addition, the method of weight sharing enables the model to map to higher dimensions. Compared with CNN, the causal convolution combined with void convolution is proposed in TCN for the forecast of sequence data, and its convolution process effectively reduces the reuse of data information.

The comparison table illustrates that although the proposed ASMNET exhibits shorter error metrics compared to all models reviewed, it excels in established models across all error metrics, thus making it the most favorable for high-quality predictions. The comparison of the forecast PV power generation is shown in Tables 2.1 and 2.2. It is observed that although the error metrics of the proposed ASMNET are only relatively shorter among all models compared, it consistently exhibits the best error metrics within the established models. The RMSE and MAE of the ASMNET method are the smallest, registering 27.91 and 13.17 for the Belgian dataset, and 0.2824 and 0.0786 for the NREL dataset, respectively. Even compared with the suboptimal model, the ASMNET approach achieved a reduction in RMSE and MAE of 0.3 MW and 0.16 MW, respectively, indicating that the forecast result is the closest to the real.

Figure 6.12 illustrates the discrepancies between the forecast result and the real at each forecast time, which is more intuitive. Compared with the traditional CNN, TCNs retain only partial historical information, demonstrating superior predictive performance on smooth, long sequences. Belgian PV results indicate that within the ultrashort term, solar power generation requires only half of the sequence information to be effective.

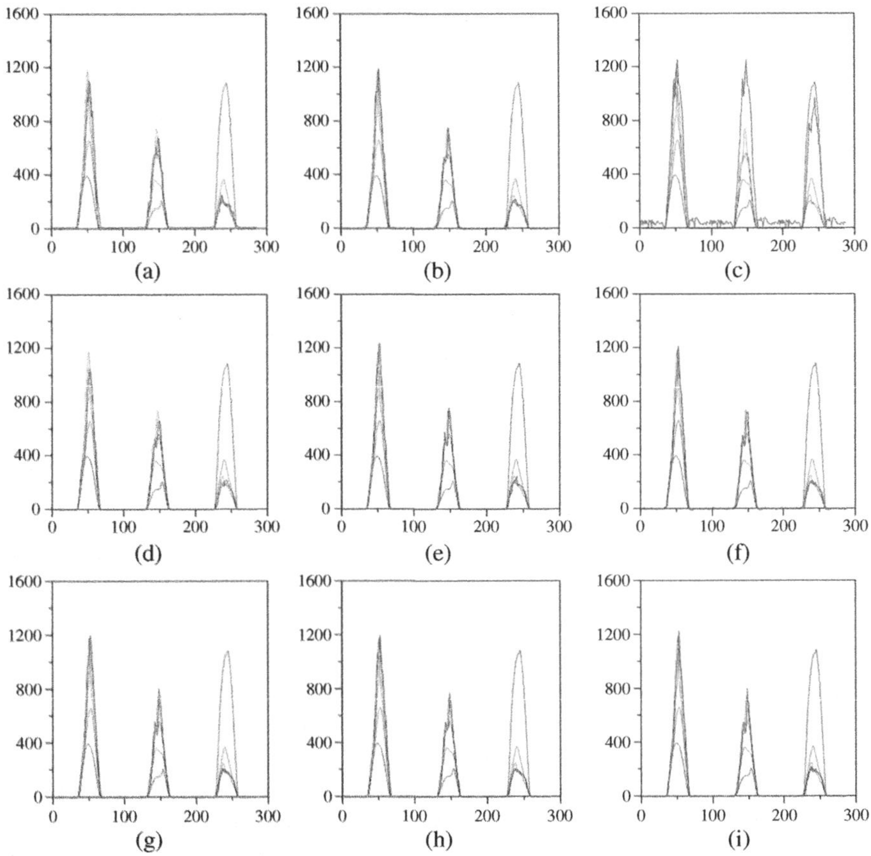

Figure 6.12 Forecast result of different models in the testing set of Belgian. (a) CNN, (b) TCN, (c) RNN, (d) LSTM, (e) GRU, (f) N-BEATS, (g) MIXER, (h) Transformer, (i) ASMNET.

In this section, we introduce a novel Adaptive Masking Network (ASMNET) designed to optimize PV power generation forecasting tasks. ASMNET incorporates a self-learning mask into the networks input structure, which effectively screens redundant information in the buffer based on correlations between input sequences. This enables the network to better extract features from high-value time steps. Within the model, the Transformer architecture translates sensitive dependencies across input data sequences into attention issues through multi-head self-attention, enhancing the accurate expression of features within the network. Our experiments, conducted using publicly available real data from Belgium, involve extensive comparative studies. These experiments demonstrate that our proposed model outperforms recent deep sequence models in forecasting abilities. Furthermore, ablation studies validate the effectiveness of our improvements. The output analysis of the masking results indicates that ASMNET can effectively process time series with strong correlations, masking redundant information to achieve superior forecasting outcomes. Additionally, due to its versatility, the proposed masking module can also be applied to other sequential tasks or natural image processing tasks. Given that our model continues to utilize the Transformer architecture for feature extraction, inherent issues such as susceptibility to overfitting and high computational demands remain unaddressed. As a next step, we plan to apply and refine our model in more domains.

References

1 Nie, F., Wang, X. and Huangm H. (2014). Clustering and projected clustering with adaptive neighbors. *Proceedings of the 20th ACM SIGKDD international conference on Knowledge discovery and data mining*, 977–986. https://doi.org/10.1145/2623330.2623726.

2 Yang, B., Li, J., Li, Y. et al. (2022). A critical survey of proton exchange membrane fuel cell system control: Summaries, advances, and perspectives. *International Journal of Hydrogen Energy* 47 (17): 9986–10020. https://doi.org/10.1016/j.ijhydene.2022.01.065. 457.

3 Fu, X. and Zhou, Y. (2022). Collaborative optimization of pv greenhouses and clean energy systems in rural areas. *IEEE Transactions on Sustainable Energy* 14 (1): 642–656. https://doi.org/10.1109/TSTE.2022.3223684.

4 K. Osmani, M. Ramadan, T. Lemenand, B. Castanier, and A. Haddad. Optimization of pv array tilt angle for minimum levelized cost of energy, *Computers & Electrical Engineering*, 96: 107474, 2021. doi:https://doi.org/10.1016/j.compeleceng.2021.107474.

5 Zhang, C., Liu, Q., Zhou, B. et al. (2022). A central limit theorem-based method for dc and ac power flow analysis under interval uncertainty of

renewable power generation. *IEEE Transactions on Sustainable Energy* 14 (1): 563–575.

6 Li, Y., Han, M., Shahidehpour, M. et al. (2023). Data-driven distributionally robust scheduling of community integrated energy systems with uncertain renewable generations considering integrated demand response. *Applied Energy* 335: 120749. https://doi.org/10.1016/j.apenergy.2023.120749.

7 Jing, Y., Yang, Y., Feng, Z. et al. (2020). Neural style transfer: a review. *IEEE Transactions on Visualization and Computer Graphics* 26 (11): 3365–3385. https://doi.org/10.1109/TVCG.2019.2921336.

8 Jin, R., Chen, Z., Wu, K. et al. (2022). Bi-LSTM-based two-stream network for machine remaining useful life prediction. *IEEE Transactions on Instrumentation and Measurement* 71: 1–10. https://doi.org/10.1109/TIM.2022.3167778.

9 Chen, C., Jin, G., Li, T.H., and Gao, W. (2023). Mitigating label noise in GANs via enhanced spectral normalization. *IEEE Transactions on Circuits and Systems for Video Technology* 33 (8): 3924–3934. https://doi.org/10.1109/TCSVT.2023.3235410.

10 Spielman, D. (2012). Spectral graph theory. *Combinatorial Scientific Computing* 18: 18. https://doi.org/10.1201/b11644-19.

11 Fan, K. (1949). On a theorem of Weyl concerning eigenvalues of linear transformations I. *Proceedings of the National Academy of Sciences* 35: 652–655. https://doi.org/10.1073/pnas.35.11.652.

12 Yang, F., Fu, X., Yang, Q., and Chu, Z. (2024). Decomposition strategy and attention-based long short-term memory network for multi-step ultra-short-term agricultural power load forecasting. *Expert Systems with Applications* 238: 122226. https://doi.org/10.1016/j.eswa.2023.122226.

13 Wu, Z. and Huang, N.E. (2009). Ensemble empirical mode decomposition: a noise-assisted data analysis method. *Advances in Adaptive Data Analysis* 1 (01): 1–41. https://doi.org/10.1142/S1793536909000047.

14 Yang, B., Wang, J., Zhang, X. et al. (2020). Comprehensive overview of meta-heuristic algorithm applications on PV cell parameter identification. *Energy Conversion and Management* 208: 112595. https://doi.org/10.1016/j.enconman.2020.112595.

15 Esteban, C., Hyland, S.L., and Rätsch, G. (2017). Real-valued (medical) time series generation with recurrent conditional gans. *arXiv preprint arXiv:1706.02633*. https://doi.org/10.48550/arXiv.1706.02633.

16 Yuan, R., Wang, B., Sun, Y. et al. (2023). Conditional style-based generative adversarial networks for renewable scenario generation. *IEEE Transactions on Power Systems* 38 (2): 1281–1296. https://doi.org/10.1109/TPWRS.2022.3170992.

17 Chen, Y., Li, P., and Zhang, B. (2018). Bayesian renewables scenario generation via deep generative networks. *2018 52nd Annual Conference on Information*

Sciences and Systems (CISS), Princeton, NJ, USA, pp: 1–6. https://doi.org/10.1109/CISS.2018.8362314.

18 Sannigrahi, S., Ghatak, S.R., and Acharjee, P. (2019). Multi-scenario based bi-level coordinated planning of active distribution system under uncertain environment. *IEEE Transactions on Industry Applications* 56: 850–863. https://doi.org/10.1109/TIA.2019.2951118.

19 Hochreiter, S. and Schmidhuber, J. (1997). Long short-term memory. *Neural Computation* 9 (8): 1735–1780. https://doi.org/10.1162/neco.1997.9.8.1735.

20 Chung, J., Gulcehre, C., Cho, K.H., and Bengio, Y. (2014). Empirical evaluation of gated recurrent neural networks on sequence modeling. *arXiv preprint arXiv:1412.3555*. https://doi.org/10.48550/arXiv.1412.3555.

21 He, K., Zhang, X., Ren, S. and Sun, J. (2016). Deep residual learning for image recognition. *Proceedings of the IEEE conference on computer vision and pattern recognition*, pp. 770–778.

22 Long, J., Shelhamer, E. and Darrell, T. (2015). Fully convolutional networks for semantic segmentation. *Proceedings of the IEEE conference on computer vision and pattern recognition*, pp. 3431–3440.

23 Mao, J., Zhang, M. and Xu, Q. (2022). CNN and LSTM based data-driven cyberattack detection for grid-connected PV inverter. *2022 IEEE 17th International Conference on Control & Automation (ICCA), IEEE*, pp. 704–709. https://doi.org/10.1109/ICCA54724.2022.9831934.

24 Pandey, A. and Wang, D. (2019). TCNN: Temporal convolutional neural network for real-time speech enhancement in the time domain. *ICASSP* 2019–*2019 IEEE International Conference on Acoustics, Speech and Signal Processing (ICASSP), IEEE*, pp. 6875–6879. doi:https://doi.org/10.1109/ICASSP.2019.8683634. 582

25 F. Wang, Z. Xuan, Z. Zhen, K. Li, T. Wang, and M. Shi. A day-ahead pv power forecasting method based on lstm-rnn model and time correlation modification under partial daily pattern prediction framework, *Energy Conversion and Management*, 212: 112766, 2020. doi:https://doi.org/10.1016/j.enconman. 2020.112766.

26 Y. Wang, W. Liao, and Y. Chang. Gated recurrent unit network-based short-term photovoltaic forecasting, *Energies*, 11(8): 2163, 2018. doi: https://doi.org/10.3390/en11082163.

27 Vaswani, A., Shazeer, N., Parmar, N. et al. (2017). Attention is all you need. *Advances in neural information processing systems* 30: 5998–6008.

28 Brown, T., Mann, B., Ryder, N. et al. (2020). Language models are few-shot learners. *Advances in Neural Information Processing Systems* 33: 1877–1901.

29 B. N. Oreshkin, G. Dudek, P. Pełka, and E. Turkina. N-beats neural network for mid-term electricity load forecasting, *Applied Energy*, 293: 116918, 2021. doi:https://doi.org/10.1016/j.apenergy.2021.116918.

7

Dynamic Reconfiguration of PV-TEG Hybrid Systems via Improved Whale Optimization Algorithm

Bo Yang, Jiarong Wang, and Yulin Li

Faculty of Electric Power Engineering, Kunming University of Science and Technology, Kunming, China

7.1 Introduction

At present, the governance and protection of the climate environment is the great challenge of our time, and the energy industry is the most important factor in climate change. In order to achieve the sustainable and healthy development of society, most countries in the world are fully promoting the construction of renewable energy, so as to establish a clean and resilient new power system. With its wide distribution, safety, and clean characteristics, solar energy has become an indispensable role in the process of energy structure transformation [1]. In the past decade, the scale of solar energy manufacturing has expanded tenfold to meet the growing global demand for clean energy. According to a report by the International Energy Agency, it is expected that the global scale of photovoltaic (PV) power generation will increase from 220 GW in 2022 to around 500 GW by 2030. This rapid expansion underscores the urgency of further improving PV conversion efficiency, a topic that continues to draw significant attention from the research community [2].

During operation, PV cells primarily convert high-energy photons into electricity, while low-energy photons are absorbed as heat. This excess heat elevates module temperature, and prolonged exposure to high thermal conditions degrades conversion efficiency. To mitigate such thermal penalties, PV-thermoelectric generator (PV-TEG) hybrid systems have emerged as a promising approach in recent years. By coupling TEG devices with PV modules, part of the waste heat can be harvested for additional power generation, while simultaneously reducing PV operating temperature [3]. A typical PV-TEG hybrid setup integrates PV panels, thermoelectric modules, and an associated heat dissipation mechanism [4]. Prior studies have investigated different coupling strategies. For instance,

Artificial Intelligence Empowered Smart Energy Systems, First Edition. Edited by Qiang Yang and Gang Huang.

Reference [5] experimentally compared electrical separation, series, and parallel connections, while also analyzing the electrothermal interactions between PV cells and TEGs under these modes. Building on this line of research, work [6] introduced an intelligent thermal management design in which the TEG subsystem not only generates electricity from waste heat but also actively contributes to lowering PV module temperature.

Meanwhile, both PV and TEG have strong modular capabilities, which give them a high degree of design flexibility and allow for a wide range of applications between several hundred W and several thousand MW. However, the PV power generation efficiency is highly dependent on the solar radiation condition. So, the TEG module will also be affected by the solar irradiance when the PV-TEG hybrid system is used for power generation. In addition, due to the particularity of the PV power generation mode, factors such as clouds, buildings, trees, and dust will partially shield the PV module, resulting in uneven radiation on PV surfaces. The system output under partial shading conditions (PSCs) will be reduced, and the power-voltage (*P-V*) characteristic is prone to multiple peaks. Severe partial shading can also induce hot-spot formation, which in extreme cases damages PV modules [7]. To alleviate mismatch losses under PSC, reference [8] analyzed different topologies of PV modules and proved that the total cross-tied (TCT) extracted more power than other topologies. Beyond topology design, two primary strategies are widely adopted to address the multi-peak characteristics of PV output curves: maximum power point tracking (MPPT) and array reconfiguration. MPPT methods adjust the input impedance to align with the global maximum power point, thereby maximizing array output. However, under irregular shading, conventional MPPT controllers often converge on local maxima rather than the true global peak, which lowers energy yield [9]. Reconfiguration, in contrast, modifies either the physical layout or the electrical wiring of shaded modules to equalize row currents. By redistributing shading effects more uniformly across the array surface, this approach reduces current imbalance and transforms the *P-V* curve into a smoother profile with a dominant single peak [10].

Reconfiguration approaches are generally categorized as static or dynamic. The former physically relocates modules within the array so that shaded units are spread more evenly. While straightforward, this technique scales poorly for large installations and complex PSC scenarios. By contrast, dynamic reconfiguration preserves the physical placement of modules and instead alters their electrical interconnections, making it more adaptable across arrays of different sizes and shading patterns. Dynamic approaches include electrical array reconfiguration (EAR), adaptive array reconfiguration (AAR), and a variety of methods driven by intelligent optimization algorithms. EAR was among the earliest techniques, employing switch matrices to modify wiring based on control rules [11]. Its simplicity comes at the cost of significant switching losses and the requirement for numerous switches, which raises implementation expense. AAR improves upon this by dividing the array into two subgroups, where a microcontroller dynamically rewires the adaptive sub-array to balance irradiance levels [7]. However, its algorithmic complexity limits its use to small or building-level PV systems. More recently, optimization-based strategies have attracted considerable attention, as they offer high flexibility and do not require precise system models. Examples include methods inspired by sociopolitical processes, such as the democratic political algorithm [12], as well as bio- and swarm-inspired heuristics like the genetic algorithm (GA) [13], particle swarm optimization (PSO) [14], improved Harris hawks optimization (HHO) [15], and the marine predator algorithm (MPA) [16]. These approaches highlight the increasing role of metaheuristics in advancing PV array reconfiguration.

Although reconfiguration strategies have been widely investigated for PV system optimization and demonstrated to be effective, their application to PV-TEG hybrid arrays remains relatively underexplored. To address this gap, this chapter develops a PV-TEG hybrid array model and introduces a reconfiguration scheme based on an improved whale optimization algorithm (IWOA) to mitigate the adverse influence of PSCs on hybrid system efficiency. The key contributions of this work are as follows:

- A detailed PV-TEG array model is established to capture both electrical and thermal interactions. The model leverages the waste heat from PV modules for thermoelectric conversion while simultaneously reducing PV operating temperature, thereby enhancing overall performance.
- An IWOA-driven method is proposed to address the discrete and nonlinear nature of dynamic reconfiguration. By incorporating a nonlinear convergence factor and a roulette-based perturbation mechanism, the algorithm achieves a stronger balance between exploration and exploitation, offering faster convergence and improved optimization capability compared to alternative metaheuristics.
- To reflect practical array configurations of different sizes and layouts (symmetric and asymmetric), the method is tested on 6×6 and 6×10 PV-TEG arrays under six representative PSC scenarios, confirming its robustness and effectiveness.

The remainder of this chapter is organized as follows: Section 7.2 presents the adopted PV-TEG hybrid system model. Section 7.3 describes the WOA framework and the improvements introduced in IWOA. Section 7.4 reports simulation studies on 6×6 and 6×10 arrays under six PSC cases. Finally, Section 7.5 provides conclusions and a detailed discussion of the results.

7.2 PV-TEG Hybrid System Model

7.2.1 PV System Model

A PV array typically consists of multiple modules arranged in series and/or parallel. Since reconfiguration studies do not demand highly detailed device-level modeling, a balance between simplicity and accuracy is sufficient. In this chapter, the single-diode equivalent circuit, a well-established and widely accepted representation, is adopted to describe PV cell behavior and evaluate output performance [17, 18]. The corresponding circuit is illustrated in Figure 7.1a, and the governing equations for the cell current are expressed as

$$I = I_{\text{pv}} - I_{\text{D}} - I_{\text{sh}} \tag{7.1}$$

$$I_{\text{pv}} = \frac{G}{G_0}[I_{\text{pvstc}} + K_{\text{sc}}(T_{\text{c}} - T_0)] \tag{7.2}$$

$$I_{\text{D}} = I_0 \left[\exp\left(q\frac{V + IR_{\text{s}}}{bT_{\text{c}}\alpha} - 1\right)\right] \tag{7.3}$$

$$I_{\text{sh}} = \frac{V + IR_{\text{s}}}{R_{\text{sh}}} \tag{7.4}$$

where I_{pv} denotes the photocurrent generated under illumination, while I_{D} is the diode current and I_{sh} represents leakage through the shunt resistance. G and G_0 refer to the actual and standard irradiance levels, respectively, with I_{pvstc} being the photogenerated current at standard test conditions. K_{sc} is the short-circuit current coefficient, and T_0 and T_{c} represent the reference and operating cell temperatures, respectively. I_0 corresponds to the reverse saturation current of the diode, q is the electron charge, V is the terminal voltage of the PV cell, b is the Boltzmann constant, α is the diode ideality factor, and R_{s} is the series resistance.

In this study, the TCT configuration is employed for PV array interconnection, as depicted in Figure 7.1b,c. Under the TCT scheme, modules in each row are first connected in series, and the resulting rows are then cross-linked in parallel across columns [19]. This arrangement improves current sharing under partial shading and reduces mismatch losses compared with simpler topologies. The overall array voltage can be expressed as

$$V_{\text{total}} = \sum_{m=1}^{M} V_m \tag{7.5}$$

where V_m denotes the maximum voltage of row mmm, and M is the total number of rows.

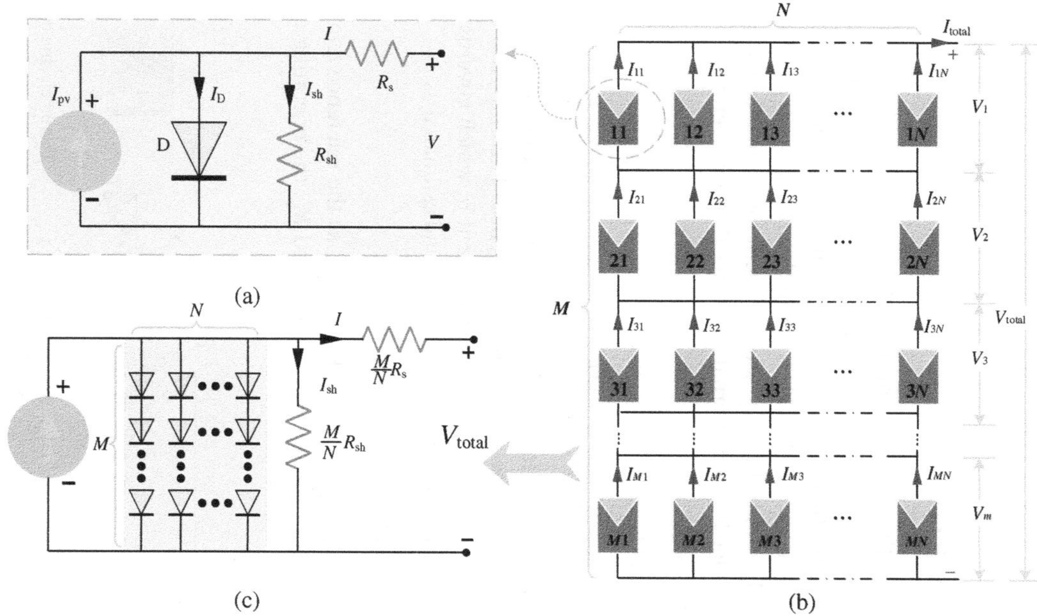

Figure 7.1 Equivalent circuit and TCT configuration of a PV array: (a) single-diode equivalent circuit of a PV cell; (b) topology of a PV array under the TCT scheme; (c) array-level equivalent circuit in TCT configuration.

Applying Kirchhoff's current law, the total array current is given by

$$I_{\text{total}} = \sum_{n=1}^{N}(I_{mn} - I_{(m+1)n}) = 0, \quad m = 1, 2, \cdots, M, \quad n = 1, 2, \cdots, N \tag{7.6}$$

where I_{mn} represents the output current of the module located in row m and column n, and N denotes the total number of columns.

7.2.2 TEG System Model

TEG directly converts thermal gradients into electrical energy. Each TEG module typically consists of two ceramic substrates enclosing a hot side and a cold side, between which multiple thermocouple pairs are linked in series by metallic connectors, as shown in Figure 7.2 [20, 21]. From an electrical perspective, the module can be represented by an ideal voltage source in series with an internal resistance. The open-circuit voltage V_{oc} and internal resistance R_{i} are temperature-dependent and can be expressed as

$$V_{\text{oc}} = \mu \Delta T = \mu(T_{\text{h}} - T_{\text{c}}) \tag{7.7}$$

$$R_{\text{in}} = \beta_1 + \beta_2 \cdot T_{\text{avg}} = \beta_1 + \beta_2 \cdot \frac{(T_{\text{h}} - T_{\text{c}})}{2} \tag{7.8}$$

where T_{h} and T_{c} denote the hot- and cold-side temperatures, respectively, T_{avg} is their average, and ΔT is the temperature difference. The empirical constants are $\beta_1 = -2.579$, $\beta_2 = 0.014$. The Seebeck coefficient μ, which governs the thermoelectric conversion, is temperature-dependent and estimated as

$$\mu(T_{\text{avg}}) = \mu_0 + \mu_1 \ln(T_{\text{avg}}/T_{\text{ref}}) \tag{7.9}$$

with $\mu_0 = 210\ \mu\text{V/K}$ and $\mu_1 = 120\ \mu\text{V/K}$, and T_{ref} as the reference temperature. When the module is connected to an external load R_{L}, the operating current is

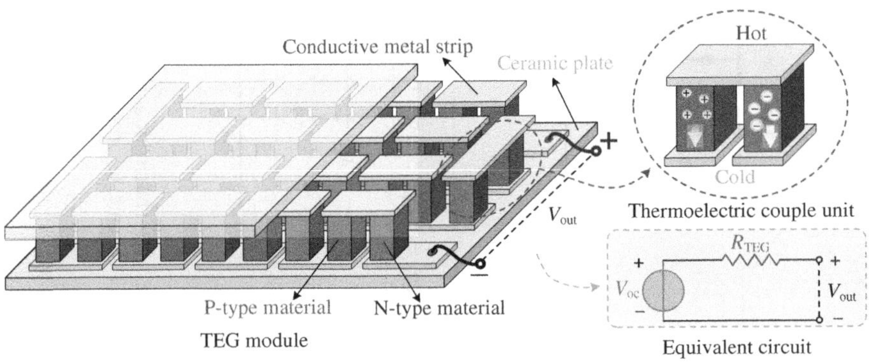

Figure 7.2 Schematic diagram of the TEG module.

determined by

$$I = \frac{V_{\text{oc}}}{R_{\text{L}} + R_{\text{in}}} \tag{7.10}$$

The delivered output power is given by

$$P_L = \left(\frac{V_{\text{oc}}}{R_{\text{L}} + R_{\text{in}}} \right)^2 R_{\text{L}} \tag{7.11}$$

An $M \times N$ TEG array is considered, as illustrated in Figure 7.3a. Based on Eqs. (7.4) and (7.5), the equivalent open-circuit voltage $V_{\text{oc}}(n)$ and internal resistance $R_{\text{in}}(n)$ for each column can be expressed as

$$V_{\text{oc}}(n) = \sum_{m=1}^{M} V_{\text{oc}}(m, n), \qquad n = 1, \ldots, N \tag{7.12}$$

$$R_{\text{in}}(n) = \sum_{m=1}^{M} R_{\text{in}}(m, n), \qquad n = 1, \ldots, N \tag{7.13}$$

where V_{oc} and R_{in} represent the column-level open-circuit voltage and resistance, respectively. Collectively, the entire array can be modeled as N voltage sources connected in parallel, as shown in Figure 7.3b.

Each column can further be represented by an equivalent current source, given by

$$I(n) = \frac{V_{\text{oc}}(n)}{R_{\text{in}}(n)}, \qquad n = 1, \ldots, N \tag{7.14}$$

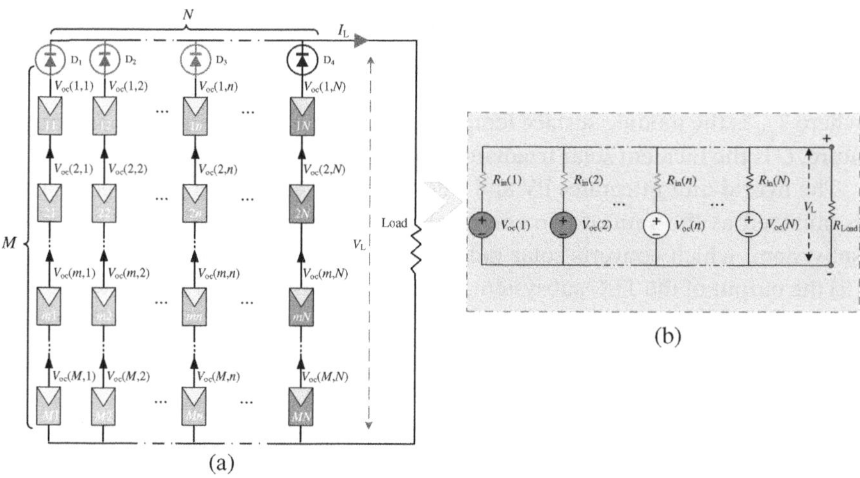

Figure 7.3 Equivalent circuit representation of a TEG array: (a) Column-level equivalent circuit of an $M \times N$ TEG array; (b) Overall equivalent circuit model of the TEG array.

where $I(n)$ denotes the output current of the nth column. The total output current I_s and total internal resistance R_{s_in} of the TEG array are formulated as

$$I_s = \sum_{n=1}^{N} I(n) \tag{7.15}$$

$$R_{s\_in} = \frac{1}{\sum_{n=1}^{N} \frac{1}{R_{in}(n)}} \tag{7.16}$$

By applying Thevenin's theorem, the current-source model can be transformed into an equivalent voltage source, with the open-circuit voltage given by

$$V_{s\_oc} = R_{s\_in} \cdot I_s \tag{7.17}$$

Finally, the maximum output power of the array is calculated as

$$P_{s\_max} = \frac{V_L^2}{R_{load}} = \frac{\left(\frac{1}{2}V_{s\_oc}\right)^2}{R_{s\_in}} = \frac{V_{s\_oc}^2}{4R_{s\_in}} \tag{7.18}$$

7.2.3 PV-TEG Hybrid System Model

In this chapter, a stacked PV-TEG hybrid configuration is considered, in which the TEG module is mounted directly on the backside of the PV module. The structural layout of the system is shown in Figure 7.4, consisting of four primary components: PV cells, TEG modules, a load circuit, and a heat sink. The heat sink facilitates thermal management by lowering the overall device temperature and sustaining a sufficient temperature gradient for thermoelectric conversion [22]. The surface temperature of the PV module is approximated by:

$$T_{pv} = 0.943T_e + 0.0195G - 0.528v_f + 0.3529 \tag{7.19}$$

where T_{pv} is the module surface temperature, T_e denotes the ambient air temperature, G is the incident solar irradiance, and v_f represents wind velocity.

The hybrid unit integrates PV and TEG modules, and its net power generation is obtained as the sum of two contributions: (i) the electrical output of the PV subsystem, which converts solar radiation into electricity via the PV effect, and (ii) the output of the TEG subsystem, which transforms part of the PV waste heat into electricity through the Seebeck effect. The total power of the PV-TEG hybrid module is therefore expressed as

$$P_{out} = P_{pv} + P_{TE} \tag{7.20}$$

where P_{pv} is the PV module output power; P_{TE} denotes the TEG contribution.

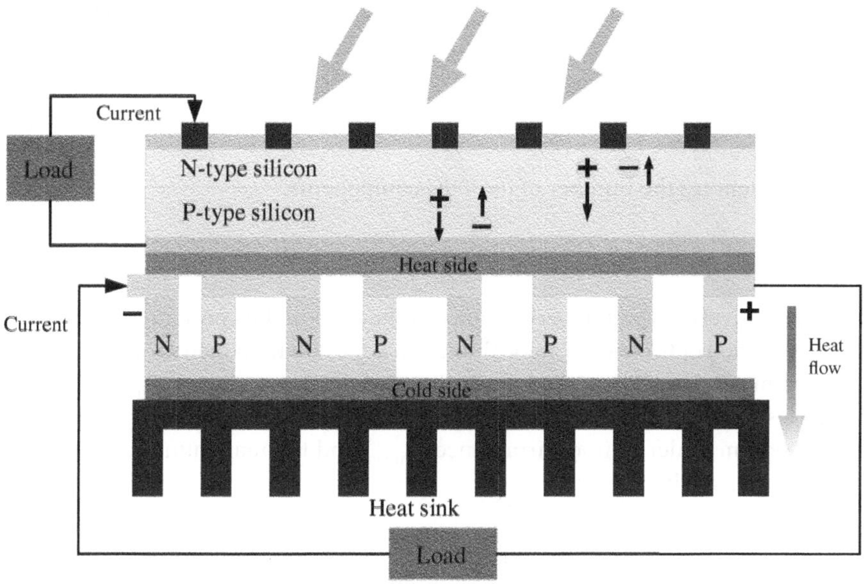

Figure 7.4 Structure of the hybrid system.

7.2.4 Objective Function

The optimization goal in this work is to maximize the total output power of the PV-TEG hybrid array, formulated as

$$F = \max(P_{\text{PV}} + P_{\text{TE}}) \tag{7.21}$$

To achieve this objective, the switching strategy must also account for system reliability and efficiency. Frequent or unnecessary switching actions shorten device lifetime and introduce additional power losses. Therefore, the reconfiguration process incorporates constraints to minimize superfluous operations of the switch matrix. Within the PV array, each module may only be exchanged with modules located in the same column, leading to the following constraint conditions:

$$\begin{cases} x_{i,j} \in \{1, 2, \dots, M\} \\ \bigcup_{i=1}^{M} x_{i,j} = \{1, 2, \dots, M\} \end{cases} \tag{7.22}$$

where $x_{i,j}$ means the number of the PV components.

Similarly, in the TEG array, modules are restricted to exchanges within the same row, and the state variables of the electrical switches are subject to

$$\begin{cases} k_{i,j} \in \{1, 2, \dots, N\} \\ \bigcup_{i=1}^{N} k_{i,j} = \{1, 2, \dots, N\} \end{cases} \tag{7.23}$$

where $k_{i,j}$ denotes the number of the TEG components.

7.2.5 Performance Evaluation

To quantitatively assess the performance of the proposed reconfiguration strategy and the applied optimization algorithms, two key evaluation metrics are employed [23]: mismatch loss (P_{mloss}) and power improvement percentage (P_{inc}). The mismatch loss is defined as the difference between the maximum output power of the hybrid system under uniform irradiance (P_{mpr}) and the maximum output power under partial shading (P_{m}):

$$P_{\text{mloss}} = P_{\text{mpr}} - P_{\text{m}} \tag{7.24}$$

The power improvement percentage evaluates the effectiveness of reconfiguration. It is calculated as the ratio of the power gain after reconfiguration to the maximum power before reconfiguration

$$P_{\text{re}} = \frac{P_{\text{Mre}} - P_{\text{Mbe}}}{P_{\text{Mbe}}} \times 100\% \tag{7.25}$$

where P_{Mre} and P_{Mbe} represent the maximum output power of the PV-TEG system after and before reconfiguration, respectively.

7.3 Improved Whale Optimization Algorithm

7.3.1 Whale Optimization Algorithm

WOA is a population-based metaheuristic introduced by Mirjalili in 2016. The method mimics the cooperative hunting strategies of humpback whales and incorporates three primary mechanisms: prey encirclement, bubble-net feeding, and random exploration [24]. WOA does not need to manually set various control parameters, and the algorithm is simple in structure, and has been outstanding in many numerical optimization and engineering problems. Therefore, WOA algorithm has been recognized and adopted by many scholars since it was proposed.

7.3.1.1 Encircling Prey

WOA assumes that the best-known position at a given iteration represents either the true prey location or an approximation of it. Other agents in the population progressively update their locations by moving closer to this reference solution [25, 26]. The update rule is given by

$$\vec{D} = |\vec{C} \cdot \vec{X}^*(t) - \vec{X}(t)| \tag{7.26}$$

$$\vec{X}(t+1) = \vec{X}(t) - \vec{A} \cdot \vec{D} \tag{7.27}$$

where $\vec{X}^*$ denotes the current best solution, t is the iteration index, and $\vec{X}$ is the current whale position. The vectors $\vec{A}$ and $\vec{C}$ are coefficient vectors defined as

$$\vec{A} = 2\vec{a} \cdot \vec{r_1} - \vec{a} \tag{7.28}$$

$$\vec{C} = 2 \cdot \vec{r_2} \tag{7.29}$$

where, $\vec{r_1}$ and $\vec{r_2}$ are random vectors in [0, 1], and $\vec{a}$ is a linearly decreasing convergence factor. As iterations proceed, $\vec{a}$ decreases from 2 to 0 according to:

$$\vec{a} = 2 - (2t)/T \tag{7.30}$$

where T is the maximum number of iterations.

7.3.1.2 Bubble-Net Attack

The bubble-net strategy is modeled using two simultaneous behaviors: (i) shrinking encirclement and (ii) spiral updating. The update of spiral position mainly refers to the whale's continuous approach to the optimal position through spiral swimming. In this process, the update of spiral position can be obtained by Eq. (7.30):

$$\vec{X}(t+1) = \vec{D'} \cdot e^{bl} \cdot \cos(2\pi l) + \vec{X}^*(t) \tag{7.31}$$

where l denotes the random number between [0, 1]; b controls the spiral shape (typically $b = 1$). $\vec{D'} = |\vec{X}^*(t) - \vec{X}(t)|$ represents the distance between the current whale position and the optimal position.

At each iteration, the algorithm selects either contraction encirclement or spiral updating with equal probability (50%):

$$\vec{X}(t+1) = \begin{cases} \vec{X}^*(t) - \vec{A} \cdot \vec{D}, p < 0.5 \\ \vec{D'} \cdot e^{bl} \cdot \cos(2\pi l) + \vec{X}^*(t), p \geq 0.5 \end{cases} \tag{7.32}$$

where p is a random number between [0, 1].

7.3.1.3 Random Search

To strengthen global search capability and avoid premature convergence, a random exploration step is introduced. When $\vec{A} \notin [-1, 1]$, the position update is based on a randomly chosen individual from the population:

$$\vec{D} = |\vec{C} \cdot \overrightarrow{X_{\text{rand}}}(t) - \vec{X}(t)| \tag{7.33}$$

$$\vec{X}(t+1) = \overrightarrow{X_{\text{rand}}}(t) - \vec{A} \cdot \vec{D} \tag{7.34}$$

where $\overrightarrow{X_{\text{rand}}}$ denotes the location of a randomly selected whale. This mechanism diversifies the search and improves the chance of locating the global optimum. $\vec{D}$ measures the positional difference between the present solution and a randomly selected whale within the swarm.

7.3.2 Design of Improved Whale Optimization Algorithm

7.3.2.1 The Roulette Mechanism

Reconfiguration of a PV-TEG hybrid array represents a high-dimensional discrete optimization problem. To improve search diversity and prevent premature convergence, this study introduces modifications to the basic WOA framework. The first enhancement is the roulette mechanism. Specifically, when the iterative process yields the same best solution for five consecutive iterations, the roulette procedure is triggered. At that point, one of three perturbation operators, i.e., swap, reversion, or insertion, is randomly applied to guide the population out of local optima.

Swap (20% Probability): Two modules are randomly chosen within the same PV column and their positions exchanged (Figure 7.5a). Similarly, in the TEG array, two modules within the same row are selected and swapped. The process is repeated across all PV columns and TEG rows.

Reversion (40% Probability): Two positions within a PV column are randomly selected, and the sequence of modules between them is reversed (Figure 7.5b).

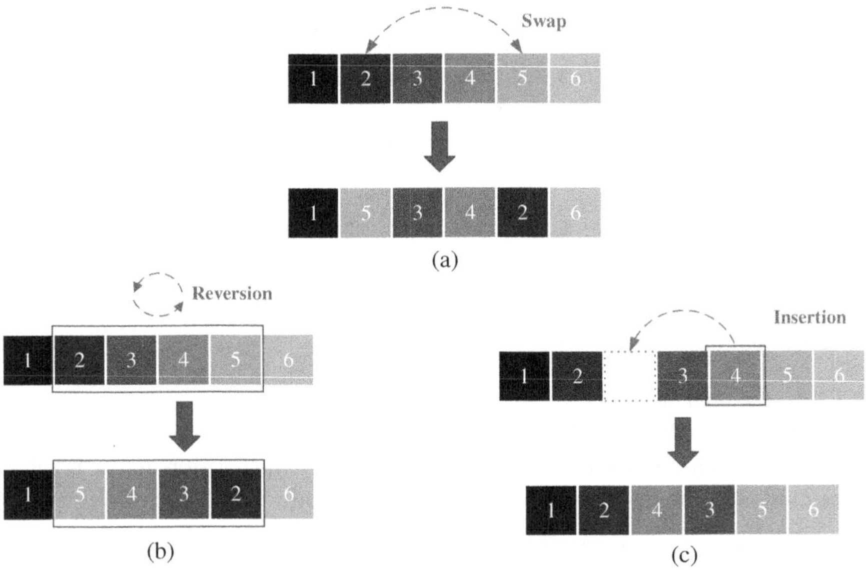

Figure 7.5 Improved mechanism: (a) Swap, (b) Reversion, (c) Insertion.

For the TEG array, two positions within the same row are chosen, and the intermediate sequence is reversed. This reordering is performed iteratively for each column and row.

Insertion (40% Probability): Two elements within a PV column are selected, and the latter is inserted immediately after the former (Figure 7.5c). The same procedure is applied in the TEG array by selecting two elements within a row. The operation is executed repeatedly for all columns and rows.

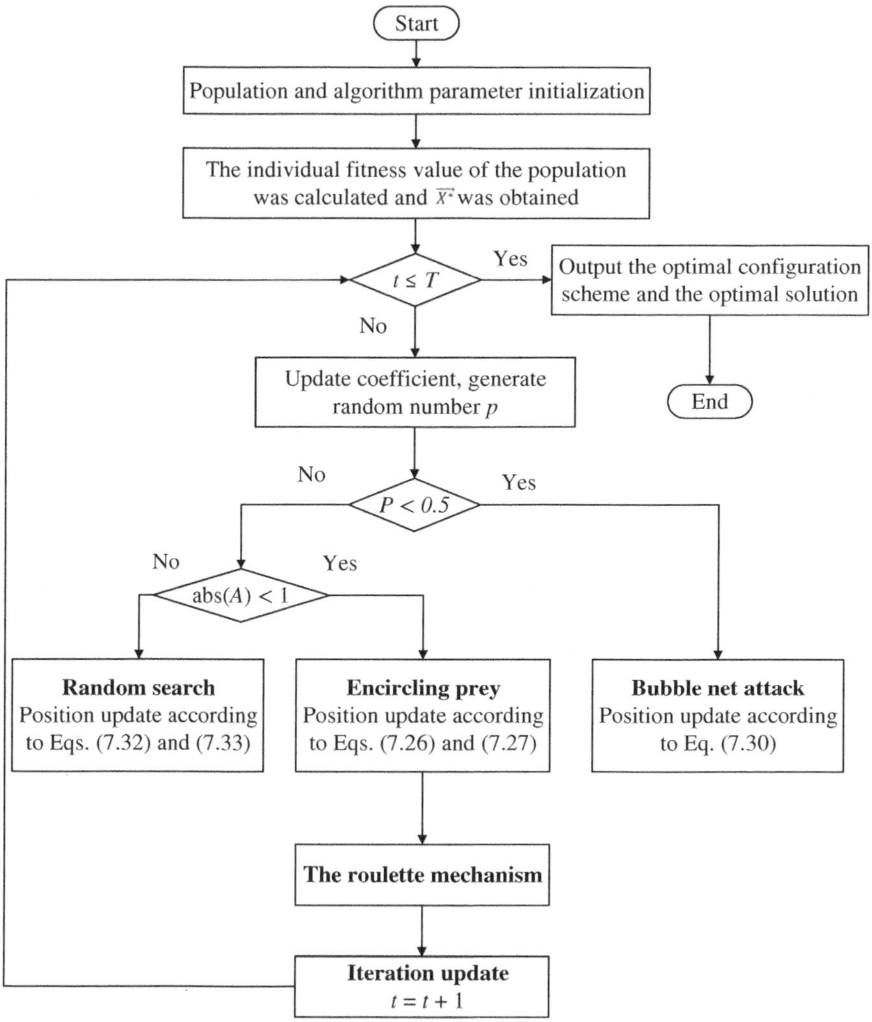

Figure 7.6 The flow chart of IWOA.

7.3.2.2 Nonlinear Convergence Factor

The second enhancement involves modifying the control parameter of the WOA. Within the algorithm, the balance between global exploration and local exploitation is governed by the vector $\vec{A}$, which itself is determined by the parameter $\vec{a}$. As a result, the behavior of $\vec{a}$ directly influences both convergence speed and solution accuracy. In the standard WOA, $\vec{a}$ decreases linearly from 2 to 0 across iterations.

To enhance optimization performance, a nonlinear convergence schedule is introduced in this study, defined as:

$$\vec{a} = 2 \cdot e^{-\left(\tan\left(1.2 \cdot \frac{t}{T}\right)^2\right)} \tag{7.35}$$

Through the above improvement measures, the IWOA algorithm can better handle the hybrid system reconfiguration problem in different sizes and PSC modes, and the solution flow of the IWOA algorithm is shown in Figure 7.6.

7.4 Case Study

In this chapter, the proposed reconfiguration strategy is validated on two PV-TEG hybrid arrays of different dimensions, i.e., one square and one non-square, under six representative PSCs. This design allows a comprehensive validation of effectiveness across array sizes and shading patterns. The electrical/thermal parameters used for the PV and TEG modules are summarized in Table 7.1 [26].

Table 7.1 Parameters of PV/TEG module.

PV component (A 10Green Technology AJ-M60-225)		TEG component (TGM199-1.4-2.0)	
Maximum power (W)	224.98	Parameter measurement condition	$T_c = 30°C$ $T_h = 200°C$
The voltage of the maximum power (V)	30.24	Size of module	40 mm × 40 mm × 40 mm
The current of the maximum power (A)	7.44	Maximum output power (W)	7.3
Open-circuit voltage (V)	36.24	Open-circuit voltage (V)	11
Open-circuit current (A)	8.04	Open-circuit current (A)	2.65
Number of batteries per component	60	Number of thermoelectric units	$n_{np} = 199$

To benchmark performance, the method is compared with several established metaheuristics, including WOA [27], GA, PSO, ACO, and BWO [28]. For fairness and reproducibility, all algorithms use identical settings: 200 maximum iterations, a population size of 30, and 20 independent runs per case.

7.4.1 6 × 6 Square Array

A 6 × 6 PV-TEG array is first examined. Six complex PSC patterns are considered, including short-narrow, long-narrow, diagonal, outer, center, and random, as depicted in Figure 7.7. Prior to reconfiguration, shading is highly localized, producing strong irradiance nonuniformity across the array (see Figure 7.7) and, consequently, severe current mismatch.

After applying the proposed IWOA-based reconfiguration, the resulting interconnection pattern (Figure 7.8) redistributes shaded modules so that irradiance is far more uniform across both the PV and TEG layers, thereby reducing mismatch risk. The corresponding *I-V* and *P-V* characteristics before and after reconfiguration for all PSCs are plotted in Figure 7.9. As the plots indicate, the characteristic multi-peak behavior observed under PSC is substantially suppressed following reconfiguration, yielding a smoother, near-single-peak profile. In turn, the maximum power P_{max} of both the PV and TEG subsystems increases, demonstrating that the IWOA-guided reconfiguration effectively mitigates the adverse impacts of partial shading on the hybrid array.

To further validate performance, the proposed IWOA method was compared against WOA, ACO, BWO, GA, and PSO. The maximum output powers obtained for all six PSC scenarios are summarized in Table 7.2. The results show that IWOA and GA consistently achieve the highest power levels across cases. Beyond average performance, algorithm stability was examined using boxplots of the 20 independent runs (Figure 7.10). The IWOA distributions exhibit much tighter quartile ranges and smaller box sizes compared with GA and other methods, indicating that the solutions are more concentrated around the optimum. The narrow gap between maximum and minimum outcomes demonstrates that IWOA more reliably converges to the global optimum, whereas GA shows higher variability. Figure 7.11 further compares the average mismatch loss and average power improvement percentage across algorithms. Before reconfiguration, the mismatch loss of the 6 × 6 PV-TEG array was 3.48 kW. After reconfiguration with IWOA, the mismatch loss decreased to 2.10 kW, while the average power output improved by 30.19%.

Overall, these results confirm that the proposed IWOA-based reconfiguration strategy provides robust, repeatable, and high-quality solutions, making it a reliable approach for dynamic reconfiguration of PV-TEG hybrid arrays under diverse PSC scenarios.

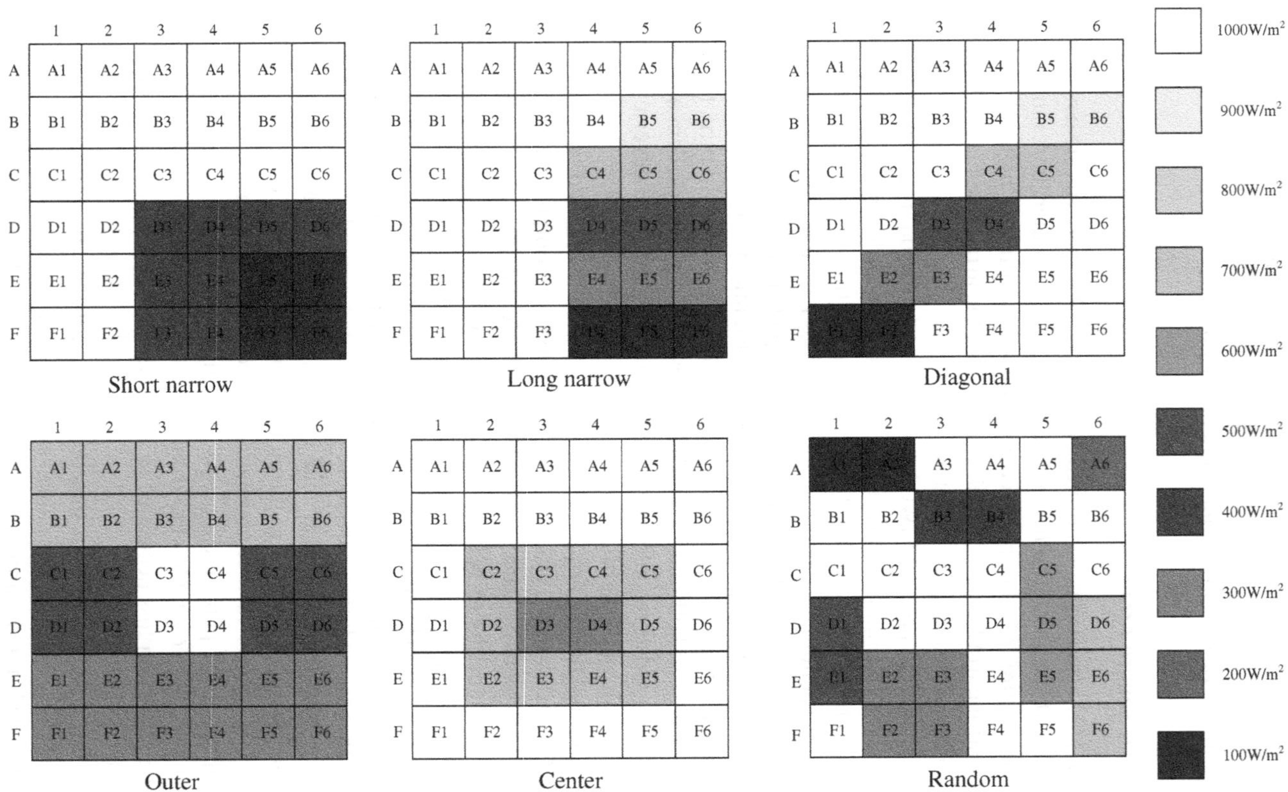

Figure 7.7 Irradiance distribution of a 6 × 6 PV-TEG array under six PSC patterns.

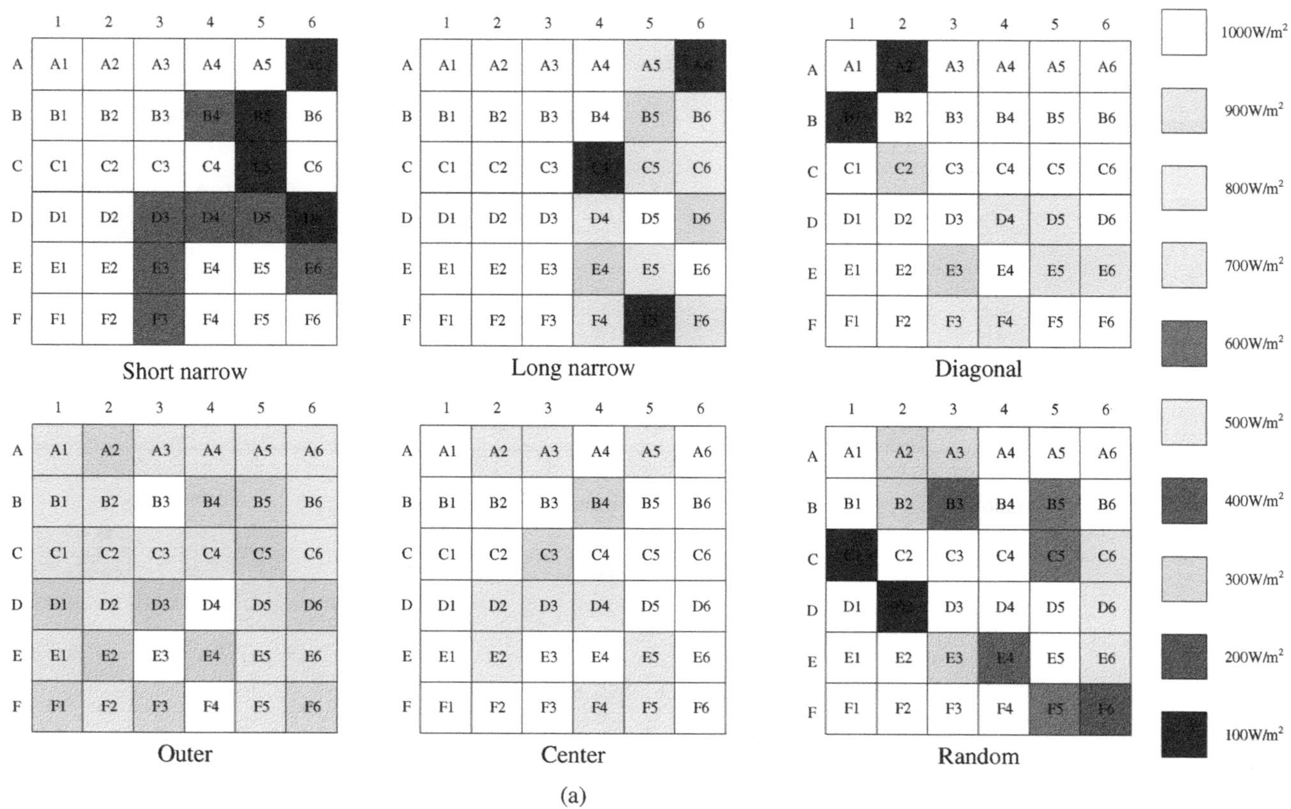

Figure 7.8 Reconfiguration schemes for the 6 × 6 PV-TEG array under six PSC conditions. (a) PV array scheme; (b) TEG array scheme.

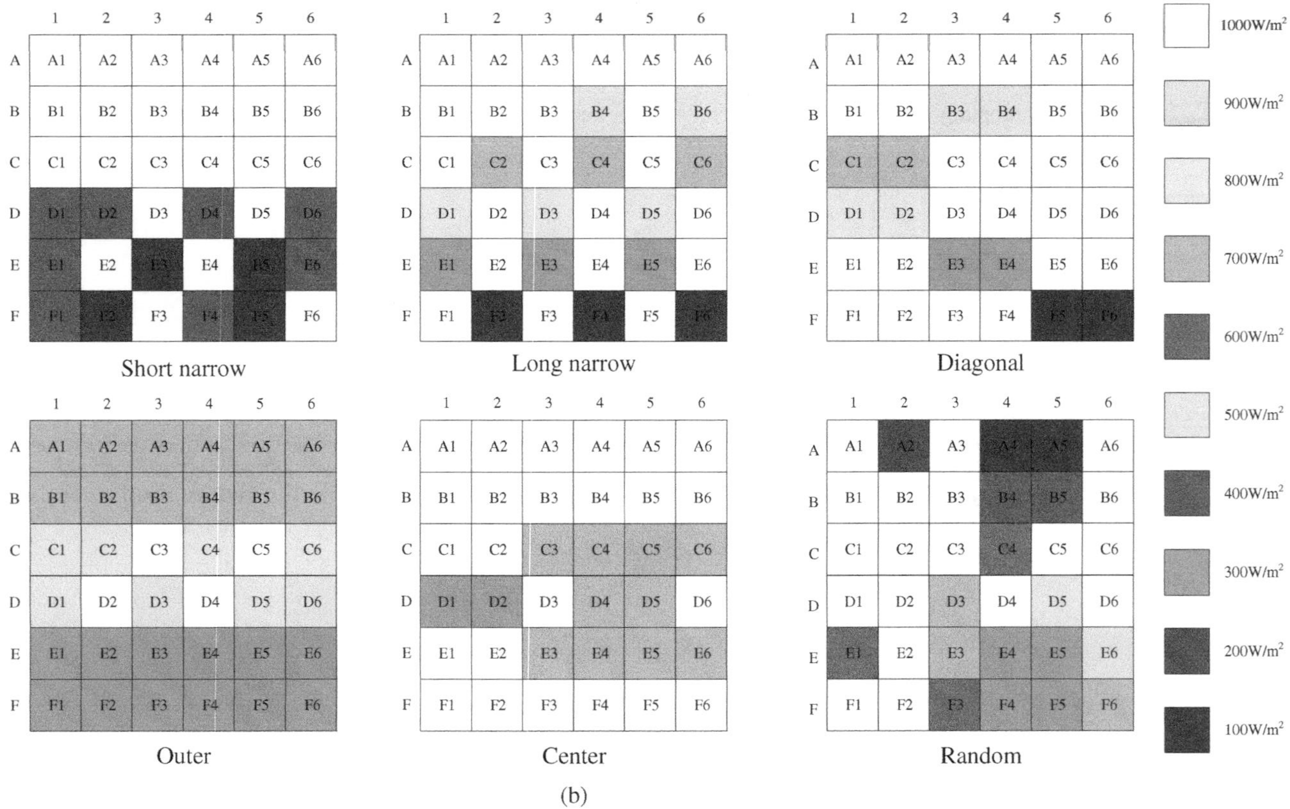

Figure 7.8 *(Continued)*

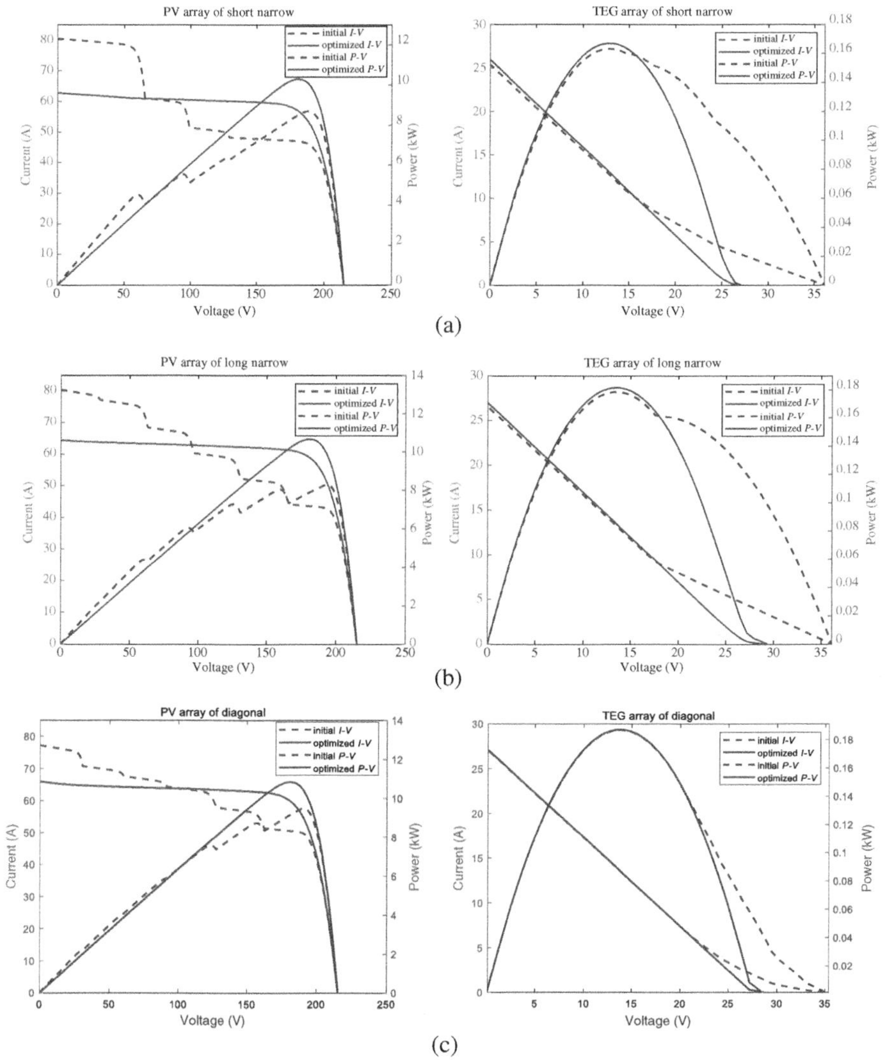

Figure 7.9 Output characteristics of the 6 × 6 PV-TEG hybrid array under six PSC cases. (a) short-narrow, (b) long-narrow, (c) diagonal, (d) outer, (e) center, (f) random.

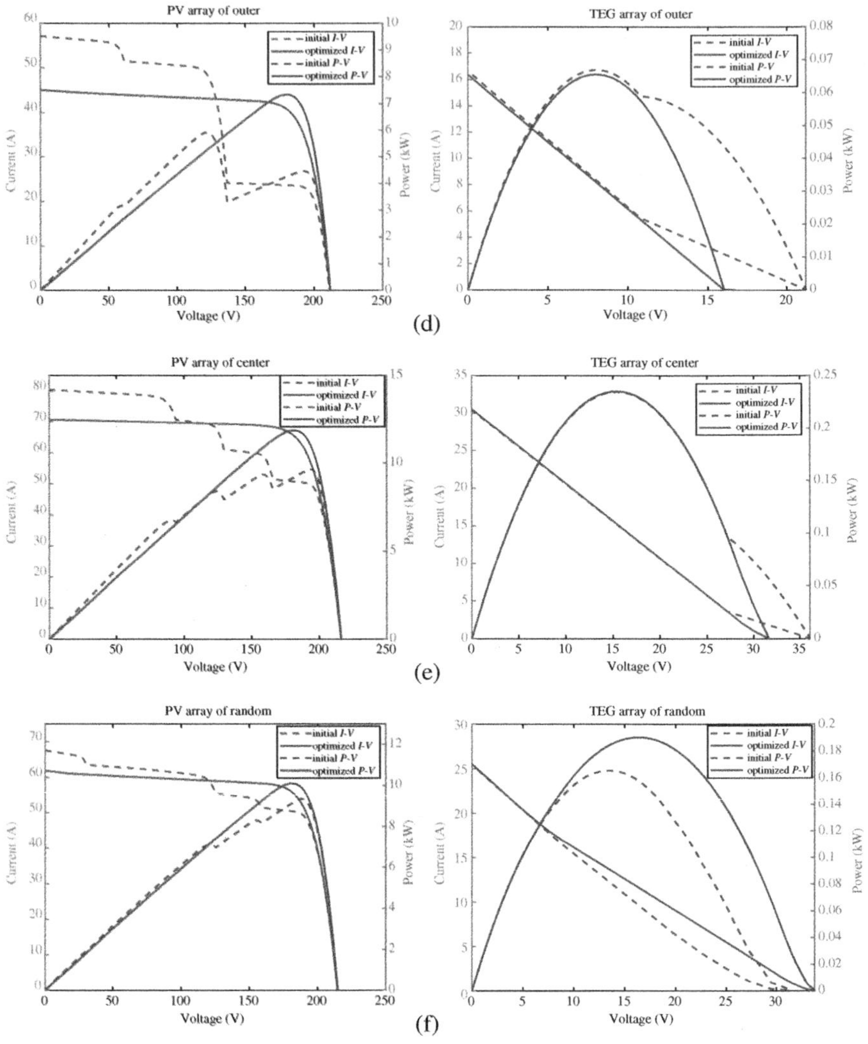

Figure 7.9 (*Continued*)

Table 7.2 Simulation results of various algorithms in case one.

Shadow pattern		Short narrow	Long narrow	Diagonal	Outer	Center	Random
Before reconfiguration (kW)		4.06	4.46	5.68	3.60	5.41	4.60
IWOA	P_{max}(kW)	5.68	6.35	6.89	4.46	6.89	5.81
	P_{avg}(kW)	5.68	6.35	6.89	4.46	6.89	5.81
WOA	P_{max}(kW)	5.68	6.35	6.76	4.46	6.89	5.81
	P_{avg}(kW)	5.68	6.24	6.61	4.34	6.79	5.62
PSO	P_{max}(kW)	5.68	6.22	6.89	4.32	6.89	5.81
	P_{avg}(kW)	5.29	6.13	6.72	4.28	6.83	5.60
GA	P_{max}(kW)	5.68	6.35	6.89	4.46	6.89	5.81
	P_{avg}(kW)	5.62	6.22	6.79	4.34	6.85	5.68
ACO	P_{max}(kW)	5.68	6.22	6.76	4.32	6.89	5.68
	P_{avg}(kW)	5.29	5.98	6.53	4.14	6.78	5.43
BWO	P_{max}(kW)	5.68	6.22	6.76	4.32	6.89	5.54
	P_{avg}(kW)	5.25	6.02	6.44	4.02	6.73	5.30

7.4.2 6 × 10 Non-Square Array

To further evaluate the robustness and generality of the proposed reconfiguration strategy, a 6 × 10 non-square PV-TEG array was constructed and tested under the same six PSC patterns. The initial irradiance distributions prior to reconfiguration are presented in Figure 7.12, while Figure 7.13 shows the distributions after applying the IWOA scheme. Following reconfiguration, shading is more evenly dispersed across the array surface, thereby mitigating excessive localized heating and reducing mismatch risks. The *I-V* and *P-V* output characteristics before and after reconfiguration under all six PSC conditions are illustrated in Figure 7.14. Similar to the 6 × 6 case, the multi-peak behavior caused by shading is significantly suppressed after reconfiguration, resulting in smoother and more stable curves. In addition, the maximum output power P_{max} of both the PV and TEG subsystems is clearly improved.

These findings demonstrate that the IWOA-based reconfiguration approach not only enhances the performance of square arrays but is also highly effective for non-square PV-TEG arrays across diverse shading scenarios, confirming its adaptability and reliability.

The objective of PSC reconfiguration is to mitigate multi-peak behavior and power loss caused by mismatch effects. When intelligent algorithms are employed to redistribute irradiance across the array, both optimization capability and solution stability become critical evaluation metrics. The detailed simulation

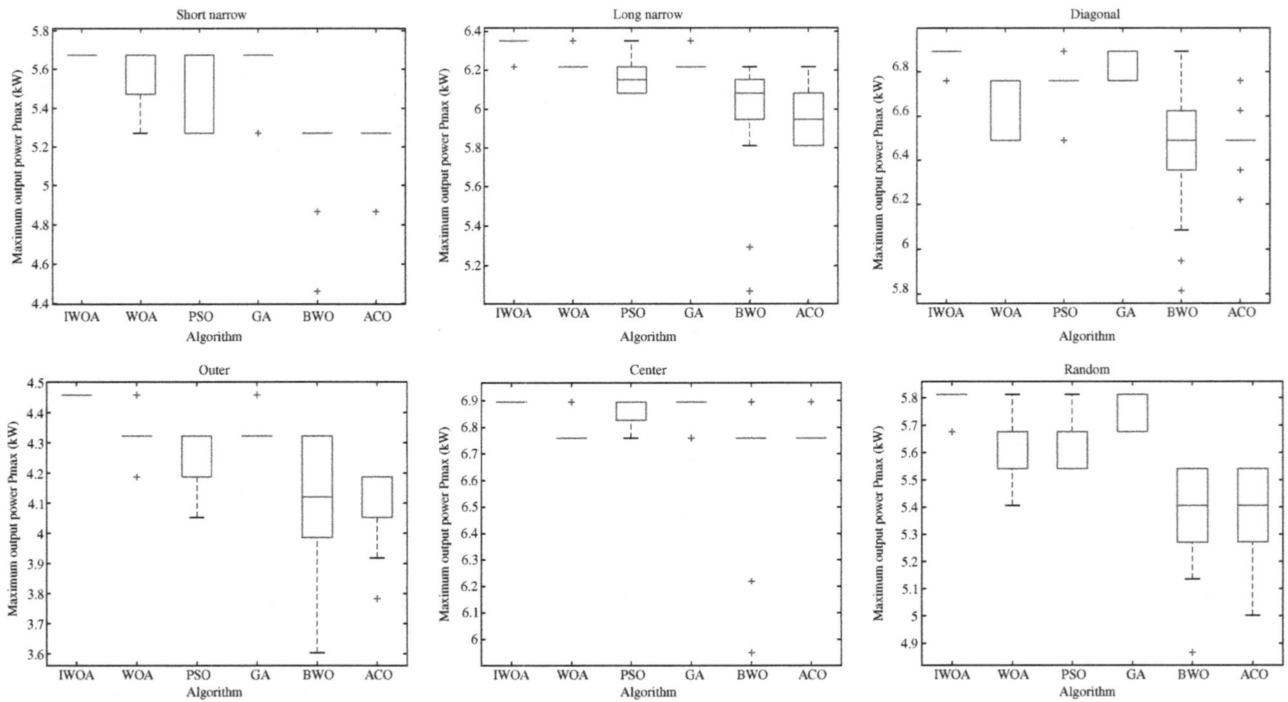

Figure 7.10 Boxplots of the 6×6 PV-TEG hybrid array after reconfiguration.

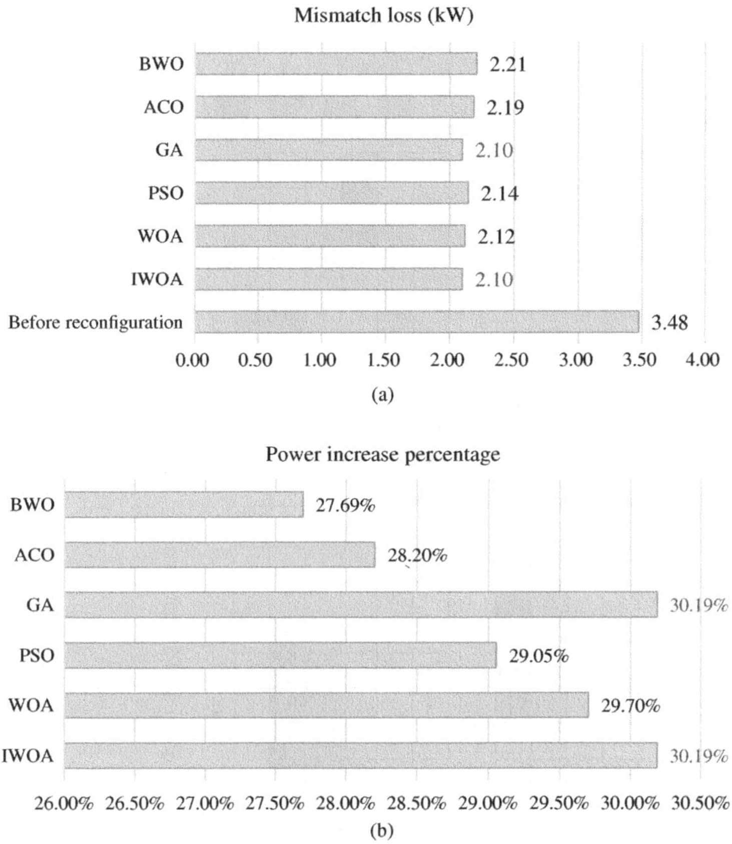

Figure 7.11 Comparative simulation results of the 6×6 PV-TEG hybrid system. (a) Average mismatch loss; (b) Average power increase percentage.

outcomes for different algorithms applied to the 6×10 non-square PV-TEG array are reported in Table 7.3. Across all six complex PSC scenarios, the proposed IWOA consistently achieves the highest maximum power output. Algorithmic robustness is further illustrated in the boxplots of Figure 7.15. The IWOA distributions are more compact, indicating greater stability and lower variability across 20 independent runs compared with alternative methods.

Additional insight is provided in Figure 7.16, which summarizes the average mismatch loss and average power improvement percentage. Prior to reconfiguration, the mismatch loss of the 6×10 array reached 5.64 kW. After IWOA-based reconfiguration, this value dropped to 3.32 kW, corresponding to a 29.80% improvement in output power. These findings confirm that the proposed method delivers reliable and scalable performance. The IWOA-based reconfiguration

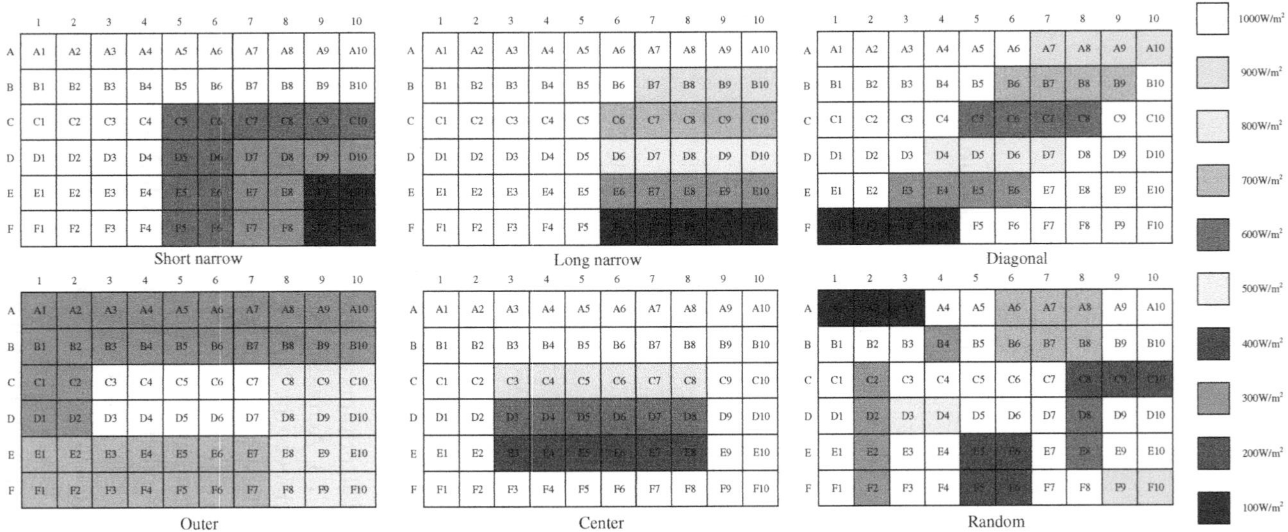

Figure 7.12 Irradiance distribution of a 6 × 10 PV-TEG array under six PSC patterns.

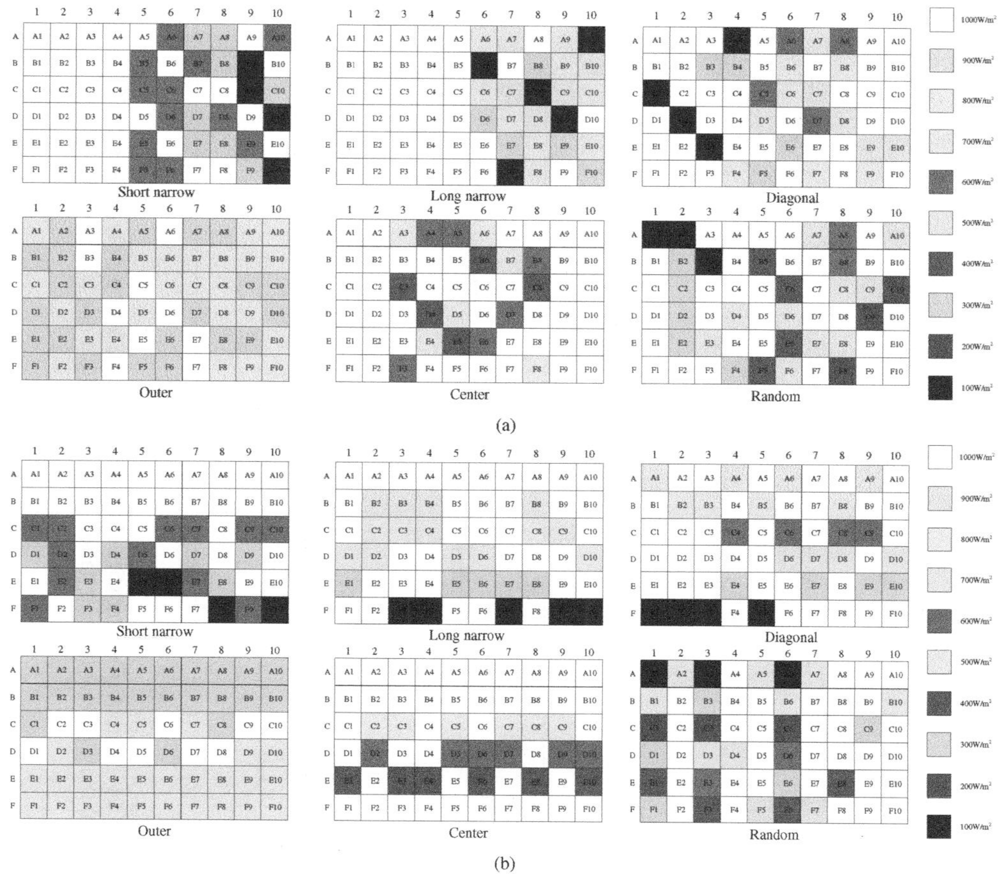

Figure 7.13 Reconfiguration schemes for the 6×10 PV-TEG array under six PSC conditions. (a) PV array scheme; (b) TEG array scheme.

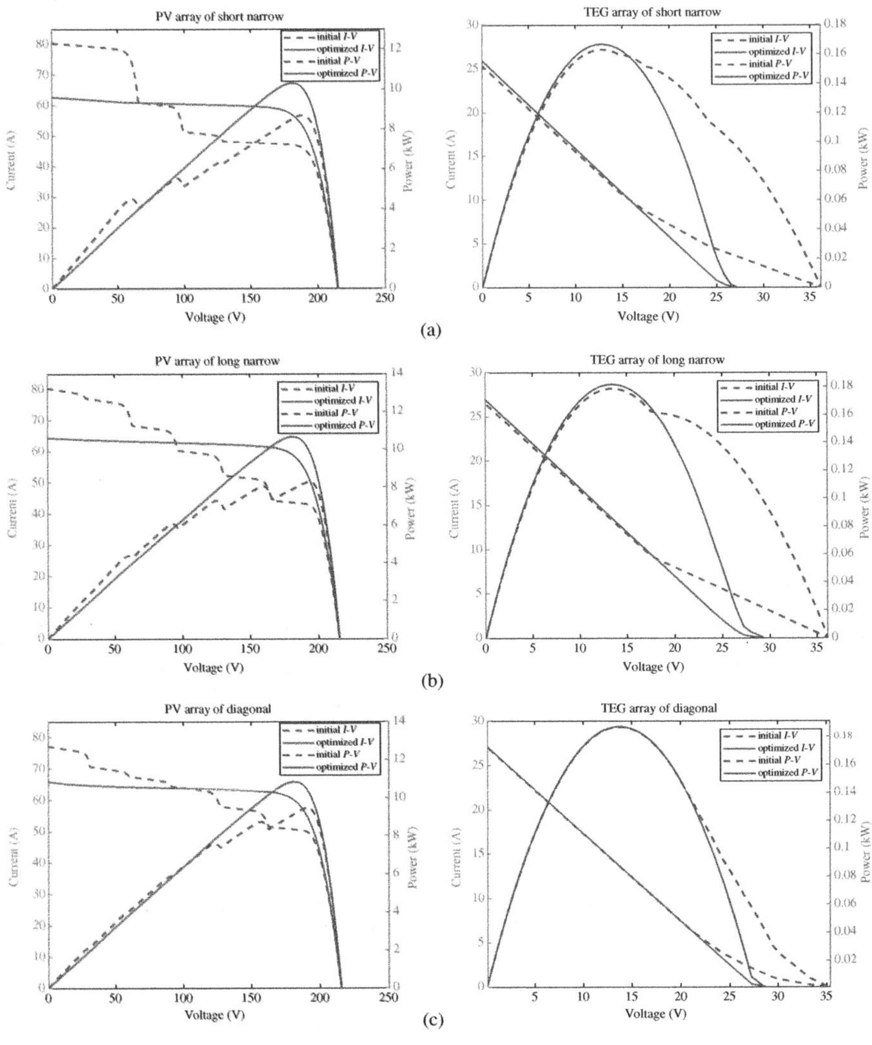

Figure 7.14 Output characteristics of the 6 × 10 PV-TEG hybrid array under six PSC cases. (a) short-narrow, (b) long-narrow, (c) diagonal, (d) outer, (e) center, (f) random.

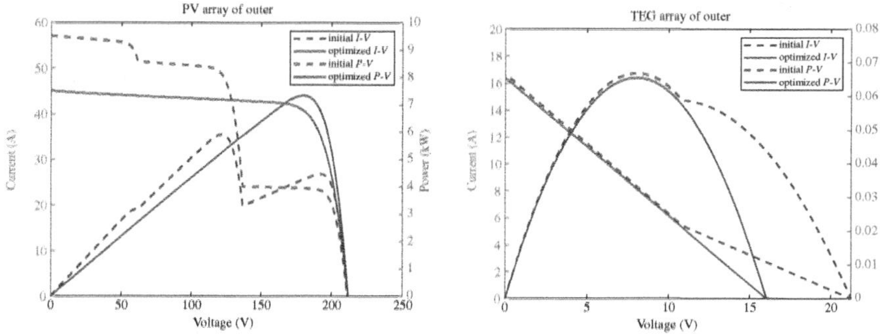

(d)

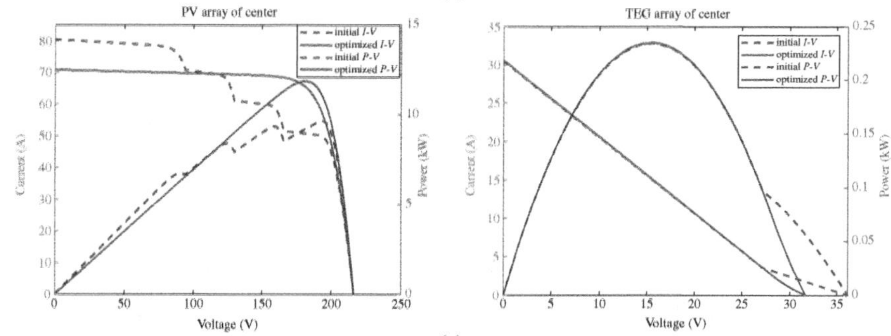

(e)

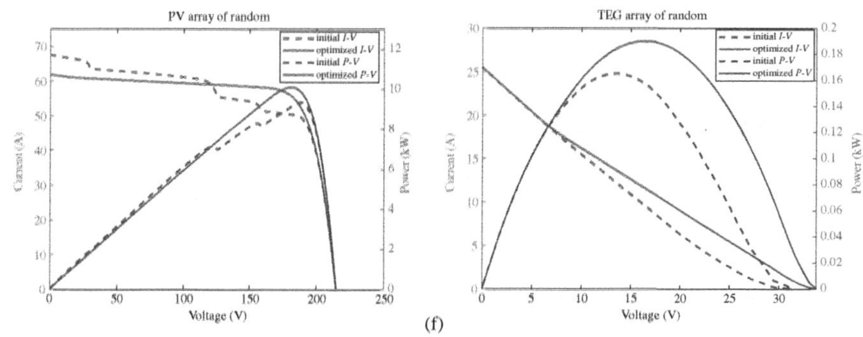

(f)

Figure 7.14 (*Continued*)

Table 7.3 Simulation results of various algorithms in case two.

Shadow pattern		Short narrow	Long narrow	Diagonal	Outer	Center	Random
Before reconfiguration (kW)		8.11	7.44	8.65	5.76	8.66	8.65
IWOA	P_{max}(kW)	10.27	10.68	10.81	7.29	11.90	10.14
	P_{avg}(kW)	10.26	10.62	10.77	7.29	11.90	10.03
WOA	P_{max}(kW)	10.27	10.54	10.68	7.29	11.63	10.00
	P_{avg}(kW)	10.15	10.50	10.37	7.05	11.53	9.83
PSO	P_{max}(kW)	10.27	10.54	10.68	7.29	11.90	10.00
	P_{avg}(kW)	10.02	10.34	10.52	7.11	11.64	9.83
GA	P_{max}(kW)	10.27	10.54	10.81	7.29	11.90	10.00
	P_{avg}(kW)	10.18	10.52	10.66	7.25	11.65	9.97
ACO	P_{max}(kW)	9.87	10.54	10.68	7.16	11.63	10.00
	P_{avg}(kW)	9.75	10.23	10.30	6.92	11.49	9.57
BWO	P_{max}(kW)	10.00	10.54	10.81	7.16	11.90	9.87
	P_{avg}(kW)	9.87	10.12	10.44	6.95	11.40	9.63

strategy not only provides effective dynamic reconfiguration for the 6×10 non-square PV-TEG array under diverse PSC conditions, but also demonstrates strong generality and adaptability for hybrid arrays of varying sizes.

7.5 Conclusion

This chapter introduced a dynamic reconfiguration strategy for PV-TEG hybrid arrays based on the IWOA to enhance energy conversion efficiency under PSC and mitigate mismatch losses. By adaptively altering module interconnections, the proposed approach reduces hotspot risk and smooths the array's electrical output characteristics. Simulation studies on 6×6 and 6×10 arrays under six PSC scenarios confirmed that IWOA delivers superior performance compared with other commonly used metaheuristics, including GA, PSO, ACO, BWO, and WOA. In terms of both power yield and solution stability, IWOA consistently achieved more reliable results. The major contributions of this work can be summarized as follows:

- A PV-TEG hybrid system model was developed, capturing both PV and thermoelectric processes. The model leverages waste heat from PV modules to generate additional power while lowering module operating temperature, thereby improving overall system efficiency.

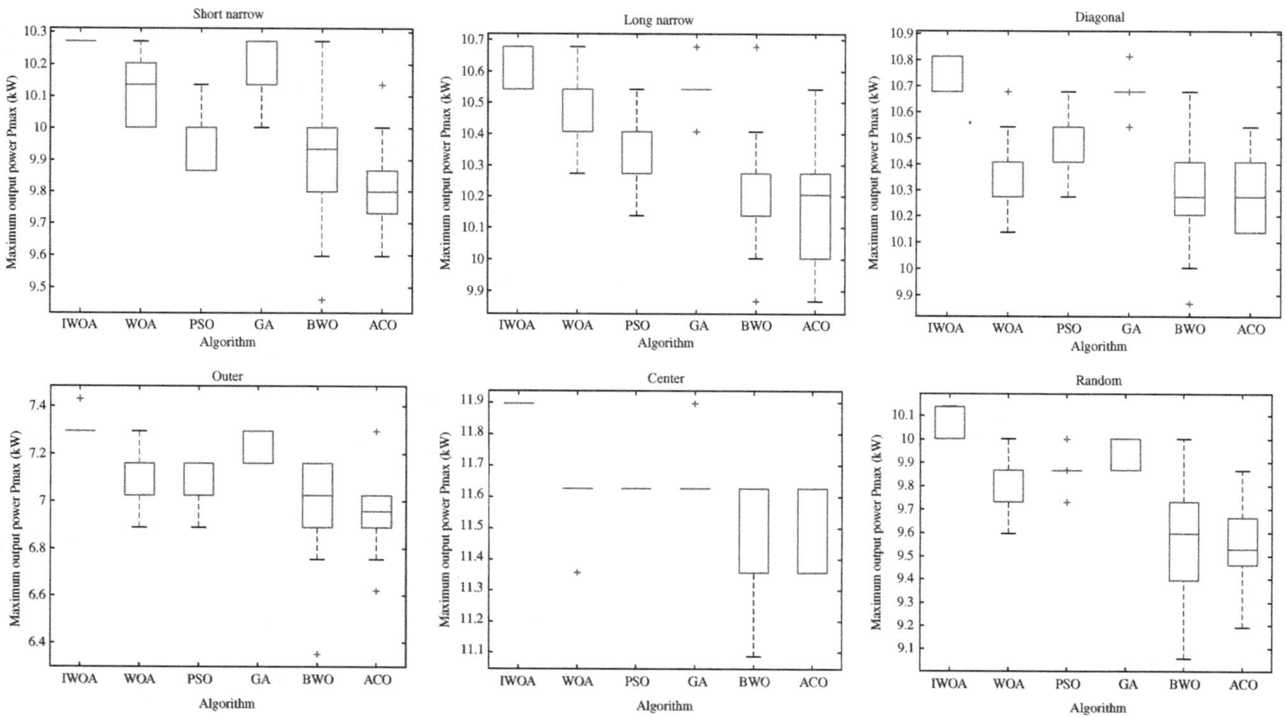

Figure 7.15 Boxplots of the 6 × 10 PV-TEG hybrid array after reconfiguration.

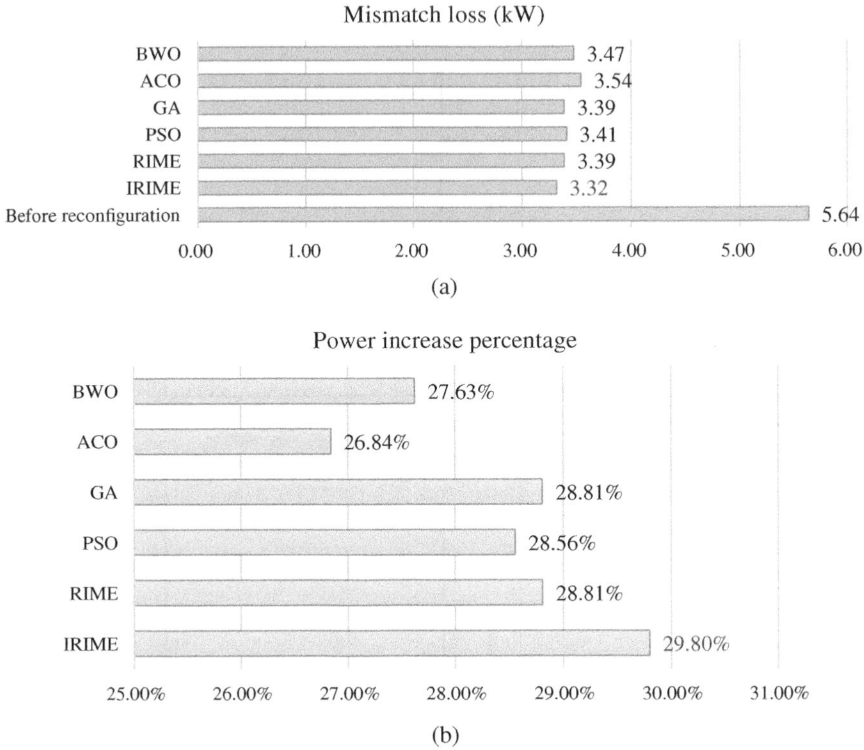

Figure 7.16 Comparative simulation results of the 6×10 PV-TEG hybrid system. (a) Average mismatch loss; (b) Average power increase percentage.

- A novel IWOA-driven reconfiguration method was proposed to handle the discrete optimization challenge of hybrid array interconnections. The algorithm demonstrated stronger stability and accuracy than other benchmark metaheuristics. On average, reconfiguration by IWOA improved the maximum power output by 30.19% for the 6×6 array and 29.80% for the 6×10 array. At the same time, mismatch loss was significantly reduced compared with competing methods.

References

1 International Energy Agency (2021). Global net-zero Emissions Roadmap for the energy sector to 2050. Website: www.iea.org.

2 International Energy Agency (2023). World Energy Outlook 2023. Website: www.iea.org.

3 Tyagi, K., Gahtori, B., Kumar, S., and Dhakate, S.R. (2023). Advances in solar thermoelectric and photovoltaic-thermoelectric hybrid systems for power generation. *Solar Energy* 254: 195–212. https://doi.org/10.1016/j.solener.2023.02.051.

4 Tang, J., Ni, H., Peng, R.L. et al. (2023). A review on energy conversion using hybrid photovoltaic and thermoelectric systems. *Journal of Power Sources* 562: 232785. https://doi.org/10.1016/j.jpowsour.2023.232785.

5 Yin, E. and Li, Q. (2022). Achieving extensive lossless coupling of photovoltaic and thermoelectric devices through parallel connection. *Renewable Energy* 193: 565–575. https://doi.org/10.1016/j.renene.2022.05.054.

6 Wang, N., Ni, L.L., Wang, A. et al. (2022). High-efficiency photovoltaic-thermoelectric hybrid energy harvesting system based on functionally multiplexed intelligent thermal management. *Energy Conversion and Management* 272: 116377. https://doi.org/10.1016/j.enconman.2022.116377.

7 Yang, B., Ye, H.Y., Wang, J.B. et al. (2021). PV arrays reconfiguration for partial shading mitigation: recent advances, challenges and perspectives. *Energy Conversion and Management* 247: 114738. https://doi.org/10.1016/j.enconman.2021.114738.

8 Bingöl, O. and Özkaya, B. (2018). Analysis and comparison of different PV array configurations under partial shading conditions. *Solar Energy* 160: 336–343. https://doi.org/10.1016/j.solener.2017.12.004.

9 Yang, B., Zhu, T.J., Wang, J.B. et al. (2020). Comprehensive overview of maximum power point tracking algorithms of PV systems under partial shading condition. *Journal of Cleaner Production* 268: 121983. https://doi.org/10.1016/j.jclepro.2020.121983.

10 Yang, B., Zhang, M.T., Guo, Z.X. et al. (2023). Adaptive evolutionary jellyfish search algorithm based optimal photovoltaic array reconfiguration under partial shading condition for maximum power extraction. *Expert Systems with Applications* 215: 119325. https://doi.org/10.1016/j.eswa.2022.119325.

11 Nguyen, D. and Lehman, B. (2008). An adaptive solar photovoltaic array using model-based reconfiguration algorithm. *IEEE Transactions on Industrial Electronics* 55 (7): 2644–2654. https://doi.org/10.1109/TIE.2008.924169.

12 Yang, B., Shao, R.N., Zhang, M.T. et al. (2021). Socio-inspired democratic political algorithm for optimal PV array reconfiguration to mitigate partial shading. *Sustainable Energy Technologies and Assessments* 48: 101627. https://doi.org/10.1016/j.seta.2021.101627.

13 Deshkar, S.N., Dhale, S.B., Mukherjee, J.S. et al. (2015). Solar PV array reconfiguration under partial shading conditions for maximum power extraction using genetic algorithm. *Renewable and Sustainable Energy Reviews* 43: 102–110. https://doi.org/10.1016/j.rser.2014.10.098.

14 Babu, T.S., Ram, J.P., Dragičević, T. et al. (2018). Particle swarm optimization based solar PV array reconfiguration of the maximum power extraction under partial shading conditions. *IEEE Transactions on Sustainable Energy* 9 (1): 74–85. https://doi.org/10.1109/TSTE.2017.2714905.

15 Yousri, D., Allam, D., and Eteiba, M.B. (2020). Optimal photovoltaic array reconfiguration for alleviating the partial shading influence based on a modified harris hawks optimizer. *Energy Conversion and Management* 206: 112470. https://doi.org/10.1016/j.enconman.2020.112470.

16 Yousri, D., Babu, T.S., Beshr, E. et al. (2020). A robust strategy based on marine predators algorithm for large scale photovoltaic array reconfiguration to mitigate the partial shading effect on the performance of PV system. *IEEE Access* 8: 112407–112426. https://doi.org/10.1109/ACCESS.2020.3000420.

17 Yang, B., Wang, J.B., Zhang, X.S. et al. (2020). Comprehensive overview of meta-heuristic algorithm applications on PV cell parameter identification. *Energy Conversion and Management* 208: 112595. https://doi.org/10.1016/j.enconman.2020.112595.

18 Zhang, X.S., Meng, D., Cai, J.H. et al. (2023). A swarm based double Q-learning for optimal PV array reconfiguration with a coordinated control of hydrogen energy storage system. *Energy* 266: 126483. https://doi.org/10.1016/j.energy.2022.126483.

19 Yang, B., Wu, S.C., Li, Q. et al. (2023). Jellyfish search algorithm based optimal thermoelectric generation array reconfiguration under non-uniform temperature distribution condition. *Renewable Energy* 204: 197–217. https://doi.org/10.1016/j.renene.2022.12.067.

20 Chen, Y.J., Yang, B., Guo, Z.X. et al. (2022). Dynamic reconfiguration for TEG systems under heterogeneous temperature distribution via adaptive coordinated seeker. *Protection and Control of Modern Power Systems* 7: 38. https://doi.org/10.1186/s41601-022-00259-6.

21 Guo, Z.X., Yang, B., Chen, Y.J. et al. (2023). Modular thermoelectric generation arrays reconfiguration under heterogeneous temperature distribution via improved cooperation search algorithm: modelling, design and HIL validation. *Applied Thermal Engineering* 219: 119323. https://doi.org/10.1016/j.applthermaleng.2022.119323.

22 Babu, C. and Ponnambalam, P. (2018). The theoretical performance evaluation of hybrid PV-TEG system. *Energy Conversion and Management* 173: 450–460. https://doi.org/10.1016/j.enconman.2018.07.104.

23 Yang, B., Li, Y.L., Huang, J.X. et al. (2023). Modular reconfiguration of hybrid PV-TEG systems via artificial rabbit algorithm: modelling, design and HIL validation. *Applied Energy* 351: 121868. https://doi.org/10.1016/j.apenergy.2023.121868.

24 Mirjalili, S. and Lewis, A. (2016). The whale optimization algorithm. *Advances in Engineering Software* 95: 51–67. https://doi.org/10.1016/j.advengsoft.2016.01.008.

25 Liu, Y.G., Zhao, Z.Z., Ma, Y. Z. (2023). Active wake control strategy of tandem wind turbine based on whale optimization algorithm. *Proceedings of the CSEE*. https://doi.org/10.13334/j.0258-8013.pcsee.222567.

26 Yang, B., Wang, J.R., Su, S. et al. (2024). Mismatch losses mitigation of PV-TEG hybrid system via improved RIME algorithm: design and hardware validation. *Journal of Cleaner Production* 434: 139957. https://doi.org/10.1016/j.jclepro.2023.139957.

27 Zhao, J.T., Luo, X.C., and Liu, J.M. (2023). Application of improved whale optimization algorithm in robot path planning. *Journal of Northeastern University (Natural Science)* 44 (8): 1065–1071. https://doi.org/10.12068/j.issn.1005-3026.2023.08.001.

28 Zhong, C.T., Li, G., and Meng, Z. (2022). Beluga whale optimization: a novel nature-inspired metaheuristic algorithm. *Knowledge-Based Systems* 251: 109215. https://doi.org/10.1016/j.knosys.2022.109215.

8

Coordinating Transactive Energy and Carbon Emission Trading Among Multi-Energy Virtual Power Plants for Distributed Learning

Peiling Chen and Yujian Ye

School of Electrical Engineering, Southeast University, Nanjing, China

8.1 Introduction

8.1.1 Background and Motivation

Global efforts to decarbonize energy systems intensify due to energy security and environmental concerns [1]. Virtual power plants (VPPs) aggregate diverse distributed energy resources (DERs), promising efficient energy management and facilitating trading [2].

Focusing solely on electricity decarbonization neglects other energy sectors, hindering global carbon reduction efforts [3]. Integrated optimization of diverse energy carriers, including electricity, heat, and natural gas, is increasingly recognized for its socioeconomic, operational, and environmental benefits [4]. Multi-Energy Virtual Power Plants (MEVPPs), integrating various energy production, conversion, storage, and consumption units, have garnered significant research attention [5]. Optimizing multiple energy forms is essential for transitioning to hybrid, environmentally friendly energy systems. This priority was evident in the EU's 5th, 6th, and 7th frameworks, which emphasized coordinated optimization of various energy forms, resulting in numerous influential multi-energy microgrid (MEMG) and MEVPP projects globally [6, 7].

DERs' flexibility depends on market involvement, particularly in transactive energy markets, fostering coordinated use of renewables and DER flexibility for sustainable development [8]. Energy sharing reduces reliance on suppliers, offering revenue opportunities and cost savings. Carbon Emission Right (CER) markets enforce carbon reduction targets within deregulated environments [9]. Microgrids (MGs) are considered specific embodiments of VPPs at the physical layer, while VPPs represent MGs at energy management and market levels [10, 11].

Artificial Intelligence Empowered Smart Energy Systems, First Edition. Edited by Qiang Yang and Gang Huang.

Meanwhile, distributed methods like the Alternating Direction Method of Multipliers (ADMM) protect participants' privacy but require careful penalty parameter selection and adaptations to fit distributed learning frameworks.

8.1.2 Review of Previous Work

Recent research focuses on energy management and trading among MEVPPs. Initially, studies optimized single entities by coordinating complementary DERs [12], emphasizing energy-shifting flexibility across different carriers, guided by exogenous energy and/or carbon prices [13]. However, a single MEVPP offers limited capacity. Integrating heterogeneous and geographically close MEVPPs into a regional system allows for local energy sharing and trading in transactive markets, enhancing socioeconomic and environmental benefits. Additionally, CER markets can be established alongside transactive energy markets.

The coordinated operation paradigm, typically examined as an optimal energy flow problem for networked MEVPPs and the upstream grid, faces practical limitations despite offering theoretically optimal solutions from the system perspective [14]. These limitations include significant communication requirements for transmitting diverse DER parameters, computational complexity with system scale, and privacy concerns regarding disclosing MEVPP information [15].

Addressing challenges, an agent-centric approach in transactive markets allows MEVPPs to control their DERs, reducing communication and computational needs and partially addressing privacy concerns [16]. Chen et al. [17] used multi-agent deep reinforcement learning for Peer-to-Peer (P2P) energy trading among MEVPPs. Siqin et al. [18] optimized joint operation and energy sharing among multiple MEVPPs using a Shapley Value-based profit allocation. Wang et al. [19] developed a P2P transaction platform for self-organized electricity trading and Xu et al. [20] introduced a P2P transactive multi-resource trading framework. However, these studies overlook network constraints, which can lead to unrealistic trading outcomes. They focus primarily on electricity trading, neglecting heat. Encouraging transactive heat trading can optimize MEVPP operations by redistributing heat efficiently from more to less efficient MEVPPs.

In addition to techno-economic analyses of MEVPP energy management, studies propose balancing economic and environmental impacts by integrating low-carbon factors into optimization to minimize emissions. Elsied et al. [21] optimized MEVPP operation considering carbon emissions. Zhou et al. [22] proposed a multi-energy coordination framework with quadratic emission constraints. Carbon tax policies promoting efficient renewable use were analyzed by Khorasany et al. [23]. However, these methods relying on external interventions may limit social efficiency compared to market-based approaches.

Emission trading schemes provide an alternative to carbon taxes, crucial for promoting emissions reduction and electricity marketization in China [24]. The EU

Emission Trading System effectively reduced emissions by 21% in 2020 compared to 2005 levels. Regulators allocate free CERs to participating entities, which must cover their emissions. Surplus or deficit CERs can be traded among entities or in the CER market. Mu et al. [25] and Hua et al. [26] integrate carbon trading into P2P trading frameworks, enhancing emission mitigation and overall MEVPP efficiency. However, coordinated trading of different energy forms and CERs among networked MEVPPs remains unexplored.

Unlike electricity, CER trading doesn't require physical delivery. Ensuring transparency is crucial, is often achieved through distributed and credible trading environments. Lu et al. [27] proposed a blockchain-based CER trading scheme, while Khaqqi et al. [28] customized CER for industries using reputation-based blockchain. Distributed coordination approaches address privacy concerns and reduce communication and computational burdens. Transactive energy markets based on decentralized optimization methods like dual sub-gradient algorithms [24] protect participant privacy, but still face a single point of failure.

ADMM is widely employed in distributed optimization across various domains, including machine learning and signal processing [29]. It is also utilized for managing P2P energy transactions among energy hubs [30] and MGs [31]. The effectiveness of ADMM relies heavily on the initial selection of penalty parameters [32]. To address this issue, adaptive penalty parameter algorithms have been developed, which dynamically adjust based on feedback and estimates of the dual function's local curvature [33]. Moreover, there is a growing interest in enhancing these adaptive methods with machine learning techniques, particularly reinforcement learning [34], to optimize penalty parameter selection. Furthermore, the integration of a consensus formulation [35] has led to the development of Consensus ADMM (C-ADMM), which requires a mechanism for distributed adaptive adjustments of penalty parameters to maintain efficiency, promote performance, and enhance suitability for distributed learning frameworks.

To address these limitations, this study introduces several novel contributions.

1) A comprehensive optimization framework coordinates electricity, heat, and CER trading among heterogeneous MEVPPs, ensuring power and thermal network constraint compliance for efficient energy shifting.
2) To protect MEVPP privacy and facilitate transparent CER trading, we developed a fully distributed AC-ADMM method that incorporates parallel computation, a leader-follower node mechanism, and adaptive penalty parameter tuning. This enhanced method accelerates convergence, outperforming traditional ADMM, and improves its applicability to distributed learning frameworks.
3) Case studies demonstrate the economic benefits of local energy and CER trading, surpassing traditional methods. Various market participation strategies are explored, showcasing the advantages of coordinating heterogeneous MEVPPs. Superior convergence is achieved compared to ADMM and C-ADMM methods.

8.2 Overall Transactive Trading Market in Heterogeneous Networked MEVPPs

We analyze the coordination problem for networked MEVPPs in a local transactive energy and CER market, depicted in Figure 8.1. MEVPPs can directly trade energy (electricity and heat) and CER within the transactive market at local trading prices. Additionally, they can engage in energy exchange with upstream energy and CER markets. Physically, diverse MEVPPs are distributed across different nodes of both the power network and heat network, interconnected with the main grid. The coordination framework incorporates network constraints to ensure the physical feasibility of MEVPPs' energy trading decisions.

To promote low-carbon energy production and utilization, MEVPPs must consume CERs for electricity and heat generating, taking responsibility for emissions. Initially allocated free CERs, MEVPPs can trade locally. Surplus CERs, stemming from instances where renewable energy sources predominantly meet a VPP's energy demand, can be sold locally or upstream for profit. When lacking sufficient CERs, MEVPPs are obligated to purchase them locally to avoid penalties for surpassing emission limits. If local markets cannot meet the demand, MEVPPs must resort to the upstream market, incurring additional costs.

Following precedent [17], we analyze three MEVPP types: residential, commercial, and industrial. Figure 8.2 illustrates their energy sources, conversion/storage devices, and demand composition.

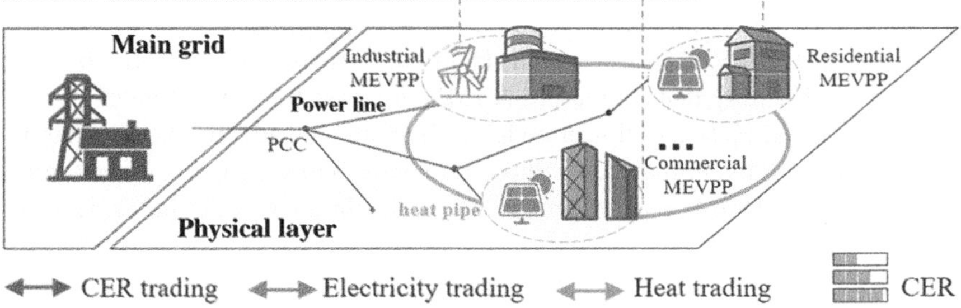

Figure 8.1 The framework of the decentralized market.

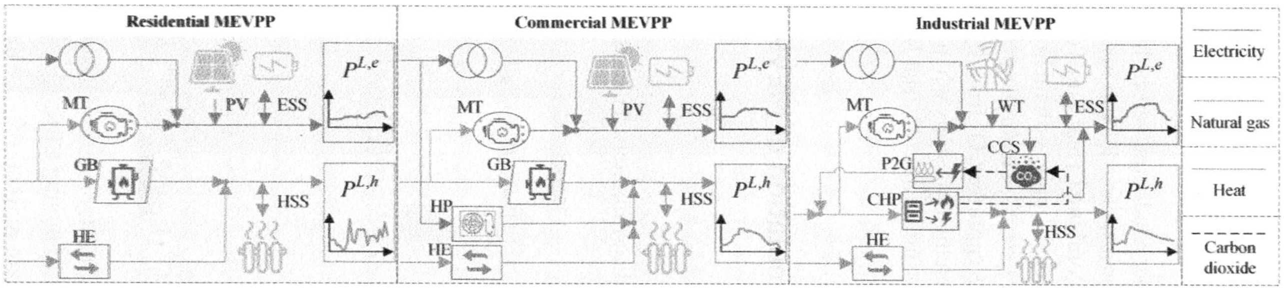

Figure 8.2 The heterogeneity of different MVPPs.

Residential MEVPPs typically feature rooftop solar PV panels, supplemented by electricity from the main grid. Micro-Turbines (MTs) act as backup generators, while Gas Boilers (GBs) provide heat from natural gas. Electrical Storage Systems (ESSs) and Heat Storage Systems (HSSs) smooth demand peaks by managing excess energy.

Commercial MEVPPs resemble residential setups but prioritize space heating. Heat Pumps (HPs) efficiently convert electricity to heat, with GB as a backup.

Industrial MEVPPs employ Combined Heat and Power (CHP) generators for simultaneous electricity and heat supply, enhancing efficiency. Power-to-Gas (P2G) and Carbon Capture Storage (CCS) technologies are integrated with CHPs to reduce carbon emissions. CCS captures CO_2 from CHP, while P2G electrolyzes water to produce hydrogen, reducing emissions when CHP is the primary. Wind Turbines (WTs) harness wind resources at industrial MEVPP sites, often offshore or rural.

8.3 Mathematical Formulation of MEVPP Coordination Problem

8.3.1 Objective Function

All MEVPPs are represented indiscriminately. The optimization problem aims to minimize total MEVPP cost.

(1) Power purchase cost:

$$C^{e}_{i,t} = \sum_{j\in I_i^{VPP}} \lambda^{+,e}_{i,j,t} P^{+,e}_{i,j,t} - \sum_{j\in I_i^{VPP}} \lambda^{-,e}_{i,j,t}(1-\alpha)P^{-,e}_{i,j,t} + \pi^{+,e}_{t} P^{+,e}_{ii,t} - \pi^{-,e}_{t} P^{-,e}_{ii,t} \tag{8.1}$$

(2) Heat purchase cost:

$$C^{h}_{i,t} = \sum_{j\in I_i^{VPP}} \lambda^{+,h}_{i,j,t} P^{+,h}_{i,j,t} - \sum_{j\in I_i^{VPP}} \lambda^{-,h}_{i,j,t}(1-\alpha)P^{-,h}_{i,j,t} + \pi^{+,h}_{t} P^{+,h}_{ii,t} - \pi^{-,h}_{t} P^{-,h}_{ii,t} \tag{8.2}$$

(3) Natural gas purchase cost:

$$C^{g}_{i,t} = \pi^{+,g}_{t} P^{+,g}_{i,t} \tag{8.3}$$

(4) CER purchase cost:

$$C^{c}_{i} = \sum_{j\in I_i^{VPP}} \lambda^{+,c}_{i,j} P^{+,c}_{i,j} - \sum_{j\in I_i^{VPP}} \lambda^{-,c}_{i,j}(1-\alpha)P^{-,c}_{i,j} + \pi^{pen,c} R^{pen,c}_{i} - \pi^{sur,c} R^{sur,c}_{i} \tag{8.4}$$

where, cost of buying electricity, heat, or natural gas of VPP i at step t is denoted by $C^{e/h/g}_{i,t}$.$C^{c}_{i,t}$ represents the cost of buying CER. $P^{+/-,e/h}_{ii,t}$ signifies electricity/heat VPP i buys/sells from/to upstream market, $P^{+/-,e/h}_{i,j,t}$ means electricity/heat VPP i buys/sells from/to VPP j. λ means the local prices and π denotes the upstream market prices. I_i^{VPP} means set of VPPs trading with VPP i. Grid compensates for network losses during power transmission by deducting transaction costs to ensure energy balance compliance. The commission charge α set at 0.05, consistent with [25].

Considering joint trading, the objective function of optimization problem is:

$$P_{i,t}^{L,e} = P_{i,t}^{T,e} + P_{i,t}^{X1,e} + P_{i,t}^{dis,e} - P_{i,t}^{ch,e} - P_{ii,t}^{-,e} + \sum_{j\in I_i^{VPP}} P_{i,j,t}^{+,e} - \sum_{j\in I_i^{VPP}} P_{i,j,t}^{-,e} \tag{8.5}$$

The obtained cost varies based on VPP's operation and transaction strategy.

8.3.2 Constraints

8.3.2.1 Energy Balance Constraint

First of all, the energy management and trading variables of each MEVPP must obey the relevant electricity and heat balance constraints, as outlined below for residential, commercial, and industrial MEVPPs:

$$P_{i,t}^{L,e} = P_{i,t}^{T,e} + P_{i,t}^{X1,e} + P_{i,t}^{dis,e} - P_{i,t}^{ch,e} - P_{ii,t}^{-,e} + \sum_{j\in I_i^{VPP}} P_{i,j,t}^{+,e} - \sum_{j\in I_i^{VPP}} P_{i,j,t}^{-,e} \tag{8.6}$$

$$P_{i,t}^{L,h} = P_{i,t}^{HE,h} + P_{i,t}^{X2,h} + P_{i,t}^{dis,h} - P_{i,t}^{ch,h} - P_{ii,t}^{-,h} + \sum_{j\in I_i^{MG}} P_{i,j,t}^{+,h} - \sum_{j\in I_i^{MG}} P_{i,j,t}^{-,h} \tag{8.7}$$

where, $P_{i,t}^{L,e/h}$ denotes VPP i's power/heat demand at step t, $P_{i,t}^{T,e}$ represents the transformer's power output, and $P_{i,t}^{HE,h}$ indicates the heat output of the heat exchanger (HE). $P_{i,t}^{ch/dis,e/h}$ signifies the charge/discharge of ESS/HSS, while $P_{i,t}^{X1,e}$ denotes generators X1's power output and $P_{i,t}^{X2,h}$ indicates heat-generated devices X2's heat output. The variation in X1 and X2 across MEVPPs reflects heterogeneity.

8.3.2.2 CER Balance Constraint

Energy firms use CERs in carbon trading to regulate emissions. Initial allocation follows baseline methods. Initial allocation of free CERs follows baseline methods. High-emission MEVPPs buy CERs to cover shortfalls, encouraging low-carbon practices. MEVPPs meet daily emission constraints, as CER trading doesn't involve physical delivery. This flexibility allows for carbon management alongside energy decisions. MEVPPs meet daily emission limits, managing carbon alongside energy decisions, as CER trading does not entail physical delivery. Given thermal plants' and coal-fired heating, initial CER allocation considers electricity, heat purchases, and gas-fueled equipment.

$$R_i^{buy,e/h} = \sum_{t\in T} P_{ii,t}^{+,e/h} \sigma^{buy,e/h} \Delta t \tag{8.8}$$

$$R_i^{MT} = \sum_{t \in T} P_{i,t}^{MT,e} \sigma^e \Delta t \tag{8.9}$$

$$R_i^{GB} = \sum_{t \in T} P_{i,t}^{GB} \sigma^h \Delta t \tag{8.10}$$

$$R_i^{CHP} = \sum_{t \in T} \left(P_{i,t}^{CHP,e} \sigma^e + P_{i,t}^{CHP,h} \sigma^h \right) \Delta t \tag{8.11}$$

where $R_i^{buy,e}/R_i^{buy,h}$ denotes initial CER (kg) for electricity and heat purchase. $\sigma^{buy,e}/\sigma^{buy,h}$ represents initial CER (kg) per unit electricity/heat purchase. $R_i^{MT}/R_i^{GB}/R_i^{CHP}$ represents initial CER (kg) for MT/GB/CHP, and σ^e/σ^h represents initial CER per electricity/heat production.

The emission factor method, widely used for emission estimation, involves multiplying activity data by emission factors, as outlined in [36].

$$e_{i,t}^{MT} = \beta^{MT} P_{i,t}^{MT,e} \tag{8.12}$$

$$e_{i,t}^{GB} = \beta^{GB} P_{i,t}^{GB,h} \tag{8.13}$$

$$e_{i,t}^{CHP} = \beta^{CHP} \left(P_{i,t}^{CHP,e} + P_{i,t}^{CHP,h} \right) \tag{8.14}$$

where e represents the emissions, β denotes carbon emission rate (kg/kWh).

In this study, CHP is coupled with CCS and P2G systems. CCS adopts post-combustion carbon capture to trap CO_2 from CHP. P2G locally utilizes this captured CO_2 to produce natural gas, reducing costs and risks associated with transmission and storage. This integration enables industrial MEVPPs to use gas for energy conversion without significantly increasing CER demand.

$$e_{i,t}^{CCS} = \eta^{CCS} e_{i,t}^{CHP}, \quad e_{i,t}^{P2G} = e_{i,t}^{CCS} \tag{8.15}$$

where $e_{i,t}^{CCS}$ is the carbon capture amount (kg), η^{CCS} is the capture rate (%) and $e_{i,t}^{P2G}$ is the amount of carbon received by P2G (kg).

The emissions from purchasing from upstream markets are calculated.

$$e_{i,t}^{buy,e/h} = \beta^{buy,e/h} P_{ii,t}^{+,e/h} \tag{8.16}$$

The initial CER allocation and carbon emission for heterogeneous MEVPPs:

(1) & (2) Residential MEVPP and commercial MEVPP:

$$R_i = R_i^{MT} + R_i^{GB} + R_i^{buy,e} + R_i^{buy,h} \tag{8.17}$$

$$e_{i,t} = e_{i,t}^{buy,e} + e_{i,t}^{buy,h} + e_{i,t}^{MT} + e_{i,t}^{GB} \tag{8.18}$$

(3) Industrial MEVPP:

$$R_i = R_i^{MT} + R_i^{CHP} + R_i^{buy,e} + R_i^{buy,h} \tag{8.19}$$

$$e_{i,t} = e_{i,t}^{buy,e} + e_{i,t}^{buy,h} + e_{i,t}^{MT} + e_{i,t}^{CHP} - e_{i,t}^{CCS} \tag{8.20}$$

If a MEVPP exceeds its daily emission limit, it must purchase the corresponding CER shortfall. If unable to do so locally, it incurs a penalty. Conversely, if emissions are below the limit, surplus CERs can be sold locally or upstream.

$$\sum_{t\in T} e_{i,t} - R_i = \sum_{j\in I_i^{VPP}} R_{i,j}^{+,c} + R_i^{pen,c}, \quad \sum_{t\in T} e_{i,t} \geq R_i \tag{8.21}$$

$$R_i - \sum_{t\in T} e_{i,t} = \sum_{j\in I_i^{VPP}} R_{i,j}^{-,c} + R_i^{sur,c}, \quad \sum_{t\in T} e_{i,t} \leq R_i \tag{8.22}$$

$$P_i^{pen,c} = \sum_{t\in T} e_{i,t} - R_i, \quad \sum_{t\in T} e_{i,t} \geq R_i \tag{8.23}$$

$$P_i^{sur,c} = R_i - \sum_{t\in T} e_{i,t}, \quad \sum_{t\in T} e_{i,t} < R_i \tag{8.24}$$

8.3.2.3 Operating Constraints of Conversion Devices

Constraints for energy conversion in MEVPP devices are outlined.

(1) & (2) Residential and Commercial MEVPPs:

$$P_{i,t}^{MT,e} = \eta_i^{MT} P_{i,t}^{+,g} v_{i,t}^{MT,g}, \quad \underline{P}^{MT} \leq P_{i,t}^{MT,e} \leq \overline{P}^{MT} \tag{8.25}$$

$$P_{i,t}^{GB,h} = \eta_i^{GB} P_{i,t}^{+,g} v_{i,t}^{GB,g}, \quad \underline{P}^{GB} \leq P_{i,t}^{GB,h} \leq \overline{P}^{GB} \tag{8.26}$$

$$v_{i,t}^{MT,g} + v_{i,t}^{GB,g} = 1, v_{i,t}^{GB,g} \geq 0, \quad v_{i,t}^{MT,g} \geq 0 \tag{8.27}$$

$$P_{i,t}^{T,e} = \eta_i^{T} P_{ii,t}^{+,e}, \quad 0 \leq P_{i,t}^{T,e} \leq \overline{P}^{T} \tag{8.28}$$

$$P_{i,t}^{HE,h} = \eta_i^{HE} P_{ii,t}^{+,h}, \quad 0 \leq P_{i,t}^{HE,h} \leq \overline{P}^{HE} \tag{8.29}$$

$$P_{i,t}^{HP,h} = \eta_i^{HP} P_{i,t}^{HP,e}, \quad \underline{P}^{HP} \leq P_{i,t}^{HP,h} \leq \overline{P}^{HP} \tag{8.30}$$

where constraints (8.25) and (8.26) express the power output of MT and GB. Constraint (8.27) ensures gas input proportions sum to 1. Constraints (8.28)–(8.30) define HE, transformer, and HP power output.

(3) Industrial MEVPP:

$$P_{i,t}^{CHP,e/h} = \eta_i^{CHP,e/h} \left(P_{g,i,t}^{+,g} + P_{g,i,t}^{P2G,g}\right) v_{i,t}^{g,CHP}, \quad \underline{P}^{CHP,e/h} \leq P_{i,t}^{CHP,e/h} \leq \overline{P}^{CHP,e/h} \tag{8.31}$$

$$P_{i,t}^{MT,e} = \eta_i^{MT} \left(P_{g,i,t}^{+,g} + P_{g,i,t}^{P2G,g}\right) v_{i,t}^{g,MT}, \quad \underline{P}^{MT} \leq P_{i,t}^{MT,e} \leq \overline{P}^{MT} \tag{8.32}$$

$$v_{i,t}^{CHP,g} + v_{i,t}^{MT,g} = 1, \quad v_{i,t}^{CHP,g} \geq 0, \quad v_{i,t}^{MT,g} \geq 0 \tag{8.33}$$

$$P_{i,t}^{CCS,e} = \chi^{CCS} e_{i,t}^{CCS}, \quad \underline{P}^{CCS} \leq P_{i,t}^{CCS,e} \leq \overline{P}^{CCS} \tag{8.34}$$

$$P_{i,t}^{P2G,e} = \chi^{P2G} e_{i,t}^{P2G}, \quad \underline{P}^{P2G} \leq P_{i,t}^{P2G,e} \leq \overline{P}^{P2G} \tag{8.35}$$

where, constraints (8.31) define CHP's electrical and heat output, using natural gas from the upstream market and P2G-generated gas. Constraints (8.34) and (8.35) detail CCS and P2G power consumption for carbon capture.

8.3.2.4 Constraints of Energy Storage Devices

MEVPPs offer flexibility in future power systems, leveraging energy storage (electricity, heat, or gas) to shift supply and demand across networks. Energy storage addresses intermittency and enhances renewable energy integration. ESS operating model constraints apply, with analogous constraints for HSS.

$$E_t^e = E_{t-1}^e + \Delta t P_t^{ch,e} \eta^{ch} + \Delta t P_t^{dis,e} / \eta^{dis}, \underline{E}^e \le E_t^e \le \overline{E}^e \tag{8.36}$$

$$0 \le P_t^{ch,e} \le \overline{P}^e u_t^e, 0 \le P_t^{dis,e} \le \left(1 - u_t^e\right) \overline{P}^e \tag{8.37}$$

where, constraint (8.36) balances ESS energy, accounting for charging/discharging losses and setting energy bounds. Constraints (8.37) limit charging/discharging based on capacity. Binary variable u_t^e prevents simultaneous charge/discharge.

8.3.2.5 Trading Constraints

The bilateral energy and CER trading activities of buyer and seller MEVPPs in the local transactive market are described as follows:

$$P_{i,j,t}^{+,e/h} = P_{j,i,t}^{-,e/h}, P_{i,j,t}^{-,e/h} = P_{j,i,t}^{+,e/h}, \forall j \in I_i^{VPP}, t \in T \tag{8.38}$$

$$R_{i,j}^{+,c} = R_{j,i}^{-,c}, R_{i,j}^{-,c} = R_{j,i}^{+,c}, \forall j \in I_i^{VPP} \tag{8.39}$$

$$P_{i,j,t}^{+,e/h} \ge 0, P_{j,i,t}^{-,e/h} \ge 0, P_{i,j,t}^{-,e/h} \ge 0, P_{j,i,t}^{+,e/h} \ge 0, R_{i,j}^{-,c} \ge 0, R_{i,j}^{+,c} \ge 0, R_{j,i}^{-,c} \ge 0, R_{j,i}^{+,c} \ge 0 \tag{8.40}$$

8.3.2.6 Network Constraints

This study considers a tree-topology radial power distribution network. The point of common coupling (PCC) is the reference bus with a voltage magnitude of 1 p.u. The linearized DistFlow model is employed, as in [22].

$$P_{h,i,t} - \sum_{m \in I_n^d} P_{i,m,t} = -p_{i,t} \tag{8.41}$$

$$Q_{h,i} - \sum_{m \in I_n^d} Q_{i,m,t} = -q_{i,t} \tag{8.42}$$

$$V_{h,t} - V_{i,t} = r_{h,i} P_{h,i,t} + x_{h,i} Q_{h,i,t} \tag{8.43}$$

$$\underline{P}_l \le P_{i,j,t} \le \overline{P}_l, \underline{Q}_l \le Q_{i,j,t} \le \overline{Q}_l, \underline{V} \le V_{i,t} \le \overline{V} \quad \forall i,j \in I \tag{8.44}$$

where, I_n^d / I_n^u denote downstream/upstream neighboring buses of node i.

This study employs a thermal network model with constant supply and return temperatures at network nodes while allowing variable water flow rates [37].

The difference between the mass of hot water flowing into a node and the mass flowing out of the node must equal the node flow:

$$\dot{m}^{s}_{h,i,t} - \dot{m}^{s}_{i,m,t} = \dot{m}^{N}_{i,t}, h \in N_i^{-}, m \in N_i^{+} \tag{8.45}$$

where, $N_i^{+/-}$ is the neighboring downstream/upstream node of node i, S/N denotes supply/demand side.

The relationship between power injection and node flow is given by:

$$P^{H}_{i,t} = c_f \dot{m}^{N}_{i,t} \left(T_i^{S} - T_i^{R}\right) \tag{8.46}$$

where, c_f is water's specific heat. When $\dot{m}^{N}_{i,t}$ is positive, it shows flow from supply to return. $T_i^{S/R}$ are fixed temperatures on supply/return side.

Pipeline water flow is unidirectional, while node flow is bidirectional.

$$0 \leq \dot{m}^{S}_{h,i,t} \leq m_p^{-S}, h \in S_i^{-} \tag{8.47}$$

$$-\underline{m}^{N}_{i,t} \leq \dot{m}^{N}_{i,t} \leq m^{-N}_{i,t} \tag{8.48}$$

Unidirectional hot water flow in pipelines restricts upstream MEVPPs from selling heat energy to downstream counterparts, adding constraints to heat trading.

$$P^{-,h}_{i,j,t} = 0, \forall j \in I_i^{h,u}, t \in T \tag{8.49}$$

$$P^{+,h}_{i,j,t} = 0, \forall j \in I_i^{h,d}, t \in T \tag{8.50}$$

8.4 Adaptive Consensus ADMM

To protect the privacy of MEVPPs and ensure transparency of CER trading, an efficient distributed optimization method is developed in this section.

8.4.1 Consensus ADMM

C-ADMM handles composite objective functions $\left(f = \sum_{i=1}^{N} f_i(x_i)\right)$ and constraints $(x_i - \widehat{x} = 0, i = 1, \ldots, N)$ by assigning each agent responsibility for its local objective (f_i) and constraints. Global variables $(\widehat{x})$ are shared among all N agents, each maintaining a local copy (x_i). When global optimality is reached, local copies converge to the common global value $(x_i = \widehat{x}, \forall i)$. This transformation from conventional ADMM to parallel processing introduces global variables for local copies to update. However, a central collector is still necessary for receiving local information and calculating global variables.

8.4.2 Adaptive Tuning of Penalty Parameters

The convergence of ADMM (and C-ADMM) methods relies heavily on the penalty parameter, typically considered a predefined hyper-parameter set beforehand and held constant throughout iterations. Employing a variable penalty parameter for each iteration can enhance convergence by reducing reliance on the predetermined value. Therefore, an adaptive tuning scheme is employed:

$$\rho^{k+1} := \begin{cases} \tau^{incr}\rho^k & \|r^k\|_2 > \mu\|s^k\|_2 \\ \rho^k/\tau^{decr} & \|s^k\|_2 > \mu\|r^k\|_2 \\ \rho^k & other \end{cases} \tag{8.51}$$

The core is to balance the primal ($\|r^k\|_2$) and dual residual norms ($\|s^k\|_2$), ensuring they both converge to zero. Choosing a large value for ρ penalizes primal feasibility violations, rapidly reducing the primal residual but increasing the dual residual. Conversely, a small ρ value reduces the dual residual but slows the decline of the primal residual. The mechanism automatically adjusts ρ for the next iteration based on the current primal and dual residuals, maintaining a balanced decline rate for both.

8.4.3 AC-ADMM-Based Energy and CER Trading for Networked MEVPPs

We implement leader and follower nodes to handle global constraints. A selected leader node gathers information from its follower nodes, updates the global variable, and sends the updated value back to the followers. This fully distributed and parallel approach significantly boosts computational efficiency. In our optimization problem, global constraints fall into two categories: trading and network constraints. The handling of these constraints varies in the proposed AC-ADMM.

(1) Trading Constraints:

Trading constraints involve local electricity, heat, and CER trading (Eq. (8.38) and (8.39)). Despite the different types, these trades involve bilateral transactions, with each transaction occurring between two VPPs. Focusing on local electricity trading, we elaborate on the treatment of trading global constraints by the AC-ADMM algorithm, as illustrated in Figure 8.3.

For local electricity trading constraints between VPP i and j at step t, VPP i and j are designated as leader and follower nodes for global variables $\widehat{P}^{+,e}_{i,j,t}$ and $\widehat{P}^{-,e}_{i,j,t}$, respectively. VPP i stores and updates local copies $P^{+,e}_{i,j,t}$, $P^{-,e}_{i,j,t}$, while VPP j does the same for $P^{+,e}_{j,i,t}$ and $P^{-,e}_{j,i,t}$. Both VPPs concurrently solve their local optimization problems. Subsequently, VPP j transmits its local information $P^{+,e}_{j,i,t}$, $P^{-,e}_{j,i,t}$, $\lambda^{-,e,k}_{j,i,t}$ and $\lambda^{+,e,k}_{j,i,t}$ to VPP i. VPP i updates global variables based on the equations in Figure 8.3's shaded area. Updated global variables are sent back to VPP j for the next iteration.

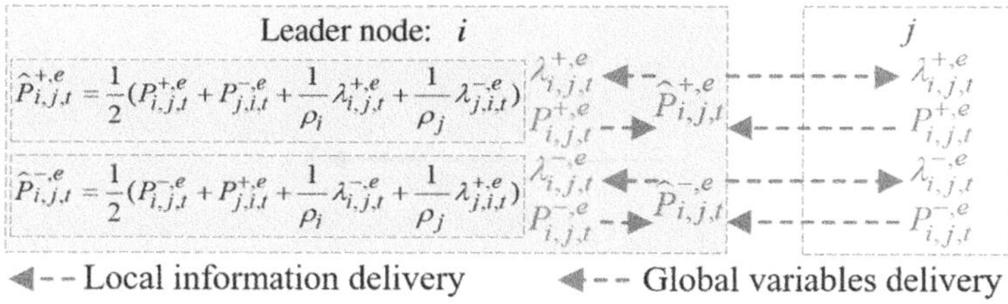

Figure 8.3 Schematic of updates for local electricity trading.

Iterations end when all local copies converge to corresponding global variables, indicating completion of local electricity trading between VPP i and VPP j at step t.

(2) Network Constraints:

Network constraints, categorized into power and heat networks, are investigated. Initially, we examine power network constraints, encompassing Eqs. (8.41)–(8.43). With a power factor of 0.8, pertinent global variables include line active power and node voltage magnitude. Adapting these global constraints for AC-ADMM ensures that local updates comply with constraints.

$$P^{(i)}_{h,i,t} - \sum_{m\in I^d_n} P^{(i)}_{i,m,t} = -p_{i,t} \tag{8.52}$$

$$V^{(i)}_{h,t} - V^{(i)}_{i,t} = r_{h,i}P^{(i)}_{h,i,t} + x_{h,i}Q^{(i)}_{h,i,t} \tag{8.53}$$

Constraints ensure equality between global variables and local copies:

$$P^{(i)}_{h,i,t} - \widehat{P}_{h,i,t} = 0, P^{(i)}_{i,m,t} - \widehat{P}_{i,m,t} = 0 \tag{8.54}$$

$$V^{(i)}_{h,t} - \widehat{V}_{h,t} = 0, V^{(i)}_{i,t} - \widehat{V}_{i,t} = 0 \tag{8.55}$$

Illustrated in Figure 8.4, for line (h,i) and (i,m), node i updates global variables for voltage magnitude ($\widehat{V}_{i,t}$) and the active power flow on line (h,i) ($\widehat{P}_{h,i,t}$). Nodes h and m are designated as follower nodes, while node i acts as the leader node. The information exchange is described in Figure 8.4.

For thermal network constraints, the global coupling constraint (Eq. (8.45)) is transformed into local constraints, similar to power network constraints.

$$\dot{m}^{S(i)}_{h,i,t} - \dot{m}^{S(i)}_{i,m,t} = \dot{m}^{N}_{i,t}, h \in S^-_n, m \in S^+_n \tag{8.56}$$

Constraints ensure equality between global variables and local copies:

$$\dot{m}^{S(i)}_{h,i,t} - \hat{\dot{m}}^{S}_{h,i,t}, \dot{m}^{S(i)}_{i,m,t} = \hat{\dot{m}}^{S}_{i,m,t} \tag{8.57}$$

In Figure 8.5, considering pipeline (h, i) and (i, m), for the update of the global variable $\hat{\dot{m}}^{S}_{h,i,t}$, node h is designated as the follower node, while node i is designated as the leader node. The information exchange is described as follows.

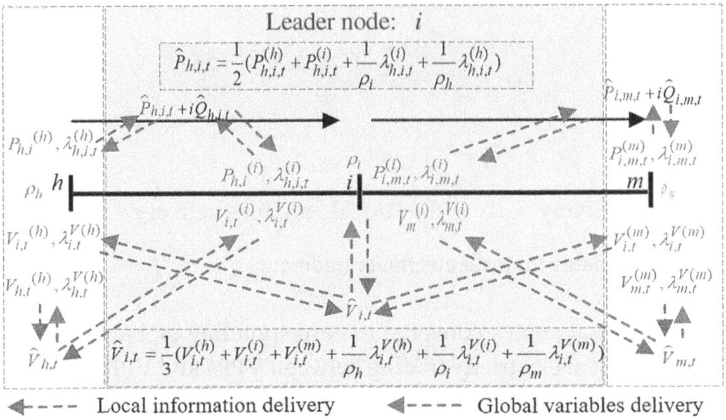

Figure 8.4 Schematic of updates for power network constraints.

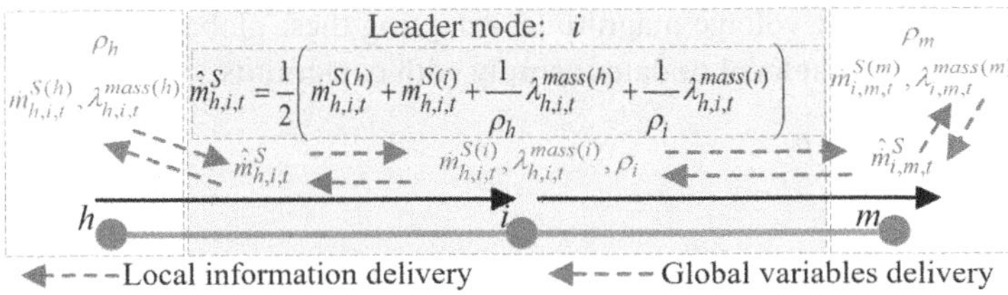

Figure 8.5 Schematic of updates for heat network constraints.

The proposed AC-ADMM treats network constraints and trading constraints similarly. However, due to variations in the entities involved in specific global variables, there are differences in the designation of leader and follower nodes.

Finally, the optimization problem is reformulated to be suitable for AC-ADMM.

$$min \text{ Cost} = \sum_{i \in I^{VPP}} \left(\sum_{t \in T} \left(C^e_{i,t} + C^h_{i,t} + C^g_{i,t} \right) + C^c_i \right)$$

$$s.t. \ (1.6)\text{–}(1.37), (1.40), (1.44), (1.46)\text{–}(1.50), (1.52)\text{–}(1.57) \tag{8.58}$$

Step 1: Solve local optimization problems in parallel at each node. Power network constraints differentiate nodes, as those not connected to VPPs also participate in global network constraints. However, the thermal network connects nodes via heat pipes, treating each node along a heat pipe similarly for thermal network constraint optimization and decomposition. Nodes are categorized into two types: those not connected to VPPs ($i \in I^{NVPP}$) and those connected to VPPs ($i \in I^{VPP}$).

For node $i \in I^{VPP}$, the following local optimization problem is solved:

$$x_i^{k+1} = \arg \min_{x_i} F_i^{VPP}\left(x_i, \lambda^k, \rho_i^k\right)$$
$$\text{s.t. } (1.6)\text{–}(1.37), (1.40), (1.44), (1.46)\text{–}(1.50), (1.52)\text{–}(1.57) \tag{8.59}$$

where, the objective function covers energy costs for purchasing energy and CER, along with penalties and regularization terms for violating global constraints.

For node $i \in I^{NVPP}$, the following local optimization problem is solved:

$$x_i^{k+1} = \arg \min_{x_i} F_i^{NVPP}\left(x_i, \lambda^k, \rho_i^k\right)$$
$$\text{s.t. } (1.44), (1.46)\text{–}(1.50), (1.52)\text{–}(1.55) \tag{8.60}$$

where, the objective function consists of penalties and regularization terms for violating power network constraints.

Step 2 involves the fully distributed update of global variables. For trading global variables, each involves only two nodes, namely the VPP nodes acting as sellers and buyers. When node i is designated as the leader node, it receives local information from follower node j and performs computations as follows:

$$\widehat{P}_{i,j,t}^{+,e,k+1} = \frac{1}{2}\left(P_{i,j,t}^{+,e,k+1} + P_{j,i,t}^{-,e,k+1} + \frac{1}{\rho_i^k}\lambda_{i,j,t}^{+,e,k} + \frac{1}{\rho_j^k}\lambda_{j,i,t}^{-,e,k}\right) \tag{8.61}$$

$$\widehat{P}_{i,j,t}^{-,e,k+1} = \frac{1}{2}\left(P_{i,j,t}^{-,e,k+1} + P_{j,i,t}^{+,e,k+1} + \frac{1}{\rho_i^k}\lambda_{i,j,t}^{-,e,k} + \frac{1}{\rho_j^k}\lambda_{j,i,t}^{+,e,k}\right) \tag{8.62}$$

The same process applies to local heat and CER variables. Node i sends the computed global variable results back to follower node j.

For power network constraints, each line active power segment involves start and end nodes, with the end node as the leader node.

$$\widehat{P}_{h,i,t}^{k+1} = \frac{1}{2}\left(P_{h,i,t}^{(h),k+1} + P_{h,i,t}^{(i),k+1} + \frac{1}{\rho_i^k}\lambda_{h,i,t}^{(i),k} + \frac{1}{\rho_h^k}\lambda_{h,i,t}^{(h),k}\right) \tag{8.63}$$

However, global variables for voltage magnitude involve multiple nodes due to their appearance in constraints related to both upstream and downstream nodes. Therefore, each relevant node retains a local copy of this variable. When node i is the leader, the calculation proceeds as follows:

$$\widehat{V}_{i,t}^{k+1} = \frac{1}{\left(1 + |I_n^u| + |I_n^d|\right)}\left(V_{i,t}^{(i),k+1} + \sum_{h \in I_n^u} V_{i,t}^{(h),k+1} + \sum_{m \in I_n^d} V_{i,t}^{(m),k+1} + \frac{1}{\rho_i^k}\lambda_{i,t}^{V(i)} + \sum_{h \in I_n^u}\frac{1}{\rho_h^k}\lambda_{i,t}^{V(h)} + \sum_{m \in I_n^d}\frac{1}{\rho_m^k}\lambda_{i,t}^{V(m)}\right) \tag{8.64}$$

For thermal network constraints, global variables represent hot water mass flow within pipes. Updating involves the two nodes at each pipe end, with the downstream node as the leader. In pipe (h, i), with global variable $\hat{\dot{m}}^{S}_{h,i,t}$, node i leads:

$$\hat{\dot{m}}^{S}_{h,i,t} = \frac{1}{2}\left(\dot{m}^{S(h)}_{h,i,t} + \dot{m}^{S(i)}_{h,i,t} + \frac{1}{\rho^k_h}\lambda^{mass(h),k}_{h,i,t} + \frac{1}{\rho^k_i}\lambda^{mass(i),k}_{h,i,t}\right) \tag{8.65}$$

Step 3 involves the local update of corresponding dual variables at each node, gradually converging values toward a consensus. Each node stores its respective dual variables locally. For instance, for updating dual variables related to local trading, taking local electricity trading as an example, the process is as follows:

$$\lambda^{+,e,k+1}_{i,j,t} = \lambda^{+,e,k}_{i,j,t} + \rho^k_i\left(P^{+,e,k+1}_{i,j,t} - \hat{P}^{+,e,k+1}_{i,j,t}\right) \tag{8.66}$$

$$\lambda^{-,e,k+1}_{i,j,t} = \lambda^{-,e,k}_{i,j,t} + \rho^k_i\left(P^{-,e,k+1}_{i,j,t} - \hat{P}^{-,e,k+1}_{i,j,t}\right) \tag{8.67}$$

The update of dual variables related to power network constraints:

$$\lambda^{(i),k+1}_{h,i,t} = \lambda^{(i),k}_{h,i,t} + \rho^k_i\left(P^{(i),k+1}_{h,i,t} - \hat{P}^{k+1}_{h,i,t}\right) \tag{8.68}$$

$$\lambda^{V(i),k+1}_{i,t} = \lambda^{V(i),k}_{i,t} + \rho^k_i\left(V^{(i),k+1}_{i,t} - \hat{V}^{k+1}_{i,t}\right) \tag{8.69}$$

Similarly, the update of dual variables related to thermal network constraints:

$$\lambda^{mass(i),k+1}_{h,i,t} = \lambda^{mass(i),k}_{h,i,t} + \rho^k_i\left(\dot{m}^{S(i),k+1}_{h,i,t} - \hat{\dot{m}}^{S,k+1}_{h,i,t}\right) \tag{8.70}$$

Step 4 involves each node i adopting the distributed adaptive penalty parameter adjustment mechanism. Notably, residuals in AC-ADMM differ from those in the conventional ADMM. Residuals can be decomposed into local agents:

$$r^k = \begin{pmatrix} r^k_1 \\ \vdots \\ r^k_N \end{pmatrix}, \quad S^k = \begin{pmatrix} S^k_1 \\ \vdots \\ S^k_N \end{pmatrix} \tag{8.71}$$

$$\left\|r^k_i\right\|_2 = \left\|x^k_i - \hat{x}^k\right\|_2, \quad \left\|s^k_i\right\|_2 = \left\|\rho^k_i(\hat{x}^k - \hat{x}^{k-1})\right\|_2 \tag{8.72}$$

Therefore, each node updates penalty parameters locally in parallel.

$$\rho^{k+1}_i := \begin{cases} 2\rho^k_i & \left\|r^{k+1}_i\right\|_2 > 10\left\|s^{k+1}_i\right\|_2 \\ \rho^k_i/2 & \left\|s^{k+1}_i\right\|_2 > 10\left\|r^{k+1}_i\right\|_2 \\ \rho^k_i & other \end{cases} \tag{8.73}$$

To ensure convergence, after a certain number of iterations, the local penalty parameter values for all nodes will be fixed to a globally uniform constant. The iteration terminates when the following stopping criteria are met:

$$\|r^k\|_2 \le \varepsilon_{pri}, \|s^k\|_2 \le \varepsilon_{dual} \tag{8.74}$$

8.5 Case Studies

8.5.1 Experimental Setup

In this study, a 4-MEVPPs system is implemented, which includes two residential MEVPPs (VPP1 and VPP4), a commercial MEVPP (VPP2), and an industrial MEVPP (VPP3). The topologies of this system in both the heat network and the distribution power network are illustrated in the Figure 8.6. The nodal temperatures on both the supply side and the return side are pre-set and specified.

The equipment configuration, hourly load and renewable energy output are derived from the real data set [38]. Table 8.1 outlines the upstream market prices. Electricity prices are divided into peak, off-peak, and flat rates, while gas and heat prices are fixed. Natural gas is priced at 0.35 RMB/kWh, based on a lower heating value of 9.7 kWh/m^3. The purchase price $\pi^{pen,c}$ of CER from the upstream market is 0.4 RMB/kg, and the selling price $\pi^{sur,c}$ is 0.05 RMB/kg.

The equipment configurations and parameters are provided in Table 8.2. The electricity consumption rate χ^{CCS} and χ^{P2G} are 0.5 and 0.7 kW/kg.

The initial CER allocation coefficients and carbon emission parameters are as shown in Table 8.3. The dispatch periods are 24 hours, while electricity and heat are balanced in real time every hour. CER trading is set to settle once a day for more flexibility because physical execution is not required. The values of α, ε_{pri}, and ε_{dual} are 5%, 0.001, and 0.001, respectively.

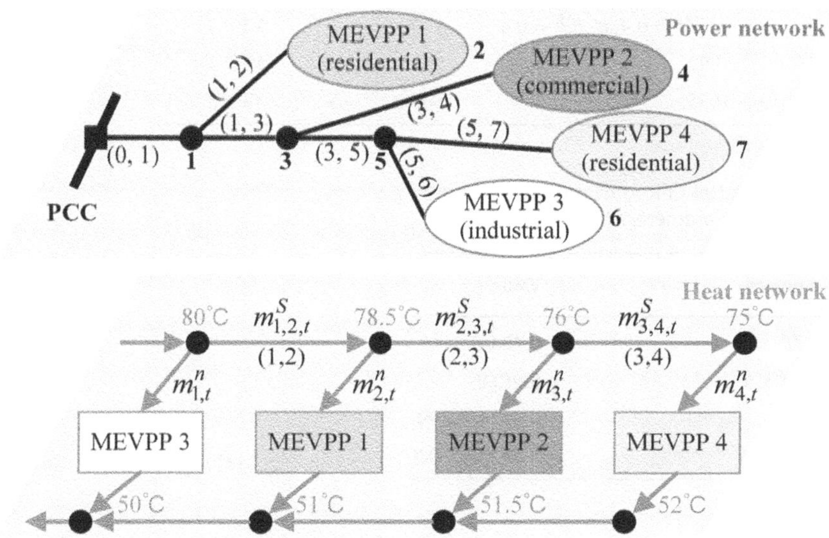

Figure 8.6 Topology of the 4-MEVPPs system.

Table 8.1 Prices of buying electricity, heat, and natural gas from the upstream markets and prices of selling electricity and heat to the upstream markets (RMB/kWh).

	Buying			Selling	
Time period	**Electricity**	**Heat**	**Gas**	**Electricity**	**Heat**
11–13; 18–22	1.16	0.42	0.35	0.23	0.1
1–6; 23–24	0.34			0.15	
7–10; 14–17	0.75			0.07	

Table 8.2 Efficiency and capacities of devices.

Devices	Efficiency	Capacity (kW)	Location
T	0.98	2500	VPP 1,2,3,4
HE	0.98	1000	VPP 1, 4
HE	0.98	2000	VPP 2,3
MT	0.6	800	VPP 1,4
MT	0.6	1000	VPP 2,3
GB	0.7	1000	VPP 1,2,4
HP	3	600	VPP 2
CHP	$\eta^{CHP,e} = 0.4$, $\eta^{CHP,h} = 0.4$	1200	VPP 3
CCS	0.92	500	VPP 3
P2G	0.55	800	VPP 3

Table 8.3 Initial CER Allocation Coefficients and Emission Parameters (kg/kWh).

Parameters	Value
$\sigma^{buy,e} / \beta^{buy,e}$	0.728/0.9419
$\sigma^{buy,h} / \beta^{buy,h}$	0.6/0.9
σ^{h}, σ^{e}	0.2124, 0.392
$\beta^{CHP}, \beta^{GB}, \beta^{MT}$	0.36, 0.65, 0.8

Various scenarios were scrutinized to assess the efficacy of electricity, heat, and CER trading in coordinating MEVPPs, while also appraising the indispensable role of network constraints. Optimization models were formulated and solved employing Gurobi in Python, executed on a computing system equipped with a 10-core 3.70 GHz Intel(R) Core (TM) i9-10900K processor and 128 GB of RAM.

8.5.2 Impact of Transactive Trading

Scenario 1 (S1): MEVPPs trade energy and CER with upstream markets in silo.

Scenario 2 (S2): MEVPPs engage in local trading of electricity, heat, and CER.

Scenario 3 (S3): MEVPPs engage in transactive trading of electricity and heat, while CER is traded with upstream markets.

Scenario 4 (S4): MEVPPs engage in transactive trading of electricity and CER, while heat is traded with upstream markets.

Table 8.4 details costs across four scenarios. Comparison between S1 and S2 shows a local market reduces overall MEVPP costs by about 7.12%, illustrating how it leverages MEVPP complementarity. After introducing local trading, costs significantly decreased for VPP1, 2, and 4, with reductions of 25.27%, 5.49%,

Table 8.4 Comparison of costs for Scenarios 1, 2, 3, and 4.

Scenario	MEVPP	Cost (RMB)	Total cost (RMB)
S1	1	32,448	133,894
	2	31,346	
	3	37,652	
	4	32,447	
S2	1	25,903	124,999
	2	29,716	
	3	43,620	
	4	25,760	
S3	1	28,161	125,239
	2	29,392	
	3	40,004	
	4	27,680	
S4	1	33,323	127,660
	2	26,801	
	3	35,745	
	4	31,791	

Table 8.5 Comparison of CER trading between S1 and S2.

		Buying CER (kg)		Selling CER (kg)	
Scenario	MEVPP	Upstream	Local	Upstream	Local
1	1	12,644	0	0	0
	2	12,317	0	0	0
	3	0	0	0	0
	4	12,644	0	0	0
2	1	10,743	0	0	0
	2	8317	4755	0	0
	3	0	0	0	4755
	4	10,950	0	0	0

and 25.96%, respectively. Conversely, the cost for the industrial MEVPP3 rose by 13.68%, due to the optimization goal focusing on minimizing total costs, rather than individual costs. It is clear that a local electricity, heat, and CER market enables heterogeneous MEVPPs to lessen their reliance on upstream markets by leveraging their mutual energy complementarities.

Further, Table 8.5 illustrates that the introduction of local markets significantly reduced CER purchases from upstream markets across all MEVPPs. Notably, VPP2 saw a reduction of 4 tons in CER purchases in S2 compared to S1, representing a 31.64% decrease. Additionally, VPP 3 employed coupled CHP-P2G-CCS systems in both scenarios to mitigate costs associated with external CER acquisitions.

8.5.2.1 Impact of Transactive CER Trading

Based on the cost comparisons in Table 8.4, it is observed that without local CER trading, the total cost would increase by approximately 0.2%. The introduction of local CER market leads to cost reductions for both commercial MEVPP2 and industrial MEVPP 3, with VPP 3 experiencing a more substantial decrease of 8.98%.

Table 8.6 shows that the local CER market reduced upstream CER purchases by 31 kg. Industrial VPP3, with unique CHP-CCS-P2G systems, sells surplus CER, offsetting costs. In contrast, without a local CER market, VPP3 regulates CHP output to match initial free CER, avoiding added costs.

Figure 8.7 illustrates CER trading in S2 and S3 with six statuses: initial CER, local trades, upstream purchases, total emissions, and CCS capture. In S2, VPP3 sells surplus CER locally, while VPP2 buys, showing the advantages of the local market.

Table 8.6 Initial allocated CER and total emissions for each VPP in S2 and S3 (kg).

	MEVPP	Initial CER	Carbon emission	From upstream
S2	1	24,029	34,772	30,011
	2	28,495	41,569	
	3	38,797	34,041	
	4	24,491	35,442	
S3	1	17,963	27,019	30,042
	2	25,091	37,085	
	3	54,827	54,827	
	4	17,880	26,872	

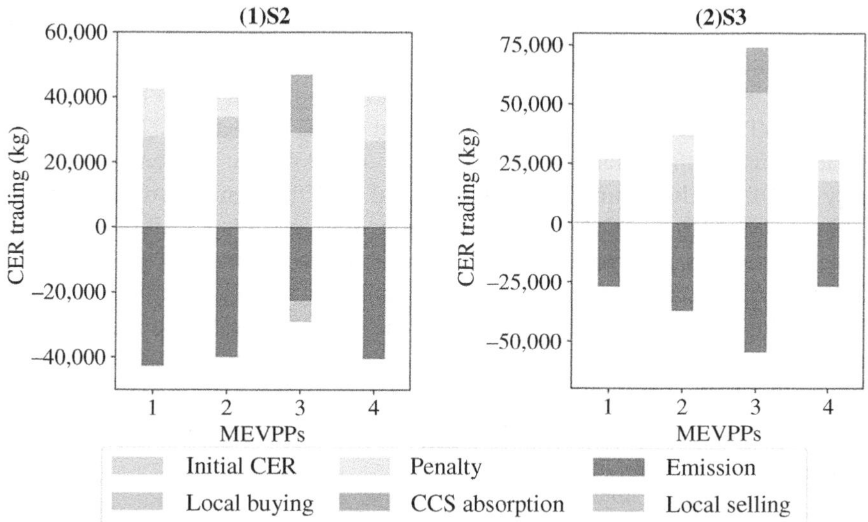

Figure 8.7 (1) CER trading in S2; (2) CER trading in S3.

The local CER market significantly influences industrial MEVPP3's optimization behavior. As depicted in Figure 8.8, equipped with CHP-CCS-P2G, VPP3 avoids selling surplus CERs at low upstream prices. In S2, it sells locally, while in S3, it buys more upstream electricity to sell locally, using CERs to offset emissions. Despite being less advantageous than direct sales, VPP3 minimizes costs, emphasizing its role in low-carbon transformations. The local market enhances decision-making flexibility, promoting economic and environmental benefits within the MEVPP cluster.

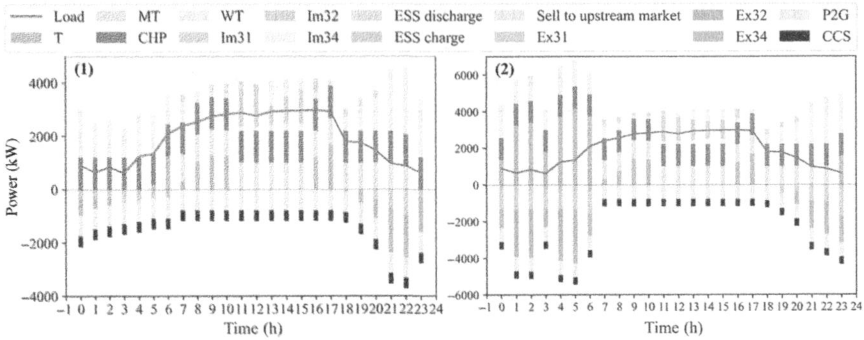

Figure 8.8 The breakdown of electricity demand–supply balance for VPP3 in (1) S2; (2) S3.

8.5.2.2 Impact of Transactive Heat Trading

Table 8.4 reveals that local heat trading in S2 and S4 reduces the total costs by approximately 2.13%. Residential MEVPP1 and MEVPP4 experience cost reductions of 9.09% and 7.03%, respectively, whereas Commercial MEVPP2 and Industrial MEVPP3 see cost increases of 4.44% and 2.62%, respectively. This variation is attributed to the differing heat production technologies: residential MEVPPs rely on GB, while commercial and industrial MEVPPs benefit from the higher efficiencies of HP and CHP. The local heat market prompts a shift toward these efficient technologies, elevating individual costs for the commercial and industrial MEVPPs but decreasing overall costs due to enhanced energy efficiency.

Combining the analysis with Figure 8.9, the impact of local heat trading on the heat load satisfaction within VPP2 is examined. It is evident that due to the high efficiency of HP, in both S2 and S4, the HP in VPP 2 is operating at full capacity. However, the key difference lies in the management of surplus heat. In S4, VPP 2

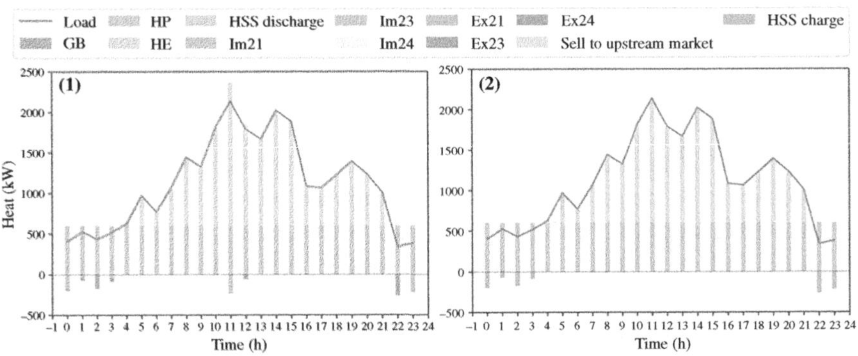

Figure 8.9 The breakdown of heat demand–supply balance for VPP2 in (1) S2; (2) S4.

opts to store excess heat generated by the HP in the HSS, which is later released during peak heat demand periods to reduce costs. In S2, VPP 2 chooses to sell this surplus heat in the local market to achieve higher profits.

The local heat market allows VPPs with lower heating costs to transfer heat to those with higher costs, reducing overall expenses. Overall, introducing a local electricity-heat-carbon market significantly lowers overall costs and carbon emissions of interconnected VPPs, leveraging their energy complementarity and flexibility.

8.5.3 Impact of Network Constraints

This section uses Scenario 2 as the baseline and introduces two new scenarios.

S5: Identical to S2, but considers power network constraints.

S6: Identical to S2, but considers both heat and power network constraints.

Table 8.7 compares total costs of three scenarios. S2, with no network constraints, has the lowest costs. However, it may lead to some transactions and energy conversions being physically inexecutable due to safety considerations. Introducing power network constraints in S5 increases costs by 2.58% compared to S2. Adding heat network constraints in S6 doesn't significantly alter costs, indicating power network constraints mainly restrict local trading in this study.

Figures 8.10 and 8.11 depict the active power and nodal voltage with (S5) and without (S2) power network constraints. After applying network constraints, nodal voltage magnitudes and line active power stability improve. From Figure 8.10(1), during 4 : 00-7 : 00, line (0,1) shows high positive active power, indicating significant electricity purchases from the upstream market. This demand arises due to low renewable energy production and high electricity needs during off-peak price hours, making upstream purchases economical. During 16 : 00-20 : 00, lines (0,1) and (1,3) exhibit increased active power corresponding to peak demand times and high upstream prices, promoting natural gas to power conversions. With network constraints, power flow is moderated for grid safety, leading to costlier but reliable load fulfillment methods. Figure 8.11 indicates similar nodal voltage stability efforts.

Table 8.7 Comparison of costs for Scenarios 2, 5, and 6.

Scenario	Total cost (RMB)
S2	124,999
S5	128,230
S6	128,231

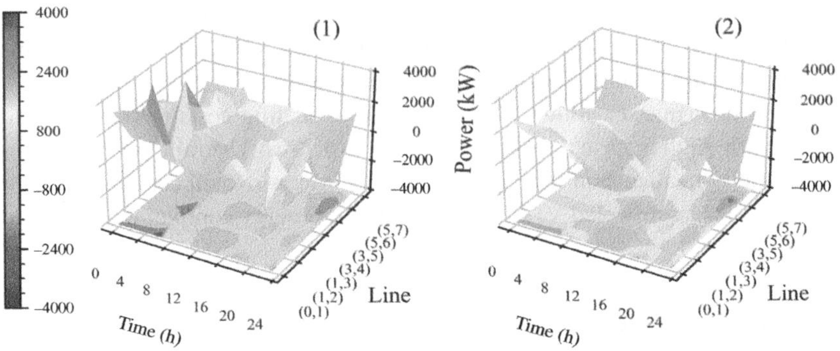

Figure 8.10 Variation of line active power throughout the day in (1) S2; (2) S5.

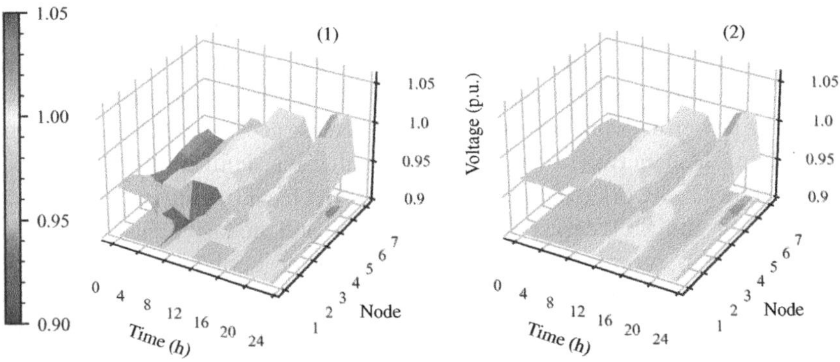

Figure 8.11 Variation of nodal voltage magnitudes throughout the day in (1) S2;(2) S5.

8.5.4 Convergence Analysis

ADMM provides enhanced privacy protection and better alignment with local transactive markets compared to centralized methods, ensuring optimal solutions to optimization problems. C-ADMM differs from traditional ADMM by employing parallel updates, saving significant time. In C-ADMM, each node stores local copies of global variables, eliminating the sequential transmission typical of ADMM. AC-ADMM, an extension, adopts a leader-follower node mechanism, enhancing VPP privacy without a central authority. Additionally, adaptive penalty parameter tuning reduces reliance on pre-selected parameters, speeding up convergence. The comparison of the three algorithms is shown in Table 8.8.

The convergence behavior of the proposed AC-ADMM is compared with that of C-ADMM using two different fixed penalty parameter values in Figure 8.12.

Table 8.8 Iterations and Computation Times for ADMM, C-ADMM, and AC-ADMM.

Algorithm	Iterations	Computation time (s)
ADMM ($\rho = 0.01$)	187	973
C-ADMM ($\rho = 0.01$)	129	525
AC-ADMM	78	309

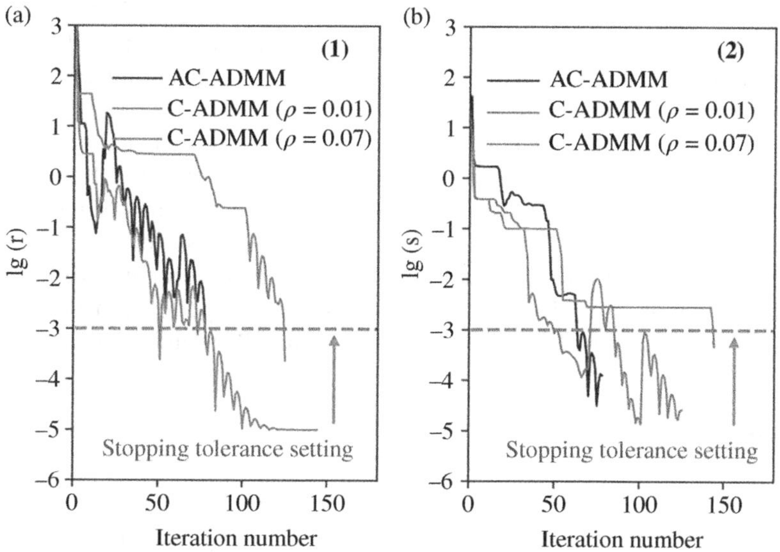

Figure 8.12 Comparison of logarithmic (a) primal and (b) dual residual.

ε_{pri} and ε_{dual} are set at 0.001. The iteration process only terminates when both the primal residual and the dual residual satisfy the stopping criteria. According to Figure 8.12(1), the primal residual of C-ADMM with a penalty factor of 0.07 quickly decreases below the stopping line. However, in Figure 8.12(2), the dual residual requires 144 iterations to fall below the stopping line. Conversely, for C-ADMM with a penalty factor of 0.01, the dual residual drops below the stopping line after about 50 iterations, but the slower decrease in primal residual leads to a total of 129 iterations. Overall, since AC-ADMM employs an adaptive penalty parameter tuning mechanism that maintains a consistent ratio between primal and dual residuals within a certain range, it achieves a more balanced reduction in both residuals. Consequently, AC-ADMM converges by the 78th iteration.

8.6 Conclusions

In this study, a comprehensive optimization framework is introduced to coordinate electricity, heat, and CER trading among heterogeneous networked MEVPPs, considering both power and thermal network constraints. The framework integrates the heterogeneity of MEVPPs, which enhances energy shifting flexibility, P2P transactive energy in various forms, and CER trading, providing a multidimensional analysis from economic and environmental perspectives while ensuring compliance with both power and thermal network constraints.

Additionally, an efficient, fully distributed AC-ADMM method is developed to ensure privacy and transparency. This method leverages C-ADMM's parallel computation, integrates a leader-follower node mechanism, and adopts adaptive penalty parameter tuning to expedite convergence and enhance its suitability for distributed learning frameworks.

Case studies demonstrate the effectiveness of the proposed method, highlighting economic benefits for MEVPPs through coordinated energy and CER trading within local markets as opposed to isolated trading with upstream energy markets, all while adhering to power and thermal network constraints. The advantages of using diverse energy forms through efficient coordination among MEVPPs are confirmed. Last, sensitivity analysis validates the superior convergence performance of the proposed AC-ADMM compared to ADMM and C-ADMM.

References

1 Hirsch, A., Parag, Y., and Guerrerxo, J. (2018). Microgrids: a review of technologies, key drivers, and outstanding issues. *Renewable and Sustainable Energy Reviews* 90: 402–411. https://doi.org/10.1016/j.rser.2018.03.003.

2 Shafiekhani, M., Ahmadi, A., Homaee, O. et al. (2022). Optimal bidding strategy of a renewable-based virtual power plant including wind and solar units and dispatchable loads. *Energy* 239: 122379. https://doi.org/10.1016/j.energy.2022.122379.

3 Dall'Anese, E., Mancarella, P., and Monti, A. (2017). Unlocking flexibility: integrated optimization and control of multienergy systems. *IEEE Power and Energy Magazine* 15: 43–52. https://doi.org/10.1109/MPE.2017.2663121.

4 Liu, J., Ma, L., and Wang, Q. (2023). Energy management method of integrated energy system based on collaborative optimization of distributed flexible resources. *Energy* 264: 125981. https://doi.org/10.1016/j.energy.2022.125981.

5 Kong, X., Xiao, J., Liu, D. et al. (2020). Robust stochastic optimal dispatching method of multi-energy virtual power plant considering multiple uncertainties. *Applied Energy* 279: https://doi.org/10.1016/j.apenergy.2020.115707.

6 Calvo-Gallardo, E., Arranz, N., and de Arroyabe, J.C.F. (2021). Analysis of the European energy innovation system: contribution of the framework programmes to the EU policy objectives. *Journal of Cleaner Production* 298: 126690. https://doi.org/10.1016/j.jclepro.2021.126690.

7 LocalRes (2023). Mevpp. [Online]. Available: https://www.localres.eu/tools-and-resources/mevpp/.

8 Adeyemi, A., Yan, M., Shahidehpour, M. et al. (2020). Transactive energy markets for managing energy exchanges in power distribution systems. *The Electricity Journal* 33: 106868. https://doi.org/10.1016/j.tej.2020.106868.

9 Strbac, G., Papadaskalopoulos, D., Chrysanthopoulos, N. et al. (2021). Decarbonization of electricity systems in Europe: market design challenges. *IEEE Power and Energy Magazine* 19: 53–63. https://doi.org/10.1109/MPE.2020.3010812.

10 Alabi, T.M., Lu, L., and Yang, Z. (2022). Data-driven optimal scheduling of multi-energy system virtual power plant (MEVPP) incorporating carbon capture system (CCS), electric vehicle flexibility, and clean energy marketer (CEM) strategy. *Applied Energy* 314: 118997. https://doi.org/10.1016/j.apenergy.2022.118997.

11 Soares, J., Ghazvini, M.A.F., Vale, Z., and de Moura Oliveira, P. (2016). A multi-objective model for the day-ahead energy resource scheduling of a smart grid with high penetration of sensitive loads. *Applied Energy* 162: 1074–1088. https://doi.org/10.1016/j.apenergy.2015.10.099.

12 Liu, Y., Jiang, C., Tan, S. et al. (2016). Optimal dispatch of virtual power plant considering risk-adjusted return on capital constraints. *Proc. CSEE* 36 (17): 4617–4627. https://doi.org/10.13334/j.0258-8013.pcsee.2016.17.004.

13 Ye, Y., Qiu, D., Wu, X. et al. (2020). Model-free real-time autonomous control for a residential multi-energy system using deep reinforcement learning. *IEEE Transactions on Smart Grid* 11: 3068–3082. https://doi.org/10.1109/TSG.2020.2983356.

14 Mashayekh, S., Stadler, M., Cardoso, G., and Heleno, M. (2017). A mixed integer linear programming approach for optimal DER portfolio, sizing, and placement in multi-energy microgrids. *Applied Energy* 187: 154–168. https://doi.org/10.1016/j.apenergy.2016.11.051.

15 Li, J., Ye, Y., Papadaskalopoulos, D., and Strbac, G. (2021). Distributed consensus-based coordination of flexible demand and energy storage resources. *IEEE Transactions on Power Systems* 36: 3053–3069. https://doi.org/10.1109/TPWRS.2020.3004013.

16 Ye, Y., Tang, Y., Wang, H. et al. (2021). A scalable privacy-preserving multi-agent deep reinforcement learning approach for large-scale peer-to-peer transactive energy trading. *IEEE Transactions on Smart Grid* 12: 5185–5200. https://doi.org/10.1109/TSG.2020.3045793.

17 Chen, T., Bu, S., Liu, X. et al. (2021). Peer-to-peer energy trading and energy conversion in interconnected multi-energy microgrids using multi-agent deep reinforcement learning. *IEEE Transactions on Smart Grid* 13: 715–727. https://doi.org/10.1109/TSG.2021.3089836.

18 Siqin, Z., Niu, D., Li, M. et al. (2022). Distributionally robust dispatching of multi-community integrated energy system considering energy sharing and profit allocation. *Applied Energy* 321: 119202. https://doi.org/10.1016/j.apenergy.2022.119202.

19 Wang, X., Liu, Y., Liu, C., and Liu, J. (2020). Coordinating energy management for multiple energy hubs: from a transaction perspective. *International Journal of Electrical Power & Energy Systems* 121: 106060. https://doi.org/10.1016/j.ijepes.2020.106060.

20 Xu, D., Zhou, B., Liu, N. et al. (2020). Peer-to-peer multienergy and communication resource trading for interconnected microgrids. *IEEE Transactions on Industrial Informatics* 17: 2522–2533. https://doi.org/10.1109/TII.2020.2967163.

21 Elsied, M., Oukaour, A., Gualous, H., and Hassan, R. (2015). Energy management and optimization in microgrid system based on green energy. *Energy* 84: 139–151. https://doi.org/10.1016/j.energy.2015.03.108.

22 Zhou, X., Ma, Z., Zou, S., and Zhang, J. (2022). Consensus-based distributed economic dispatch for multi micro energy grid systems under coupled carbon emissions. *Applied Energy* 324: 119641. https://doi.org/10.1016/j.apenergy.2022.119641.

23 Khorasany, M., Najafi-Ghalelou, A., Razzaghi, R., and Mohammadi Ivatloo, B. (2021). Transactive energy framework for optimal energy management of multi-carrier energy hubs under local electrical, thermal, and cooling market constraints. *International Journal of Electrical Power & Energy Systems* 129: 106803. https://doi.org/10.1016/j.ijepes.2021.107297.

24 Zhang, F., Fang, H., and Song, W. (2019). Carbon market maturity analysis with an integrated multi-criteria decision-making method: a case study of EU and China. *Journal of Cleaner Production* 241: 118296. https://doi.org/10.1016/j.jclepro.2019.118296.

25 Mu, C., Ding, T., Zhu, S. et al. (2022). A decentralized market model for a microgrid with carbon emission rights. *IEEE Transactions on Smart Grid* 14 (2): 1388–1402. https://doi.org/10.1109/TSG.2022.3153421.

26 Hua, W., Jiang, J., Sun, H., and Wu, J. (2020). A blockchain-based peer-to-peer trading framework integrating energy and carbon markets. *Applied Energy* 279: 115539. https://doi.org/10.1016/j.apenergy.2020.115539.

27 Lu, Y., Li, Y., Tang, X. et al. (2022). A blockchain-enabled smart emission cap restrictive and carbon permit trading system. *Applied Energy* 313: 118787. https://doi.org/10.1016/j.apenergy.2022.118787.

28 Khaqqi, K.N., Sikorski, J.J., Hadinoto, K., and Kraft, M. (2018). Incorporating seller/buyer reputation-based system in blockchain-enabled emission trading application. *Applied Energy* 209: 8–19. https://doi.org/10.1016/j.apenergy.2017.10.058.

29 Boyd, S., Parikh, N., Chu, E. et al. (2011). Distributed optimization and statistical learning via the alternating direction method of multipliers. *Foundations and Trends in Machine learning* 3: 1–122. https://doi.org/10.1561/2200000016.

30 Najafi, A., Pourakbari-Kasmaei, M., Jasinski, M. et al. (2022). The role of EV-based peer-to-peer transactive energy hubs in distribution network optimization. *Applied Energy* 319: 119267. https://doi.org/10.1016/j.apenergy.2022.119267.

31 Wei, C., Shen, Z., Xiao, D. et al. (2021). An optimal scheduling strategy for peer-to-peer trading in interconnected microgrids based on RO and Nash bargaining. *Applied Energy* 295: 117024. https://doi.org/10.1016/j.apenergy.2021.117024.

32 Mhanna, S., Verbič, G., and Chapman, A.C. (2019). Adaptive ADMM for distributed AC optimal power flow. *IEEE Transactions on Power Systems* 34 (3): 2025–2035. https://doi.org/10.1109/TPWRS.2018.2876734.

33 Xu, Z., Mario, F., and Tom, G. (2017). Adaptive ADMM with spectral penalty parameter selection. In: *Artificial Intelligence and Statistics*, 718–727. PMLR.

34 Zeng, S., Kody, A., Kim, Y. et al. (2022). A reinforcement learning approach to parameter selection for distributed optimal power flow. *Electric Power Systems Research* 212: 108546. https://doi.org/10.1016/j.epsr.2022.108546.

35 Mateos, G., Bazerque, J.A., and Giannakis, G.B. (2010). Distributed sparse linear regression. *IEEE Transactions on Signal Processing* 58: 5262–5276. https://doi.org/10.1109/TSP.2010.2053028.

36 Zhong, X., Zhong, W., Liu, Y. et al. (2022). Optimal energy management for multi-energy multi-microgrid networks considering carbon emission limitations. *Energy* 246: 123428. https://doi.org/10.1016/j.energy.2022.123428.

37 Frölke, L., Sousa, T., and Pinson, P. (2022). A network-aware market mechanism for decentralized district heating systems. *Applied Energy* 306: 117956. https://doi.org/10.1016/j.apenergy.2022.117956.

38 Chen, P., Ye, Y., Wang, H. et al. (2024). Holistic coordination of transactive energy and carbon emission right trading for heterogeneous networked multi-energy microgrids: a fully distributed adaptive consensus ADMM approach. *Sustainable Energy Technologies and Assessments* 64: 103729. https://doi.org/10.1016/j.seta.2024.103729.

9

Cluster-Based Heuristic Algorithm for Collection System Topology Generation of a Large-Scale Offshore Wind Farm

Jincheng Li[1,2], Zhengxun Guo[1,2], and Xiaoshun Zhang[1,2]

[1] *Foshan Graduate School of Innovation, Northeastern University, Guangdong, Foshan, China*
[2] *College of Information Science and Engineering, Northeastern University, Liaoning, Shenyang, China*

9.1 Introduction

Over the past decade, wind energy has increasingly attracted attention, becoming a hot topic in both industry and academia [1]. Offshore wind energy, in particular, is undergoing a significant growth and expansion phase. This surge is fueled by technological advancements, decreasing costs, and supportive policy frameworks worldwide. According to projections from the International Energy Agency (IEA), renewable energy sources are expected to contribute 98% of the 2518 Terawatt hours (TWh) of new electricity generation capacity slated for global addition between 2022 and 2025. Within this renewable energy spectrum, wind energy, especially offshore wind, is poised to become a leading force.

9.1.1 Background and Importance

The excessive use of fossil fuels has severely harmed ecosystems and significantly accelerated climate change [2]. Recently, sustainability has become a critical issue in global energy discussions. Wind energy has notably risen as a key renewable source, contributing significantly to electricity production worldwide.

Data from the US Department of Energy's Wind Energy Technologies Office (WETO) [3] suggests an immense potential for wind power within the United States, including its federal waters and the Great Lakes. Estimates indicate the presence of over 4000 Gigawatts (GWs) of wind capacity, capable of producing around 13,500 TWh of electricity annually. Similarly, China, with its extensive coastline of over 18,000 km and maritime domain exceeding 3 million square kilometers, stands out for its exceptional offshore wind resource potential [4].

Artificial Intelligence Empowered Smart Energy Systems, First Edition. Edited by Qiang Yang and Gang Huang.

This potential positions China as a global leader in the wind energy market. The sector has witnessed significant growth, with record-breaking increases in offshore wind capacity. In the year 2022 alone, the global offshore wind sector saw an addition of 27.7 GW in grid-connected installations [5]. Forecasts predict that by the year 2030, the cumulative installed capacity of offshore wind could surpass 65 GW [6]. These developments underscore the pivotal role of wind energy in transitioning toward a more sustainable and environmentally friendly global energy framework.

The wind power industry is marked by its dynamic growth, driven by policy support, technological innovation, and market expansion [7]. These factors together create a conducive environment for its advancement. For identifying potential sites and formulating development plans for Offshore Wind Farms (OWFs), understanding the spatial aspects and technical viability of offshore wind energy is essential [8]. The development dynamics and challenges differ notably between onshore and offshore wind power. OWFs offer distinct advantages over their onshore counterparts, such as higher resource density, greater site selection flexibility, more consistent wind directions, and lower turbulence, which collectively enhance operational efficiency and energy production potential [9, 10].

9.1.2 Research Status

To lower the construction expenses of OWF, extensive research has been conducted on optimizing the internal electrical and submarine cable connection system [11]. The capital cost of electrical infrastructure accounts for 15–30% of the total initial cost of an OWF [12]. Specifically, the Collection System Topology (CST) is identified as a crucial component of the electrical infrastructure within an OWF, underscoring its role in optimizing the overall efficiency and cost-effectiveness of the wind farm's operation [13].

9.1.2.1 Radial Topologys Optimization Methods

Radial topology is recognized as a more cost-effective approach than ring topology for connecting wind turbines in OWFs, primarily due to its requirement for shorter cable lengths without the need for any loops. Addressing this optimization challenge, the most favored methodologies have been developed based on principles such as the Minimum Spanning Tree (MST), Genetic Algorithm (GA), Fuzzy C-Means (FCM), K-means clustering, and Particle Swarm Optimization (PSO). These techniques are celebrated for their ability to find efficient solutions that minimize cable lengths and reduce costs. To further enhance the benefits offered by these methods, several innovative approaches have been introduced through the integration of these techniques. Examples of such hybrid methods include combinations of GA and MST, MST and FCM, and K-means clustering and

MST [14]. These integrated approaches leverage the strengths of each individual technique to achieve more refined and economically viable solutions for the electrical infrastructure of OWFs.

9.1.2.2 Graph Theory-Based Methods

Graph theory-based approaches are utilized to create the topology using principles from graph theory. The most commonly applied techniques are the MST and the Dynamic MST (DMST), which are instrumental in designing the CST. MST is effective in minimizing the total cable length required for connections but may not always achieve the lowest capital costs due to varying prices among different cables. DMST addresses this issue by incorporating the cost variations of cables into its calculations [15]. Both MST and DMST can quickly provide viable CST designs, yet they may result in elevated capital expenses for a large-scale OWF (LSOWF) project unless integrated with broader global optimization strategies.

9.1.2.3 Meta-Heuristic Optimization Methods

Meta-heuristic optimization techniques offer flexible solutions for the optimal design of the CST, due to their adaptability to various optimization mathematical models. [16–18] include several variations of GA, which have been specifically tailored for CST design. Despite these methods having the ability to handle high-dimensional optimization tasks for LSOWF, these GAs can encounter issues such as premature convergence and failure to meet cross-constraints [17].

For instance, an enhanced GA [18] was developed to tackle the CST design by framing the optimization challenge as a classical multiple Traveling Salesman Problem (TSP), yet it struggled with cross-constraint [18] compliance in large-scale OWFs. A binary GA [15] approach was utilized to optimally categorize all wind turbines (WTs), subsequently determining each group's connected topology via the MST method. Moreover, combining DMST with GA has been shown to improve CST's optimum quality, thereby reducing capital costs further [19].

The optimization efficiency of GAs can be enhanced by integrating them with other meta-heuristic algorithms, such as merging GA's cross-operators with the vaccine extraction technique from the Immune Algorithm (IA) [20]. PSO, another meta-heuristic method, offers advantages in global search capabilities and parallel processing. Variations of PSO, including Binary PSO (BPSO) [21] and Adaptive PSO (APSO) [22], have been proposed for CST optimization, demonstrating their ability to achieve lower capital costs and quicker convergence times compared to other algorithms like the Non-Dominated Sorting GA-III (NSGA-III).

In the context of CST's optimal design, which shares similarities with the TSP, Ant Colony Optimization (ACO) has been applied, considering various constraints [23]. To enhance optimization efficiency, MST was integrated with ACO [24], significantly reducing the dimensionality of optimization tasks.

This multifaceted approach, combining meta-heuristic methods with classical optimization strategies, underscores the complexity and innovation in optimizing CST design for OWFs.

9.2 Mathematical Model for CS Optimization in LSOWFs

This section begins by detailing the components of LSOWFs. Subsequently, it draws an analogy between the optimization of CS in LSOWFs and the MST problem within graph theory. Furthermore, an in-depth explanation of the cost model pertinent to LSOWFs is provided. The section concludes by outlining the constraints associated with the CS in LSOWFs.

9.2.1 Composition of the LSOWF

An LSOWF principally comprises four key components: the Power CS, Offshore Substations (OSSs), the CS, and onshore substations. Figure 9.1 depicts these components.

In LSOWFs, WTs are fundamental in harvesting wind energy and converting it into electrical power, which is then conveyed to the OSS via the CS [25]. The planning and design of the OWF's CS are of utmost importance, as adjustments in the cabling topology can lead to significant reductions in the CS's construction costs. Broadly, the CS is divided into two configurations: single-sided ring topology and double-sided ring topology, as explored in reference [26]. Although the double-sided ring topology is favored for its higher reliability, it is associated with increased construction costs. Consequently, the single-sided ring topology is utilized to minimize construction expenses. Furthermore, the effective optimization of the CS is pivotal in decreasing the overall construction costs of the LSOWF.

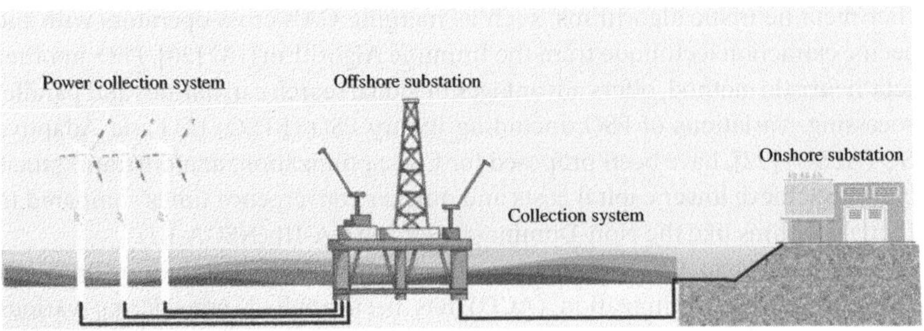

Figure 9.1 Components of the LSOWF.

9.2.2 Objective Function

In this study, the formulation of the objective function for optimizing CST design emphasizes the reduction of overall capital costs. Such costs are comprehensive, encompassing the expenses related to the procurement, shipment, and installation of submarine cables, as well as considering the potential costs associated with system failures. Accordingly, the objective function is articulated as follows [11]:

$$\min f = C_\text{p} + C_\text{s} + C_\text{f} \tag{9.1}$$

Herein, f delineates the aggregate capital cost targeted for minimization. C_p specifies the purchase cost of the submarine cables. C_s denotes the total expenses for transporting and setting up the submarine cables, it is posited to be directly proportional to the total length of the undersea cables utilized. Lastly, C_f is the cost related to any failures in the CS, highlighting the financial impact of potential downtimes or inefficiencies.

The purchase cost C_p of submarine cables is influenced by both the type and the length of the cables used. Typically, the type of submarine cables is chosen based on the transmission capacity needed for the WTs they connect. In an OWF, the total purchase cost can be determined by the following formula [19]:

$$C_\text{p} = \sum_{i=1}^{N_\text{f}} \sum_{j=1}^{n_i} L_i^j C_i^j, i = 0, 1, 2, \ldots, N_\text{f};\ \ j = 0, 1, 2, \ldots, n_i \tag{9.2}$$

In this equation, N_f stands for the total number of feeders that connect to the OSS. The term n_i indicates the number of segments within the ith feeder. L_i^j represents the length of the jth segment in the ith feeder, and C_i^j is the cost coefficient for the jth segment in the ith feeder. This way of calculating C_p allows us to account for the various types and lengths of cables needed throughout the OWF, ensuring that the purchase cost reflects both the physical characteristics of the cables and the specific requirements of the OWF's design.

The failure opportunity cost C_f is essentially the cost incurred when a submarine cable failure prevents WTs from generating electricity. This can be mathematically expressed as [27]:

$$C_\text{f} = \pi K_\text{rw} E_\text{e} \tag{9.3}$$

Here, K_rw stands for the reliability weight of the CST, influenced by factors such as the wake effect and ambient temperature. It's calculated by the equation:

$$K_\text{rw} = \left(\lambda T \sum_{v=0}^{V_d} P_v F_v \right) \Big/ P_{WT}, V_d = 0, 1, 2, \ldots, d \tag{9.4}$$

The Expected Energy Not Supplied (EENS) of the OWF, denoted as E_e, represents the total power outage impact due to CST failures within a year, assuming normal operation of WTs throughout. It is determined by:

$$E_e = P_{OWF}E_a \tag{9.5}$$

The annual availability loss, E_a is calculated as:

$$E_a = 8760\left(1 - \frac{E_x}{P_{OWF}}\right) \tag{9.6}$$

In these equations, λ signifies the efficiency coefficient, which for optimally designed wind farms typically ranges between 0.9 and 0.95. The term d represents different wind speeds, while V_d indicates the distribution of these speeds. P_v and F_v are the output power of a WT and the frequency of the wind speed, respectively. P_{OWF} denotes the total installed capacity of the OWF, and E_a calculates the total hours the OWF is nonoperational due to cable failures.

E_x represents the expected value of the output power for the topology considering the failure of CST. The E_x can be written as:

$$E_x = \left(1 - \prod_{k=0}^{N}(1 - q_L)\right)\left(\sum_{l=0}^{N}(N - l)P_{WT}C_N^l q^l(1 - q)^{N-l}\right) \tag{9.7}$$

where, N represents the number of WTs; q_L represents the failure rate of submarine cables, and q denotes the failure rate of WTs.

9.2.3 Constraints

In designing the CST for OWFs, certain constraints must be considered, particularly regarding the selection and connection of submarine cables to WTs. These constraints ensure the system's efficiency, reliability, and capacity to meet energy production goals. They are as follows:

9.2.3.1 Constraint on the Number of WTs per Feeder

The number of WTs connected by each feeder is subject to two main limitations. First, it's constrained by the number of feeders that the OSS can accommodate, ensuring each feeder connects more than a minimum required number of WTs. Second, the capacity of WTs each feeder can support is limited by the maximum transmission capacity of the submarine cables used. This can be mathematically expressed as [28]:

$$\begin{cases} n^{\min} \leq n_i \leq n^{\max} \\ n^{\min} = \frac{P_t}{n^{\max} P_{WT}} \\ n^{\max} = \frac{S_C}{P_{WT}} \\ S_C = \sqrt{3}UI_i^{\max}\cos\varphi \end{cases} \tag{9.8}$$

Here, n^{min} and n^{max} are the minimum and maximum number of WTs that can be connected by each feeder. P_t represents the rated power of the substation, and P_{WT} denotes the rated power of each WT. S_C is the maximum transmission capacity of the submarine cables, where U is the rated voltage of the CS, I_i^{max} is the maximum current capacity of the submarine cables in the *ith* feeder, and *cos* φ is the power factor, indicating the efficiency with which the electrical power is converted into useful work output.

9.2.3.2 Constraint on the Load Capacity of Submarine Cable

To ensure efficient and reliable operation, the chosen submarine cable for each segment must have a maximum transmission capacity that can accommodate the combined load of all downstream connected WTs. This requirement is formalized as follows.

$$S_{i,j} \leq S_{i,j}^{max}, i = 0, 1, 2, \ldots, N_f; \; j = 0, 1, 2, \ldots, n_i \tag{9.9}$$

where, $S_{i,j}$ denotes the capacity sum of downstream connected WTs behind the jth segment in the ith feeder; and $S_{i,j}^{max}$ is the maximum transmission capacity of the submarine cable for the jth segment in the ith feeder.

9.2.3.3 Non-Crossing Constraint of Submarine Cables

A critical design consideration for OWF is the avoidance of submarine cable crossings. As depicted in Figure 9.2, cable crossings can introduce significant complications, including increased construction costs for laying one cable over

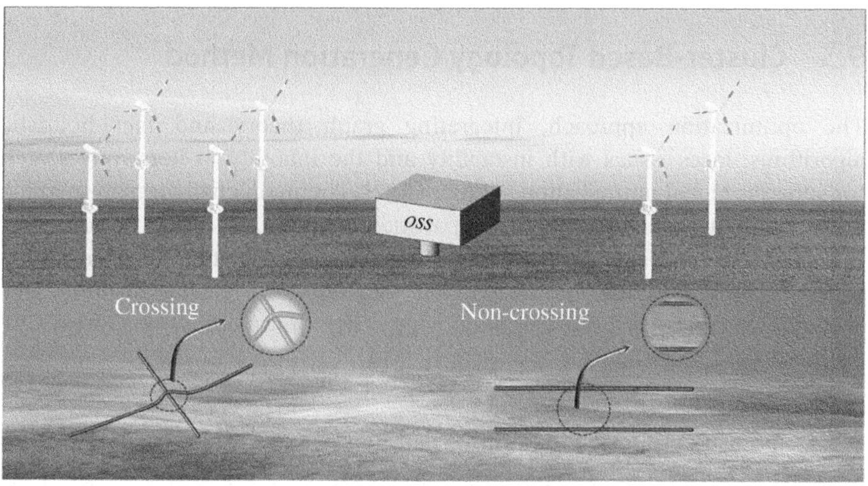

Figure 9.2 Illustration of cross constraint.

another, heightened reactive power loss, and elevated risks of damage [28]. To mitigate these issues, it is essential that pairs of submarine cables adhere to the non-crossing constraint. This constraint is applied to each segment of the feeder, ensuring that cables do not physically intersect.

$$\begin{cases} (M_{\bar{i}} - P_i) \times (Q_i - P_i) \cdot (N_{\bar{i}} - P_i) \times (Q_i - P_i) < 0 \\ (P_i - M_{\bar{i}}) \times (N_{\bar{i}} - M_{\bar{i}}) \cdot (Q_i - M_{\bar{i}}) \times (N_{\bar{i}} - M_{\bar{i}}) < 0 \end{cases}, \quad i = 0, 1, 2, \ldots, N_f, \bar{i} \neq i \tag{9.10}$$

In this formula, P_i, Q_i, M_i, and N_i denote the location coordinates of the connection points by the two submarine cables. The symbol × represents the cross product, while denotes the dot product.

9.2.3.4 Voltage Constraint

Maintaining the voltage within specified limits at each node in the CS is essential for the stable and efficient operation of an OWF. This requirement is encapsulated in the voltage constraint, which stipulates that the voltage magnitude at every node must stay within its predetermined lower and upper bounds [29].

$$V_{i,j}^{\min} \leq V_{i,j} \leq V_{i,j}^{\max}, i = 0, 1, 2, \ldots, N_f; \ j = 0, 1, 2, \ldots, n_i \tag{9.11}$$

where $V_{i,j}^{\min}$ and $V_{i,j}^{\max}$ denote the minimum and maximum allowable voltage magnitudes at each node of the feeder, ensuring the system operates within safe electrical limits. $V_{i,j}$ represents the actual voltage magnitude at the jth node in the ith feeder.

9.3 Cluster-Based Topology Generation Method

The optimization approach, integrating graph theory and meta-heuristic algorithms, faces issues with instability and the inability to iteratively search for superior global optimization solutions [27]. The main challenges in merging these techniques include efficiently partitioning the CST design optimization into discrete processes for individual algorithm application, and maximizing the strengths while minimizing the weaknesses of each algorithm. This section presents two cluster-based topology generation methods to address CST optimization for OWF.

9.3.1 Polar Coordinate Clustering

OWFs adopt the method of Polar Coordinate Clustering in order to effectively manage and optimize the power transmission system of the OWF. In this method,

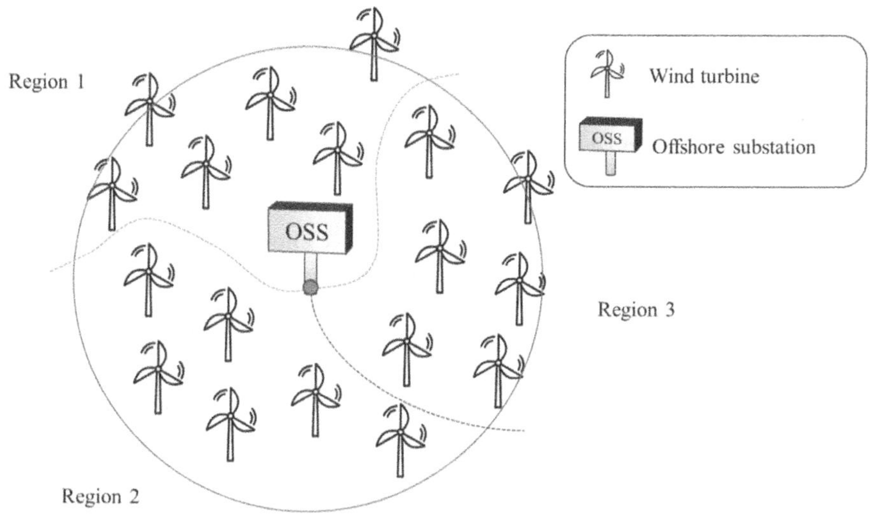

Figure 9.3 Polar Coordinate Clustering process.

the OSS is regarded as the center of the polar coordinate system, while the WTs are distributed at various polar angles and radii relative to this center. The primary aim of Polar Coordinate Clustering is to group the wind turbines based on their positions relative to the OSS, in order to optimize cable layout and power transmission efficiency.

To implement distributed optimization for the WTs connection system, WTs are ordered and segmented into groups based on polar coordinates, using the OSS as the origin. Each group contains an equal number of WTs, except potentially the final group, which may have fewer. The WT count varies between $n^{\min}$ and $n^{\max}$. Additionally, any WT can serve as a starting point, indicating that the total number of WTs determines the group division, as illustrated in Figure 9.3. The overall group count is thus described as:

$$G = N(n^{\max} - n^{\min} + 1) \tag{9.12}$$

where G is the total number of groups, and N is the number of WTs.

9.3.2 Radial Clustering

The FCM algorithm is a method of fuzzy clustering that assigns data points to multiple clusters, each with varying degrees of membership. This approach provides flexibility in grouping data points but introduces a challenge in the context of WT grouping for OWF. In particular, FCM tends to organize WTs into a loose, block-style distribution as a result of each clustering operation producing a cluster center. Such a configuration is less than ideal, as it fails to guarantee a straightforward, unambiguous path for submarine cables connecting WTs to the OSS. This limitation can lead to inefficiencies in cable routing and potential increases in the cost and complexity of the OWF's CS.

9.3.2.1 Fuzzy C-Means Algorithm

The FCM algorithm is a clustering method that allows one piece of data to belong to two or more clusters. This method is an extension of the more widely known k-means clustering technique, with the key difference being its use of fuzzy logic to assign membership levels to data points relative to different clusters. Instead of assigning a data point to a single cluster, as in k-means, FCM assigns a degree of membership to each cluster for each data point, indicating the strength of association between that data point and a particular cluster.

The algorithm iteratively updates the cluster centroids and the membership levels of each data point to the clusters until the objective function converges to a minimum. The objective function of the FCM algorithm is described as follows:

$$L = \sum_{i=1}^{N_{\text{sub}}} \sum_{j=1}^{N_{\text{total}}} u_{ij}^{m} d_{ij}^{2} \tag{9.13}$$

$$u_{ij} = \frac{1}{\sum_{r=1}^{c} \left(\frac{d_{ij}(k)}{d_{rj}(k)} \right)^{\frac{2}{m-1}}} \tag{9.14}$$

where, L represents the objective function of the FCM algorithm; m stands as the weighting coefficient; u_{ij} signifies the membership degree, ranging between 0 and 1; d_{ij} indicates the Euclidean distance from each turbine to the clustering centroid; N_{sub} signifies the number of partitions; N_{total} stands for the total number of WTs in LSOWFs; k denotes the iteration number, and c signifies cluster center number.

9.3.2.2 Radial Fuzzy C-Means Algorithm

In the Radial Fuzzy C-Means (RFCM) algorithm, the traditional reliance on Euclidean distance for clustering membership is replaced with a novel approach based on the angular orientation of WTs relative to the OSSs. The pivotal metric for cluster assignment is the angle α which is determined by the intersection of two vectors: one from the WTs to the OSSs, and another from the clustering centroid c to the OSSs. As delineated in Eq. 9.15 and illustrated in Figure 9.4, the distance d_{ij} adopts the Euclidean distance d_{wo} from a WT to the OSSs, multiplied by $\sin\alpha$, for angles α less than 90°. For angles exceeding 90°, d_{ij} is assigned an effectively

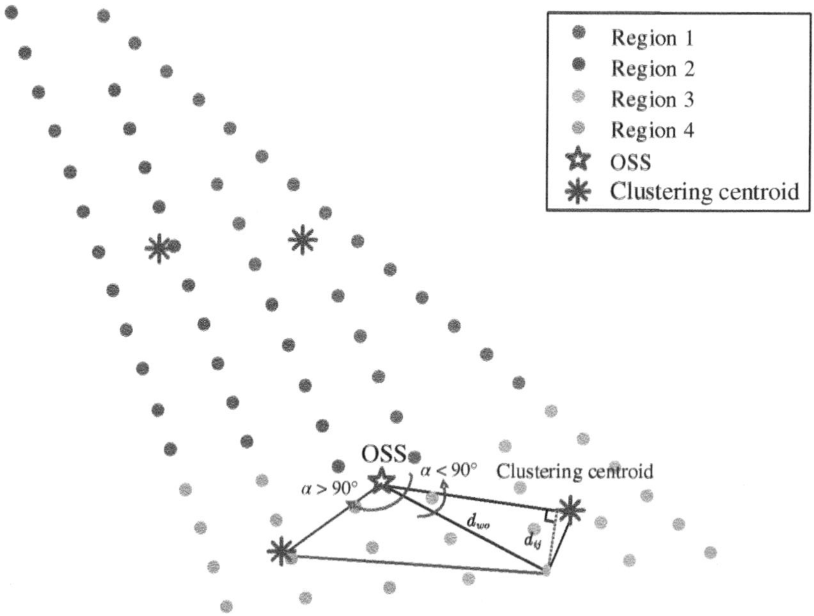

Figure 9.4 RFCM classification schematic diagram.

infinite value. This innovative modification ensures that WTs markedly divergent in angular positioning from the OSSs are excluded from clustering by the RFCM algorithm, facilitating a radial organization of turbines around the OSSs.

$$d_{ij} = \begin{cases} d_{wo} \cdot \sin\alpha, \alpha \leq 90° \\ \infty, \alpha > 90° \end{cases} \tag{9.15}$$

This strategy ensures that each feeder creates a radial distribution area surrounding the OSS, effectively isolating it from other feeders and enabling fan-shaped clustering in LSOWFs. The contrast between the two clustering approaches is depicted in Figure 9.5.

As illustrated in Figure 9.5 (a), the FCM algorithm overlooks the OSS's location, leading to a grid-like clustering pattern in LSOWFs. This arrangement inadvertently creates a barrier between the OSS and the WTs in yellow Region 3, implying that these WTs cannot be seamlessly connected to the OSS without the cables intersecting with others. Remarkably, Figure 9.5 (b) demonstrates that the RFCM algorithm can create a radial block area, effectively resolving the connectivity issues between turbine clusters and the OSS.

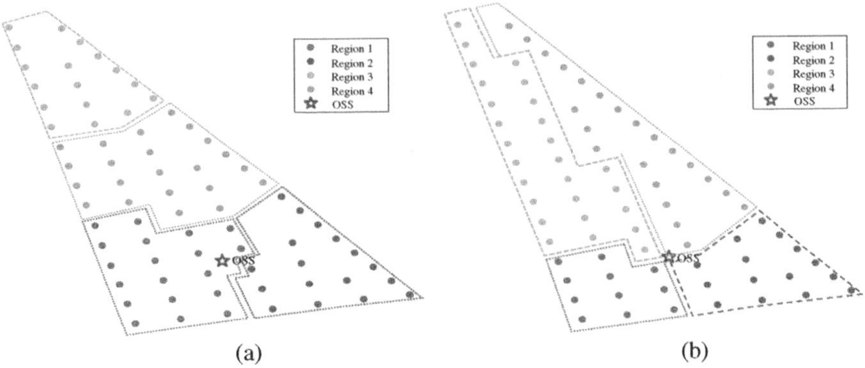

Figure 9.5 Comparison chart of the two clustering algorithm results: (a) FCM clustering schematic diagram and (b) RFCM clustering schematic diagram.

9.3.3 Dynamic Minimum Spanning Tree

Designing an optimal topology within the CS can be conceptualized as a variant of the MST problem in mathematics. This approach entails treating the OSS and each WT as vertices. Meanwhile, submarine cables connecting two WTs are considered edges, with the cable lengths serving as the edge weights in the weighted graph. Consequently, this allows for the representation of the LSOWFs' topological structure through eq. 9.16.

$$G_s = (V, E_s, M_s), G_s \in G, E_s \in E, W_s \in W \tag{9.16}$$

To identify the optimal connection strategy for a single group of WTs, the optimization model leverages an enhanced version of the DMST. The configuration of upstream submarine cables can be adjusted with the incorporation of additional downstream WTs. Consequently, each addition of WTs necessitates a revision of the upstream edge factors in the optimization model's cost calculations. The Improved DMST builds upon the Prim algorithm by incorporating additional factors to construct an MST. By modifying the Prim algorithm, the Improved DMST facilitates the creation of a directed graph, allowing for the iterative updating of edge factors with each edge addition, as illustrated in Figure 9.6.

The problem should be found with a few initial solutions while satisfying all constraints. All WTs with the same power respectively and divided into groups by polar coordinates can avoid the non-crossing constraint effectively. Besides, optimizing the initial solutions with a heuristic algorithm significantly enhances the grouping derived from simple division and connection processes based on polar coordinates. These preliminary solutions, while adhering to basic constraints, often fall short of the optimal grouping strategy due to their straightforward generation method. The heuristic algorithm introduces a dynamic optimization

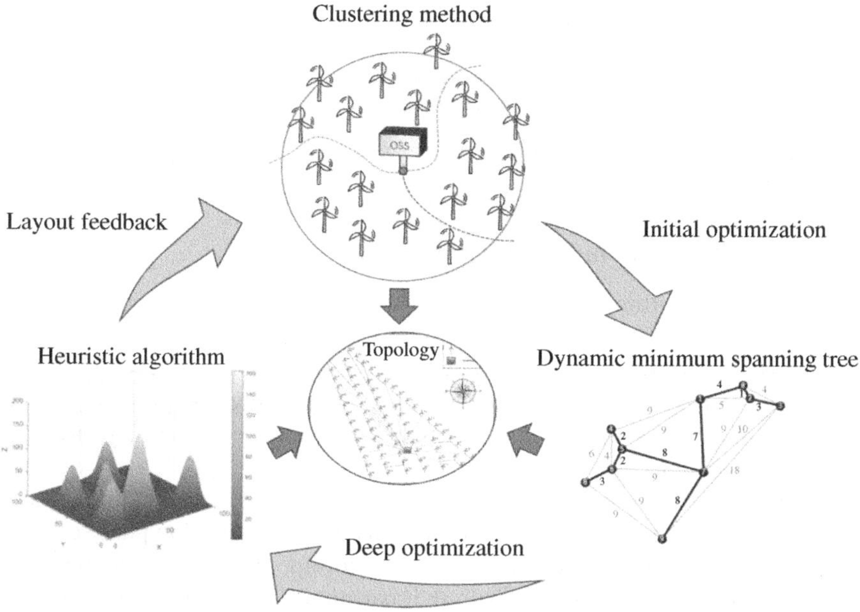

Figure 9.6 Operation framework of the proposed approach for CST design.

layer by enabling the exchange of WTs between groups. It's possible to systematically refine and evolve these groupings toward a more optimal configuration, thereby addressing and potentially surpassing the limitations of the initial solution set. This method ensures that as new WTs are added, the model dynamically updates to reflect changes in upstream cable configurations, optimizing the connection layout for efficiency and cost-effectiveness.

In the Improved DMST method, the process is initiated by establishing two distinct sets: $\widetilde{M} = \{0\}$, representing the OSS and WTs already connected to the tree, and $\check{N} = \{i \mid i \in n_l\}$, denoting the WTs yet to be connected. The procedure unfolds as follows:

- **Initial Setup:** Create the sets $\widetilde{M} = \{0\}$ for the OSS and already connected WTs, and $\check{N}$ for WTs awaiting connection.
- **WT Ranking:** Order the WTs $(i \in n_l)$ based on their distance to the OSS, preparing for sequential connections.
- **Connection Process:** ① Select the foremost WT from $\check{N}$ for potential connection to each member in $\widetilde{M}$. ② Evaluate the connection feasibility and efficiency using a specified factor to determine the optimal connection target. ③ Transfer the selected WT from $\check{N}$ to $\widetilde{M}$, marking its connection to the network. ④ Record

the established connection in a new set, $\widetilde{L} = \{\emptyset\}$, designed to track the connectivity relations.

- **Repetition and Completion:** Iterate the third step for each WT in $\check{N}$ until all WTs are included in $\widetilde{M}$, ensuring $\widetilde{N} = \{\emptyset\}$ and $\widetilde{M} = \{0, i \mid i \in n_l\}$.
- **Output:** Generate the final topology tree and calculate the capital cost associated with the CST.

9.3.4 Firefly Algorithm-Based Optimization

The Firefly Algorithm (FA) draws its inspiration from the luminescent and alluring behavior of fireflies, who use their glow to attract mates. This natural mechanism is mirrored in the FA, where the attraction between fireflies is harnessed as a strategy for solution-seeking. In tackling complex, large-scale issues, FA has been shown to have an edge over conventional heuristic methods such as GA and PSO [29].

To leverage the FA for optimizing the CS in LSOWFs, the preliminary step involves initializing the population of solutions. This initialization phase consists of generating an n-dimensional random vector for each partition, where n corresponds to the number of WTs in that partition. The elements of this vector range between 0 and 1 and are determined based on the quantity of WTs. An element within this vector that is 0.5 or above is converted to 1, denoting the establishment of a connection between two nodes. If an element falls below 0.5, it is changed to 0, indicating the absence of a connection. Subsequently, this vector is transformed into an $n \times n$ adjacency matrix to represent these connections explicitly. For every swarm size specified, a unique set of adjacency matrices is created, forming the basis for the optimization process. For objectives centered on minimizing costs, the luminous intensity $I(x)$ of a firefly is defined as inversely proportional to the value of the objective function $f(x)$. This relationship is expressed through the formula 9.17:

$$I(x) \propto \frac{1}{f(x)} \tag{9.17}$$

where, the objective function $f(x)$ is explicitly defined as the total cost C_{all} of the CS. Consequently, when the total cost C_{all} is relatively low, the luminous intensity $I(x)$ of a firefly is correspondingly high. This designates a firefly with a more robust luminosity, which, in turn, attracts other fireflies toward it, effectively guiding the FA progressively toward the optimal solution.

Furthermore, within the framework of the FA, fireflies exhibiting weaker luminosity are inherently attracted toward those with stronger luminosity, embodying a fundamental aspect of the algorithm's search dynamics. This behavior is mathematically encapsulated as follows:

$$\beta = \beta_0 e^{-\gamma r^2} \tag{9.18}$$

In this equation, the parameter β signifies the level of attraction between fireflies, where β_0 is the base attraction strength observed in the absence of distance ($r = 0$). The variable r represents the Euclidean distance between two WTs, serving as a proxy for the distance between two solutions in the search space. As r increases, indicating that two fireflies are further apart, the attraction strength β undergoes a rapid decay, modeled by the exponential term $e^{-\gamma r^2}$, where γ delineates the rate of this decay.

By fine-tuning the three core parameters—β_0, the initial attraction; γ, the decay rate; and the distance metric r the FA is endowed with the flexibility to efficiently navigate and optimize the CS topology within LSOWFs. This tailored approach allows for the adaptive adjustment of the algorithm's sensitivity to distance, ensuring that the search process is both dynamic and responsive to the evolving landscape of solutions. The FA's ability to simulate the natural attraction phenomenon among fireflies toward more optimal positions enables it to effectively minimize costs and refine the CS topology, demonstrating its prowess in addressing complex optimization challenges.

9.4 Case Study

In order to demonstrate the capabilities and application of the proposed technique, two OWFs with varying quantities (i.e., 75, 250) and layouts of WTs are optimized on three heuristic algorithms: PSO, ICA, and FA. To ensure objectivity and consistency across evaluations, we standardize the computational experiments by setting both the maximum number of iterations and the population size to 100 and 50, respectively, for three heuristic algorithms tested. The experiments were performed on a personal computer equipped with an Intel(R) Core i7-10700 CPU, operating at a frequency of 2.90 GHz, and supported by 8 GB of RAM. This hardware setup provided the necessary computational power and memory capacity to efficiently handle the demands of the optimization algorithms and the complexities of the OWF simulations.

9.4.1 Test Case #1

In the analysis of Test Case #1, the configuration includes 75 WTs, each with a generating capacity of 8 MW, leading to a collective installed capacity of 600 MW. The connectivity among these WTs is facilitated by three distinct types of submarine cables. The selection of these submarine cables is influenced by several critical factors, including the rated capacity of each cable type in MW. This allows for the calculation of the maximum number of 8 MW WTs that each cable type can support: Type #1 cable is suitable for connecting up to 3 WTs, Type #2 cable can accommodate 4 WTs, and Type #3 cable is capable of supporting 5 WTs.

Table 9.1 The parameters of submarine cables in Test Case #1.

Type	Loadable quantity of WTs	Ampacity (A)	Resistance AC 90°C (Ω/km)	Reactance (Ω/km)	Price (¥/km×10⁶/km)	Represent color
#1	2	278	0.246	0.153	2.106	
#2	3	398	0.127	0.138	2.623	
#3	4	512	0.078	0.127	3.231	

Further, essential cable characteristics such as standard current carrying capacity (measured in Amperes), AC resistance (considering a reference temperature of 90°C, in Ohms per kilometer), initial costs (expressed in tens of thousands of RMB), reactance, and inductance are meticulously evaluated. These critical parameters are systematically presented in Table 9.1.

For Test Case #1, we applied the aforementioned clustering methods to classify the OWF into four distinct areas. An initial optimization was performed using the MST approach, followed by a secondary optimization using heuristic algorithms to obtain the CS layout of the OWF. The number of iterations in FA is 100. The convergence process of the algorithms indicates that FA can break through the local optimal solution. Additionally, the layout of Test Case #1 and optimal CST graphs derived by clustering methods with FA under this OWF are shown in Figure 9.7. By integrating the computation time and capital cost obtained by all methods, Polar Coordinate Clustering with heuristic algorithms can require high-quality solutions to the CST problem in OWF.

9.4.2 Test Case #2

For Test Case #2, the infrastructure comprises WTs, each with an 8 MW capacity, culminating in a total generation capacity of 2000 MW. The design and planning phase involves the consideration of seven distinct submarine cable types, all rated for a 66 kV operational voltage. The capacity of these cables to carry current varies significantly, impacting the number of turbines they can effectively connect. For example, the Type#1 cable is designed to support up to 3 WTs, whereas the Type#7 cable extends its capability to connect up to 9 WTs. The detailed specifications, including current carrying capacity and other relevant parameters for these submarine cables, are meticulously outlined in Table 9.2.

We adopted the same clustering method as Test Case#1 to partition the OWF, reducing the complexity of the problem. Given that Test Case#2 comprises 250 wind turbines, we utilized a clustering algorithm to divide it into 6 distinct regions. The optimized CS topology diagrams obtained by FA are illustrated in Figure 9.8.

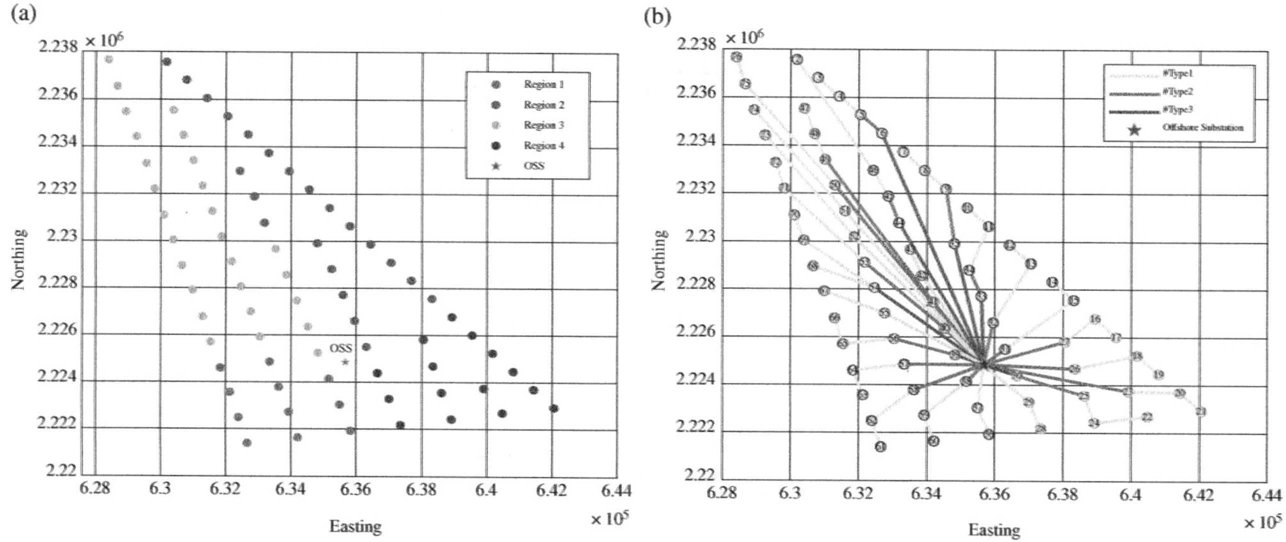

Figure 9.7 Test Case #1 connection graphs obtained by Clustering method with FA: (a) Clustering results; (b) OWF topology optimization results.

Table 9.2 The parameters of submarine cables in Test Case #2.

Type	Loadable quantity of WTs	Ampacity (A)	Resistance AC 90°C (Ω/km)	Reactance (Ω/km)	Price (¥/km × 10^6/km)	Represent color
#1	3	230	0.342	0.160	0.195	
#2	4	278	0.246	0.153	0.211	
#3	5	398	0.127	0.138	0.262	
#4	6	512	0.078	0.127	0.323	
#5	7	642	0.046	0.118	0.421	
#6	8	715	0.036	0.115	0.365	
#7	9	789	0.027	0.112	0.580	

9.4.3 Cost Comparison

To demonstrate the FA's effectiveness in optimization, Table 9.3 offers a detailed analysis of performance metrics, such as total cost and execution time, for Region 3 in Test Case #1, comparing FA against other algorithms. Despite FA's longer computation time, it achieves the most cost-efficient outcome, with a total expense of 57,104,007 RMB. Notably, this figure is significantly lower than the costs incurred by the PSO and Imperialist Competitive Algorithm (ICA), by 72,393 RMB and 70,562,370 RMB, respectively. This reduction represents a substantial cost efficiency improvement of 12.35%, underscoring FA's potential to optimize financial resources in OWF development projects more effectively than its counterparts.

Figure 9.9 shows the convergence paths for different algorithms. Remarkably, the FA finds the best solution by the 4th iteration. The total costs for ICA and PSO are also higher compared to FA, indicating FA's superior speed in convergence and its effectiveness in global search, making it the most efficient among the compared algorithms.

Table 9.3 Optimal cost by different algorithms in the subregion.

Method	Total cost (¥10 K)	Running time (s)
FA	5710.4007	101.2
PSO	5717.6400	59.1
ICA	6416.0244	40.0

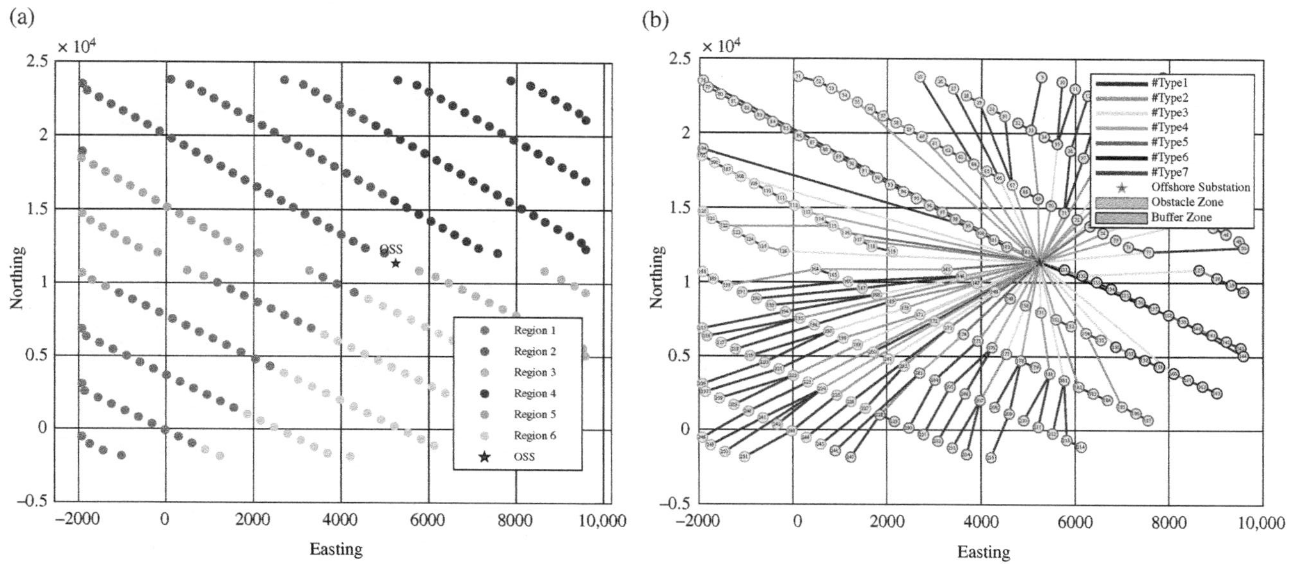

Figure 9.8 Test Case #2 connection graphs obtained by Clustering method with FA: (a) Clustering results; (b) OWF topology optimization results.

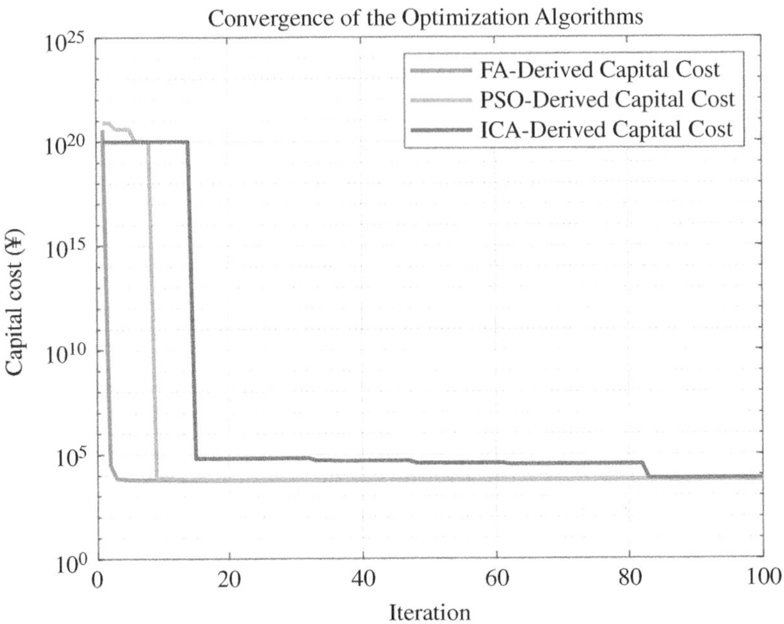

Figure 9.9 Convergence curves of different algorithms.

9.5 Conclusion

In summary, this chapter has detailed the development and comparative analysis of two distinct methods for the topology generation of LSOWF CSs, namely Polar Coordinate Clustering and radial grouping. Each method employs FA approach to optimize the design, focusing on the efficiency, cost reduction, and reliability enhancement of OWF. These clustering methods not only optimize the structural layout for better load management but also contribute to the durability and reliability of the OWF's CS.

Future research should aim at enhancing these methodologies, possibly through the integration of more advanced optimization techniques or the exploration of hybrid models that combine the strengths of both approaches. Further, the application of emerging technologies, such as artificial intelligence and machine learning, could offer dynamic and adaptive solutions to the challenges of OWF design. The ongoing refinement of these methods will be crucial in advancing the cost-effectiveness and reliability of offshore wind energy, thus playing a vital role in the global shift toward sustainable energy solutions.

References

1 Fischetti, M. and Pisinger, D. (2018). Mixed Integer Linear Programming for new trends in wind farm cable routing. *Electronic Notes in Discrete Mathematics* 64: 115–124. https://doi.org/10.1016/j.endm.2018.01.013.

2 Kim, A., Kim, H., Choe, C., and Lim, H. (2023). Feasibility of offshore wind turbines for linkage with onshore green hydrogen demands: A comparative economic analysis. *Energy Conversion and Management* 277: 116662. https://doi.org/10.1016/j.enconman.2023.116662.

3 Offshore Wind Research and Development. Available: https://www.energy.gov/eere/wind/offshore-wind-research-and-development. (accessed 23 July 2023).

4 Li, P., Lian, J., Ma, C., and Zhang, J. (2023). Complementarity and development potential assessment of offshore wind and solar resources in China seas. *Energy Conversion and Management* 296: 117705. https://doi.org/10.1016/j.enconman.2023.117705.

5 Global Wind Report 2023. Available: https://gwec.net/globalwindreport2023/ (accessed 23July 2023).

6 Zhang, S. and Li, X. (2021). Future projections of offshore wind energy resources in China using CMIP6 simulations and a deep learning-based downscaling method. *Energy* 217: 119321. https://doi.org/10.1016/j.energy.2020.119321.

7 Wu, F., Gao, R., Li, C., and Liu, J. (2023). A comprehensive evaluation of wind-PV-salt cavern-hydrogen energy storage and utilization system: A case study in Qianjiang salt cavern, China. *Energy Conversion and Management* 277: 116633. https://doi.org/10.1016/j.enconman.2022.116633.

8 Gao, Q., Hayward, J.A., Sergiienko, N. et al. (2024). Detailed mapping of technical capacities and economics potential of offshore wind energy: a case study in South-eastern Australia. *Renewable and Sustainable Energy Reviews* 189: 113872. https://doi.org/10.1016/j.rser.2023.113872.

9 Liu, Z., Fan, S., Wang, Y., and Peng, J. (2021). Genetic-algorithm-based layout optimization of an offshore wind farm under real seabed terrain encountering an engineering cost model. *Energy Conversion and Management* 245: 114610. https://doi.org/10.1016/j.enconman.2021.114610.

10 Li, X., Zhou, S., and Zhao, Y. (2023). Onshore and offshore wind power generation forecasting using a novel flexible time-varying fractional nonlinear grey model. *Energy Conversion and Management* 297: 117695. https://doi.org/10.1016/j.enconman.2023.117695.

11 Gonzalez-Longatt, F.M., Wall, P., Regulski, P., and Terzija, V. (2011). Optimal electric network design for a large offshore wind farm based on a modified genetic algorithm approach. *IEEE Systems Journal* 6 (1): 164–172. https://doi.org/10.1109/JSYST.2011.2163027.

12 Fischetti, M. and Pisinger, D. (2018). Optimizing wind farm cable routing considering power losses. *European Journal of Operational Research* 270 (3): 917–930. https://doi.org/10.1016/j.ejor.2017.07.061.

13 El Mokhi, C. and Addaim, A. (2020). Optimal substation location of a wind farm using different metaheuristic algorithms. *2020 IEEE 6th International Conference on Optimization and Applications (ICOA). IEEE*, 1–6. https://doi.org/10.1109/ICOA49421.2020.9094469.

14 Chen, R., Zhang, Z., Hu, J. et al. (2024). Grouping-based optimal design of collector system topology for a large-scale offshore wind farm by improved simulated annealing. *Protection and Control of Modern Power Systems* 9 (1): 94–111. https://doi.org/10.23919/PCMP.2023.000151.

15 Dahmani, O., Bourguet, S., Machmoum, M. et al. (2014). Optimization of the connection topology of an offshore wind farm network. *IEEE Systems Journal* 9 (4): 1519–1528. https://doi.org/10.1109/JSYST.2014.2330064.

16 Li, D.D., He, C. and Shu, H. Y. (2008). Optimization of electric distribution system of large offshore wind farm with improved genetic algorithm. *2008 IEEE Power and Energy Society General Meeting-Conversion and Delivery of Electrical Energy in the 21st Century*. IEEE, 1–6. https://doi.org/10.1109/PES.2008.4596194

17 González, J.S., Payán, M.B., and Santos, J.R. (2013). A new and efficient method for optimal design of large offshore wind power plants. *IEEE Transactions on Power Systems* 28 (3): 3075–3084. https://doi.org/10.1109/TPWRS.2013.2251014.

18 Gonzalez-Longatt, F.M., Wall, P., Regulski, P., and Terzija, V. (2011). Optimal electric network design for a large offshore wind farm based on a modified genetic algorithm approach. *IEEE Systems Journal* 6 (1): 164–172. https://doi.org/10.1109/JSYST.2011.2163027.

19 Zhang, Y., Qiao, Y., Lu, Z., and Sun, W. (2017). Optimisation of offshore wind farm collection systems-based on modified genetic algorithm. *The Journal of Engineering* 2017 (13): 1045–1049. https://doi.org/10.1049/joe.2017.0488.

20 Li, D.D., He, C. and Fu, Y. (2008). Optimization of internal electric connection system of large offshore wind farm with hybrid genetic and immune algorithm. *2008 Third International Conference on Electric Utility Deregulation and Restructuring and Power Technologies*. IEEE, 2476–2481. https://doi.org/10.1109/DRPT.2008.4523827.

21 P. Hou, G. Yang, W. Hu, C. Chen, M. Soltani, and Z. Chen. Cable connection scheme optimization for offshore wind farm considering wake effect. *2018 IEEE Congress on Evolutionary Computation (CEC)*. IEEE, 2018: 1–8. https://doi.org/10.1109/CEC.2018.8477866.

22 Tao, S., Xu, Q., Feijóo, A., and Zheng, G. (2020). Joint optimization of wind turbine micrositing and cabling in an offshore wind farm. *IEEE Transactions on Smart Grid* 12 (1): 834–844. https://doi.org/10.1109/TSG.2020.3022378.

23 Hou, P., Hu, W., and Chen, Z. (2016). Optimisation for offshore wind farm cable connection layout using adaptive particle swarm optimisation minimum spanning tree method. *IET Renewable Power Generation* 10 (5): 694–702. https://doi.org/10.1049/iet-rpg.2015.0340.

24 Srikakulapu, R. and Vinatha, U. (2016). Optimal design of collector topology for offshore wind farm based on ant colony optimization approach. *2016 IEEE International Conference on Power Electronics, Drives and Energy Systems (PEDES)*. IEEE, 1–6. https://doi.org/10.1109/PEDES.2016.7914458.

25 Neagu, B.C. and Georgescu, G. (2014). Wind farm cable route optimization using a simple approach. *2014 International Conference and Exposition on Electrical and Power Engineering (EPE)*. IEEE, 1004–1009. https://doi.org/10.1109/ICEPE.2014.6970060.

26 Shen, X., Wu, Q., Zhang, H., and Wang, L. (2023). Optimal planning for electrical collector system of offshore wind farm with double-sided ring topology. *IEEE Transactions on Sustainable Energy*. https://doi.org/10.1109/TSTE.2023.3241357.

27 Huang, L. and Fu, Y. (2010). Reliability evaluation of the offshore wind far. *2010 Asia-Pacific Power and Energy Engineering Conference*. IEEE, 2010: 1–5. https://doi.org/10.1109/APPEEC.2010.5449007.

28 Zuo, T., Meng, K., Tong, Z. et al. (2019). Offshore wind farm collector system layout optimization based on self-tracking minimum spanning tree. *International Transactions on Electrical Energy Systems* 29 (2): e2729. https://doi.org/10.1002/etep.2729.

29 Wang, W., Qiao, Y., Dou, F. et al. (2019). Optimization of offshore wind farm collector systems based on improved genetic algorithm. *Electric Power* 52 (01): 63–68. https://doi.org/10.11930/j.issn.1004-9649.20180604.

10

Transmission Line Multi-Fitting Detection Method Based on Implicit Space Knowledge Fusion

Qianming Wang, Congbin Guo, Xuan Liu, and Yongjie Zhai

School of Control and Computer Engineering, North China Electric Power University, Baoding, Hebei, China

10.1 Introduction

As a crucial infrastructure component of the energy Internet, transmission lines are tasked with the vital role of transporting electric energy, and their operational status is intricately linked to societal production and livelihoods. Due to the necessity of electric energy transportation in remote regions, transmission lines are predominantly constructed in environments with harsh natural conditions, such as forests, snowfields, or deserts [1]. fittings [2, 3], as essential electrical components on transmission lines, are particularly susceptible to defects like corrosion, soiling, and breakage due to complex environmental conditions and adverse weather. These factors can ultimately lead to transmission line failures and significant economic losses [4]. Hence, the regular inspection of fittings is of paramount importance in transmission line maintenance. It stands as a critical task, serving as a necessary prerequisite to ensure the safety and stability of transmission line operations [4]. Consequently, it is imperative to implement regular and systematic inspection practices to effectively mitigate risks and ensure the robust performance and integrity of the infrastructure.

The high-risk nature inherent in traditional manual inspections has inevitably resulted in a sluggish growth in the number of frontline operation and inspection personnel for transmission lines. When coupled with the inherent inefficiencies and limitations of manual methods, this shortfall has rendered traditional inspection techniques increasingly inadequate for meeting the expanding and complex demands of transmission line inspections. Consequently, there is an urgent and pressing need for the development and implementation of high-efficiency, low-risk inspection methods to bridge this gap. With the widespread adoption

Artificial Intelligence Empowered Smart Energy Systems, First Edition. Edited by Qiang Yang and Gang Huang.

of Unmanned Aerial Vehicle (UAV) inspection technology on transmission lines [5, 6], the offline burden on operation and inspection personnel has significantly decreased, alleviating them from many physically demanding and hazardous tasks [7]. Nevertheless, while the use of drones for inspecting transmission line components, particularly the vast array of transmission line fittings, has proven to be beneficial, it generates a substantial volume of corresponding inspection images. Relying solely on manual visual inspection of these images is not only labor-intensive and time-consuming but also highly susceptible to fatigue-induced errors. This inefficiency underlines the necessity for the introduction and integration of intelligent inspection technologies for transmission line components. Among these advanced technologies, the accurate detection of fitting targets stands out as a crucial prerequisite for conducting fitting state analysis and fault detection. Therefore, the study and development of accurate and effective transmission line fitting detection technology remain of paramount importance.

Traditional fitting detection methods predominantly rely on artificially designed features, such as the morphological characteristics of fittings [8, 9], and color and texture features [10], to identify fittings or their defects. However, these manually designed features impose inherent limitations on traditional detection methods, rendering them typically effective only for specific types of fittings or defects. This approach lacks scalability and often results in low recognition accuracy. Furthermore, these methods are highly susceptible to environmental complexities, thus diminishing their practical applicability in real-world scenarios. In stark contrast, detection methods based on deep learning have revolutionized the field by enabling autonomous learning of image feature extraction processes through neural networks. These methods yield deep image features that are better suited for the intended detection task, offering superior anti-interference capabilities and significantly higher recognition accuracy. As a result, deep learning-based approaches can adapt to different types of fittings and defects more efficiently and robustly than traditional methods. Given these advantages, there is a growing body of scholarly research focusing on the development and refinement of deep learning-based fitting inspection methods. Researchers are increasingly exploring how these advanced techniques can be harnessed to improve the precision, reliability, and overall effectiveness of fitting detection processes, further cementing the importance of deep learning in the evolution of intelligent inspection technologies.

For instance, leveraging the Single-Shot Detector (SSD) model, literature [11], achieved notable results in multi-class gadget detection tasks on transmission lines by employing techniques such as rotation, cropping for sample augmentation, Intersection over Union (IoU) enhancements for scale-sensitive prediction frames, and introducing repulsive loss to address small-target and dense-obscuration issues in complex gadget backgrounds. Literature [12], based on the YOLOX model, augmented and broadened aerial image datasets of fittings and further improved model robustness with online enhancement strategies like Mosaic and Mixup. Subsequently, the model was trained using transfer learning and cosine annealing methods to realize gold tool detection on transmission lines. Addressing the challenge of multi-angle demand and limited sample size in a vibration-proof hammer and wire clip detection, literature [13], explored the optimal number of migration layers through experiments dividing the network into deep and shallow layers. This enabled the incorporation of transfer learning techniques to augment the YOLOv3 model, thereby improving its effectiveness in detecting vibration-proof hammers and wire clips. Additionally, literature [14], proposed an advanced method utilizing an improved Cascade R-CNN, which incorporates a unique six-layer feature pyramid network to address small-target detection challenges. The method employs Soft Non-Maximum Suppression (Soft-NMS) to manage occlusion issues and Focal Loss to counteract the imbalance between positive and negative samples. This approach effectively facilitates multi-target detection on transmission lines. Finally, literature [15], primarily utilizing the Faster R-CNN model, incorporated multi-scale training to address the issue of significant insulator size variations. The approach included adjusting candidate frame proportions to better adapt to insulator characteristics and introducing adversarial generation strategies to mitigate sample occlusion problems. This comprehensive method significantly enhanced the detection accuracy of insulators on transmission lines.

While the aforementioned fitting detection methods often utilize image enhancement techniques or adjust corresponding models based on specific gadget characteristics to enhance feature extraction capabilities and detection accuracy, they primarily focus on features within proposed bounding boxes to detect targets individually. This approach often overlooks the rich contextual information present among fittings and neglects the extensive expertise embedded in the electric power domain, which derives from standardized connection rules. The integration of contextual information and domain-specific expertise could lead to significantly improved detection accuracy. By considering the relationships and contextual cues among various fittings, alongside the specialized knowledge within the power field, more accurate and reliable detection can be achieved. Therefore, leveraging such contextual information and expert knowledge to enhance detection methods holds substantial research value and practical importance. This approach aims to bridge current gaps in detection accuracy and make meaningful advancements in the field of intelligent inspection technologies for power transmission systems.

The term "transmission line fittings" encompasses a collection of components within a transmission line system that serve various essential functions, including fixing, protecting, and connecting. These fittings are categorized into different types, each playing a distinct role and functioning in unique ways within the transmission line. For instance, connectors such as Adjusting Boards (ABs), Hanging Boards (HBs), and U-type hanging rings primarily facilitate connections, whereas protective components like Grading Rings (GRs) and Shielded Rings (SRs) are designed to safeguard the system. While a single category of fitting performs a specific function, the holistic operation of a transmission line must satisfy multiple requirements across different domains to ensure overall safety and stability. Consequently, the assembly and design of these fittings often necessitate adherence to established combination rules founded on industry experience and engineering principles. This standardization has led to the development of numerous common and similar structures, as depicted in Figure 10.1. In the figure, the orange box signifies the true box identified before detection. It can be observed that the fittings are densely packed, and their

(a) (b) (c) (d)

Figure 10.1 Example diagrams of fitting structures. (a) Structure example 1, (b) Structure example 2, (c) Structure example 3, and (d) Structure example 4.

connection combinations and operational modes result in varying degrees of mutual occlusion among the fittings. This, in turn, leads to similar characteristics among the fittings. For example, in Figure 10.1b, the SR provides protection by enveloping the entire structure, leading to candidate features that resemble those of the enclosed fittings during detection. Similarly, in Figure 10.1c, Yoke Plates (YPs), ABs, and HBs, which function as connectors, experience obstruction at their connection points. This results in a high degree of feature similarity between these fittings and the surrounding components during the detection process. The abundance of similar features can easily induce recognition biases within the current detection paradigm, which heavily relies on strong perceptual intelligence but lacks robust logical reasoning. Therefore, enhancing the logical reasoning capabilities within the detection paradigm may be advantageous in effectively addressing the aforementioned issue. This would enable a more accurate and reliable identification process by mitigating the biases stemming from feature similarities.

While the standardized combination structure leads to dense occlusion of fittings, causing the object features to be highly similar and significantly impacting detection results, it concurrently offers a wealth of implicit spatial knowledge. This duality is evident in Figure 10.1b–d, where YPs and HBs are frequently used together, implying a higher likelihood of HBs existing in various spatial orientations relative to the YPs. These spatial relationships manifest as additional spatial information, observed through the connection structures between YPs and HBs and the visual information surrounding the YPs. The specific spatial context varies across different images. For instance, Figure 10.1a illustrates that influenced by factors such as the type of YP, there are primarily HBs and bag-type suspension clamps in its vicinity. Consequently, the connection structure between the YP and the HB may exhibit slight variations when compared to Figure 10.1b–d. This regularization of fitting combinations results in extensively similar patterns among fittings within transmission lines. These patterns can be summarized by humans to form critical knowledge that aids recognition. Through cognitive adjustments, humans can adapt this summarized knowledge to specific practical applications efficiently. However, current detection paradigms often overlook this ability to learn and summarize the knowledge hidden within these spatial patterns, as well as to utilize the specificities inherent in them. These paradigms predominantly focus on perceptual intelligence, which is limited without incorporating logical reasoning derived from pattern-based knowledge. By neglecting to harness this implicit spatial information, existing detection methods miss out on opportunities to enhance detection accuracy and performance, thereby highlighting the necessity to develop methods that can effectively learn and apply the spatial patterns and their specificities in practical scenarios.

Therefore, to enable the current detection paradigm to adequately learn and leverage implicit spatial knowledge, enhancing the model's logical reasoning capabilities and comprehensively considering the specificity of such knowledge across various scenarios, this chapter proposes a multi-fitting detection method for transmission lines based on the fusion of implicit spatial knowledge. This proposed method allows the model to mine and utilize the implicit spatial knowledge inherent in the fitting detection process through meticulous module design and structural improvements. The learning and summarization of implicit spatial knowledge are grounded in the spatial context information derived from different spatial positions of fitting objects. This approach enables the model to fully consider the specificity of implicit spatial knowledge and make effective adjustments during the recognition process.

Specifically, this chapter primarily employs the Double-Head model as the foundational framework. Initially, it designs a spatial box setting module and a spatial context extraction module to acquire precise spatial context information regarding fitting objects. Additionally, a spatial context memory module is designed to facilitate the retention of correct spatial context information and the elimination of incorrect spatial context information affected by factors such as shooting angles. This process aids in summarizing and fusing implicit spatial knowledge while fully considering potential specificities. Moreover, given the inherent disadvantage of the model's post-processing structure in dealing with dense occlusion situations, Soft-NMS [13] is introduced. This enhancement aims to improve the post-processing structure and boost the model's adaptability to dense occlusion scenarios. By integrating Soft-NMS, the method mitigates the issues arising from dense occlusion, thereby improving detection accuracy and robustness. Through these comprehensive improvements, the model effectively leverages implicit spatial knowledge, enhancing both logical reasoning abilities and performance in varied practical situations.

10.2 Overall Overview of Methods

The overall network framework discussed in this chapter is depicted in Figure 10.2. Its specific components can be categorized into two primary structures: the Implicit Spatial Knowledge Fusion (ISKF) structure and the enhanced post-processing structure:

1) The ISKF structure is the main component. Specifically, this structure is based on the Double-Head model [12]. The aerial images of transmission line fittings are first processed through a ResNet-101 backbone network combined with a feature pyramid network for extracting image features [16]. Subsequently, the region proposal network is employed to generate R (regions of interest).

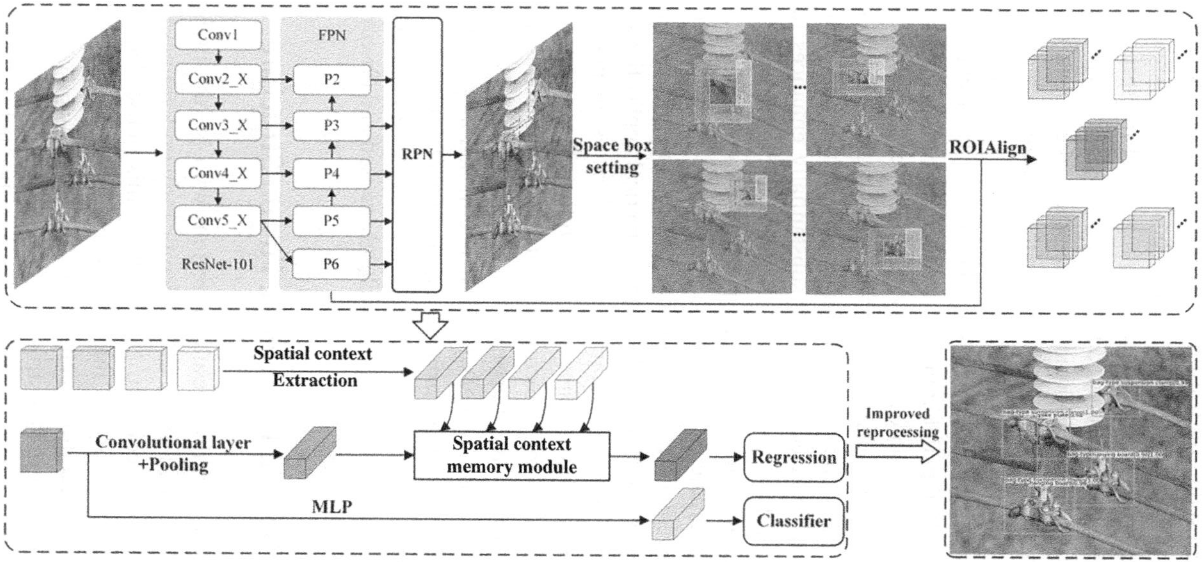

Figure 10.2 Overall network framework diagram of multi-fitting detection method for transmission lines based on implicit spatial knowledge fusion.

Based on this, reasonably set regions with rich spatial context information as spatial boxes. These regions of interest and spatial boxes undergo ROI Align operations at corresponding feature levels, extracting M_r (fixed-size feature maps of regions of interest) and M_s (spatial box features). Afterward, in the classification stage, the feature maps of the regions of interest are processed in accordance with the baseline model, going through fully connected layers and a classification detection head to accomplish the object classification task. In the regression stage, M_r refines F_r (the feature vectors of the fitting candidate boxes) through convolution and pooling, while the feature map of the spatial box is processed by a spatial context information extraction module to extract F_s (spatial context information). Building upon F_r as the foundation, the spatial context memory module effectively remembers and selectively forgets spatial context information from all four different spatial directions. This approach not only facilitates the learning and summarization of implicit spatial knowledge but also substantially enhances the model's capability to accurately complete the task of bounding box regression. By effectively capturing and integrating spatial context, the model becomes more proficient in predicting precise bounding boxes. This, in turn, leads to a significant improvement in its overall object detection performance, ensuring more accurate and reliable results in various practical applications.

2) The enhanced post-processing framework functions as an integral complementary component. This advanced structure refines the model's redundancy suppression technique for detection outcomes by employing a Soft-NMS methodology. This refinement operates in synergy with the ISKF framework, significantly boosting the model's performance in handling dense occlusion object detection tasks. By optimizing the suppression of redundant predictions, this Improved Post-Processing (IPP) structure ensures greater accuracy and robustness in scenarios characterized by closely packed and obscured objects.

10.3 Implicit Spatial Knowledge Fusion Structure

The dense distribution of fittings results in regions with highly similar features, particularly within occluded areas, which significantly undermines the model's accuracy in localization prediction. Nevertheless, this dense distribution also offers a substantial repository of implicit spatial knowledge that the model can exploit and utilize. Building upon this premise, the chapter introduces the concept of an ISKF structure. This structure serves as a sophisticated mechanism for the effective learning and application of the embedded spatial knowledge. The primary aim of this approach is to augment the model's capacity to integrate and apply spatial information, thereby enhancing its overall performance in localization prediction and related tasks. To adequately address the specificity of implicit spatial knowledge in practical applications, the designed structure assimilates and learns from this knowledge through spatial context information associated with various candidate boxes. This methodology ensures that the model can precisely capture and utilize spatial relationships and patterns inherent in different candidate regions, thus improving its overall performance and efficacy in real-world scenarios. The overarching objective is to leverage the abundance of implicit spatial information derived from the dense and variable spatial distribution of fittings. By doing so, the aim is to bolster the model's comprehension and utilization of spatial relationships and patterns, ultimately enhancing its predictive accuracy and resilience in complex environments. To realize these objectives, the proposed structure incorporates several critical components, including the spatial box setting module, the spatial context extraction module, and the spatial context memory module. This integrated approach ensures a comprehensive learning mechanism, enabling the model to harness and apply intricate spatial relationships and patterns effectively, thereby improving its localization prediction capabilities and overall performance.

10.3.1 Spatial Box Setting Module

Spatial context information is the foundation for implementing the ISKF structure. So, it is necessary to pre-locate the areas where spatial context information may exist to form a spatial box, in order to extract specific spatial context information within the spatial box. The fitting detection of two-stage models often relies on the fitting candidate boxes generated by the region proposal network, which can be represented in the form of angular coordinates:

$$R = \left\{ r^i = \left(x^i_{\text{min}}, y^i_{\text{min}}, x^i_{\text{max}}, y^i_{\text{max}} \right) \right\}_{i=1}^{N} \tag{10.1}$$

In the equation, $(x^i_{\text{min}}, y^i_{\text{min}})$ represents the top-left coordinate of the fitting candidate box, $(x^i_{\text{max}}, y^i_{\text{max}})$ represents the bottom-right coordinate, and N represents the number of fitting candidate boxes.

For potential objects within the candidate fitting boxes, the regularized combination structure between fittings and the partial visual information of adjacent fittings provide the model with a wealth of information necessary for learning implicit spatial knowledge. This intricate arrangement ensures that the model can effectively leverage the relational and contextual cues present in the environment, thereby enhancing its ability to accurately identify and predict the locations of fittings within complex visual scenes. This approach significantly contributes to the model's overall understanding of spatial dynamics, ultimately improving its performance in object detection tasks. Building on this concept, the detailed configuration of the spatial boxes must encompass information about the four distinct spatial orientations of the fitting candidate boxes. This ensures that the model can accurately capture and utilize the spatial relationships and orientations. This information varies among objects in the same category across different images, thereby forming distinct spatial context information for the fitting candidate boxes. Such variability captures the unique spatial configurations and interactions present in each image, enriching the model's understanding and enhancing its ability to generalize across diverse visual scenarios. This provides a foundation for the model to learn implicit spatial knowledge while also taking into account its specificity. By doing so, the model can effectively incorporate nuanced spatial relationships and context-specific details, thereby enhancing its overall learning and predictive capabilities. Therefore, the process of setting spatial boxes is shown in Figure 10.3, and the specific setting formula for the different orientation spatial boxes corresponding to each fitting candidate box can be expressed as follows:

$$\begin{cases} s_u^i = \left(x_{\min}^i - \alpha w^i, y_{\min}^i - \alpha h^i, x_{\max}^i - \alpha w^i, y_{\max}^i - (1-\alpha)h^i\right) \\ s_d^i = \left(x_{\min}^i + \alpha w^i, y_{\min}^i + (1-\alpha)h^i, x_{\max}^i + \alpha w^i, y_{\max}^i + \alpha h^i\right) \\ s_l^i = \left(x_{\min}^i - \alpha w^i, y_{\min}^i + \alpha h^i, x_{\max}^i - (1-\alpha)w^i, y_{\max}^i + \alpha h^i\right) \\ s_r^i = \left(x_{\min}^i + (1-\alpha)w^i, y_{\min}^i - \alpha h^i, x_{\max}^i + \alpha w^i, y_{\max}^i - \alpha h^i\right) \end{cases} \tag{10.2}$$

In the equation, s_u^i, s_d^i, s_l^i and s_r^i represent the spatial boxes corresponding to the top, bottom, left, and right positions of the i-th fitting candidate box, and α is the predetermined spatial box coefficient. w^i and h^i represent the width and height of the i-th fitting candidate box respectively. w^i and h^i are obtained by the following calculations:

$$\begin{cases} w^i = x_{\max}^i - x_{\min}^i \\ h^i = y_{\max}^i - y_{\min}^i \end{cases} \tag{10.3}$$

In summary, the primary advantage of implementing spatial box settings through the aforementioned method is its ability to accurately determine regions rich in dense spatial context information based on their orientation, while concurrently avoiding overlap. This strategy ensures the efficient extraction of spatial context information and prevents the redundant extraction of information from overlapping regions in subsequent processes. Additionally, given the generation characteristics of candidate boxes in two-stage detection models, there is potential for spatial boxes corresponding to fitting candidate boxes located at the image periphery to extend beyond the image boundary. Since such extensions encompass no useful information, it is crucial to restrict the spatial boxes to a reasonable range during the setting process and eliminate any portions that exceed this boundary. This ensures that the spatial boxes remain within the valid confines of the image, preserving the integrity and relevance of the extracted spatial context information. By carefully managing the boundaries of the spatial boxes and ensuring they do not extend beyond the image limits, the method maintains a high level of precision in capturing useful spatial context

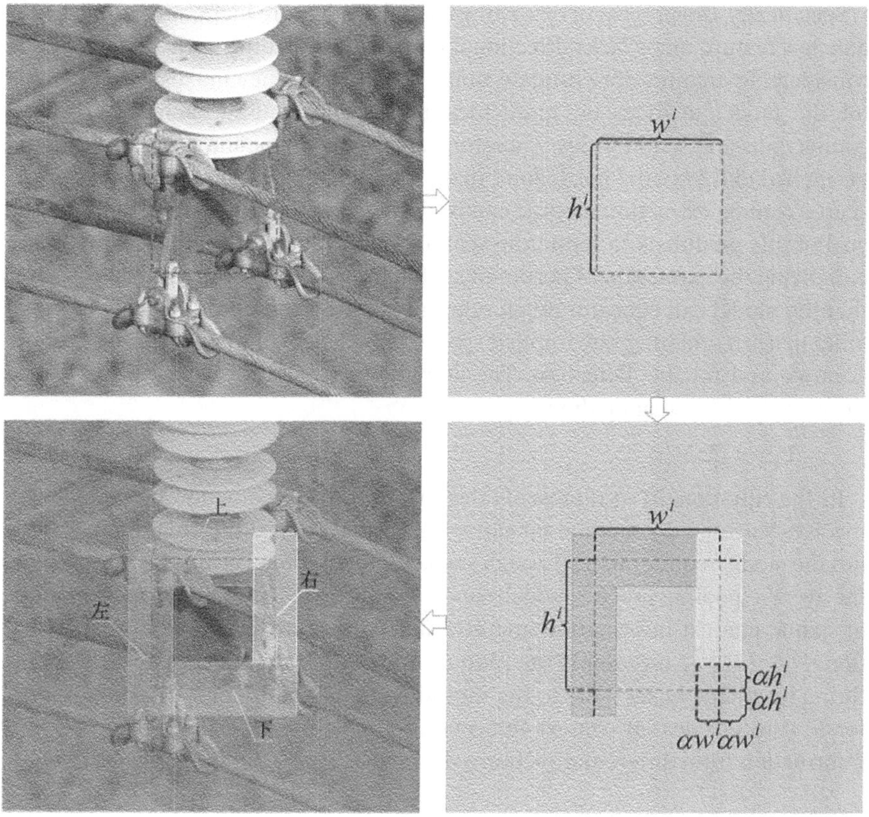

Figure 10.3 Process diagram of space box setting.

information. This approach not only enhances the efficiency and accuracy of the spatial information extraction process but also upholds the overall integrity and applicability of the spatial context data, thereby optimizing the subsequent analytical and predictive tasks performed by the model.

10.3.2 Spatial Context Extraction Module

Given the defining characteristics of spatial boxes, the image area they delineate contains an abundance of information. To enable the model to autonomously learn implicit spatial knowledge that can assist in detection, it is essential to first extract spatial context information. This information serves as the source of implicit spatial knowledge for model learning and must be derived from the spatial feature maps corresponding to different spatial boxes.

Specifically, the process of extracting spatial box feature maps and fitting candidate box feature maps lacks directional information, leading to the inclusion of a considerable amount of redundant noise. This redundancy makes it challenging for the model to focus on more meaningful information, such as structural connections and spatial rules. Therefore, additional extraction processing should be applied to the spatial box feature maps to mitigate this issue. To address this, a spatial context extraction module can be designed within the ISKF structure. The goal of this module is to refine the spatial box feature maps, reducing noise and enhancing the extraction of pertinent spatial information. This approach ensures that the model can concentrate on relevant features, thereby improving its capability to learn and integrate implicit spatial knowledge, ultimately aiding in more accurate and reliable detection. The aforementioned task can be accomplished through the following methodology:

$$F_s = \Gamma(\mathrm{T}(M_s)) \tag{10.4}$$

In the equation, $\mathrm{T}(\sim)$ represents the spatial context extraction process, which involves learning the method for extracting spatial context information that satisfies the model requirements through stacked convolutional layers. $\Gamma(\sim)$ represents the average pooling process, which serves to eliminate redundant representations of spatial context information and effectively integrate spatial context information. $F_s = \left\{F_s^i\right\}_{i=1}^{N}$ represents the spatial context information corresponding to N fitting candidate boxes. Due to the multidirectional nature of spatial box setting, for F_s^i (the i-th spatial context information), it will specifically contain various information from different directions, denoted as f_{su}^i, f_{sd}^i, f_{sl}^i, and f_{sr}^i respectively.

10.3.3 Spatial Context Memory Module

In aerial imagery, certain fitting objects may display spatial context information biases or errors in specific directions due to factors such as the drone's shooting angle. These inaccuracies have the potential to mislead the model's inferential detection processes. Additionally, various factors such as fitting type and structural context can further complicate the situation. Rigidly instructing the model based on fixed spatial knowledge – for instance, assuming that the presence of one type of fitting necessitates the presence of another type of fitting in a specific direction – can also lead to erroneous model detections. To mitigate the potential issues that arise from the rigid application of spatial knowledge, it is imperative for the model to autonomously learn and integrate implicit spatial knowledge while taking into account the specificity of different scenarios. This enables the model to correctly apply the learned knowledge based on the varying circumstances it encounters.

With this objective in mind, a spatial context memory module has been designed. This module utilizes the feature vectors of the regions of interest as carriers and employs Gated Recurrent Units (GRUs) [17]. The inclusion of GRUs in the model encourages the learning and integration of implicit spatial knowledge by memorizing effective information from diverse directional spatial contexts while discarding irrelevant information. This design ensures that the model remains adaptive, enhancing its ability to deal with dynamic and complex spatial relationships. Moreover, during the detection phase, it is essential to apply the same methodology to allow for the rational adjustment and specific application of implicit spatial knowledge. By doing so, the model can consider the uniqueness of each object within the image and effectively utilize the implicit spatial knowledge to assist in inference detection. This approach helps guide the model in making more accurate and context-aware detections, thereby improving its overall performance and reliability in various scenarios. The autonomous learning and integration of implicit spatial knowledge facilitated by the spatial context memory module and GRUs not only address the biases and errors inherent in aerial imagery but also enhance the model's adaptability and precision. This sophisticated handling of spatial context information ensures that the model can make informed and accurate detections, even in the presence of complex spatial relationships and occlusions.

The GRU is an integral component within the spatial context memory module, playing a pivotal role in enabling selective remembering and forgetting functions. The operational principle of the GRU is depicted in Figure 10.4. Its distinctive dual-gate mechanism imparts it with superior long-term memory capabilities, which are crucial for maintaining and recalling relevant context over extended sequences. The GRU's design incorporates an update gate, which plays a critical role during the memory process. This design ensures that the memory process

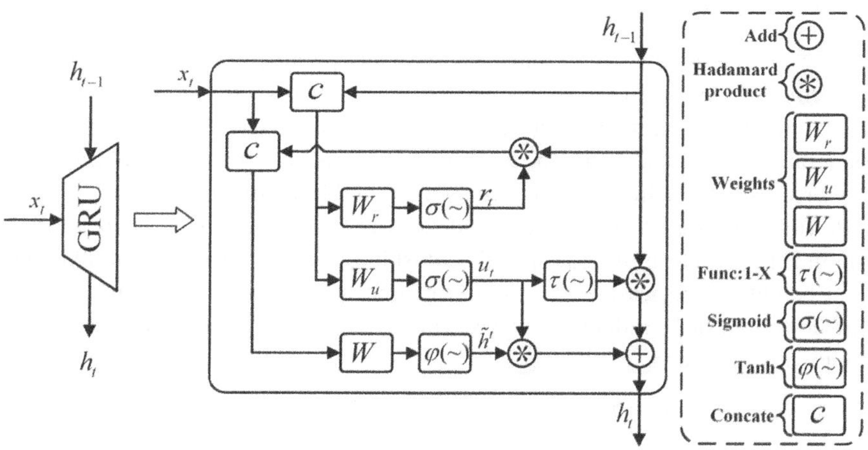

Figure 10.4 Working principle diagram of gate recurrent unit.

within the GRU is controllable, allowing for the selective retention of pertinent information while simultaneously eliminating redundant information. This functionality is essential for optimizing the model's performance by focusing on relevant features and discarding extraneous data.

Within the spatial context memory module, the GRU leverages the fitting candidate box features as the initial hidden state. By employing the corresponding temporally sequenced spatial context information as input, the GRU effectively facilitates the transmission of information and the integration of features over time. This process allows the model to maintain a coherent understanding of the spatial context, thereby enhancing its ability to accurately locate and identify objects within a given scene. The use of the GRU within the spatial context memory module significantly contributes to the model's ability to manage long-term dependencies and context information. The controllable memory process, coupled with the effective integration of spatial context information, underscores the GRU's importance in ensuring the model's overall efficacy in complex object detection tasks. This strategic utilization of the GRU enhances the model's capability to handle dense occlusion and other challenging scenarios, thereby improving detection performance across various fitting types. Specifically, the calculation formula for r_t (the reset gate) is as follows:

$$r_t = \sigma(W_r \cdot [h_{t-1}, x_t]) \tag{10.5}$$

In the equation, W_r represents the learnable parameters of the reset gate, h_{t-1} represents the previous time step's hidden state, x_t represents the current time step's input, and $[\sim, \sim]$ represents vector concatenation. Furthermore, the calculation formula for u_t (the update gate) is as follows:

$$u_t = \sigma(W_u \cdot [h_{t-1}, x_t]) \tag{10.6}$$

In the equation, W_u represents the learnable parameters of the update gate. The final processing of the above two gates requires $\sigma(\sim)$ (the Sigmoid activation function), which sets their values between 0 and 1. Therefore, the GRU can generate two gates to define the retention or exclusion of information, providing a prerequisite for achieving selective memory and forgetting functions.

$\widetilde{h}_t$ (the candidate-hidden state) used for transitions in the gate recurrent unit should efficiently capture the spatial context information at the current time step, and be able to determine the amount of information to retain from h_{t-1} (the previous time step's hidden state) based on r_t. Therefore, the specific construction method is as follows:

$$\widetilde{h}_t = \varphi(W \cdot [r_t * h_{t-1}, x_t]) \tag{10.7}$$

In the equation, W is a learnable parameter, * represents element-wise multiplication, and $\varphi(\sim)$ denotes the Tanh activation function. The parameter r_t, which ranges between 0 and 1, controls the information flow from h_{t-1} to $\widetilde{h}_t$ in a manner defined by $r_t * h_{t-1}$. h_{t-1} contains historical information that has been recorded up to the previous moment, so Eq. (10.7) can effectively store new input spatial context information through the reset gate while ensuring the effective preservation of the information of h_{t-1}, in order to define the combination form of historical information and new input information in $\widetilde{h}_t$.

Subsequently, through the coordination of the update gate, h_t (the current time step's hidden state) can be computed:

$$h_t = (1 - u_t) * h_{t-1} + u_t * \widetilde{h}_t \tag{10.8}$$

From Eq. (10.8), it can be observed that the update gate primarily further enhances the selective memory and forget functions of the gate recurrent unit. To complete the update of h_t, it is essential to selectively retain part of the information contained in h_{t-1} based on the selective memory and forget requirements during model learning, while clearly specifying the new information to be incorporated. Therefore, using u_t to selectively choose between the information contained in h_{t-1} and $\widetilde{h}_t$, the final fusion is achieved to obtain h_t that has memorized effective information and forgotten invalid information.

In summary, the h_t obtained at the final moment is the result of selective memory of different spatial context information in different orientations achieved through the multi-temporal implementation of the candidate box features of fittings. The final output is used for regression prediction within the model, where the predicted results are evaluated against the ground truth through the calculation of a loss function. This process enables the model to quantify the discrepancy between its predictions and the actual values, thereby facilitating iterative improvements through optimization techniques to enhance predictive accuracy. Through backpropagation, the model parameters are updated, especially the updates of W_r, W_u, and W within the spatial context memory module. This process enables the model to induce and synthesize implicit spatial knowledge throughout effective memory retention. As a result, the model becomes capable of autonomously learning and integrating this implicit spatial knowledge, thereby improving its inference and detection capabilities within the framework. By leveraging such knowledge, the model enhances its ability to perform more accurate and robust predictions and detections across various scenarios.

10.4 Improved Post-Processing Structure

Many anchor-based object detection models face intrinsic challenges, particularly the scenario where multiple candidate bounding boxes are generated for the same detection object. Consequently, these models necessitate post-processing techniques to resolve and eliminate redundant predictions on the same object. These post-processing methods are crucial for refining the model's output, ensuring that only the most accurate and nonoverlapping predictions are retained, which ultimately enhances the precision and reliability of the object detection process. The most fundamental post-processing method is the traditional Non-Maximum Suppression (NMS) technique, which follows these steps: first, arrange all candidate bounding box detection results of a certain class in descending order based on their classification confidence; then, based on V (the box with the highest confidence level in the current classification), delete the same type of detection results adjacent to V (i.e., the same type of detection results whose IoU with V is higher than a certain threshold); and last, iterate the aforementioned procedure for the remaining unprocessed segments of the detection results within this category. Continue this process until all the detection results belonging to this category have been thoroughly processed.

The objective is to obtain a comprehensive set of post-processed results for this specific category, thereby ensuring accuracy and consistency across the entire detection framework. Therefore, while the ISKF structure facilitates the learning and practical application of implicit spatial knowledge within models, thereby enhancing detection precision, it remains ineffective in addressing the adaptability issues associated with traditional NMS methods. Specifically, from a post-processing standpoint, these traditional NMS techniques struggle to manage the complexities inherent in dense object detection tasks. Consequently, the integration of implicit spatial knowledge alone does not fully mitigate these limitations in scenarios requiring the detection of closely packed objects. The underlying reason is that traditional NMS methods struggle to adequately account for and balance the following two distinct scenarios. In the first scenario, if the selected IoU threshold is set too low, objects of the same type that are situated in densely populated areas may experience erroneous suppression. Specifically, the object with the lower confidence score is likely to be incorrectly eliminated. Conversely, in the second scenario, if the selected IoU threshold is set too high, redundant detection results for the same object are prone to being misclassified as separate, actual objects. This failure to effectively suppress redundant detections leads to an increased occurrence of duplicate detection phenomena. Traditional NMS methods, therefore, face substantial challenges in achieving an optimal balance between these conflicting needs, particularly in complex environments characterized by dense object distributions.

Therefore, to address the inherent challenges posed by the post-processing structure in dense occlusion tasks, this chapter introduces an improvement through the implementation of the Soft-NMS method. This enhancement is designed to significantly boost the model's adaptability and effectiveness in detecting objects within densely occluded environments. By integrating

Soft-NMS, the post-processing framework is refined to manage overlapping objects more efficiently, thereby reducing the likelihood of erroneous suppression. This methodological adjustment not only op0timizes the detection process but also ensures a more sophisticated and accurate interpretation of detection confidence scores, ultimately leading to more reliable results in complex detection scenarios. Unlike traditional NMS structures that directly remove adjacent boxes with an IoU higher than a certain threshold from V, the key concept of Soft-NMS is to gradually decay their confidence scores. Given that a higher IoU between detection boxes indicates a greater degree of overlap, and boxes with high overlap are more likely to correspond to the same object, it is essential to suppress redundant detection boxes. Consequently, the penalty for the decay of confidence scores should be directly proportional to the IoU. By establishing this positive correlation, the model can more effectively reduce false positives by penalizing overlapping detection boxes in accordance with the extent of their overlap, ensuring that only the most relevant detections are retained. Additionally, if the penalty function could be represented as a continuous function, it can avoid sudden changes in confidence ranking caused by discontinuous penalty functions. So, the penalty function for improving the post-processing structure is:

$$o_i = o_i e^{-\frac{iou(V, b_i)^2}{\sigma}} \tag{10.9}$$

In the equation, o_i represents the classification confidence of the penalty box, σ is the standard deviation of the Gaussian function, V represents the box with the highest classification confidence, and b_i represents the penalty box.

Therefore, based on Eq. (10.9), the following deductions can be made: When there is no overlap between boxes V and b_i (i.e., the IoU is 0), it signifies that the two boxes are not detecting the same object. In this scenario, o_i remains unchanged, and no penalty is required. Conversely, when there is an overlap between boxes V and b_i, the likelihood that the two boxes are detecting the same object increases with the degree of overlap (i.e., a larger IoU). Consequently, the penalty function should be continuous and monotonically decreasing with respect to the IoU value. As the IoU increases, the value of the penalty function decreases (corresponding to a higher degree of penalty), thereby causing o_i to diminish.

10.5 Experimental Results and Analysis

10.5.1 Experimental Dataset and Environment

The dataset images utilized in the experiment for this chapter were selected from those captured during unmanned aerial vehicle inspections of transmission lines. Due to the chapter's primary focus on the problem of dense occlusion of fittings, images containing densely packed objects were specifically chosen. The training set is composed of 1442 images, while the test set includes 348 images, resulting in a total of 1790 images. The LabelImg tool was used to selectively annotate 14 classes of fittings objects, resulting in a total of 25,675 annotated fittings objects. These 14 types of fittings are Pre-Twisted Suspension Clamp (PT), Bag-type Suspension Clamp (BT), Compression-Type Strain Clamp (CT), Wedge-Type Strain Clamp (WT), HB, U-Type Hanging Ring (UT), YP, Parallel Groove Clamp (PG), Shockproof Hammer (SH), Spacer (SP), GR, SR, Weight (WE) and AB, respectively.

The operating system version employed in this chapter is Ubuntu 18.04.6 LTS. The specialized accelerator card utilized is the NVIDIA RTX3090Ti. The programming language version used is Python 3.8, with Pytorch serving as the foundational framework for model development. The maximum number of iterations chosen is 18, and the batch size is set to 2. The stochastic gradient descent method is used as the optimizer, with a learning rate of 0.005, a momentum parameter of 0.9, and a WE decay rate of 0.0001.

10.5.2 Comprehensive Comparative Experiment

In order to comprehensively evaluate the progressiveness of the model presented in this chapter, an extensive comparison experiment was conducted. The evaluation was performed on the fitting dataset utilized in this chapter, using the COCO evaluation metrics and the time required to test each image as criteria for assessing model detection performance. The models subjected to this comparative evaluation include SSD [18] (designated as M1), RetinaNet [19] (designated as M2), YOLOF [20] (designated as M3), Dynamic R-CNN [21] (designated as M4), Libra R-CNN [22] (designated as M5), and the baseline model [12] (designated as M6). Additionally, the model proposed in this chapter is referred to as M7 in the comparative analysis. The results of this comprehensive comparison are detailed in Table 10.1, highlighting the detection performance across different models. This comparative study rigorously benchmarks the performance of the proposed model (M7) against various state-of-the-art detection models, thereby providing a holistic view of its effectiveness and efficiency in object detection tasks within the specific dataset context. By juxtaposing the detection metrics and the processing times of these models, the evaluation provides valuable insights into the advancements introduced by the proposed model. This methodology ensures a robust and transparent comparison, further validating the progressive improvements achieved through the innovations detailed in this chapter.

The experimental results presented in Table 10.1 unequivocally demonstrate that the model developed in this chapter markedly outperforms other models in terms of both detection accuracy and recall. Specifically, when examining the Average Precision (AP) metrics, the model exhibits impressive improvements

Table 10.1 Performance comparison results of different detection models.

	M1	M2	M3	M4	M5	M6	M7
AP^{50-95}(%)	33.5	35.5	43.5	43.7	45.3	42.1	**46.6**
AP^{50}(%)	65.8	66.7	77.1	75.1	80.1	77.5	**81.0**
AP^{75}(%)	30.4	33.6	43.4	45.4	45.7	41.9	**47.6**
AP^{S}(%)	13.9	9.1	14.1	19.6	14.1	14.9	**24.1**
AP^{M}(%)	14.0	25.7	27.1	34.0	32.3	32.9	**37.1**
AP^{L}(%)	37.5	38.6	47.6	46.8	48.3	45.0	**49.7**
AR^{1}(%)	22.4	22.5	26.5	26.4	**27.6**	25.7	27.3
AR^{10}(%)	41.5	46.8	51.1	51.8	53.9	50.1	**56.9**
AR^{100}(%)	42.6	48.1	51.7	52.2	54.9	50.9	**59.6**
AR^{S}(%)	20.6	15.4	14.0	19.6	17.1	15.1	**24.4**
AR^{M}(%)	22.1	32.4	34.0	39.2	39.0	39.4	**46.8**
AR^{L}(%)	46.8	51.8	56.4	55.9	58.4	54.0	**63.1**
T(s)	**0.008**	0.050	0.033	0.041	0.042	0.069	0.190

of 4.5%, 3.5%, and 5.7% on the AP50-95, AP50, and AP75 metrics, respectively. Notably, the enhancement on the more stringent AP75 metric is particularly pronounced. This significant advancement is attributed to the model's capability to leverage implicit spatial knowledge during the localization phase of the detection task. This enables the generation of prediction boxes that are more accurately positioned. Additionally, the IPP structure ensures that these accurately positioned prediction boxes are not erroneously discarded due to factors such as dense object distribution, ultimately preserving the most accurately located results in the final output. Furthermore, the detection accuracy of the model in this chapter has seen varying degrees of enhancement across three different scales: small, medium, and large.

Among these, the most substantial improvement is observed in the detection accuracy for small-scale objects (APS), which increased by 9.2%. This is primarily due to the model's original design objective of addressing the challenge of dense occlusion. Small fittings often occupy a limited number of pixels within an image; consequently, when occlusion occurs in densely distributed areas due to factors like connection structures and shooting angles, the occluded region constitutes a larger proportion of the small fitting area. This results in higher feature similarity and, consequently, poorer detection accuracy by conventional models. Therefore, by integrating implicit spatial knowledge to assist in differentiating objects with highly similar features and addressing the issue of dense occlusion, the model achieves a more significant improvement in detecting small fittings under such challenging conditions. In terms of the Average Recall (AR) metrics, the model introduced in this chapter demonstrates substantial improvement across all AR metrics compared to the baseline model. For instance, the AR1, AR10, and AR100 metrics have increased by 1.6%, 6.8%, and 8.7%, respectively. This further exemplifies the effectiveness of the model's combined improved structures in tackling the problem of dense occlusion. Although these enhancements may result in a certain degree of detection speed loss, this trade-off is justified by the significant gains in detection accuracy and recall rates. Such improvements are crucial for meeting the stringent requirements for high detection accuracy necessitated by the current demands for high safety and stability in the maintenance of transmission lines. Thus, the advancements embodied in the model present a substantial contribution toward achieving more reliable and accurate detection in complex environments.

Additionally, to facilitate a more granular comparison, this study employed the widely recognized AP50 evaluation index and the more stringent AP75 evaluation index as the primary metrics for assessing detection accuracy. Comparative experiments were conducted to evaluate the detection accuracy of various types of fittings using both the model proposed in this chapter and the baseline model. Based on the experimental findings, a comparative chart illustrating the detection accuracy for different types of fittings was generated, as shown in Figure 10.5. The results reveal that, for most types of fittings, the model proposed in this chapter exhibits notable improvements in both the AP50 and AP75 metrics. These enhancements are particularly pronounced in fittings that are typically challenging to detect and are prone to dense occlusion issues, such as HBs, U-type hanging rings, and CT strain clamps. This further underscores the superiority of the proposed model and its effectiveness in addressing the complexities posed by

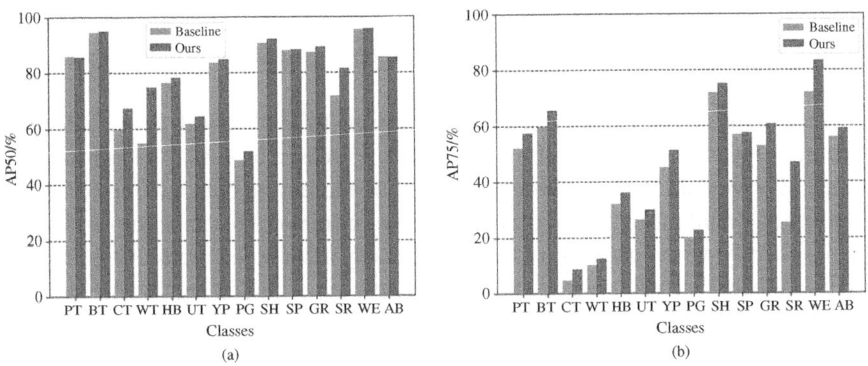

Figure 10.5 Comparison of detection accuracy for various fitting types. (a) Comparison of AP50 detection accuracy, (b) Comparison of AP75 detection accuracy.

dense occlusion scenarios. The improvements observed in the detection accuracy of these difficult-to-detect fittings highlight the robustness and adaptability of the proposed model's architecture. By effectively leveraging implicit spatial knowledge and implementing an IPP structure, the model is adept at distinguishing objects with high feature similarities and navigating densely occluded environments. This intrinsic capability significantly elevates the detection performance, particularly for fittings that present greater challenges in traditional detection frameworks. The comparative analysis provided in this chapter not only reinforces the efficacy of the proposed model but also demonstrates its advantageous application in solving dense occlusion problems. The enhancements in detection accuracy for various types of fittings affirm the model's potential for deployment in practical scenarios where high precision and reliability are paramount.

10.5.3 Ablation Experiment

1) **Model Structure Ablation Experiment:** The model proposed in this chapter principally comprises two critical components: the ISKF structure and the IPP structure. To ascertain the efficacy of these distinct structures within the proposed method, a series of ablation experiments were meticulously designed and executed. The results of these experiments are presented in Table 10.2. When the baseline model was augmented exclusively with the ISKF structure, notable enhancements were observed, with improvements of 3.5% and 3% in the AP50 and AP75 metrics, respectively. This advancement can be attributed to the unique design of the ISKF, which allows the model to adaptively assimilate implicit spatial knowledge during the training phase. This is achieved by extracting and effectively encoding the spatial context information pertinent to the objects to be detected. During the testing phase, the features of the fitting candidate box can leverage this memorized knowledge to fully integrate effective information. Consequently, the model can accurately adjust its detection criteria based on the implicit spatial knowledge acquired during training, along with the differing information of the objects and their surroundings, thus achieving precise detection outcomes.

Table 10.2 Results of model structure ablation experiment.

Method	ISKF	IPP	AP^{50}(%)	AP^{75}(%)
Baseline			77.5	41.9
	√		81.0	44.9
		√	78.2	45.0
Ours	√	√	**81.0**	**47.6**

Conversely, when the baseline model was solely enhanced with the IPP structure, there were increments of 0.7% and 3.1% in the AP50 and AP75 metrics, respectively. The rationale behind this improvement lies in the architecture of the IPP structure, which aligns with the core principles of Soft-NMS. Unlike the traditional NMS approach, Soft-NMS introduces a degree of confidence attenuation for detection results with lower confidence levels in adjacent objects of the same class. This methodology replaces the original strategy of outright deletion, thereby preventing the erroneous removal of accurate yet relatively low-confidence detection results. Finally, the method outlined in this chapter, which integrates both the implicit spatial knowledge structure and the IPP structure within the baseline model, resulted in substantial improvements. Specifically, the AP50 and AP75 metrics experienced boosts of 3.5% and 5.7%, respectively. This significant enhancement is a consequence of the synergistic benefits derived from combining these two augmented structures. The ISKF component compensates for the baseline model's deficiencies in logical reasoning capacity, while the Soft-NMS-based IPP structure addresses the limitations of the traditional NMS post-processing capabilities, especially in dense occlusion scenarios. This comprehensive approach underscores the efficacy of employing a dual-structured enhancement strategy to optimize the detection performance of the model, thereby demonstrating its superior adaptability and accuracy in complex detection tasks.

Ablation experiment on the spatial box coefficient: Within the ISKF structure of the model, the spatial box setting module includes a crucial parameter, specifically denoted as α (the spatial box coefficient). The impact of α on the spatial box paradigm is pivotal in determining the quantity of effective content within the subsequent spatial context information. This, in turn, has a significant influence on the model's detection results. Therefore, design ablation experiments about α to demonstrate its impact on model detection accuracy. Table 10.3 shows the results of the ablation experiment, and it can be seen that when α is set to 0.25, the comprehensive detection performance of the model is optimal due to the

Table 10.3 Results of spatial box coefficient ablation experiment.

α	AP^{50-95}(%)	AP^{50}(%)	AP^{75}(%)	AR^{1}(%)	AR^{10}(%)	AR^{100}(%)
0.10	45.4	80.1	45.7	26.5	56.0	59.1
0.15	45.3	80.0	46.1	26.3	56.2	58.9
0.20	46.1	80.1	**47.9**	27.3	56.3	59.1
0.25	**46.6**	**81.0**	47.6	**27.3**	**56.9**	**59.6**
0.30	45.6	80.2	47.5	26.8	55.6	58.7

spatial box coefficient being appropriately set. If the spatial box coefficient is too small, the box will not cover sufficient information, limiting the model's ability to learn implicit spatial knowledge and thereby reducing its effectiveness in aiding detection. Conversely, if the spatial box coefficient is set too large, the spatial box becomes excessively large, resulting in the corresponding area covering not only effective spatial context information but also an excess of background noise. This background noise introduces redundant information, which detracts from the model's detection capabilities and is not conducive to accurate model detection.

10.5.4 Visual Comparison Experiment

To qualitatively analyze the effectiveness of the model discussed in this chapter, a visual comparison experiment was designed. This experiment aimed to contrast the baseline model with the newly proposed model introduced in this chapter. The results of the comparison are presented in Figure 10.6. This approach allows for a clear visual assessment of the improvements and advantages of the model, providing a comprehensive understanding of its performance characteristics.

In Figure 10.6a, the obstructed bag-type suspension clamp provides limited visual information, making it difficult for the baseline model, which relies on direct perception, to detect it effectively. The model proposed in this chapter, however, utilizes implicit spatial knowledge as a supplementary driving force. In scenarios where visual information is constrained, this learned implicit spatial knowledge – when combined with contextual information surrounding the object – enables the model to engage in a higher degree of logical reasoning. As a result, the model can successfully detect occluded objects, as illustrated in Figure 10.6d. In Figure 10.6b, the baseline model often struggles with accurately predicting the precise positioning of SHs. Consequently, the post-processing phase may unintentionally suppress some true positive detections or fail to effectively eliminate redundant ones. In contrast, the model discussed in this chapter achieves significantly more accurate positioning of SHs, as seen in Figure 10.6e. This improvement is attributable to the integration of implicit spatial knowledge within the model, coupled with enhancements in the post-processing structure. These adaptations ensure more effective suppression of potential redundant detection results, thereby improving overall detection accuracy. Figure 10.6c showcases the grading shielding structure, characterized by the dense distribution of numerous fittings. Such density complicates the visual scene and exacerbates occlusion, limiting the baseline model's capacity to detect fittings such as U-type hanging rings and HBs reliably. The baseline model's post-processing method also falls short, struggling to adequately suppress detections of densely distributed, similar fittings. This issue leads to redundant detections of fittings like YPs and SRs. By contrast, the model introduced in this

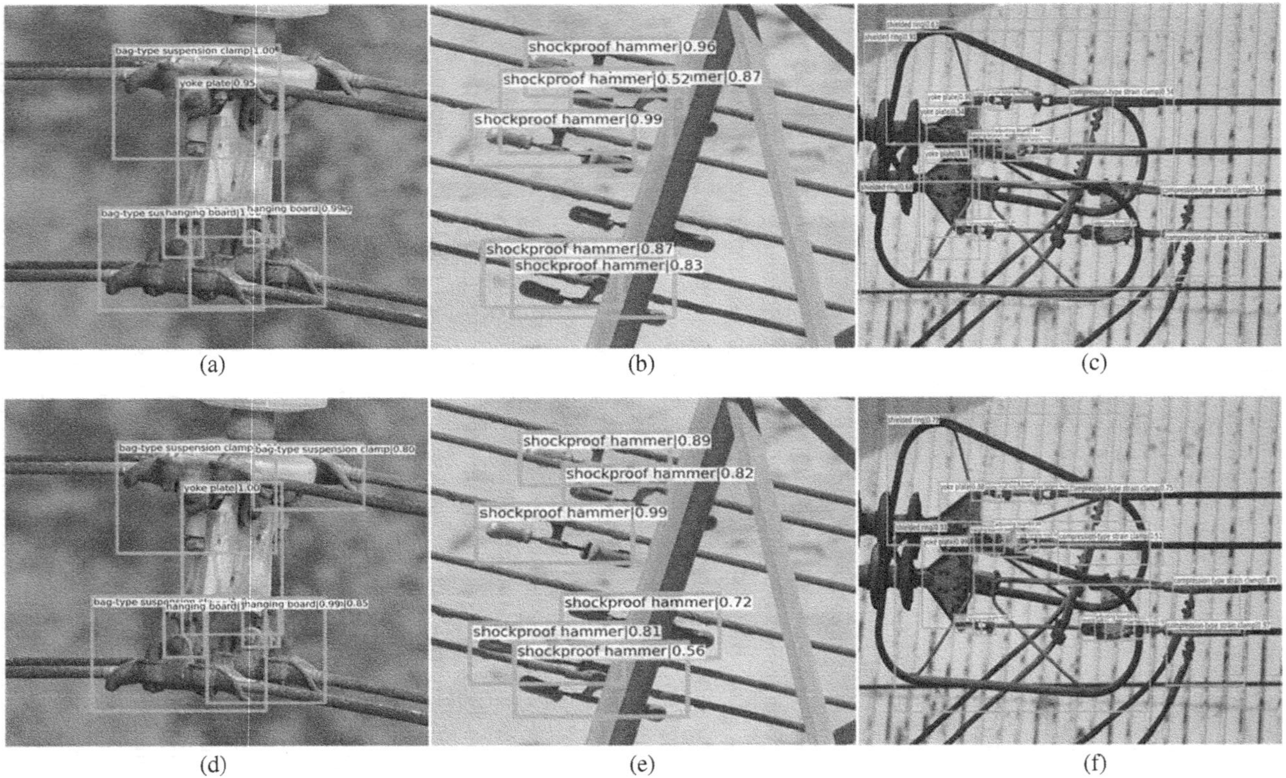

Figure 10.6 Visual comparison of detection results. (a) Baseline result 1, (b) Baseline result 2 (c) Baseline result 3, (d) Ours result 1 (e) Ours result 2 (f) Ours result 3.

chapter benefits from the combined assistance of two structural improvements, allowing it to detect previously missed U-type hanging rings and HBs, as shown in Figure 10.6f. Furthermore, it can effectively suppress redundant detections of YPs and SRs, thus offering a comprehensive solution to the detection deficiencies observed with the baseline model.

Based on the outcomes derived from the visual comparison experiment, it can be conclusively determined that the model developed in this chapter significantly enhances the logical reasoning capabilities by incorporating implicit spatial knowledge as an auxiliary driving force. This strategic integration allows the model to more effectively detect fitting objects that are prone to occlusion, even under challenging visual conditions. Specifically, the model's ability to infer and utilize spatial relationships implicitly enables it to excel in identifying objects that are partially or wholly obscured, which represents a notable improvement over traditional baseline models. Furthermore, the model exhibits remarkable adaptability in scenarios characterized by dense occlusion, where multiple objects overlap or are closely situated. These conditions often pose substantial challenges to conventional detection models, leading to a higher likelihood of missed detections and false positives. However, the model presented in this chapter leverages its enhanced logical reasoning, driven by implicit spatial knowledge, to navigate these complexities with impressive efficacy. In summary, the findings from the visual comparison experiment provide robust evidence that the model introduced in this chapter not only improves the detection accuracy of occluded fitting objects but also demonstrates exceptional adaptability to environments with significant occlusion. This advancement underscores the model's potential as a highly effective tool for object detection in complex and visually restricted settings, thereby highlighting a critical progression in the field of model-based detection systems.

10.6 Summary

Although the regularized structure and connection method between fittings contribute to dense occlusion, they also encapsulate rich implicit spatial knowledge. As a result, the methodology proposed in this chapter seeks to effectively harness this knowledge to assist the model in achieving more precise detection outcomes. The specific approach can be summarized as follows: (1) Designing a Spatial Box Setting Module and a Spatial Context Extraction Module: These modules facilitate the setting of spatial boxes and the specific extraction of spatial context information, respectively; (2) Developing a Spatial Context Memory Module: This module is responsible for filtering and retaining spatial context information, eliminating the impact of erroneous information, accurately learning implicit spatial knowledge, and ensuring its practical application in the actual detection process; (3) Enhancing the Model's Post-Processing Structure: Instead of relying on the original direct suppression method, this approach imposes varying degrees of punishment on the confidence scores of detection results. This improvement further augments the model's adaptability to tackle the issue of dense occlusion among fittings. Moreover, comprehensive experiments were conducted on the dataset previously mentioned. The experimental results demonstrate that the proposed model effectively addresses the problem of dense occlusion among fittings. Compared to the baseline model, it shows an improvement in the primary evaluation metrics, AP50 and AP75, by 3.5% and 5.7%, respectively. Additionally, its overall performance significantly surpasses that of other models. This novel approach provides a fresh perspective on the integration of implicit knowledge as a secondary driving force in intelligent inspection technology for transmission lines and establishes a foundation for further advancements in fault diagnosis.

References

1 Reddy, M.J.B., Karthik Chandra, B., and Mohanta, D.K. (2013). Condition monitoring of 11 kV distribution system insulators incorporating complex imagery using combined DOST-SVM approach. *IEEE Transactions on Dielectrics and Electrical Insulation* 20 (2): 664–674. https://doi.org/10.1109/TDEI.2013.6508770.

2 Choi, H., Koo, G., Kim, B.J., and Kim, S.W. (2021). Weakly supervised power line detection algorithm using a recursive noisy label update with refined broken line segments. *Expert Systems with Applications* 165: 113895. https://doi.org/10.1016/j.eswa.2020.113895.

3 Deng, F., Xie, Z., Mao, W. et al. (2022). Research on edge intelligent recognition method oriented to transmission line insulator fault detection. *International Journal of Electrical Power & Energy Systems* 139: 108054. https://doi.org/10.1016/j.ijepes.2022.108054.

4 Pazoki, M. (2017). A new fault classifier in transmission lines using intrinsic time decomposition. *IEEE Transactions on Industrial Informatics* 14 (2): 619–628. https://doi.org/10.1109/TII.2017.2741721.

5 Pournaras, E. and Espejo-Uribe, J. (2016). Self-repairable smart grids via online coordination of smart transformers. *IEEE Transactions on Industrial Informatics* 13 (4): 1783–1793. https://doi.org/10.1109/TII.2016.2625041.

6 Yang, L., Fan, J., Liu, Y. et al. (2020). A review on state-of-the-art power line inspection techniques. *IEEE Transactions on Instrumentation and Measurement* 69 (12): 9350–9365. https://doi.org/10.1109/TIM.2020.3031194.

7 Wang, S., Chen, W., Kuang, X.B. et al. (2020). Insulator area detection and localization algorithm based on multi-feature salient fusion. *Application Research of Computers* 37 (S2): 351–353.

8 Qi, Y.C., Jiang, A.X., Zhao, Z.B. et al. (2019). Fittings detection method in patrol images of transmission line based on improved SSD. *Electrical Measurement and Instrumentation* 56 (22): 7–12. https://doi.org/10.19753/j.issn1001-1390.2019.022.002.

9 Zhang, Y.X., Wu, G.P., Liu, Z.Y. et al. (2020). Transfer learning of transmission line damper and clamp detection based on YOLOv3 network. *Journal of Computer Applications* 40 (S2): 188–194. https://doi.org/10.11772/j.issn.1001-9081.2020020276.

10 Li, X., Liu, S.N., Yang, Z., and Wang, K.K. (2021). Multi-target detection of transmission lines based on improved cascade R-CNN. *Journal of Electronic Measurement and Instrument* 35 (10): 24–32. https://doi.org/10.13382/j.jemi.B2104058.

11 Yi, J.Y., Chen, C.F., and Gong, G.Q. (2021). Aerial insulator detection of transmission line based on improved Faster RCNN. *Computer Engineering* 47 (06): 292–298. https://doi.org/10.19678/j.issn.1000-3428.0059872.

12 Wu, Y., Chen, Y. P., Yuan, L. et al. (2020). Rethinking classification and localization for object detection. *Proceedings of the IEEE/CVF Conference on Computer Vision and Pattern Recognition*, 10186–10195. https://doi.org/10.48550/arXiv.1904.06493.

13 Bodla, N., Singh, B., Chellappa, R., and Davis, L.S. (2017). Soft-NMS–improving object detection with one line of code. *Proceedings of the IEEE International Conference on Computer Vision*, 5561–5569. https://doi.org/10.48550/arXiv.1704.04503.

14 Ren, S.Q., He, K., Girshick, R., and Sun, J. (2016). Faster R-CNN: Towards real-time object detection with region proposal networks. *IEEE Transactions on Pattern Analysis and Machine Intelligence* 39 (6): 1137–1149. https://doi.org/10.1109/TPAMI.2016.2577031.

15 He, K., Zhang, X.Y., Ren, S.Q., and Sun, J. (2016). Deep Residual Learning for Image Recognition. *Proceedings of the IEEE Conference on Computer Vision and Pattern Recognition* 770–778. https://doi.org/10.1109/cvpr.2016.90.

16 Lin, T.-Y., Dollár, P., Girshick, R. et al. (2017). Feature pyramid networks for object detection. *Proceedings of the IEEE Conference on Computer Vision and Pattern Recognition*, 2117–2125. https://doi.org/10.48550/arXiv.1612.03144.

17 Cho, K., van Merrienboer, B., Gulcehre, C. et al. (2014). Learning phrase representations using RNN encoder-decoder for statistical machine translation. *arXiv preprint arXiv:1406.1078*. https://doi.org/10.48550/arXiv.1406.1078.

18 Liu, W., Anguelov, D., Erhan, D. et al. (2016). Ssd: Single shot multibox detector. *Computer Vision–ECCV 2016: 14th European Conference, Amsterdam,*

The Netherlands, October 11–14, 2016, Proceedings, Part I 14, 21–37. https://doi.org/10.48550/arXiv.1512.02325.

19 Lin, T.-Y., Goyal, P., Girshick, R. et al. (2017). Focal loss for dense object detection. *Proceedings of the IEEE International Conference on Computer Vision*, 2980–2988. https://doi.org/10.48550/arXiv.1708.02002.

20 Chen, Q., Wang, Y., Yang, T. et al. (2021). You only look one-level feature. *Proceedings of the IEEE/CVF Conference on Computer Vision and Pattern Recognition*, 13039–13048. https://doi.org/10.48550/arXiv.2103.09460.

21 Zhang, H., Chang, H., Ma, B. et al. (2020). Dynamic R-CNN: Towards high quality object detection via dynamic training. *Computer Vision–ECCV* 2020: *16th European Conference, Glasgow, UK, August 23–28, Proceedings, Part XV 16*, 260–275. https://doi.org/10.48550/arXiv.2004.06002.

22 Pang, J., Chen, K., Shi, J. et al. (2019). Libra r-cnn: Towards balanced learning for object detection. *Proceedings of the IEEE/CVF Conference on Computer Vision and Pattern Recognition*, 821–830, 2019. https://doi.org/10.48550/arXiv.1904.02701.

Index

Artificial Intelligence Empowered Smart Energy Systems, First Edition. Edited by Qiang Yang and Gang Huang.

i

k

l

IEEE Press Series on Power and Energy Systems

Series Editor: Ganesh Kumar Venayagamoorthy, Clemson University, Clemson, South Carolina, USA.

The emphasis of the IEEE Press Series on Power and Energy Systems is to publish leading-edge books that cover a broad spectrum of current and forward-looking technologies in the fast-moving area of power and energy systems including smart grid, renewable energy systems, electric vehicles and related areas. Our target audience includes power and energy systems professionals from academia, industry and government who are interested in enhancing their knowledge and perspectives in their areas of interest.

1. *Electric Power Systems: Design and Analysis, Revised Printing*
   Mohamed E. El-Hawary
2. *Power System Stability*
   Edward W. Kimbark
3. *Analysis of Faulted Power Systems*
   Paul M. Anderson
4. *Inspection of Large Synchronous Machines: Checklists, Failure Identification, and Troubleshooting*
   Isidor Kerszenbaum
5. *Electric Power Applications of Fuzzy Systems*
   Mohamed E. El-Hawary
6. *Power System Protection*
   Paul M. Anderson
7. *Subsynchronous Resonance in Power Systems*
   Paul M. Anderson, B.L. Agrawal, and J.E. Van Ness
8. *Understanding Power Quality Problems: Voltage Sags and Interruptions*
   Math H. Bollen
9. *Analysis of Electric Machinery*
   Paul C. Krause, Oleg Wasynczuk, and S.D. Sudhoff
10. *Power System Control and Stability, Revised Printing*
    Paul M. Anderson and A.A. Fouad
11. *Principles of Electric Machines with Power Electronic Applications, Second Edition*
    Mohamed E. El-Hawary
12. *Pulse Width Modulation for Power Converters: Principles and Practice*
    D. Grahame Holmes and Thomas Lipo
13. *Analysis of Electric Machinery and Drive Systems, Second Edition*
    Paul C. Krause, Oleg Wasynczuk, and S.D. Sudhoff

14. *Risk Assessment for Power Systems: Models, Methods, and Applications*
    Wenyuan Li

15. *Optimization Principles: Practical Applications to the Operations of Markets of the Electric Power Industry*
    Narayan S. Rau

16. *Electric Economics: Regulation and Deregulation*
    Geoffrey Rothwell and Tomas Gomez

17. *Electric Power Systems: Analysis and Control*
    Fabio Saccomanno

18. *Electrical Insulation for Rotating Machines: Design, Evaluation, Aging, Testing, and Repair*
    Greg C. Stone, Edward A. Boulter, Ian Culbert, and Hussein Dhirani

19. *Signal Processing of Power Quality Disturbances*
    Math H.J. Bollen and Irene Y.H. Gu

20. *Instantaneous Power Theory and Applications to Power Conditioning*
    Hirofumi Akagi, Edson H. Watanabe, and Mauricio Aredes

21. *Maintaining Mission Critical Systems in a 24/7 Environment*
    Peter M. Curtis

22. *Elements of Tidal-Electric Engineering*
    Robert H. Clark

23. *Handbook of Large Turbo-Generator Operation and Maintenance, Second Edition*
    Geoff Klempner and Isidor Kerszenbaum

24. *Introduction to Electrical Power Systems*
    Mohamed E. El-Hawary

25. *Modeling and Control of Fuel Cells: Distributed Generation Applications*
    M. Hashem Nehrir and Caisheng Wang

26. *Power Distribution System Reliability: Practical Methods and Applications*
    Ali A. Chowdhury and Don O. Koval

27. *Introduction to FACTS Controllers: Theory, Modeling, and Applications*
    Kalyan K. Sen and Mey Ling Sen

28. *Economic Market Design and Planning for Electric Power Systems*
    James Momoh and Lamine Mili

29. *Operation and Control of Electric Energy Processing Systems*
    James Momoh and Lamine Mili

30. *Restructured Electric Power Systems: Analysis of Electricity Markets with Equilibrium Models*
    Xiao-Ping Zhang

31. *An Introduction to Wavelet Modulated Inverters*
    S.A. Saleh and M.A. Rahman

32. *Control of Electric Machine Drive Systems*
    Seung-Ki Sul

33. *Probabilistic Transmission System Planning*
Wenyuan Li

34. *Electricity Power Generation: The Changing Dimensions*
Digambar M. Tagare

35. *Electric Distribution Systems*
Abdelhay A. Sallam and Om P. Malik

36. *Practical Lighting Design with LEDs*
Ron Lenk and Carol Lenk

37. *High Voltage and Electrical Insulation Engineering*
Ravindra Arora and Wolfgang Mosch

38. *Maintaining Mission Critical Systems in a 24/7 Environment, Second Edition*
Peter Curtis

39. *Power Conversion and Control of Wind Energy Systems*
Bin Wu, Yongqiang Lang, Navid Zargari, and Samir Kouro

40. *Integration of Distributed Generation in the Power System*
Math H. Bollen and Fainan Hassan

41. *Doubly Fed Induction Machine: Modeling and Control for Wind Energy Generation Applications*
Gonzalo Abad, Jesús López, Miguel Rodrigues, Luis Marroyo, and Grzegorz Iwanski

42. *High Voltage Protection for Telecommunications*
Steven W. Blume

43. *Smart Grid: Fundamentals of Design and Analysis*
James Momoh

44. *Electromechanical Motion Devices, Second Edition*
Paul Krause Oleg, Wasynczuk, and Steven Pekarek

45. *Electrical Energy Conversion and Transport: An Interactive Computer-Based Approach, Second Edition*
George G. Karady and Keith E. Holbert

46. *ARC Flash Hazard and Analysis and Mitigation*
J.C. Das

47. *Handbook of Electrical Power System Dynamics: Modeling, Stability, and Control*
Mircea Eremia and Mohammad Shahidehpour

48. *Analysis of Electric Machinery and Drive Systems, Third Edition*
Paul C. Krause, Oleg Wasynczuk, S.D. Sudhoff, and Steven D. Pekarek

49. *Extruded Cables for High-Voltage Direct-Current Transmission: Advances in Research and Development*
Giovanni Mazzanti and Massimo Marzinotto

50. *Power Magnetic Devices: A Multi-Objective Design Approach*
S.D. Sudhoff

51. *Risk Assessment of Power Systems: Models, Methods, and Applications, Second Edition*
Wenyuan Li

52. *Practical Power System Operation*
Ebrahim Vaahedi

53. *The Selection Process of Biomass Materials for the Production of Bio-Fuels and Co-Firing*
Najib Altawell

54. *Electrical Insulation for Rotating Machines: Design, Evaluation, Aging, Testing, and Repair, Second Edition*
Greg C. Stone, Ian Culbert, Edward A. Boulter, and Hussein Dhirani

55. *Principles of Electrical Safety*
Peter E. Sutherland

56. *Advanced Power Electronics Converters: PWM Converters Processing AC Voltages*
Euzeli Cipriano dos Santos Jr. and Edison Roberto Cabral da Silva

57. *Optimization of Power System Operation, Second Edition*
Jizhong Zhu

58. *Power System Harmonics and Passive Filter Designs*
J.C. Das

59. *Digital Control of High-Frequency Switched-Mode Power Converters*
Luca Corradini, Dragan Maksimoviæ, Paolo Mattavelli, and Regan Zane

60. *Industrial Power Distribution, Second Edition*
Ralph E. Fehr, III

61. *HVDC Grids: For Offshore and Supergrid of the Future*
Dirk Van Hertem, Oriol Gomis-Bellmunt, and Jun Liang

62. *Advanced Solutions in Power Systems: HVDC, FACTS, and Artificial Intelligence*
Mircea Eremia, Chen-Ching Liu, and Abdel-Aty Edris

63. *Operation and Maintenance of Large Turbo-Generators*
Geoff Klempner and Isidor Kerszenbaum

64. *Electrical Energy Conversion and Transport: An Interactive Computer-Based Approach*
George G. Karady and Keith E. Holbert

65. *Modeling and High-Performance Control of Electric Machines*
John Chiasson

66. *Rating of Electric Power Cables in Unfavorable Thermal Environment*
George J. Anders

67. *Electric Power System Basics for the Nonelectrical Professional*
Steven W. Blume

68. *Modern Heuristic Optimization Techniques: Theory and Applications to Power Systems*
Kwang Y. Lee and Mohamed A. El-Sharkawi

69. *Real-Time Stability Assessment in Modern Power System Control Centers*
Savu C. Savulescu

70. *Optimization of Power System Operation*
Jizhong Zhu

71. *Insulators for Icing and Polluted Environments*
Masoud Farzaneh and William A. Chisholm

72. *PID and Predictive Control of Electric Devices and Power Converters Using MATLAB®/Simulink®*
Liuping Wang, Shan Chai, Dae Yoo, Lu Gan, and Ki Ng

73. *Power Grid Operation in a Market Environment: Economic Efficiency and Risk Mitigation*
Hong Chen

74. *Electric Power System Basics for Nonelectrical Professional, Second Edition*
Steven W. Blume

75. *Energy Production Systems Engineering Thomas*
Howard Blair

76. *Model Predictive Control of Wind Energy Conversion Systems*
Venkata Yaramasu and Bin Wu

77. *Understanding Symmetrical Components for Power System Modeling*
J.C. Das

78. *High-Power Converters and AC Drives, Second Edition*
Bin Wu and Mehdi Narimani

79. *Current Signature Analysis for Condition Monitoring of Cage Induction Motors: Industrial Application and Case Histories*
William T. Thomson and Ian Culbert

80. *Introduction to Electric Power and Drive Systems*
Paul Krause, Oleg Wasynczuk, Timothy O'Connell, and Maher Hasan

81. *Instantaneous Power Theory and Applications to Power Conditioning, Second Edition*
Hirofumi, Edson Hirokazu Watanabe, and Mauricio Aredes

82. *Practical Lighting Design with LEDs, Second Edition*
Ron Lenk and Carol Lenk

83. *Introduction to AC Machine Design*
Thomas A. Lipo

84. *Advances in Electric Power and Energy Systems: Load and Price Forecasting*
Mohamed E. El-Hawary

85. *Electricity Markets: Theories and Applications*
Jeremy Lin and Jernando H. Magnago

86. *Multiphysics Simulation by Design for Electrical Machines, Power Electronics and Drives*
Marius Rosu, Ping Zhou, Dingsheng Lin, Dan M. Ionel, Mircea Popescu, Frede Blaabjerg, Vandana Rallabandi, and David Staton

87. *Modular Multilevel Converters: Analysis, Control, and Applications*
Sixing Du, Apparao Dekka, Bin Wu, and Navid Zargari

88. *Electrical Railway Transportation Systems*
Morris Brenna, Federica Foiadelli, and Dario Zaninelli

89. *Energy Processing and Smart Grid*
James A. Momoh

90. *Handbook of Large Turbo-Generator Operation and Maintenance, 3rd Edition*
Geoff Klempner and Isidor Kerszenbaum

91. *Advanced Control of Doubly Fed Induction Generator for Wind Power Systems*
Dehong Xu, Dr. Frede Blaabjerg, Wenjie Chen, and Nan Zhu

92. *Electric Distribution Systems, 2nd Edition*
Abdelhay A. Sallam and Om P. Malik

93. *Power Electronics in Renewable Energy Systems and Smart Grid: Technology and Applications*
Bimal K. Bose

94. *Distributed Fiber Optic Sensing and Dynamic Rating of Power Cables*
Sudhakar Cherukupalli and George J. Anders

95. *Power System and Control and Stability, Third Edition*
Vijay Vittal, James D. McCalley, Paul M. Anderson, and A.A. Fouad

96. *Electromechanical Motion Devices: Rotating Magnetic Field-Based Analysis and Online Animations, Third Edition*
Paul Krause, Oleg Wasynczuk, Steven D. Pekarek, and Timothy O'Connell

97. *Applications of Modern Heuristic Optimization Methods in Power and Energy Systems*
Kwang Y. Lee and Zita A. Vale

98. *Handbook of Large Hydro Generators: Operation and Maintenance*
Glenn Mottershead, Stefano Bomben, Isidor Kerszenbaum, and Geoff Klempner

99. *Advances in Electric Power and Energy: Static State Estimation*
Mohamed E. El-hawary

100. *Arc Flash Hazard Analysis and Mitigation, Second Edition*
J.C. Das

101. *Maintaining Mission Critical Systems in a 24/7 Environment, Third Edition*
Peter M. Curtis

102. *Real-Time Electromagnetic Transient Simulation of AC-DC Networks*
Venkata Dinavahi and Ning Lin

103. *Probabilistic Power System Expansion Planning with Renewable Energy Resources and Energy Storage Systems*
Jaeseok Choi and Kwang Y. Lee

104. *Power Magnetic Devices: A Multi-Objective Design Approach, Second Edition*
Scott D. Sudhoff

105. *Optimal Coordination of Power Protective Devices with Illustrative Examples*
Ali R. Al-Roomi

106. *Resilient Control Architectures and Power Systems*
Craig Rieger, Ronald Boring, Brian Johnson, and Timothy McJunkin

107. *Alternative Liquid Dielectrics for High Voltage Transformer Insulation Systems: Performance Analysis and Applications*
Edited by U. Mohan Rao, I. Fofana, and R. Sarathi

108. *Introduction to the Analysis of Electromechanical Systems*
Paul C. Krause, Oleg Wasynczuk, and Timothy O'Connell

109. *Power Flow Control Solutions for a Modern Grid using SMART Power Flow Controllers*
Kalyan K. Sen and Mey Ling Sen

110. *Power System Protection: Fundamentals and Applications*
John Ciufo and Aaron Cooperberg

111. *Soft-Switching Technology for Three-phase Power Electronics Converters*
Dehong Xu, Rui Li, Ning He, Jinyi Deng, and Yuying Wu

112. *Power System Protection, Second Edition*
Paul M. Anderson, Charles Henville, Rasheek Rifaat, Brian Johnson, and Sakis Meliopoulos

113. *High Voltage and Electrical Insulation Engineering, Second Edition*
Ravindra Arora and Wolfgang Mosch

114. *Modeling and Control of Modern Electrical Energy Systems*
Masoud Karimi-Ghartemani

115. *Control of Power Electronic Converters with Microgrid Applications*
Armdam Ghosh and Firuz Zare

116. *Coordinated Operation and Planning of Modern Heat and Electricity Incorporated Networks*
Mohammadreza Daneshvar, Behnam Mohammadi-Ivatloo, and Kazem Zare

117. *Smart Energy for Transportation and Health in a Smart City*
Chun Sing Lai, Loi Lei Lai, and Qi Hong Lai

118. *Wireless Power Transfer: Principles and Applications*
Zhen Zhang and Hongliang Pang

119. *Intelligent Data Mining and Analysis in Power and Energy Systems: Models and Applications for Smarter Efficient Power Systems*
Zita Vale, Tiago Pinto, Michael Negnevitsky, and Ganesh Kumar Venayagamoorthy

120. *Introduction to Modern Analysis of Electric Machines and Drives*
Paul C. Krause and Thomas C. Krause

121. *Electromagnetic Analysis and Condition Monitoring of Synchronous Generators*
Hossein Ehya and Jawad Faiz

122. *Transportation Electrification: Breakthroughs in Electrified Vehicles, Aircraft, Rolling Stock, and Watercraft*
Ahmed A. Mohamed, Ahmad Arshan Khan, Ahmed T. Elsayed, and Mohamed A. Elshaer

123. *Modular Multilevel Converters: Control, Fault Detection, and Protection*
Fuji Deng, Chengkai Liu, and Zhe Chen

124. *Stability-Constrained Optimization for Modern Power System Operation and Planning*
Yan Xu, Yuan Chi, and Heling Yuan

125. *Interval Methods for Uncertain Power System Analysis*
Alfredo Vaccaro

126. *Practical Partial Discharge Measurement on Electrical Equipment*
Greg Stone, Andrea Cavallini, Glenn Behrmann, and Claudio Angelo Serafino

127. *Graph Database and Graph Computing for Power System Analysis*
Renchang Dai and Guangyi Liu

128. *The Power of Artificial Intelligence for the Next-Generation Oil and Gas Industry: Envisaging AI-inspired Intelligent Energy Systems and Environments*
Pethuru Raj Chelliah, Venkatraman Jayasankar, Mats Agerstam, B. Sundaravadivazhagan, and Robin Cyriac

129. *Microgrids: Theory and Practice*
Peng Zhang

130. *Electric Machinery and Drives: An Electromagnetics Perspective*
Nabeel A. O. Demerdash, JiangBiao He, and Hao Chen

131. *Smart Cyber-Physical Power Systems, Volume 1: Fundamental Concepts, Challenges, and Solutions*
Ali Parizad, Hamid Reza Baghaee, and Saifur Rahman

132. *Smart Cyber-Physical Power Systems, Volume 2: Solutions from Emerging Technologies*
Ali Parizad, Hamid Reza Baghaee, and Saifur Rahman

133. *Analysis of Electric Machinery and Drive Systems, Fourth Edition*
Paul C. Krause, Oleg Wasynczuk, Scott D. Sudhoff, and Steven D. Pekarek

134. *Understanding Electromagnetic Transients in Power System*
Luiz Cera Zanetta Jr.

135. *Communities for Clean Energy Justice and Equity in Grid Modernization*
Mohammadreza Daneshvar, Behnam Mohammadi-Ivatloo, and Amjad Anvari-Moghaddam

136. *Modern Power System*
Arindam Ghosh

137. *Hidden Risk in Occupational Electrical Safety: Through the Lens of System Safety*
H. Landis "Lanny" Floyd

138. *Artificial Intelligence Empowered Smart Energy Systems*
Qiang Yang and Gang Huang

Printed and bound by CPI Group (UK) Ltd, Croydon, CR0 4YY
17/09/2026
14964329-0006

Microplastic Monitoring Using Artificial Intelligence

Scrivener Publishing
100 Cummings Center, Suite 541J
Beverly, MA 01915-6106

Publishers at Scrivener
Martin Scrivener (martin@scrivenerpublishing.com)
Phillip Carmical (pcarmical@scrivenerpublishing.com)

Microplastic Monitoring Using Artificial Intelligence

Edited by

Abhishek Kumar

Pooja Dixit

Pramod Singh Rathore

Arun Lal Srivastav

and

Ashutosh Kumar Dubey

Scrivener Publishing

WILEY

This edition first published 2026 by John Wiley & Sons, Inc., 111 River Street, Hoboken, NJ 07030, USA and Scrivener Publishing LLC, 100 Cummings Center, Suite 541J, Beverly, MA 01915, USA

For more information about Scrivener publications please visit www.scrivenerpublishing.com.

Wiley Global Headquarters
111 River Street, Hoboken, NJ 07030, USA

For details of our global editorial offices, customer services, and more information about Wiley products visit us at www.wiley.com.

The manufacturer's authorized representative according to the EU General Product Safety Regulation is Wiley-VCH GmbH, Boschstr. 12, 69469 Weinheim, Germany, e-mail: Product_Safety@wiley.com.

Library of Congress Cataloging-in-Publication Data

ISBN 9781394450084

Cover image: Generated with AI using Adobe Firefly
Cover design by Russell Richardson

Set in size of 11pt and Minion Pro by Manila Typesetting Company, Makati, Philippines

Printed in the USA

10 9 8 7 6 5 4 3 2 1

Contents

Preface

The accelerating crisis of microplastic pollution has emerged as a critical environmental and public health challenge in the 21st century. From the depths of our oceans to the air we breathe, microplastics have permeated ecosystems, food chains, and human bodies, demanding urgent attention and innovative solutions. This book, *Microplastic Monitoring Using Artificial Intelligence*, brings together a diverse collection of chapters that explore how emerging technologies—particularly Artificial Intelligence (AI), Machine Learning (ML), and advanced sensor systems—can transform our understanding, detection, and control of microplastic pollution.

The book begins by establishing the foundational context of microplastic pollution, with **Chapter 1** offering a comprehensive overview of microplastics in air and the environment, exploring their sources, transport mechanisms, and ecological impacts. This sets the stage for the integration of AI technologies in the subsequent chapters.

Chapter 2 builds on this foundation by providing a detailed analysis of AI-driven technologies in microplastic pollution mitigation, outlining how AI can enhance detection accuracy, processing efficiency, and real-time decision-making.

Chapter 3 introduces Explainable AI (XAI) as a transformative approach to improving transparency and trust in AI-driven microplastic monitoring systems. The chapter illustrates how interpretable models can support risk assessment, stakeholder engagement, and policy formulation.

Chapter 4 delves deeper into the specific role of AI in microplastic detection, showcasing how AI algorithms, image processing techniques, and predictive models surpass traditional methods in accuracy, scalability, and reliability.

Chapter 5 transitions into the realm of AI-driven optical imaging and spectroscopic techniques, detailing advanced methods like Raman and FTIR spectroscopy, and demonstrating their potential when coupled with AI models for precise and automated detection.

Chapter 6 expands the scope by exploring the integration of AI with advanced sensor technologies for real-time monitoring of microplastics. It presents a multi-dimensional perspective on how AI and sensor systems can map microplastic hotspots, support rapid mitigation, and inform policy frameworks.

Chapter 7 presents a focused study on using machine learning for microplastic source and pathway prediction, highlighting how ML models can simulate complex environmental dynamics, identify pollution origins, and assist in targeted mitigation strategies.

Chapter 8 shifts the lens to big data analytics, emphasizing how large-scale datasets from diverse sources—satellite imagery, citizen science, IoT sensors—can be processed using AI tools to map the global distribution of microplastics and uncover patterns and trends.

Chapter 9 explores the synergy between automation, robotics, and AI in microplastic sampling and processing. This chapter underscores how intelligent systems can enhance the efficiency, consistency, and scalability of environmental monitoring and remediation efforts.

Chapter 10: This chapter presents real-world case studies demonstrating the use of AI in microplastic detection, monitoring, and analysis. It highlights successful implementations across diverse ecosystems, showcasing the adaptability and scalability of AI technologies. Readers will gain practical insights into integrating AI into environmental projects.

Chapter 11: AI systems for environmental monitoring raise critical ethical considerations, including data privacy, algorithmic bias, and the societal impact of technology adoption. This chapter explores these issues, advocating for transparent, inclusive, and responsible AI deployment in microplastic research.

Chapter 12: This chapter examines the legal and policy landscape surrounding AI-driven microplastic monitoring. It discusses global standards, data governance, and frameworks that guide responsible AI use while ensuring environmental protection and public trust.

Chapter 13: Emerging AI technologies offer groundbreaking potential for environmental monitoring, including microplastic detection. This chapter explores future trends, from quantum computing integration to decentralized AI frameworks, highlighting pathways for next-generation environmental intelligence.

Chapter 14: This chapter focuses on the role of Explainable AI (XAI) in enhancing transparency and trust in microplastic detection systems. It emphasizes model interpretability, stakeholder engagement, and the importance of explainable models in environmental decision-making.

Chapter 15: This chapter provides a forward-looking perspective on AI's potential to revolutionize microplastic detection and control. It highlights collaborative efforts, cross-disciplinary approaches, and policy integration as key drivers for achieving a sustainable, AI-powered environmental future.

Chapter 16: This chapter presents frameworks for combining AI, advanced sensors, and automated data analysis in large-scale microplastic monitoring. It outlines real-time detection systems, scalable architectures, and best practices for AI adoption in environmental surveillance.

1

Introduction to Microplastic and the Role of AI

Pooja Dixit[1]*, Shaloo Dadheech[2], Priya Batta[3] and Neeraj Bhargava[4]

[1]*Department of Computer Science, Shri Ratanlal Kanwarlal Patni Girls' College, Kishangarh, India*
[2]*Department of Computer Applications, JIET, Jodhpur, India*
[3]*Department of Computer Science, Amity University Mohali, Punjab, India*
[4]*Department of Computer Application, Amity University Jaipur, Rajasthan, India*

Abstract
Microplastic pollution has become a global environmental challenge of today that directly threatens human health, biodiversity and sustainability. Microplastics – plastic particles smaller than 5 mm – have spread everywhere; in oceans, rivers, soil and air. They also reach the human body through the food chain, where they create health risks through ingestion, inhalation and toxic leaching. Traditional detection techniques such as microscopy, FTIR and Raman spectroscopy are useful but are time-consuming, costly and have limited scalability due to which their efficiency is low. Artificial Intelligence (AI) fills this gap. AI-based methods such as machine learning and deep learning make automated detection, classification and predictive modeling of microplastics possible. AI with sensors and IoT integration also enables real-time monitoring and pollution hotspot mapping. But challenges such as data availability, high computational cost and ethical concerns need to be addressed. In the future, through explainable AI, interdisciplinary collaborations and policy integration, AI could become a game-changer for microplastic management. Overall, this study establishes a strong bridge between microplastics and AI that can provide sustainable solutions for both environment and human health.

***Keywords*:** Microplastics, AI, FTIR, Raman spectroscopy

**Corresponding author*: poojadixit565@gmail.com

Abhishek Kumar, Pooja Dixit, Pramod Singh Rathore, Arun Lal Srivastav and Ashutosh Kumar Dubey (eds.) Microplastic Monitoring Using Artificial Intelligence, (1–22) © 2026 Scrivener Publishing LLC

1.1 Introduction

1.1.1 Background and Importance of the Study

Today, environmental pollution has become a global concern, in which a relatively "new" and hidden pollutant—microplastics—has become a major challenge for scientists, policymakers and the general public. Microplastics means those small particles of plastic which are smaller than 5 millimeters. Their size is so small that they easily spread in every part of the environment - whether it is in ocean, rivers, soil or air. In the last 50-60 years, plastic production has increased exponentially. Plastic was initially considered a "wonder material" because it was durable, lightweight and cheap. But now this durability has become an environmental disaster. When this plastic degrades, it breaks into small pieces and forms microplastics. These particles persist in the environment for centuries.

This study is important because microplastics are not just a pollution problem but also a threat to human health, biodiversity and global sustainability. Microplastics affect marine species in oceans, reduce soil fertility and even reach the human body through the food chain. The role of AI (Artificial Intelligence) becomes important in this field because detection and monitoring of microplastics using conventional methods is time-consuming and expensive. AI-based models help with automatic detection, classification, and prediction. This chapter, therefore, forms a foundation for understanding microplastics and connecting the role of AI [1, 2].

1.1.2 Definition of Microplastics

Scientists have used different definitions to define microplastics, but the consensus is that they are plastic particles smaller than 5 mm in size. Some researchers refine this further, such as:

- Large Microplastics: 1–5 mm
- Small Microplastics: 1 μm–1 mm

And now there is the concept of nano plastics, which are particles smaller than 1 μm.

Microplastics are generally divided into two categories:

- Primary Microplastics: These are produced directly in small sizes, such as microbeads in cosmetics, personal care products, industrial abrasives.

- Secondary Microplastics: These are formed after the breakdown of larger plastics, such as plastic bags, bottles or fishing nets, which degrade and convert into microplastics.

Another important aspect of the definition is their composition and chemical nature. Plastics are polymer-based, with additives, stabilizers, and dyes mixed in. These additives make microplastics more toxic when they are released into the environment [3].

1.1.3 Sources and Types of Microplastics

The sources of microplastics are so diverse that they can originate from practically everywhere. Some major sources are listed below:

• Industrial and Consumer Products:
Microbeads in cosmetics and toothpaste.
Fibers derived from synthetic textiles (polyester, nylon) when we use washing machines.

• Degradation of Larger Plastics:
Sunlight (UV radiation), physical abrasion and chemical weathering break down larger plastic items to form microplastics.

• Wastewater Treatment Plants:
This is a major pathway. The microplastics we generate for domestic or industrial use are released into the environment through wastewater.

• Tire Wear Particles:
When vehicles run on roads, the friction of their tires releases small rubber particles, which is a hidden source of microplastics.

Types of Microplastics:

- Fibres (synthetic clothes, ropes, nets)
- Fragments (breakdown pieces of large plastics)
- Films (thin plastic layers like plastic bags)
- Beads/Pellets (cosmetic products or plastic industry raw material)

It is very important to identify their type and source, as this knowledge helps in creating AI-based classification systems [4, 5].

1.1.4 Environmental and Health Impacts

The impacts of microplastics are multidimensional. It is a "silent pollutant" that is not directly visible, but gradually damages both ecosystem and human health.

(A) Environmental Impacts
• Marine Ecosystems:
Oceans have become the biggest sink for microplastics. Marine organisms such as fish, plankton, and seabirds consume microplastics as food. This has a negative impact on their growth, reproduction and survival rates.

• Freshwater Systems:
Rivers and lakes become carriers of microplastics. These particles disturb aquatic biodiversity and ultimately reach the oceans.

• Soil Systems:
Sewage sludge used as fertilizer in agriculture also contains microplastics. They reduce soil fertility and microbial activity.

• Airborne Pollution:
Microplastics are also suspended in the air. They circulate with dust particles and are capable of long-distance transport.

(B) Human Health Impacts
• Ingestion:
Humans ingest microplastics through contaminated seafood, drinking water, and even packaged food.

• Inhalation:
Airborne microplastics reach the lungs, which can cause respiratory problems.

• Toxicity and Chemical Leaching:
Microplastics not only create physical blockage, but they also contain additives and absorbed pollutants (like pesticides, heavy metals), which release toxic chemicals.

• Potential Health Risks:
Endocrine disruption
Inflammation
Carcinogenic effects (long-term exposure)

These impacts are still the subject of ongoing research, but evidence strongly suggests that microplastics remain an emerging public health crisis [6].

1.2 Microplastic Distribution and Pathways

Once microplastic particles are released into the environment, they circulate in different mediums—water (marine and freshwater), soil (soil and agriculture), air (airborne particles), and finally accumulate in the bodies of living organisms through the food chain. This section explains in detail how microplastics are distributed and their major pathways.

1.2.1 Marine and Freshwater Systems

Marine and freshwater ecosystems are the largest sinks for microplastics. The majority of global plastic waste ultimately reaches oceans and rivers. When plastic items fall into oceans, they degrade and convert into microplastics by mechanical abrasion, UV radiation, and wave action.

Marine Systems (Oceans, Seas):
Research studies show that microplastics are present in surface waters, deep sea sediments, and even in Arctic ice cores.
Ocean currents carry them for thousands of kilometers, causing microplastics to be found even in remote and pristine areas.
Marine organisms such as plankton, mollusks, fish, and seabirds accidentally ingest these particles. This ingestion disturbs their growth, reproduction, and metabolic activities.

Freshwater Systems (Rivers, Lakes):
Rivers act as transport highways for microplastics. Microplastics generated from cities and industries enter rivers as wastewater and finally reach the oceans.
Lakes accumulate microplastics due to stagnant water.
Studies have shown that freshwater organisms such as crustaceans and small fish also consume microplastics, which changes their food chain dynamics.

Important Point:
Marine and freshwater systems form an interconnected cycle in which once plastic enters, it becomes almost impossible to completely remove it [7].

1.2.2 Soil and Agricultural Environments

When we talk about microplastics, the focus is usually on oceans. But soil and agriculture are equally vulnerable.

Sources in Soil Systems:
• Sewage Sludge Application:
Wastewater treatment plants that produce sludge contain large amounts of microplastics. When this sludge is used as fertilizer in fields, microplastics are directly deposited in the soil.

• Agricultural Practices:
Plastic mulching (used to cover crops) degrades and releases small particles into the soil.
Greenhouses and irrigation pipes are also a source.

• Atmospheric Deposition:
Airborne microplastics are also deposited on the soil surface.

• Impacts on Soil Health:
Soil microorganisms such as bacteria and fungi are directly affected by the presence of microplastics.
Soil porosity and water-holding capacity also change.
Agricultural productivity may be indirectly reduced as soil fertility is disturbed.

Agriculture and Food Safety:

- Crops that grow in the soil can uptake microplastics (especially nano plastics).
- This makes food safety a major concern as humans and livestock consume contaminated crops.

1.2.3 Airborne Microplastics

The presence of microplastics in air is a relatively recent discovery, but its impact is very serious.

Sources of Airborne Microplastics:

- Synthetic textiles: When we wash or wear clothes, microfibers are released that easily get suspended in the air.

- Urban dust: Roads, tire wear particles and construction materials also release microplastics.
- Industrial emissions: Factories and industrial sites are major contributors of airborne particles.

Transport Mechanisms:

- Airborne microplastics are very lightweight, so they can be transported long-distance. For example, studies have detected them in remote regions such as the Himalayas and the Arctic.
- Atmospheric currents carry them from one continent to another.

Effects of Airborne Microplastics:

- Human exposure *via* inhalation is a major health concern.
- Accumulation in the lungs can lead to respiratory problems, asthma, and long-term chronic diseases.
- Airborne deposition also contaminates soil and water systems, meaning an interconnected pathway is formed.

1.2.4 Bioaccumulation in the Food Chain

The most dangerous aspect of microplastics is their bioaccumulation and biomagnification.

Bioaccumulation:

- When an organism continuously ingests microplastics, the particles accumulate in its body tissues.
- For example, when small fish and plankton eat microplastics, they are stored in their digestive tract.

Biomagnification:

- When a predator eats contaminated prey, the microplastics and their associated toxins are transferred upward in the food chain.
- Higher trophic level species (such as big fish, marine mammals, and finally humans) are most at risk.

Food Chain Examples:
Marine Food Chain:

- Plankton → Small Fish → Large Fish → Humans
- At each step, the concentration of microplastics and associated chemicals increases.

Terrestrial Food Chain:

- Soil → Crops → Livestock → Humans
- Particles absorbed from the soil go into crops and are then passed on to consuming animals and humans.

Impacts on Human Health:

- Humans have chronic low-dose exposure through the food chain.
- Additives and absorbed pollutants (such as pesticides and heavy metals) disrupt the body's metabolism and immune system.
- This exposure can enhance the risk of long-term diseases such as endocrine disorders and cancer.

1.3 Current Methods of Microplastic Detection

The detection and analysis of microplastics is a very challenging task. The reason is that their size is extremely small, shape and color are diverse, and they are present in different environmental matrices (water, soil, air, organisms). Therefore, researchers have developed a number of sampling, collection and analytical methods that identify and quantify microplastics [8].

1.3.1 Sampling and Collection Techniques

The first step in detecting microplastics is their sampling and collection. These techniques depend on which medium of the environment the sample is being taken from—water, soil, air, or organisms.

(A) Water Sampling (Marine and Freshwater):
Surface Water Sampling:

- The most commonly used tool is neuston nets or manta trawls, with a mesh size of around 300 μm.
- These are dragged along by ships or boats to collect microplastics from the surface layer.

Sub-Surface Sampling:

- Water bottles or pumping systems are used to collect water from different depths.

(B) Sediment and Soil Sampling:

- Sediment Cores: Samples are taken from riverbeds, beaches or ocean floors using core samplers.
- Density Separation Technique: Soil or sediment sample is placed in a high-density liquid (such as NaCl, ZnCl2 solution). Plastics are light, so they float and can be collected.

(C) Air Sampling:

- High-volume air samplers fitted with filters are used to collect airborne microplastics.
- These filters trap particles which are later used for microscopic and spectroscopic analysis.

(D) Biota Sampling (Organisms):

- Microplastics are extracted by dissecting the digestive tracts of marine and terrestrial organisms (fish, mollusks, birds).
- Plastics are separated by chemically digesting the tissue of the organism.

1.3.2 Conventional Analytical Methods (Microscopy, FTIR, Raman Spectroscopy)

The most important step after sampling is analysis and identification. Till date, the most widely accepted methods are microscopy-based and spectroscopy-based techniques.

(A) Microscopy Techniques:

- Visual Microscopy (Stereo/Optical Microscope):
- A stereo microscope is used to visually identify microplastics.
- Researchers observe their size, color, shape (fibers, fragments, films, beads).
- Limitation: This method is subjective and requires an experienced observer.

Scanning Electron Microscopy (SEM):

- SEM shows detailed surface morphology of microplastics.
- It can be used to study the texture and degradation level of the particle.

(B) Fourier Transform Infrared Spectroscopy (FTIR):

- FTIR is a spectroscopic technique that identifies the chemical composition of microplastics.
- When infrared light falls on a microplastic, it generates a specific absorbance spectrum that acts as a fingerprint of the polymer.

Types of FTIR methods:

- ATR-FTIR (Attenuated Total Reflectance): for large particles.
- μ-FTIR (Micro-FTIR): for small particles (down to 20 μm).
- Advantage: Non-destructive and reliable polymer identification.

(C) Raman Spectroscopy:

- Raman spectroscopy is another powerful tool based on the principle of laser light scattering.
- Raman provides more detailed chemical information than FTIR and can identify smaller particles (up to 1 μm).
- It is particularly useful for the analysis of colored microplastics.
- Limitation: Fluorescence interference sometimes complicates the analysis.

(D) Thermal Analysis (Supporting Technique):

- Differential Scanning Calorimetry (DSC) and Pyrolysis-GC/MS (Gas Chromatography/Mass Spectrometry) are also used.
- These methods heat plastics and analyze their breakdown products.
- This helps identify the polymer type and additives of microplastics [9].

1.3.3 Limitations of Traditional Approaches

While conventional methods are widely accepted for studying microplastics, they have some critical limitations that make research slow and expensive.

(A) Time-Consuming and Labor-Intensive:

- Sampling, density separation, and microscopic identification are very tedious and slow processes.
- Large-scale monitoring is practically difficult.

(B) Size Detection Limitations:

- The effective detection limit of stereo microscopy and FTIR is around 20 μm.
- Nano plastics (those <1 μm) are extremely difficult to detect by these techniques.

(C) Subjectivity in Visual Identification:

- Identification of particles on microscopy depends on the experience of the researcher.
- This can lead to "observer bias" and inconsistent results.

(D) High Cost and Equipment Requirements:

- FTIR and Raman spectrometers are costly instruments and may not be available in every lab.
- Skilled manpower and regular calibration are required.

(E) Contamination Risks:

- There is a high risk of plastic contamination during sampling and lab work (e.g. lab coat fibers, plastic labware).
- If contamination control protocols are not strictly followed, results may be unreliable.

(F) Lack of Standardized Protocols:

- There are no globally standardized methods available for microplastic detection yet.
- Different researchers follow different protocols, making data comparison difficult.

1.4 Role of Artificial Intelligence (AI) in Microplastic Research

Detecting, monitoring and predicting microplastic pollution using conventional methods is very slow, expensive and limited. This is the point where Artificial Intelligence (AI) and Machine Learning (ML) have emerged as a game-changer. Through AI, we can do automated detection, faster classification, pollution modeling and real-time monitoring for microplastics.

1.4.1 Introduction to AI and Machine Learning Concepts

AI means making machines "intelligent" – that is, training them to make human-like decisions. This includes Machine Learning (ML) and Deep Learning (DL), which learn patterns from large datasets and make predictions and classifications.

Machine Learning (ML): Algorithms such as Support Vector Machines (SVM), Random Forest (RF), and k-Nearest Neighbor (kNN) analyze features (size, shape, color) of microplastics and make classification [10].

Deep Learning (DL): Neural networks (CNNs – Convolutional Neural Networks) are very powerful especially in image-based tasks. They automatically perform feature extraction and detect microplastics with high accuracy.

Strengths of AI are:

- Automation – Reduces the need for manual identification.
- Speed – Quickly analyzes thousands of images.
- Accuracy – Reduces human error and subjectivity.
- Scalability – Large datasets and continuous monitoring are possible.

1.4.2 AI for Image-Based Microplastic Identification

- Traditional microscopy is a slow and subjective method. With AI, microplastic identification can be automated.

Steps in AI-Based Image Analysis:

- Image Acquisition: Sample images are collected through a microscope or camera.
- Pre-Processing: Noise removal, contrast enhancement, and background subtraction are performed.
- Feature Extraction: AI models extract features of size, shape (fiber, bead, fragment), color, and texture.
- Classification: CNNs and ML classifiers divide microplastics into categories.

Example Applications:

- CNN models differentiate microplastics and non-plastic particles from microscope images.
- Through transfer learning, pre-trained models (ResNet, VGGNet) can be fine-tuned for microplastic datasets.
- Automated systems analyze thousands of particles to produce dataset-ready results that would take weeks with the manual method.

Advantage:
This method improves both speed and objectivity, and nano plastics that are difficult to detect with the human eye can be more easily identified with AI.

1.4.3 AI for Predictive Modeling of Microplastic Pollution

AI is not limited to just detection, but is also used in predictive modeling. Matlab, AI models predict future pollution trends based on past data and environmental parameters.

Inputs for Predictive Models:

- Environmental factors: wind, water current, rainfall, temperature.
- Human activities: plastic waste generation, urbanization, industrial activity.
- Geographic factors: river discharge, coastal regions, soil type.

Modeling Techniques:

- Regression Models and Random Forests: To predict spatial and temporal patterns of pollution levels.
- Deep Learning Models: To handle complex non-linear relationships, such as ocean currents and plastic transport modeling.
- GIS + AI Integration: Geographic Information Systems can be used to map microplastic hotspots.

Example:
AI can predict which coastal areas will be at the highest risk of microplastic pollution in the next 10 years. This becomes an early warning system for policymakers.

1.4.4 AI in Real-Time Monitoring and Sensing

- Real-time monitoring of microplastics is very important, especially for wastewater treatment plants and marine systems.

AI + Sensors Combination:

- Optical Sensors: Optical sensors are used for real-time water monitoring. AI detects microplastic particles by analyzing their signals.

- Spectral Sensors: AI models with Raman/FTIR sensors instantly analyze spectral fingerprints to identify microplastics.
- Autonomous Underwater Vehicles (AUVs): Drones and robotic vehicles collect water samples and analyze them with onboard AI models.

Benefits:

- Instant feedback for pollution control.
- Continuous monitoring without manual intervention.
- Early detection → better management strategies.

Example:
In a wastewater treatment plant, AI-enabled sensors detect the amount of microplastic present in treated water, and based on that, filtration efficiency can be improved.

1.4.5 Integration of AI with IoT and Remote Sensing

The power of AI is further enhanced when it is combined with IoT (Internet of Things) and Remote Sensing technologies.

IoT Integration:

- IoT devices (sensors, cameras, drones) collect real-time data continuously.
- This data is fed to AI models that provide instant analysis and predictions.

Example: Smart buoys installed in oceans to continuously monitor microplastic density.
Remote Sensing Integration:

- Satellites and drones collect high-resolution images and spectral data.
- AI algorithms analyze these data to detect coastal and riverine microplastic accumulation.

This is a game-changer for large-scale monitoring, as the limitations of ground sampling are overcome.

Advantage of Integration:

- Scalable Monitoring – regional to global level tracking is possible.
- Smart Decision-Making – Actionable insights are available to governments and agencies.
- Cost-Effective – Costs of long-term manual sampling are reduced.

1.5 Case Studies and Applications

AI's role in microplastic research appears strong when we look at its real-world use cases and case studies. These applications show how AI has made detection, monitoring, and prediction more effective. Some representative case studies are discussed below.

1.5.1 AI-Driven Microplastic Detection in Marine Systems

Marine ecosystems are most vulnerable to microplastic pollution. Traditional sampling and lab-based analysis are time-consuming. AI-based approaches fill this gap.

Case Example – CNN-Based Image Recognition in Oceans

- Researchers developed a Convolutional Neural Network (CNN) that automatically identifies microplastics from microscope images.
- The model achieved 80–90% accuracy in classifying fibers, fragments and beads in marine water samples.
- While manual analysis might take hours to study a single sample, the AI-based system can analyze a large dataset in a few minutes.

Case Example – Autonomous Underwater Vehicles (AUVs):

- There was a pilot project in Europe where underwater drones with AI-enabled cameras were used.
- These drones collected images at different depths of the ocean and onboard AI algorithms performed real-time microplastic detection.

- This created spatial distribution maps that indicate which oceanic regions are most polluted.

Key Takeaway:
AI-driven marine detection allows us to perform large-scale and near real-time monitoring that was not possible with conventional sampling.

1.5.2 AI for Wastewater Treatment Monitoring

Wastewater treatment plants (WWTPs) are a critical checkpoint for microplastic pollution. Many microplastics enter the environment *via* domestic and industrial wastewater. If their removal is monitored and optimized at the treatment plant itself, pollution can be significantly reduced.

Case Example – AI + Sensor-Based Monitoring

- An AI-driven optical sensor system was implemented at a wastewater treatment facility in Japan.
- Sensors continuously analyze treated water and AI models detect and quantify microplastic particles.
- If removal efficiency is low, the system instantly sends an alert and process parameters are adjusted.

Case Example – Predictive Maintenance in WWTPs

- AI models trained with historical data (flow rate, filter efficiency, seasonal variations) are used to predict microplastic removal efficiency.
- Predictive analytics suggests plant operators when filter replacement or system optimization is needed.

Key Benefits:

- Real-time monitoring prevents untreated microplastics from entering the environment.
- AI reduces operational costs by reducing the need for manual lab analysis.
- Predictive systems help maintain water quality regulations.

1.5.3 Predictive Analytics for Microplastic Pollution Hotspots

A big challenge of microplastic research is to predict which areas are going to be more polluted in the future. Here the role of predictive analytics + AI is very critical.

Case Example – Coastal Hotspot Mapping (GIS + AI):

- In a study of the Mediterranean Sea, researchers integrated GIS (Geographic Information Systems) and machine learning algorithms.
- Input Data: Ocean currents, population density, river discharge, fishing activities.
- Output: Pollution Hotspot Maps that clearly show which coastal regions are at high risk of microplastic accumulation.

Case Example – River Transport Prediction:

- ML-based models were used in the Great Lakes region of the US to predict microplastic flow from rivers to lakes.
- The model used rainfall and industrial discharge data to accurately estimate how microplastic loads would increase during the flood season.

Benefits of Predictive Analytics:

- Policymakers receive an early warning system.
- Targeted cleanup and prevention strategies are deployed to high-risk areas.
- Limited resources can be efficiently allocated.

1.6 Challenges and Limitations

The role of AI in microplastic research is very promising, but it also comes with some serious challenges and limitations. If these are not understood, the efficiency and reliability of AI systems can be compromised.

1.6.1 Data Availability and Quality Issues

- There is a lack of standardized datasets for microplastic research. Each lab collects data in its own way, due to which models cannot generalize.
- There is a lot of variation in images and sensor data (lighting, magnification, background noise), which makes consistent training for AI difficult.
- Annotated datasets, where microplastic particles have been properly marked by experts, are very limited.
- Result: Models overfit and performance drops in real-world samples.

1.6.2 Technical and Computational Challenges

- High computational cost: Deep learning models like CNNs require GPUs and large computing clusters to train. This is not feasible for every lab.
- Interference problem: Natural samples contain organic matter, algae, and sediments that look like microplastics. Differentiating them is tough even for AI.
- Scalability: It is easy to train a model on a small dataset, but making the algorithm scalable and robust for large-scale global monitoring is challenging.

1.6.3 Ethical and Policy Considerations

- Data privacy and sharing: Many countries do not openly share their environmental datasets, which slows down collaborative AI research.
- Bias in policy support: If an AI model is trained on incomplete data, its predictions can be biased and lead to incorrect policy recommendations.
- Accountability: If an AI-based monitoring system makes incorrect predictions and causes environmental damage, who will be responsible – the researcher, the policymaker, or the system developer?

1.7 Future Directions

The combination between AI and microplastic research is still in its infancy. More advancements are possible in the future that will revolutionize environmental monitoring.

1.7.1 Advancements in AI Models for Environmental Applications

- Explainable AI (XAI): In the future, AI models will be created that will explain their decisions, which policymakers will trust more.
- Hybrid Models: Faster and accurate detection will be possible by combining traditional spectroscopy + AI.
- Lightweight Models: AI models running on mobile and edge devices will be developed that will do real-time monitoring on-site without heavy computing.

1.7.2 Cross-Disciplinary Research Opportunities

- Environmental scientists + computer scientists + policymakers will create interdisciplinary frameworks.
- Materials science researchers will design new sensors that generate AI-friendly data.
- Collaboration can create a global standardized dataset that will boost the accuracy of AI.

1.7.3 AI for Policy Support and Decision-Making

- AI-based simulation models will show policymakers what impact pollution levels will have if plastic ban or recycling laws are implemented.
- Predictive analytics provides governments with early warning systems so they can take proactive decisions.
- Evidence-based policy making will be possible on the basis of AI reports.

1.7.4 Towards Sustainable Microplastic Management

- With AI we can design end-to-end solutions:
- Detection at source (industries, wastewater plants).

- Real-time monitoring (rivers, oceans, soils).
- Predictive hotspot mapping (to stop pollution before it spreads).
- Long-term goal: Circular economy where plastics are properly recycled and AI ensures that there is no leakage into the environment.

1.8 Conclusion

Microplastic pollution is a global environmental challenge today that poses a threat to human health, biodiversity and ecosystem sustainability. Traditional detection methods are useful, but their speed and scalability are limited. This is where Artificial Intelligence (AI) comes in as a powerful tool – making detection, monitoring and prediction more accurate and efficient. But AI adoption also requires addressing challenges such as data availability, computational load and ethical concerns. In the future, interdisciplinary research, explainable AI and policy-level integration could make AI a game-changer for microplastic management.

Ultimately, AI is not just a technology but a bridge between science and sustainability, which can take us towards a cleaner and healthier planet.

AI disclosure statement: The author confirms that no artificial intelligence (AI)-generated content or AI-assisted writing tools were used in the preparation of this chapter (including summaries and the abstract). All text, analysis, and conclusions are the result of the author's original work.

References

1. Andrady, A.L., Microplastics in the marine environment. *Mar. Pollut. Bull.*, 62, 8, 1596–1605, 2011, https://doi.org/10.1016/j.marpolbul.2011.05.030.
2. Cole, M., Lindeque, P., Halsband, C., Galloway, T.S., Microplastics as contaminants in the marine environment: A review. *Mar. Pollut. Bull.*, 62, 12, 2588–2597, 2011, https://doi.org/10.1016/j.marpolbul.2011.09.025.
3. Li, J., Liu, H., Chen, J.P., Microplastics in freshwater systems: A review on occurrence, environmental effects, and methods for microplastics detection. *Water Res.*, 137, 362–374, 2018, https://doi.org/10.1016/j.watres.2017.12.056.
4. Wang, W., Ndungu, A.W., Li, Z., Wang, J., Microplastics pollution in inland freshwaters of China: A case study in urban surface waters of Wuhan, China. *Sci. Total Environ.*, 575, 1369–1374, 2017, https://doi.org/10.1016/j.scitotenv.2016.09.213.

5. Cowger, W., Gray, A., Christiansen, S.H., DeFrond, H., Deshpande, A.D., Rochman, C.M., Critical review of processing and classification techniques for images and spectra in microplastic research. *Appl. Spectrosc.*, 74, 9, 989–1010, 2020, https://doi.org/10.1177/0003702820929064.
6. Duarte, C.M., Pitt, K.A., Lucas, C.H., Purcell, J.E., Uye, S., Robinson, K., Is global ocean sprawl a cause of jellyfish blooms? *Front. Ecol. Environ.*, 11, 2, 91–97, 2013, https://doi.org/10.1890/110246.
7. Deng, H.W., Li, W.W., Wu, M.H., Zhang, L.Y., Wang, C., Yang, J., Artificial intelligence in environmental science: Applications and challenges. *Environ. Sci. Technol.*, 55, 22, 14783–14797, 2021, https://doi.org/10.1021/acs.est.1c04987.
8. Priya, A., Ojha, V.P., Dutta, M., Artificial intelligence for environmental sustainability: A review of applications in monitoring and management of microplastic pollution. *J. Clean. Prod.*, 370, 133576, 2022, https://doi.org/10.1016/j.jclepro.2022.133576.
9. da Costa, J.P., Santos, P.S.M., Duarte, A.C., Rocha-Santos, T., (Nano)plastics in the environment – Sources, fates and effects. *Sci. Total Environ.*, 566–567, 15–26, 2016, https://doi.org/10.1016/j.scitotenv.2016.05.041.
10. Verma, P., Sharma, N., Rathore, P., Leveraging machine learning for microplastic detection: Current trends, challenges, and future directions. *Environ. Monit. Assess.*, 195, 6, 741, 2023, https://doi.org/10.1007/s10661-023-11079-3.

2

A CNN-ViT Hybrid Deep Learning Architecture for Accurate Microplastic Detection

B. Dhanalaxmi[1*], B. Saritha[2], P. Punitha[3], G. Jagan Naik[4] and B. Anupama[5]

[1]Department of CSE, Geethanjali College of Engineering and Technology, Hyderabad, Telangana, India
[2]Department of CSE, Maturi Venkata Subbarao (MVSR) Engineering College, Hyderabad, Telangana, India
[3]Department of CSE (Data Science), Vignana Bharathi Institute of Technology, Hyderabad, Telangana, India
[4]Department of CSE (Data Science), CMR Institute of Technology, Hyderabad, Telangana, India
[5]Department of ECE, B V Raju Institute of Technology Engineering College, Hyderabad, Telangana, India

Abstract
Accurate and efficient microplastic detection systems have become vital since this environmental hazard continues to endanger both nature and human health. The proposed research presents a combination of deep learning frameworks which merge Convolutional Neural Networks and Vision Transformers to improve microplastic detection within complicated settings. The CNN part provides efficient local feature recognition and the ViT section brings global context analysis through attention mechanism. The model acquired its skills by using publicly accessible datasets consisting of SEM and holographic images for both training and validation purposes. The preprocessing techniques of images focused on input quality enhancement to maximize model operational performance. The hybrid CNN-ViT model demonstrated better performance than standalone models when evaluated because it reached validation accuracy of 91.87% with a validation loss of 0.1164. The researchers tested this framework with Raspberry Pi edge devices

**Corresponding author*: dinnu18@gmail.com

Abhishek Kumar, Pooja Dixit, Pramod Singh Rathore, Arun Lal Srivastav and Ashutosh Kumar Dubey (eds.) Microplastic Monitoring Using Artificial Intelligence, (23–36) © 2026 Scrivener Publishing LLC

which demonstrated its functional capabilities. This design establishes itself as a promising method for scalable time-sensitive microplastic identification which further strengthens its status as an important environmental surveillance and public health defines system.

Keywords: Feature selection, ensemble, healthcare, machine learning, prediction, medical, hyperparameter tuning

2.1 Introduction

The environmental challenge of microplastics poses a global crisis due to their tiny status of less than 5 millimeters wide. The environmental contaminant microplastics exist throughout oceans and rivers and soil and air which threaten marine life and human health and ecosystems. Accurate microplastic detection stands essential for the comprehension of their spread patterns and the reduction of their effects. The current detection methods consume excessive effort along with showing unreliable results [1]. The continuous growth of imaging data makes deep learning present an encouraging solution path. The research operates with a new deep learning system which unites convolutional neural networks (CNN) and vision transformers (ViT) to achieve better detection results and operational speed. The proposed approach delivers an enhanced environmental monitoring technology through automatic microplastic detection because it effectively combines the strengths of its two primary modeling components [2].

Many sources contribute to microplastic formation which includes plastic degradation as well as industrial activities and usage of personal care items. Small-sized microplastics infiltrate the food chain effectively while absorbing toxic substances, thus they impose potential dangers to animals and humans [3]. The numerous detection challenges of microplastics emerge from their diverse dimensions, body shapes and chemical substances present. Although precise, the laboratory techniques FTIR and Raman spectroscopy need technical expertise but they prove expensive and time-consuming to perform. Digital microscopy and imaging technologies produced considerable dataset information which supports automated detection methods. The recognition of microplastics from different backgrounds proves successful using deep learning techniques and particularly image-based models. This research combines environmental science capabilities with computer vision approaches to establish an integrated system for ecological protection and monitoring because of rising artificial intelligence implementation for these purposes [4].

The prevalence of microplastics throughout environmental surroundings creates immediate as well as future consequences toward human wellbeing and supermarket security with ongoing impacts to ecological preservation methods. Humans obtain microplastics by consuming seafood and drinking water while also inhaling them through the air which can potentially cause health problems including inflammation together with hormone disruption and toxicity effects [5]. The current inability to detect microplastics efficiently along with the inability to remove them leads to further environmental damage that weakens vital marine ecosystems. This damage then impacts biodiversity and carbon regulation functions. The development of advanced detection methods helps both science research and regulatory organizations to create powerful environmental policies. Technological innovation about microplastics enables industrial transformation of material usage patterns and waste management practices. The development of advanced detection technology delivers a safer environment which started from today and includes future generations in this benefit [6].

Microplastic detection throughout history depended on manual collection methods as well as FTIR and Raman spectroscopy analyses done in laboratories. Research-quality analytical tools provide exact results but their application faces various significant constraints which include both substantial costs and need for complex technology alongside human interpretive and error-based deficiencies. Visual sorting under a microscope, another common method, lacks consistency and reproducibility [7]. Large-scale environmental monitoring projects cannot be scaled through these current detection methods. The expanding volume of environmental data makes manual processing inadequate, thus generating time delays and restricted observational capacity. Speed and automated solutions have reached an urgent priority status. Fundamental traditional technologies do not match the level of performance required during real-time monitoring activities in various demanding operational environments. The present situation requires AI integration to enhance detection capabilities as well as increase operational precision and system performance rates [8].

A hybrid deep learning system that joins convolutional neural networks (CNNs) and vision transformers (ViTs) exists as a solution to the problems of microplastic identification. The design of CNNs enables them to identify irregular microplastic particles through extraction of local spatial features including shape texture and edges from images. The ability of ViTs to grasp wide contextual information allows the model to recognize relationships between different regions of the image [9]. Through the combination of these two architectures the model achieves stronger performance because it gains capabilities for understanding both the detailed aspects locally and maintaining a broad view of the entire image. The hybrid model learns from microplastic images with labels before being optimized to tackle different background difficulties during training. This structural design both raises detection quality while suppressing false alarms and enables the system to scale for time-sensitive environmental surveillance duties.

The worsening issue of plastic pollution alongside its extensive environmental impact served as the main reason for conducting this study. An awareness of microplastics exists broadly but we lack feasible solutions to detect these pollutants in real-time. The environmental world demands deep learning technology for image recognition because this advancement has become essential and relevant. The detection needs to adapt for microplastics due to their irregular shapes in addition to their transparency and variable color properties [10]. This combination of CNN-ViT technology provides a solution which answers both basic image processing requirements and the most advanced artificial intelligence capabilities through its adaptive nature toward natural patterns. The research team works to contribute to environmental sustainability through sophisticated monitoring tools which drive better ecological protection research by making ecological monitoring more efficient [21–23].

2.2 Literature Review

This section focusses on the previous research done on Architecture for Accurate Microplastic Detection. This section consists of all the information that are needed to check the previous research done.

Citations	Technology used	Characteristics	Limitations	Outcomes	Future scope
[11]	CNN (Convolutional Neural Network), MATLAB	Accurate detection and quantification of microplastics in high-resolution tissue images	Focused only on tissue images; may not generalize to other samples	Validated CNN model effective in identifying microplastics in tissue	Further research into other biological tissues and real-time analysis
[12]	FTIR, ML (SVC, KNN), DL (LeNet5)	Classification of microplastics with high accuracy despite noise	Some DL models underperformed; FTIR spectral noise	LeNet5 achieved 96.93% accuracy; ML also showed strong results	Enhancement of DL models and noise filtering techniques
[13]	Object detection algorithms, SEM imaging, Open-source dataset	Broad classification by polymer types and sizes; diverse dataset	Simulated images; real-world performance needs testing	Dataset supports robust microplastic detection and benchmarking	Expanding dataset and refining detection under varied conditions
[14]	Deep learning, Polarization holographic imaging system	Portable, cost-effective, real-time MP analysis	Complex setup; may require calibration for different environments	Automated classification and dynamic MP monitoring	Broader deployment and integration into filtration systems
[15]	YOLOv5, Mobile phone microscopy, ATR-FTIR, FE-SEM	Low-cost, portable, accurate detection of MPs in products	Limited to visible MPs; relies on external confirmation tools	98% accuracy in MP identification using deep learning	Scaling for consumer use and extending to more sample types

(Continued)

(Continued)

Citations	Technology used	Characteristics	Limitations	Outcomes	Future scope
[16]	Digital holography, Deep learning (CompNet), Transfer learning	Effective on small, imbalanced datasets; prevents overfitting	Requires precise optical setup and calibration	Improved classification and generalization accuracy	Deployment for real-time *in situ* environmental monitoring
[17]	Digital holography, Deep learning	Robust classification regardless of MP morphology	Requires holography equipment and DL training expertise	Reliable in mixed samples with various contaminants	Development of automated, field-ready classification systems
[18]	IoT (Raspberry Pi), Deep learning	Affordable ($75), scalable prototype for freshwater analysis	Prototype may require adaptation for other water types	Accurate MP classification, detection, and quantification	Scalable citizen science data collection and real-time monitoring
[19]	Raspberry Pi 3, Object detection, NIR sensor, Arduino BLE Sense 33	Efficient shoreline detection; uses spectral reflectance and DL	Limited to shoreline areas; sensitive to lighting conditions	Effective low-cost solution for nurdle detection and mapping	Future field deployments and multi-material plastic tracking
[20]	Digital holography, Deep learning	Not dependent on morphological characteristics; high sensitivity	Complex optics; not yet field-deployable	Strong detection in mixed biological samples	Robust field-deployable system for real-world use

2.3 Proposed Mythology

The data acquisition and model deployment procedures structured systematically to establish a whole Architecture for Accurate Microplastic Detection use as shown in Figure 2.1.

Data Collection and Preprocessing
This study obtained its data from Kaggle by compiling labeled pictures showing microplastics alongside non-microplastic particles that were acquired with different illumination setups and background variations. All images received a uniform dimension before researchers normalized pixel values. The model eliminated background noise through the use of image enhancement tools. The researchers checked and cleaned all label information to establish high-quality training data. Such an early preprocessing step proves essential because it helps simplify computational operations while building a stronger prediction model. A split proceeded for the data into training and testing subsets while following an 80:20 split ratio. This created an environment for model training and evaluation purposes.

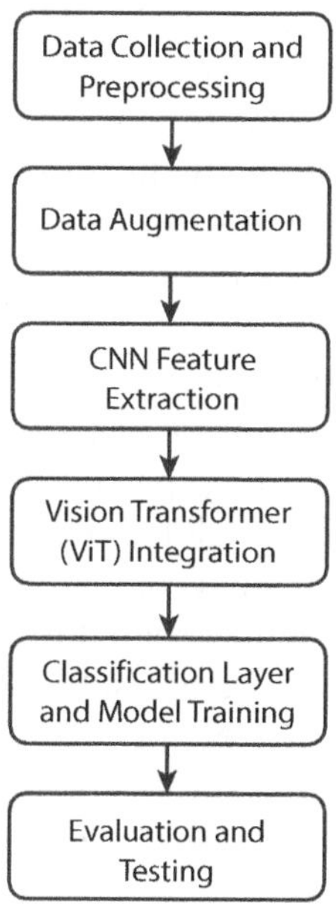

Figure 2.1 Proposed methodology.

Data Augmentation

The model achieved better robustness and generalization through data augmentation processing that was applied to its training dataset. Random rotations and horizontal and vertical flipping techniques combine with zooming adjustments and brightness modifications and slight positional changes to the data. The modeling process uses augmentation to replicate alterations which occur in genuine data samples while maintaining consistent elemental components. Such dataset characteristics need particular attention because of their limited sample number or uneven distribution. Training capabilities and overfitting prevention both receive substantial benefits from the application of data augmentation techniques. The CNN-ViT hybrid model becomes equipped with the ability to detect microplastics across different angles, lighting situations and environmental conditions through this step.

CNN Feature Extraction

Through a convolutional neural network (CNN) the program acquires localized features like edges and textures and image patterns within input data. The identification of physical microplastic characteristics depends heavily on these low- to mid-level features because microplastics display irregular forms that are both shape and surface irregular. A CNN structure uses several convolutional operations followed by pooling transformations with activation components which compress important image data. The next stage of processing receives the feature maps extracted from the last CNN layer as input. Through this process the model gains capability to detect subtle visual variations that occur in microplastic particles.

Vision Transformer (ViT) Integration

A ViT receives flattened CNN feature maps in the form of patch sequences as input. The ViT operates on image patches through its self-attention processing stage while it maintains global awareness across the image content. While CNNs specialize in recognizing nearby patterns ViT offers the model the ability to recognize spatial organization and this becomes essential to detect dispersed microplastics within different image areas. This dual approach between CNN and ViT delivers refined and comprehensive image interpretable models. The merged architecture delivers considerably better detection results particularly when observing challenging background situations.

Classification Layer and Model Training

The dense classification head receives ViT output to perform fully connected transformations and activate the output using a softmax function. The layer determines through prediction whether microplastics exist along with their specific types present within the image. The system achieved training with cross-entropy loss together with the Adam optimizer running at 0.001 learning rate. The monitoring process included observing training and validation loss over multiple epochs to guarantee proper convergence as well as to prevent overfitting. The model displayed continuous reduction of training and testing loss values which proved its capability to learn and generalize well. The combination architecture demonstrated high performance when detecting intricate visual patterns present in microplastic images.

Evaluation and Testing

The CNN-ViT hybrid model received testing data evaluation through the assessment of essential measurements including accuracy, precision, recall, F1-score and loss curves. Testing results showed that the model maintained low testing loss level with high accuracy that proved effective generalization on unknown data. All test metrics indicated that the combination of CNN and ViT models provided superior performance than the baseline models that used only CNN or only ViT. Grad-CAM was implemented as a visualization technique to identify which parts of images affected the model prediction process. The evaluation test results proved the model to be reliable for use in future real-world microplastic detection applications.

2.4 Result and Discussion

The experimental testing of the proposed CNN-ViT hybrid model focused on microplastic detection over 25 successive training iterations. The recorded metrics at each epoch included training and validation accuracy and training and validation loss to evaluate how the model learned and its generalization ability. The model demonstrated robust microplastic detection capabilities when the system achieved maximum accuracy together with minimized loss measurements between the two datasets.

Both training and validation accuracy steadily rise because of the progressive increase in training epochs. At the beginning of training the model exhibited 70.56% validation accuracy which reached its peak at 92.10% when it entered the 25th epoch. The CNN-ViT hybrid architecture demonstrates strong ability to extract spatial and contextual information

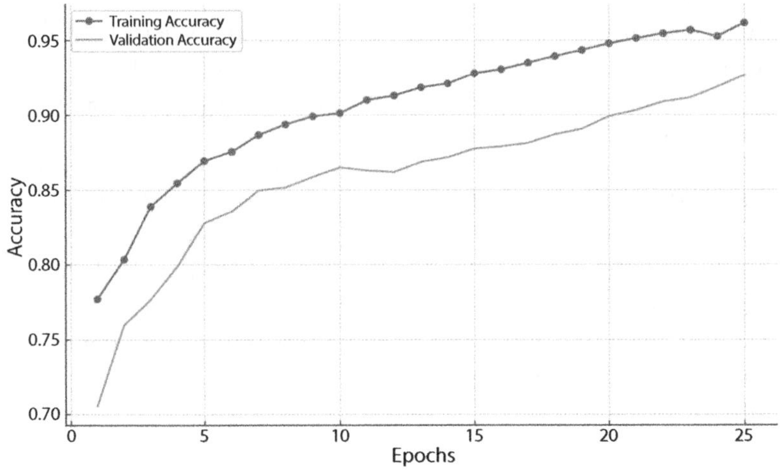

Figure 2.2 Model accuracy and validation accuracy.

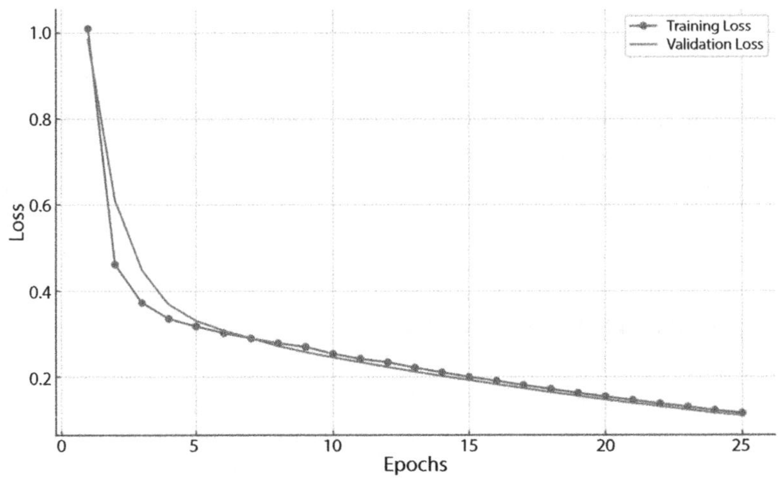

Figure 2.3 Model loss and validation loss.

from microplastic images through this observed trend. The steadily rising training accuracy demonstrates proper learning occurs without overfitting because it matches the validation accuracy values closely. Performance stability of the model became possible through stop loss implementation as well as controlled learning rate adjustments shows in Figure 2.2 and Figure 2.3.

A decline in training and validation losses shows through the loss graph because the model demonstrates its competence in minimizing errors throughout learning. Loss values start at more than 1.0 yet descend steeply in the beginning epochs and then slowly decline before reaching the minimum value near 0.11. The CNN-ViT hybrid achieves effective pattern recognition by showing this convergence trend in its learning process. During the complete process the minimal difference between training and validation loss signifies the excellent ability of the model to recognize patterns in new datasets. This evidence shows the effectiveness of the architecture to detect microplastics because it delivers both superior accuracy performance and minimal false positive results.

2.5 Concluding Remarks and Future Scope

The hybrid CNN-ViT framework proved effective at detecting microplastics because validation accuracy reached above 92% during minimal loss conditions according to training results. Both Convolutional Neural Networks and Vision Transformers used together enabled the model to acquire small-scale spatial attributes and worldwide structural relationships effectively. The model demonstrates good generalization ability because of its robust learning performance alongside minimal validation loss which makes it fitting for environmental monitoring systems deployment.

The upcoming framework development plans include an integration of multi-spectral or hyperspectral imaging data components which will enhance the detection capabilities across various environmental conditions. The deployment of real-time video handling techniques alongside federated learning protocols would provide enhanced scalability alongside protection of privacy during distributed detection system operations. The research should include more studies about transferring learning models combined with lightweight edge deployment methods to enable real-time detection of microplastics in water bodies and field monitoring stations.

AI disclosure statement: The author acknowledges the use of AI-based tools for minor language editing or grammatical correction. However, all intellectual contributions, analyses, and conclusions presented in this chapter are entirely the author's own.

References

1. B, H.C., Challa, C., G, A.R., T, G.M., T, D., Detection of Microplastic Ingestion in the Human Body Using Deep Learning Technique. *2024 International Conference on Intelligent Systems for Cybersecurity (ISCS)*, pp. 1–6, 2024.
2. Yurtsever, U. and Yurtsever, M., Object detection for microplastic using deep learning on microscopic images, 2018.
3. Thar, A.S., Laitrakun, S., Deepaisarn, S., Opaprakasit, P., Somnuake, P., Athikulwongse, K., A Comparative Study of Machine Learning and Deep Learning Models for Microplastic Classification using FTIR Spectra. *2023 18th International Joint Symposium on Artificial Intelligence and Natural Language Processing (iSAI-NLP)*, pp. 1–6, 2023.
4. Rezvani, H., Zarrabi, N., Mehta, I., Kolios, C., Jaafar, H.A., Kao, C., Saeedi, S., Yousefi, N., Morphological Detection and Classification of Microplastics and Nanoplastics Emerged from Consumer Products by Deep Learning. ArXiv, abs/2409.13688, 2024.
5. Li, Y., Zhu, Y., Huang, J., Ho, Y., Fang, J.K., Lam, E.Y., High-throughput microplastic assessment using polarization holographic imaging. *Sci. Rep.*, 14, 2024.
6. Bin, M.Z., Arju, Z., Hridi, N.A., Dewan, L., Suhaila, Amin, M.N., Rashid, T.U., Azad, A.K., Rahman, S., Hossain, M., Habib, A., Deep-learning enabled rapid and low-cost detection of microplastics in consumer products following on-site extraction and image processing. *RSC Adv.*, 15, 10473–10483, 2025.
7. Zhu, Y., Yeung, C.H., Lam, E.Y., Digital holographic imaging and classification of microplastics using deep transfer learning. *Appl. Opt.*, 60, 4, A38–A47, 2020.
8. Zhu, Y., Yeung, C.H., Lam, E.Y., Digital holographic microplastics detection and characterization in heterogeneous samples *via* deep learning. Other Conferences, 2021.
9. Sundar, S., A Novel Low-Cost Approach For Detection, Classification, and Quantification of Microplastic Pollution in Freshwater Ecosystems using IoT devices and Instance Segmentation. *2022 IEEE MIT Undergraduate Research Technology Conference (URTC)*, pp. 1–5, 2022.
10. Zhang, K. and Zhang, C.C., Designing a Low-Cost Microcontroller-Based Rover for Microplastic Detection Using Deep-Learning Image Detection and Near-Infrared Spectroscopy. *2023 Congress in Computer Science, Computer Engineering, & Applied Computing (CSCE)*, pp. 957–962, 2023.
11. Sarmiento, J., Anaya, M., Tibaduiza, D.A., Microplastic Identification Using Impedance Spectroscopy and Machine Learning Algorithms. *Int. J. Distrib. Sens. Netw.*, 2024.
12. Abhishek, K., Rathore, P.S., Dubey, A.K., Srivastav, A.L., Dutt, V., Ananth Kumar, T., *Sustainable Management of Electronic Waste*, John Wiley & Sons, 2024.

13. Rathore, P.S., Sustainable smart homes and buildings with internet of things. John Wiley & Sons, Inc., Scrivener Publishing LLC, 2025a.
14. Munagala, N.S., Kumar, V., Srividhya, S., Usha, G., Jeyalakshmi, R., Micro-Objects Classification for Microplastic Pollution Detection using Holographic Images. *2024 International Conference on Advances in Computing, Communication and Applied Informatics (ACCAI)*, pp. 1–7, 2024.
15. Rathore, P. S., Ahuja, S., Burri, S. R., Khunteta, A., Baliyan, A., Kumar, A., Deep Learning Techniques for Automation and Industrial Applications, John Wiley & Sons, Inc. Scrivener Publishing LLC, 2024a.
16. Rathore, P.S., Kumar, A., Bhatia, S., Mashat, A., Gadekal, T.R., Sustainable Smart Homes and Buildings with Internet of Things, John Wiley & Sons, 2025.
17. Valentino, M.R., Běhal, J., Pirone, D., Memmolo, P., Miccio, L., Bianco, V., Ferraro, P., On the use of machine learning for microplastic identification from holographic phase-contrast signatures. *Opt. Metrol.*, 2023.
18. Sasso, D., Merone, M., Nicolai, E., Vollero, L., Sabatini, A., A Benchmarking on Optofluidic Microplastic Pattern Recognition: A Systematic Comparison Between Statistical Detection Models and ML-Based Algorithms. *IEEE Access*, 12, 20091–20102, 2024.
19. Valentino, M.R., Běhal, J., Bianco, V., Itri, S., Mossotti, R., Fontana, G.D., Stella, E., Miccio, L., Ferraro, P., Synthetic microfibers discriminated by AI-enabled polarization resolved Digital Holography. *2022 IEEE International Workshop on Metrology for the Sea; Learning to Measure Sea Health Parameters (MetroSea)*, pp. 511–515, 2022.
20. Bhatt, E., Gauba, P., Sarethy, I.P., Assessing Recent Technologies for Addressing Microplastic Pollution and Pushing the Case of Bioremediation as an Attractive Approach, in: *Recent Advances in Food, Nutrition & Agriculture*, 2024.
21. Dimaongon, N.G., Araña, K.N., Bacosa, H.P., Emerging Microplastic Contamination in the Food Industry: the Case of Commercial Table Salts in Iligan, Philippines, in: *Philippine Journal of Science*, 2024.
22. Choi, D., Jung, S., Lee, J., Kwon, E.E., Analysis of microplastics distributed in the environment: Case studies in South Korea. *Energy Environ.*, 2024.
23. Shokunbi, O.S., Jegede, D.O., Shokunbi, O.S., A study of the microplastic contamination of commercial table salts: A case study in Nigeria, in: *Environmental Health Engineering and Management*, 2023.

3

XAI for Decision Support in Microplastic Pollution Management

Srinibas Pattanaik[1]*, Sachin Ahuja[1], Sartajvir Singh Dhillon[1], Jasneet Chawla[1], Deeksha Sonal[1] and Alessandro Vinciarelli[2]

[1]*Department of Computer Science & Engineering, University Institute of Engineering, Chandigarh University, Mohali, India*
[2]*Computational Social Intelligence (Computing Science), University of Glasgow, Glasgow, United Kingdom*

Abstract
The Microplastic pollution management is the instructive and influential environmental issue that needs sophisticated decision support systems for efficient and significant management of microplastic pollution and very risks and poses for serious and decisive with and to the human health and aquatic ecosystems. The management systems are essential for the human beings and additionally automated solutions are frequently viewed with suspicion due to the intricacy of the ecosystems and the opaqueness of AI models. By improving the ability to be explained and detailed of AI-driven decision support systems, XAI Artificial Intelligence offers a remedy by empowering environmental scientists and policymakers to comprehend and verify model predictions. With an emphasis on how interpretable machine learning approaches might enhance monitoring, risk assessment and risk management and security for the microplastic pollution frameworks, and remediation strategies and integrated with AI frameworks, this study examines the use of XAI with integrated AI support in managing microplastic pollution. We go over several XAI techniques to improve model transparency, including and using Shapley and Cognitive computing and Generative transform XAI and Additive Explanations with Local Interpretable Paradigm-agnostic explanations. We also provide case studies that demonstrate how XAI is being used in real-world environmental sustainability applications. Explainability and comprehensibility can be included into machine learning-driven decision support to foster trust,

**Corresponding author*: srinibas.21bcs7837@gmail.com

Abhishek Kumar, Pooja Dixit, Pramod Singh Rathore, Arun Lal Srivastav and Ashutosh Kumar Dubey (eds.) Microplastic Monitoring Using Artificial Intelligence, (37–54) © 2026 Scrivener Publishing LLC

assist well-informed policymakers, and create more potent mitigation plans for microplastic contamination and potential mitigation and support of XAI AI and collected data by Remote Sensing and XAI with Data collections and analytics.

Keywords: Machine learning, predictive analytics, explainable AI, AI-integrated policy making, AI framework, sustainable waste management, remote sensing, image processing integrated AI

3.1 Introduction

The microplastic pollution with XAI is the contamination and accumulation of microplastics and presence of microgenres. Contaminated microplastic with its adverse impact on freshwater bodies, the underwater world, and even terrestrial settings, microplastic contamination has become an important problem for the environment [1]. These tiny plastic particles, which are frequently less than 5 mm, come from consumer goods, and trade-offs, integrated with industrial operations, and the breakdown of larger plastic debris [2]. Because of their tiny dimension and pervasiveness, microplastics are challenging to monitor, evaluate, and reduce, which makes managing them a challenging undertaking requiring sophisticated decision-making techniques [3]. In environmental research, AI has shown itself to be a useful tool, providing data-driven solutions for risk assessment, remediation planning, and pollutant monitoring [4]. However, the stakeholders like researchers, and other resources, persons insights with all of the details and environmental organizations want openness and confidence in the decision-making process, the black-box character of many AI models and advisory systems and migration and integration with is a major obstacle to their implementation. Defensible Artificial Intelligence and Big-Data are essential in this situation because it is the collection of all the necessary data to predict the input and monitoring the XAI approaches. By improving model interpretability, XAI approaches enable users to comprehend the logic underlying AI-driven suggestions and predictions and decision analysis.

Individuals can increase the dependability followed by acceptability of AI-based solutions by including XAI into decision-making engines for managing microplastic pollution. Stakeholders may more confidently, adopt mitigation strategies and targeted mitigation initiatives, forecast pollution hotspots, and evaluate the detrimental ecological effects of microplastics— thanks to transparent and recognizable AI models. Additionally,

XAI promotes ownership in AI applications by guaranteeing that choices are founded on morally and scientifically valid principles and valid fundamentals of microplastic pollution with description and insights [5, 6].

The integration of Artificial Intelligence (AI) enables automated analysis of large-scale primary and secondary environmental datasets. Through AI-driven systems, massive volumes of data can be visualized, interpreted, and represented graphically with improved precision and effectiveness. In the context of microplastic pollution, AI can automatically classify plastic particles based on their source and chemical composition, predict the dispersion of pollutants and micro-contaminants, and identify images of microplastics in water samples [7, 8].

Furthermore, drones and other smart, sustainable data collection devices—integrated with AI and sensor technologies—can enhance real-time pollution tracking. These systems provide valuable insights and meaningful data that support prompt remedial actions for environmental management.

By integrating AI and sensor-based connectivity, researchers and policymakers can make data-driven decisions that lead to more effective pollution control strategies and environmentally sustainable governance frameworks [9, 10].

Automation systems and sustainable AI-based technologies play a crucial role in enhancing the effectiveness of microplastic pollution detection. These technologies can recognize and analyze microplastic particles in environmental samples. Since microplastic contamination poses serious threats to both human health and the environment, it necessitates the use of advanced monitoring and management techniques [11]. Through the integration of various data sources, predictions, and analysis with integrated with AI, decision support systems are essential in directing pollution control efforts. By providing scalable and effective methods for pollution assessment, the incorporation of AI into environmental monitoring has greatly enhanced the detection, categorization and initial data representation with enhanced technologies, and analysis of contaminants. But as AI-driven systems proliferate, it is imperative to conform with regulations, trust, and accountability, and identify all data that decision-making processes be transparent and explainable. By tackling these issues, more informed and efficient responses will be possible, supporting long-term pollution control and environmental preservation [12, 13].

3.2 Causes and Consequences and Effects of Microplastic Pollution

Microplastics come from a variety of sources, which can be broadly divided into four categories: airborne, marine, home, and industrial. The production, processing, and disposal of plastic waste are industrial operations that greatly increase the pollution caused by microplastics [14, 15]. Tiny plastic particles are frequently released into the environment and waterways by factories that make polymers manufactured from synthetic debris other ingredients, and microbeads for cleaning and cosmetic items. The issue is also exacerbated by domestic sources, such as household goods like detergents, synthetic textiles, and personal hygiene products. When synthetic clothes are washed, microfibers are released into wastewater, which frequently makes its way past treatment facilities and into natural environments. Fishing gear, plastic waste, and tire wear particles are examples of marine sources of microplastics [16, 17]. These materials break down over time because of ocean currents, UV light, and microbial activity and with examples [18]. The extensive usage and disposal of those who leave plastic materials have resulted in the growing environmental problem of microplastic contamination. Primary microplastics, which are produced as tiny particles, and secondary microplastics, which are produced when bigger plastic objects break down, are the two main sources. Toothpaste, body washes, face scrubs, and the cosmetics which includes microbeads are examples of personal care products that purposefully create and discharge primary microplastics into the environment. Additionally, industrial techniques are utilized in the production of synthetic textile and processing with fibers that are released when washing. Microplastics are also employed in research and medical settings for things like industrial abrasives and drug delivery systems. However, larger plastic debris gradually decomposes under the influence of environmental factors such as UV radiation, mechanical abrasion, and chemical reactions, leading to the formation of secondary microplastics. AI- and IoT-based sensor systems, combined with remote sensing and satellite imagery, can be utilized to monitor and collect data on these pollutants. Such integrated technologies provide valuable decision-support tools for effective management and mitigation of microplastic pollution [19, 20].

The implications of microplastic pollution remain far-reaching, damaging many kinds of ecosystems and the well-being of maritime habitats; microplastics are swallowed by aquatic creatures including fish, seagulls, and plankton out getting them for food. This leads to internal damage, malnutrition, and eventually death of many species. Moreover, microplastics assemble and biomagnified along the food loop, causing greater levels in larger predators, involving humans. Coral reefs, critical to the wave's biodiversity, are also impacted when microplastics bring deadly bacteria and block required sunlight. On land, microplastics are found in agricultural soils due to the usage of sewage sludge and compost, altering the productivity of soil, water retention, and population sizes of bacteria. In the atmosphere, tiny plastic pieces can be breathed, posing dangers to respiratory health. These tiny crumbs have also been identified in drinking water, shellfish, salt, and possibly the blood of adults and placenta, raising doubts over their long-term health impacts. They have been shown to induce problems with hormonal balance, inflammatory conditions, oxidative stress, and probable cell damage because to the harmful compounds they commonly hold like in the BPA and phthalates and remote sensing [21, 22].

In terms of economics microplastic pollution affects various sectors. The current fishing and aquaculture enterprises suffer because of the contaminated seafood items resulting in sacrifices in livelihoods and food security. Coastal tourism is disproportionately harmed by unclean beaches and decreased ecosystems, while agriculture confronts lower crops from plastic-contaminated soil. Besides that, the general price of water treatment methods has risen on account to the demand for enhanced filter systems for purging microplastics. They can originate from both drinking water and stormwater sources [23]. Around the globe microplastics have been detected in those that is most faraway areas of the Earth, which might include deep-sea tunnels and high summits like Everest, showing their pervasive prevalence and persistence in the surrounding ecosystem. Microplastic contamination in ecosystem poses a major difficulty to marine being healthy, natural ecosystems, and human well-being. Addressing this issue involves a multi-faceted approach comprising reduced plastic manufacture and usage, growth of biodegradable alternatives, effective administration of waste, democratic requirements, and public awareness. Only by concerted and persistent worldwide actions can we limit the detrimental impacts of microplastics and safeguard the survival of our beloved planet for further generations to come [24, 25].

3.3 The Application of AI in Management of the Environment

Artificial Intelligence in managing the environment. is playing a transformational role in addressing many ecological concerns. One of the key uses of artificial intelligence comprises environmental data going through processing. With the use inside learning techniques such as machine learning (ML), the method of deep learning was (DL), while information metal extraction, AI enables efficient examination of large-scale, complicated environmental datasets. It aids in spotting patterns that incorporate shifts in conditions, precipitation trends, and climate anomalies, thus enabling proactive decision-making. Forecasting models based on time series are applied to predict natural catastrophes such as disasters, droughts, and storms, even though methods for detecting anomalies alert authorities about rapid changes in pollutant levels, contaminants in water, or illegal deforestation. Lastly, artificial intelligence automates compliance tracking as a result of extracting important facts from raw sensor signals. A famous example, as part of such an approach is Google's AI algorithm which expects urban atmospheric pollutants with astonishing precision.

An important area where AI aids is in the identification and anticipation of microplastics and tiny plastic bits that penetrate oceans and food chains. Utilizing convolutional neural network networks AI can examine microscopic or spectra images and identify microplastic particles. ML models also distinguish plastic kinds by the analysis of spectrometry techniques data, including FTIR or Raman signals. Predictive analytics further assesses their group and concentration of these plastics using data linked to turbulent ocean currents, temperature, in and salinity. This is why AI-powered infrared technology imaging devices are now capable of spotting microplastics that are inside marine life and lakes, thus facilitating prompt intervention efforts.

AI-based decision-making technologies are currently being applied in pollution management. Air quality monitoring techniques apply ML models to model the Air Quality Index, also known as the AQI, and identify damage hotspots, helping monarchies manage transport in cities, industrial combustion, and consumer health alerts. In water planning, AI determines chemical spillages before optimizing waste water and sewage flow operations, with smart infrastructure for treatment altering settings automatically. Waste management has also adopted AI, with computer vision infrastructures classifying waste into recyclable portions and improving garbage pickup route using predictive algorithms described in Figure 3.1.

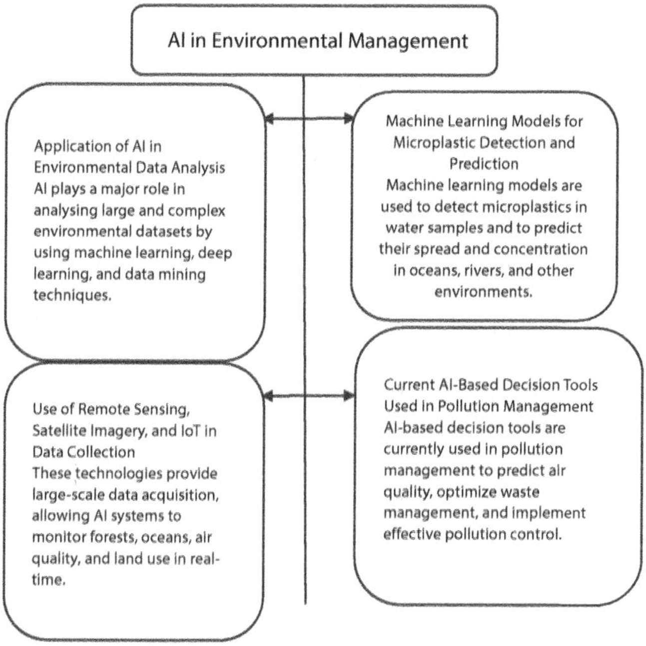

Figure 3.1 AI integrated with environmental management roles.

AI-guided urban planning technologies help evaluate the environmental impacts of new developments, thereby supporting the creation of low-emission zones and sustainable infrastructure. A notable example is IBM's Green Horizons initiative, which demonstrates how AI can aid in pollution control by providing real-time recommendations for urban development, energy management, and microservice-based environmental monitoring.

3.4 XAI Frameworks are Flexible and for the Micro Plastic Environmental Management and the Summary to Explainable Artificial Intelligence

Simple to understand, artificial intelligence encapsulates a set of techniques and approaches intended to render the mathematical results of computerized learning algorithms intelligible to humans. The basic purpose of XAI would be to make AI systems visible, believable, and accountable by offering detailed explanations of their judgments or forecasts. XAI attempts to bridge the gap within complicated machine learning algorithms and human usability, particularly in high-stakes use cases such as the health sector observation of the environment, and decisions, where understanding the logic behind a choice is vital. A major contrast belongs to interpretable and simple to comprehend models. Interpretable models are intrinsically visible, such as decision trees, regression models such as linear, and logistic regression, the state of the procedure for making choices is easy and theoretically tractable. In contrast, explicable models usually refer to specialized "black-box" mathematical models like complex neural networks or a set of approaches, that are *post-hoc* reasons modern methods (like LIME or the SHAP method.) make the assumptions they make intelligible. While interpretable models offer fundamental transparency, explainable models strive to acquire interpretability through extra layers or methods without compromising performance.

Some inscription of utilizing artificial intelligence with geographical designation information systems (GIS) gives a strong concentrating tool for environment analysis. With the help of overlaying computational forecasts and logic on geospatial maps, other interested parties may visually analyze microplastic dispersion around regions, think about contributing causes, and make geographically-aware choices as well shows in Figure 3.2. For instance, heatmaps derived from mapping saliency or SHAP values can be placed on the coastal regions; GIS data, displaying locations by the contamination by microplastic and their associated objects likely sources,

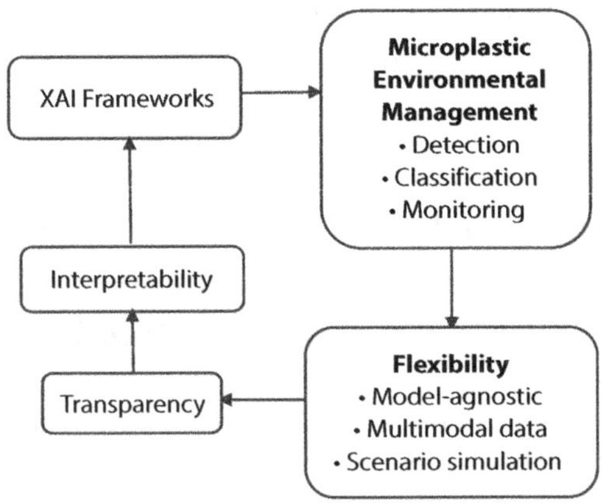

Figure 3.2 Explainable artificial intelligences applications and frameworks.

thereby contributing to easier to organize cleaning up efforts or preventive measures. This spatially shaped explanation integrates technical ideas with daily environmental understanding and trust and verify insights with AI.

XAI platforms comprise the neighbourhood apparent model-agnostic explanations), the SHAP method (Shapley Additive explanations), Hooks, and attention-based precedents communicate flexibility and a modular structure, making them changeable to a wide range of hypothetical environmental datasets, encompassing imagery collected from satellites chemical readings from water examples and data from geographic areas. These frameworks help environmental researchers to analyze model projections and explain, for instance, why a certain coastal zone is classified as a contaminating hotspot or which sources from manufacturing are most definitely involved in microplastic contamination. The reason behind this ability to be understood supports better judgment and targeted their choices, allowing authorities to institute initiatives in controlling airborne pollutants and connected with Satellite imaginary and remote sensing.

3.5 Application and Case Studies of XAI Microplastic Pollution Management

The usefulness of XAI in this field of expertise is exemplified through various case studies. In the first scenario, the SHAP function was taken into account spectroscopy-based microplastic categorization utilizing FTIR (Fourier-Transform Infrared) The method, collected data from marine samples. The model recognized critical wavelength bands that separated between different debris kinds, such as polyethylene resin additionally polypropylene which renders it more interpretable. The second scenario involves the implementation of Gradient Computer Assisted in a CNN-based imaging system for microplastics. By offering visual documentation of the deep learning model's judgments, Grad-CAM assisted environmental scientists comprehend the rationale behind some misclassifications in microscope images. The third case study addressed the application of LIME (Local Capable of being understood Model-agnostic Explanations) for modeling with predictive accuracy of plastic spread in maritime ecosystems. LIME was used, as it clarifies local model results based on the marine environment parameters,revealing knowledge concerning how factors like alternating currents and temperature result in the circulatory process of microplastics. In a collaborative effort, these applications and research investigations convince themselves how XAI enhances the dependability, accountability, and usefulness of AI-driven solutions for managing the proliferation problem of pollution caused by micro plastic.

Explainable Artificial Intelligence involves a crucial role in boosting the degree of transparency while participating interpretability of the machine learning forms administered in microplastic pollution management. One significant use of utilizing artificial intelligence features the weight analysis, where approaches like SHAP (SHapley Analytic exPlanations) are used to find whichever features—such as chemical characteristics, the number of particles, and geographic location—have an especially significant influence over small plastics identification. The outcome allows researchers to better comprehend the particular drivers driving model projections. Another key application is visual explanation, notably utilizing Gradient Computer Assisted (Gradient-weighted Class Activation Diagramming), which shows the specific regions in microscope pictures that assist most to the classification judgments that were created by convolution neural networks (CNNs). This visual knowledge is crucial for confirming and trusting the model's principal behavior. Additionally, XAI supports in model the debugging process and acknowledgment by helping researchers assess erroneous beneficial consequences throughout counterfeit negatives, thereby completing the purpose of determining the presence models for improved accuracy. Furthermore, the policy assistance provided by XAI ensures that choices based on AI models are visible, understood. and reproducing, giving environmental governing bodies the confidence to execute data-driven interventions.

Case Studies 1: In this 1st case study there will be the use of integrated Machine learning and in this scenario, (SHapley Additive exPlanations) were utilized to an algorithm such as machine learning model built using FTIR (Fourier-Transform Infrared) spectroscopic data acquired from marine samples. The model's purpose was to classify many other kinds of microscopic plastics such as polyethylene resin and polypropylene. SHAP enabled the discovery of the most instructive wavelength companies that contributed to discriminating between the material in question types. This provided researchers with a comprehensive understanding of the chemical signals responsible for classification, making the model more interpretable and reliable for scientific experts and regulatory applications. It also strengthened decision-support capabilities through the use of supervised machine learning and artificial intelligence.

This study featured the application of Gradient Computer Assisted (Gradient-weighted Class). The stimulation mapping according to a model powered by convolutional neural networks (C created for tiny pieces of plastic detection in microscope pictures. Grad-CAM developed heatmaps to display which portions of a picture invested most to the template's classification judgments. These renderings enabled environmental scientists to evaluate and validate the model's purpose and lifestyle choices, especially in cases of a mistake in classification. By identifying imaging characteristics responsible for both maintain and bad anticipated developments, the model could be more fully modified, and domain experts acquired trust in the approach known as deep learning.

In the subsequent scenario (area's subject comprehensible model-agnostic explanations) was applied to explain predictions generated by a model predicting facilitating the movement of microplastics in the ocean.

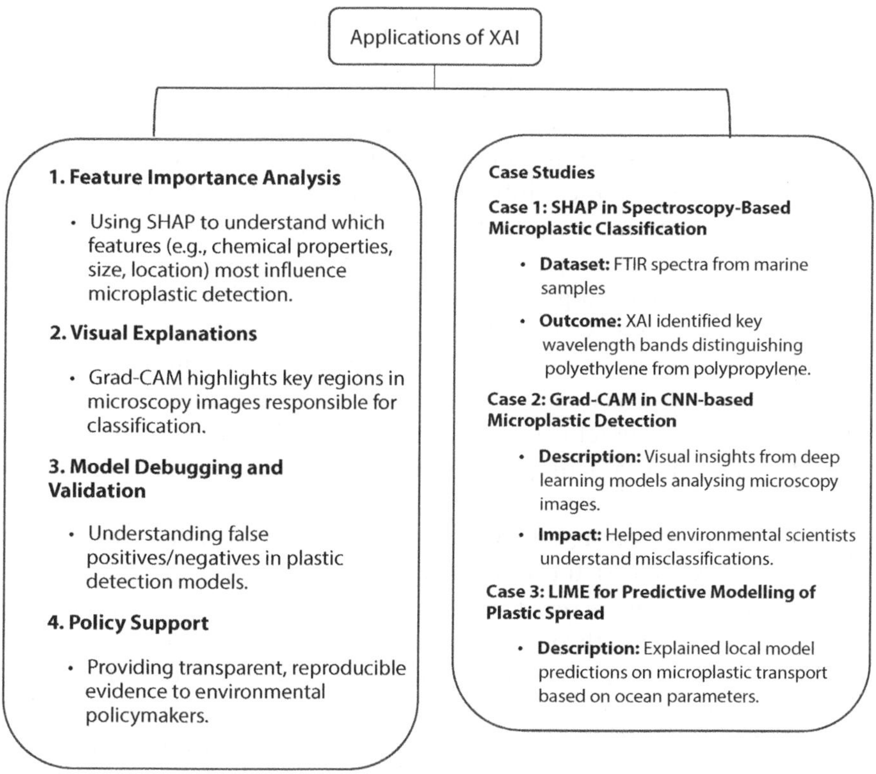

Figure 3.3 Applications with case studies for XAI.

The model employed different oceanographic characteristics, such as currents in the ocean temperature is desirable, and salt concentration, to forecast possible the variation in the distribution of plastic atomic particles shows in Figure 3.3. LIME gave localized detailed explanations for each individual predictions, helping researchers to recognize how unique environmental conditions dictated the transport of how they act microplastics. This kind of interpretability has been helpful for constructing forecasting tools that might alert mitigation efforts and maritime policy decisions.

3.6 The Utilization of Machine Learning with Explainable AI (XAI) Regarding Decision Support Systems

The encoding of integrating Explainable Artificial Intelligence with DSS (Decision Support Systems) gives a powerful, transparent framework for addressing microplastic contamination in aquatic ecosystems. This XAI + DSS architecture begins with the input of environmentally related information obtained through multiple sources, such as oceanographic sensors, geographic information systems satellites that transmit information and marine sample stations. These data streams contain factors including water temperature, salinity, shifting trends, and small plastic particles content levels. The information contained in the data is next run through a comprehensive ML/XAI pipeline, since robotics versions have received instruction on detect, organize, and prevent microfibers distribution of the component, while XAI models like SHAPED, LIME, and Grad-CAM provide apparent explanations for each anticipated. The output that comes out from this pipeline's completion consists of valuable knowledge complemented by clear, practical explanations that enable both domain leaders and specialists in evaluating the reliability and rational thinking behind each proposal.

For them to encourage involvement from users as well as strengthen accessibility as well as the integrated system incorporates visualization dashboards that provide real-time data, predictions of the model, and their correlated with clarity through maps with interactive features, graphs, and stacked charts. These electronic dashboards allow stakeholders—such marine scientists, lawmakers and policymakers and environmental regulators—to study data patterns and predict behavior intuitively. In addition, the software is equipped with warning mechanisms that cause explainable advertisements in response to key environmental thresholds. For example, when accumulations of microplastic surpass safe limits from a marine zone, the method that can send alerts outlining what was the idea behind the hint, the underlying environmental conditions, and possible mitigation strategies. By combining interpretability and accepting automation, the XAI-DSS the integration not only promotes data-driven maritime protection strategies but also develops confidence as well as transparency in the use of AI technology in environmental monitoring and integrated with transparency.

3.7 Futures Directions and Challenges of Explainable AI with Microplastic Pollution

The importance of the flourishing production capacity of utilizing artificial intelligence in environmental science, various hurdles must be solved to properly exploit its potential in tiny pieces of plastic pollution management. One significant difficulty signifies the development of XAI models to handle vast, heterogeneous environmental datasets. As information garnered from satellites that transmit information sensors, and laboratory tests continue to rise in volume and complexity, it has turned increasingly challenging to use current XAI methodologies efficiently while preserving performance. Another key problem lies in managing their interpretability with precision. Highly interpretable models, from decision trees alongside linear models, may lack a capacity for forecasting needed for complex systems within the environment, while more specific deep learning models typically compromise transparency. Striking the correct balance of these two attributes is key for assignment reliable and intelligible AI systems and integrated with Machine learning and deep learning and Generative AI.

Notwithstanding the changes that have taken place, there are several significant obstacles to overcome. Plastics made from micro datasets are generally noisy, inconvenient, and insufficient necessitating XAI systems to handle uncertainty while remaining useful. A major trade-off arises between their comprehensible and accuracy; while less complicated algorithms like decision trees give better communication, they could fail to work and additionally murky deep learning models and integrated models. Additionally, the lack of defined XAI techniques in environmental science inhibits reproducibility and comparison between investigations. The restricted being able to labeled microscopic facts again negatively impacts the concept training and evaluation. Bias and fairness difficulties develop owing to specific places analysis, which can contribute to biased or inaccurate explanations. Lastly, trust and acceptance among key stakeholders—such as researchers, policymakers, and local communities—remain limited. This highlights the urgent need for user-friendly interfaces and improved understanding of Explainable AI (XAI) technologies. Additionally, better integration of sampling site data with remote sensing information is essential to enhance transparency and reliability in environmental monitoring.

Several AI-integrated and semi-automated systems play a vital role in addressing these challenges. Deep neural networks, enhanced with AI-based interpretability tools such as Grad-CAM, SHAP, and LIME, help in understanding and visualizing model decisions. Moreover, convolutional neural networks (CNNs) have demonstrated significant success in analyzing microscopic images of plastic particles. The processing of natural language with concurrent integrated processing, and natural language processing (NLP), enables in extracting knowledge acquired from the academic community, while Bayes and integrated with reinforcement learning integrated with techniques with XAI provide a probabilistic interpretation of risks with explicit rationale. Auto-ML systems improve with explainability features speed model building while ensuring transparency. A well-defined strategy is essential to advance the accessibility and application of Explainable AI (XAI) in microplastic research. In the short term, efforts should focus on developing high-quality, well-labeled datasets and integrating interpretability tools such as SHAP and LIME directly into operational machine learning pipelines. In the latter stages, hybrid models that combine physical modeling of plastic behavior with AI-driven analytical approaches should be developed. In the long term, establishing a global, open-access framework for predicting and assessing microplastic distribution will be essential. Such a platform would promote greater collaboration, transparency, and data sharing in policymaking, while integrating advances in deep learning, explainable AI (XAI), and microplastic research.

In conclusion, Explainable AI holds great potential to alter how we battle the introduction of microplastic, thereby transforming semi-transparent AI models directly into what appears to be transparent, trustworthy tools. However, fulfilling this promise requires a deliberate effort incorporating research that spans disciplines high-quality data, that include stakeholder involvement. By addressing technological and social difficulties jointly, utilizing artificial intelligence can convince every single the academia and the public help them make knowledgeable, decent choices in the global campaign against microplastic contamination with Machine learning predictions and integrated with Agentic AI integration.

3.8 Conclusion

The implementation of clarified AI (artificial intelligence and agentic AI) and tiny plastic particles pollution management presents a disruptive thought about environmental decision-making. By boosting transparency, accountability, and confidence in AI-driven models, XAI helps elected officials, research, and constituents comprehend more completely the rationale behind forecasts and recommendations. This is particularly relevant in the complex realm of the proliferation of microplastic where interdisciplinary ideas and data-driven tactics are necessary for effective intervention. As the industry grows in importance, the acceptance of XAI not only guarantees more informed decisions and accountable judgments but also encourages confidence from the public in the use of modern technology for environmental sustainability. Continuous studies and teamwork are necessary to develop XAI frameworks, increase ability to be interpreted and eliminate the disparity between AI models and actual solutions to the environment and microplastics pollution and it can help with integrated AI and agentic AI models to integrate and remove the microplastic detection using earth observation with remote sensing imagery.

AI disclosure statement: The author confirms that no artificial intelligence (AI)-generated content or AI-assisted writing tools were used in the preparation of this chapter (including summaries and the abstract). All text, analysis, and conclusions are the result of the author's original work.

References

1. Montavon, G., Samek, W., Müller, K.R., Methods for interpreting and understanding deep neural networks. *Digital Signal Process.*, 73, 1–15, 2018.
2. Liu, F.T., Ting, K.M., Zhou, Z.-H., Isolation Forest, in: *Proc. IEEE Int. Conf. Data Mining*, pp. 413–422, 2008.
3. Baehrens, D., Schroeter, T., Harmeling, S., Kawanabe, M., Hansen, K., Rasmussen, C., How to Explain Individual Classification Decisions. *J. Mach. Learn. Res.*, 11, 1803–1831, Jun. 2010.

4. Ribeiro, M.T., Singh, S., Guestrin, C., "Why Should I Trust You?": Explaining the Predictions of Any Classifier, in: *Proc. 22nd ACM SIGKDD Int. Conf. Knowledge Discovery and Data Mining*, pp. 1135–1144, 2016.
5. Rathore, P., Sustainable smart homes and buildings with internet of things. John Wiley & Sons, Inc., Scrivener Publishing LLC, 2025.
6. Samek, W., Wiegand, T., Müller, K.-R., Explainable Artificial Intelligence: Understanding, Visualizing and Interpreting Deep Learning Models. *ITU J. ICT Discov.*, 1, 1, 39–48, 2018.
7. Bhandari, P.V., *et al.*, Explainable Artificial Intelligence for Environmental Applications: A Review. *IEEE Access*, 10, 128270–128295, 2022.
8. Holzinger, A., *et al.*, What Do We Need to Build Explainable AI Systems for the Medical Domain?, arXiv preprint, arXiv:1712.09923, 2017.
9. Verma, S., Dickerson, J., Hines, K., Fairness and Explainability in Artificial Intelligence: A Survey, in: *Proc. 2020 IEEE/ACM Conf. on AI, Ethics, and Society*, 2020.
10. Garcia-Garcia, D., Orts-Escolano, A., Oprea, S., Villena-Martinez, V., Garcia-Rodriguez, J., A Review on Deep Learning Techniques Applied to Semantic Segmentation, arXiv preprint, arXiv:1704.06857, 2017.
11. Lecun, Y., Bengio, Y., Hinton, G., Deep learning. *Nature*, 521, 436–444, 2015.
12. Hahladakis, M., Velis, E., Weber, R., Iacovidou, E., Purnell, P., An overview of chemical additives present in plastics: Migration, release, fate and environmental impact during their use, disposal and recycling. *J. Hazard. Mater.*, 344, 179–199, 2018.
13. Carr, S.A., *et al.*, Transport and fate of microplastic particles in wastewater treatment plants. *Water Res.*, 91, 174–182, 2016.
14. Andrady, A.L., Microplastics in the marine environment. *Mar. Pollut. Bull.*, 62, 8, 1596–1605, 2011.
15. Boucher, A. and Friot, M., Primary Microplastics in the Oceans: A Global Evaluation of Sources, IUCN, Gland, Switzerland, 2017.
16. Van Cauwenberghe, L., Claessens, M., Vandegehuchte, M.B., Janssen, C.R., Microplastics are taken up by mussels (Mytilus edulis) and lugworms (Arenicola marina) living in natural habitats. *Environ. Pollut.*, 199, 10–17, 2015.
17. Li, Y., Li, H., Zhang, L., Zhang, Y., A Deep Learning-Based Method for Detection of Microplastics in Marine Environments. *IEEE Access*, 8, 92962–92970, 2020.
18. Sharma, R. and Choudhury, D., Artificial Intelligence and IoT Based Decision Support System for Marine Debris Monitoring, in: *Proc. IEEE Int. Conf. Comput. Sustain. Glob. Dev.*, pp. 227–232, 2019.
19. Zhang, D., Huang, W., Wang, C., AI-Based Monitoring of Plastic Pollution in River Systems. *IEEE Trans. Geosci. Remote Sens.*, 59, 8, 6441–6451, 2021.
20. Gade, M.G., Belyaev, A.V., Dankert, M., Niedzielski, F., Optical Remote Sensing of Marine Litter and Microplastics: A Review. *Remote Sens.*, 14, 10, 2354–2375, 2022.

21. Nasir, J.A., *et al.*, Machine learning and deep learning approaches for environmental monitoring: A review. *Environ. Sci. Pollut. Res.*, 29, 38613–38639, 2022.
22. Zubiaga, A.Z., Machine Learning-Based Framework for River Plastic Debris Detection Using Satellite Imagery. *IEEE J. Sel. Top. Appl. Earth Obs. Remote Sens.*, 14, 5450–5461, 2021.
23. Cook, F.L., *et al.*, Data-Driven Policy Making Using AI for Environmental Governance. *IEEE Trans. Eng. Manage.*, 70, 1, 144–153, 2023.
24. Singh, S.K., *et al.*, Design of a Decision Support System for Marine Pollution Management Using AI Techniques, in: *Proc. 2022 IEEE Int. Conf. Systems, Man, and Cybernetics (SMC)*, 2022.
25. Gamboa, L.T., *et al.*, A Framework for Explainable AI in Environmental Monitoring Systems. *IEEE Access*, 9, 155473–155487, 2021.

4

AI-Driven Technologies in Mitigation of Microplastic Pollution

Lata Rani[1], Hurmat[1], Deepa Singh[1], Babu Bharman[1], Arun Lal Srivastav[2], Jyotsna Kaushal[3,4]*, Komal Thapa[1] and Neha Kanojia[1]

[1]*Chitkara University School of Pharmacy, Chitkara University, Himachal Pradesh, India*
[2]*Chitkara University School of Engineering and Technology, Chitkara University, Baddi, Himachal Pradesh, India*
[3]*Centre for Water Sciences, Chitkara College of Pharmacy, Chitkara University, Punjab, India*
[4]*Chitkara University Institute of Engineering & Technology, Chitkara University, Punjab, India*

Abstract

The influence of microplastics (MPs) is increasing nowadays on human health as well as on natural environment. There are different sources of these compounds i.e., personal hygiene products, soaps, and the decomposition of larger plastic stuffs. Microplastics are responsible for various detrimental effects specifically on marine life and other ecosystems. As an illustration, disturbance of reproductive systems of marine animals, ingestion of marine animals, and even mortality. Microplastics generate different pecuniary effects as well, because their presence in the environment can have negative impact on various sectors i.e., tourism and fishing. Moreover, human health is also influenced negatively owing to microplastic exposure i.e., various metabolic disorders (obesity and diabetes), sleep disturbances, disease related to respiratory system. Different programs and policies have been implemented to deal with issues arising due to microplastic pollution. But there are some other problems which include, lack of funding, less awareness among the public, and inefficient legislation. In the following chapter, the writer had discussed about the role of AI in the estimation of MPs in the ecosystem.

**Corresponding author*: Jyotsna.kaushal@chitkara.edu.in

Abhishek Kumar, Pooja Dixit, Pramod Singh Rathore, Arun Lal Srivastav and Ashutosh Kumar Dubey (eds.) Microplastic Monitoring Using Artificial Intelligence, (55–82) © 2026 Scrivener Publishing LLC

It was found that AI-assisted detection techniques are beneficial to cope up with shortcomings of the traditional detection methods i.e., expensive, low efficiency, less accuracy and laborious. Additionally, this chapter discussed the role of AI to mitigate MPs pollution effects.

***Keywords*:** Environment, artificial intelligence, microplastics, pollution, control

4.1 Introduction

There are different detrimental effects of MPs on human wellbeing, environment sustainability, marine life etc. Marine animals are very prone to side effects of marine organisms, including fish and turtles, which are MPs because they consume these particles as food. These particles have the ability to damage ecosystems by altering nutrient cycles. Additionally, MPs have a plethora of bad effects on aquatic communities owing to transport of pollutants that ultimately leads to the accumulation of toxins in water. There are various reports suggesting that MPs can permeate the food chain by the route of seafood either by consumption or drinking. The entering pathways of MPs in the environment is shown in Figure 4.1. These parameters result in various drawbacks like posing risks such as inflammation, reproductive issues, growth and development problems, poor immune

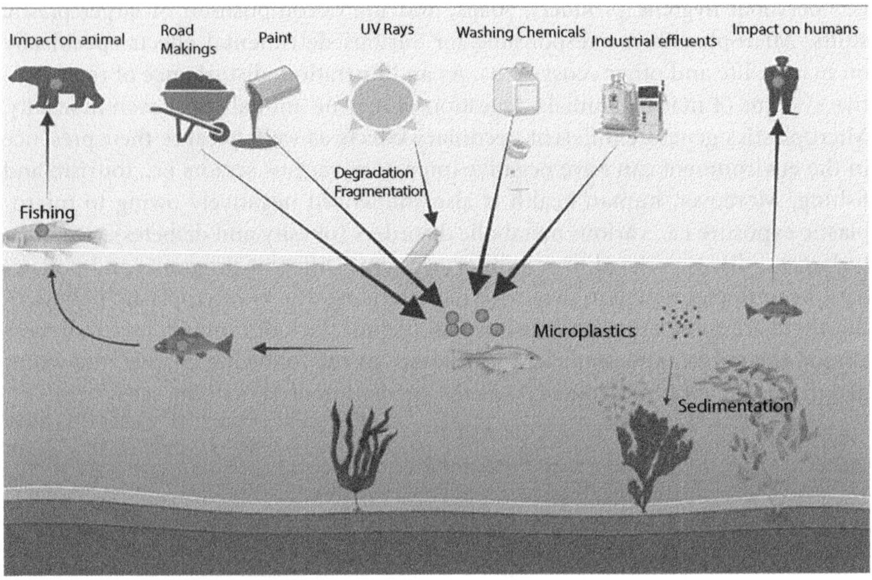

Figure 4.1 Introduction of MPs in environment.

responses etc. Moreover, MPs may damage filters and pipes of water purification plants by various ways that ultimately affects the efficiency of this plants. This factor also leads to higher operation and maintenance (O&M) costs. To conclude, evaluation to the distribution and accumulation of MPs is a crucial task globally due to its detrimental effects.

AI has become a significant component of different environmental monitoring projects in these days owing to its accessibility and efficiency. AI is a recent and magical division of computer science that deals with various computer programs i.e., learning, detection or analyzing, decisions making process and reasoning [1–3]. AI can be utilized to detect large data sets, which ease and facilitate to conclude patterns and precise predictions [4]. In addition, AI has been employed in various fields of environmental monitoring i.e., prediction of natural catastrophe, water quality assessment, pollution identification etc. [5]. It was revealed that AI can significantly advance environmental monitoring's process as well as efficiency of these techniques. It also provides different long-term benefits related to cost and reproducibility [6, 7].

AI has introduced a significant revolution in countless fields, owing to its specific role in different processes related to environmental monitoring and pollution control [8]. AI suggests different techniques to deal with palliate environmental problems with great ease and efficiency. One of the primary AI benefactions to environmental monitoring is the fact that it has the capacity to dissect enormous amounts of data from a variety of sources, including satellite imagery, remote detectors, and downfall stations [9, 10]. Machine knowledge (ML) algorithms are suitable of assaying these data sets to identify patterns and anomalies and deliver information on environmental changes with unknown perfection. Machine knowledge (ML) algorithms can exercise these data sets to fete patterns and anomalies and give perceptivity into environmental changes with unknown delicacy. AI-tested models read pollution situations in air quality and soil operation by adding up inputs from several monitoring stations and meteorology [11]. The models help minimize pollution and save public health by furnishing timely interventions. AI has been effective in detecting and relating adulterants. For illustration, ML and deep knowledge(DL) styles are employed to prognosticate PM 2.5 attention, soil heavy substance pollution, the influence of microplastics on the heavy substance uptake by shops, and mass attention values of nitrogen oxides (NO2), essential carbon (EC), and organic carbon (OC) in business- released PM10 patches [12–15]." Quality monitoring of water has also come more with the advancements in AI. Traditional styles of assessing water quality include field slice and lab analysis, which consume time and are limited in operation. AI platforms can continuously cover water bodies using detector networks and dissect the data entered to descry pollutants and prognosticate pollution events [16, 17]. This real-time monitoring capability is essential to icing safe drinking water and securing submarine ecosystems from poisonous adulterants. AI in waste operation is another area of utmost significance that needs development. Intelligent waste operation systems employ AI to maximize waste collection routes, minimizing energy use and emigrations from scrap exchanges [18]. AI-enabled sorting technologies also enhance recovering effectiveness by duly relating and separating colorful forms of waste paraphernalia [19, 20]. These inventions affect in greener waste operation practices and reduce the overall environmental footmark. In recent times, AI technology has also been employed to address the critical issue of microplastic pollution and give new results for discovery, analysis, and control [21]. Microplastics are small plastic fractions that irruption ecosystems and have come as an important environmental concern since they are not fluently biodegradable and could impact environmental organisms, biodiversity, and mortal health. Currently, the primary methodologies employed for identifying and characterizing microplastic substances (MPS) involve techniques such as visual inspection through microscopy and analytical assessment *via* spectroscopy. Visual identification is generally the original system to roughly identify MPs rested on their shape and size. Microplastics with some shape characteristics may be manually or microscopically linked. The process is total, heavily dependent on the chops of the bystander, and is associated with a high rate of recognition error [22]. While the optic microscopes give advanced exaggeration at the cost of their inferior resolution, scanning electron microscopy (SEM) can be employed to depict the morphology of microplastic patches, therefore allowing experimenters to assess their size and shape, and face characteristics to dissect face morphology [23].

SEM's high exaggeration allows for demarcation among colorful MP types according to their distinct face textures, easing correct identification. Still, this fashion is inversely time- consuming and involves complicated sample pretreatment [24]. Spectroscopic ways generally employed to descry MPs include FTIR and Raman spectroscopy. "FTIR spectroscopy is commonly utilized as a primary technique for detecting and identifying microplastics. It's used extensively for detecting and relating MPs in environmental samples, particularly to identify the polymer factors. It can fete and measure MPs with high perfection and acquire specific information of polymers by detecting characteristic spreads of plastic patches. It is, still, fluently intruded with by water or organic contaminations on the sample face, leading to disordered spectral immersion peaks [25, 26]. The analysis process is also labor-ferocious, and the flyspeck size of distinguishable microplastic patches is confined. Raman spectroscopy is a non-destructive fashion, in comparison to FTIR, that's rested on inelastic scattering of monochromic light to engage with the vibrational modes of molecular bonds in microplastic patches. This enables it to have a broader wavelength range, be resistant to water hindrance, and have an advanced resolution (100 nm) [27]. Nonetheless, the presence of complements and organic matter in the sample can lead to serious crimes in the discovery results, and the analysis process is still time- consuming [28]. Conventional discovery ways are time-consuming, labor-ferocious, and narrow in compass [29]; therefore, AI technology integration is a promising area to achieve more accurate and effective discovery and identification of microplastics, which can contribute to diving global microplastic pollution. Formerly, over 1600 exploration papers on the discovery of microplastics and over 100 exploration papers on AI-supported discovery of microplastics have been published (hunt tool, Web of Science; hunt date, Jul 31, 2024; hunt subject captions,' microplastics and discovery," microplastics and discovery, machine knowledge or deep knowledge'). Research has revealed that ways in combination with computer vision are more suitable for detecting and measuring microplastics. Partial least places in combination with interval selection models is applied in this exploration to identify images of microplastics, and it assists in bridling time destruction and escaping mortal bias in analysis [30]. Deep CNN and U- Net neural networks can be employed to pre- process and classify bitsy microbead images, member and count MPs in specified images under complicated conditions automatically, lessening primer intervention, and enjoying the capability to fete images in batches [31, 32]. Presently, different DL models like VGG16, UNet, and MultiResUNet have been employed to prop SEM microplastic image identification, all demonstrating bettered vaticination delicacy [33]. ML technology is excellent in data type and processing. It's applied for batch processing of FTIR data, pulling out unique information of colorful polymers, which can identify types of polymers in environmental MPs directly and efficiently [34]. Machine learning is also employed in Raman spectroscopy to enhance signal strength, thereby improving the effectiveness of microplastic (MP) detection and analysis. Together with ML, Raman spectroscopy can quicken exploration into MPs in factual samples [35]. Presented with a wide range of spectral data, ML is arising as an effective process suitable of streamlining the classifying process and making environmental samples analysis easier. AI technology may also be employed in managing microplastic impurity. Utilizing big data to analyze pollution sources and monitor pollutant migration provides an effective method for enhancing traceability of long-term environmental trends and addressing the current lack of scientific rigor and responsibility of current microplastic traceability ways [36].

Within the constant pursuit of a pollution-free and sustainable future, the use of AI for microplastic discovery and pollution operation has surfaced as a vital technology. Conventional discovery styles of microplastics have multitudinous limitations and downsides, precluding them from directly measuring the degree of microplastic impurity. The AI driven process allows for rapid-fire and automated discovery of MPs across all environmental matrices, ranging from submarine systems to soil and indeed the atmosphere. In addition, this technology enables the aggregation and analysis of huge data sets, yielding perceptivity into patterns, sources, and goods of pollution that were not preliminarily possible. The significance of the integration of AI in microplastic discovery goes far beyond technological advancement. This composition offers a detailed review of the crossroad of AI and microplastic discovery, demonstrating the synergistic part of ML styles in the fight against plastic pollution. We shouldered a detailed study of ways for AI-supported discovery, identification, and pollution operation, shoveled the possibilities of exercising colorful ways in real- world operations, and anatomized how data pre- processing processes can be bettered by AI. Use of AI for detecting microplastics AI-supported image processing fashion. Conventional image identification and discovery procedures for MPs also have some essential limitations that limit their delicacy and responsibility.

4.2 AI Assisted Detection Techniques for the Microplastic

The detection techniques used for the MPs in the environment is represented in Figure 4.2 and details of the technique given below.

4.2.1 AI-Assisted Image Processing Technology

A significant limitation is the devilish reliance on visual interpretation, which tends to be private and susceptible to mortal error [37]. This subjectivity occurs from the homemade separation process, whereby individualities have to separate MPs from other patches on the base of physical characteristics. Misapprehensions are current, particularly in the case of environmental samples with a combination of organic and inorganic substances. The diversity in the size, shape, and color of MPs also makes

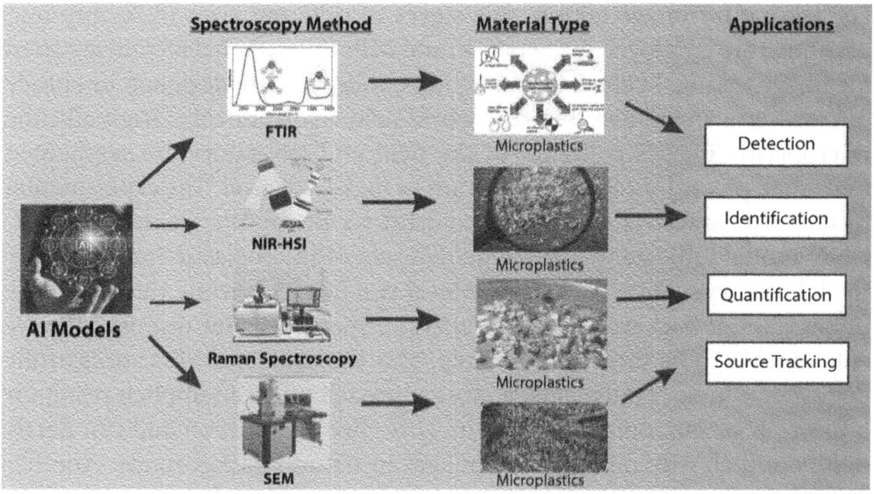

Figure 4.2 AI assisted detection techniques for MPs.

identification more complex [38]. In addition, conventional imaging bias similar as optic microscopy are generally not sufficient in terms of resolution to identify the lowest MPs, performing in major underestimation of their circumstance [39]. Conventional image recognition styles also fail due to the diversity of MPs, since these patches change physically and chemically with time [1]. In fact, riding can make MPs fracture into sharp irregular shapes and lose color, and hence be more delicate to separate from natural debris like beach or organic material [40]. Traditional system sample medicine is time-consuming and requires important trouble, adding analysis time and introducing implicit for crimes. The need for homemade intervention decreases effectiveness in general and makes scale-up of large-scale environmental monitoring systems clumsy [41]. The detection of microplastics has become significantly more efficient with the help of AI algorithms, which are well-suited for analyzing large volumes of data extracted from microscopic images . Machine knowledge and deep knowledge, which are subdomains of AI, are essential in this terrain. Through training AI models on different datasets with microplastic images, the models learn to fete different types of MPs, similar as fractions, fibers, and microbeads, with high perfection and delicacy. In fact, Han *et al.*, suggested a deep knowledge-rested approach, Mask R- CNN, which can descry, classify, member, and count MPs, and classify them automatically into orders like filaments, fractions, and patches [42]. The set of microplastic images is attained from sand field photography and laboratory image taking.

For laboratory image taking, two sieves of different sizes (4.76 mm and 0.85 mm) are employed for network, and MPs lower than 0.85 mm are tried for image taking. "Laboratory imaging conditions are designed to closely mirror those encountered in real-world field environments. originally, the microplastic images collected in the field and laboratory are segmented into small pieces by a Python program to make a dataset. The experimental results of characterizing MPs in colorful scripts by the trained Mask R-CNN model indicate that the type delicacy of microplastic images under the laboratory's white background level reaches 93.3, while the sensitivity under natural background conditions is 15% lower." But utmost MPs are still suitable to be directly deposited, and the orders of MPs successfully deposited can be rightly labeled. This approach offers an implicit result for out- of- door microplastic rapid-fire- fire- fire disquisition and can fleetly descry and classify microplastic images in natural backgrounds. Another model coupled with DL can be used to negotiate rapid-fire- fire- fire qualitative and quantitative identification of colorful types of MPs. It uses vector- quantized variational machine encoder (VQVAE2) tone- supervised point knowledge combined with optic resolution print audial microscopy to identify and descry microplastic images. originally, the collected MPs need to be screened and pulled, and the separated patches are observed and characterized under a photoacoustic microscope (PAM) to gain sample images [43]. The DL process is employed for the discovery of microplastic photoacoustic imaging data to enhance the processing speed of big data and determine MPs of varied shapes and kinds. The pre- recovering step in this process remains the same as in conventional ways; however, accelerating and refining the process of analyzing microplastic images is essential. Zhu *et al.*, designed an online digital holographic imaging system with a light DL network, CompNet. CompNet is rested on the holographic classifier convolutional neural network (HC-CNN) armature and stoked by analysis and type modules [44]. The system employs an incompletely coherent light source to image MPs, producing a hologram on the imaging airplane, captured by a CMOS camera with a discovery range of 100 μm to 5 mm. The experimental samples correspond of low- complexity synthetic samples and high- complexity in samples from factual strands. The experimental results indicate that the CompNet system can effectively determine the number of MPs in the samples, with a saving of around 72 of the training time compared to other networks. Nevertheless, it's limited to unstable and small data sets and is unfit to be used to define the type, shape, and size of microplastic patches. With the operation of AI technology, automatic recognition and analysis of microplastic images can be achieved. Fully automated image processing, combined with a flexible timber classification system, can significantly improve the efficiency and accuracy of Nile red staining, luminescence imaging, microscopy, microplastic detection, and overall automated image analysis (RF) [45]. Vitali *et al.*, set 20 samples of bottled water for examination, carried out original filtration, stained them with Nile red, and took their photos under a microscope. They employed their suggested fashion to train an ML supervised model. The imaging element may be completely automated to calculate precisely the number and sizes of the patches and roughly the weight of a single microplastic and as a result, it effectively assessed the distribution of mass within the sample. Such an interactive AI tool would be suitable to distinguish MPs between colorful marketable bottle waters and arrange the workflows like making pretreatment simpler and better. It can be employed in the discovery of MPs in more complicated food sample matrices. One system combining 3D coherent imaging and ML can automatically identify MPs in pretreated water samples at colorful scales with high delicacy [46]. The pre-treatment removes sediments and summations larger than the analysis size, builds holographic features with holographic microscopy to distinguish MPs from algae, and uses Pearson correlation measure matrix to find the sample with lesser delicacy. Incipiently, H- support vector machine (H- SVM) identifies colorful MPs in pretreated seawater with a rate of delicacy up to 99. The algorithm demonstrates performance equal to completely developed deep neural networks with efficiently zero training time. But in the case of larger algae, beach, or sediments lesser than 1 mm in the target pretreated water sample, delicacy diminishes, and thus filtering out larger contaminations during pretreatment is necessary. Another semi-automatic fashion grounded on conventional image processing is used to identify MPs. Microplastic Lab (MUPL) applies automated image segmentation to fleetly and directly identify and quantify MPs in swash and delve sediments at an average processing speed of 10 s per image [47]. They insulated MPs from swash water and delve sediments, characterized and imaged them in the lab, and entered two kinds of microplastic images (original images and Nile red stained images) to estimate MPs' cornucopia and characterization. Shape and size of MPs were enumerated by MUPL, and average divagation from ImageJ was below 10, as low as 8 when Nile red color was applied. In discrepancy to ML ways, MUPL does not involve supervised literacy of enormous sets and can break the issue of MP identification in grueling conditions; still, it detects and quantifies fibers of a specific length range and cannot guarantee discovery of small MPs.

AI-assisted detection techniques are used for the identification and classification of microplastics (MPs). The various AI-based approaches are illustrated in Figure 4.2, and their detailed descriptions are provided below.

4.2.2 AI-Assisted FTIR

FTIR spectroscopy is the most frequent tool for relating and determining the microplastic polymer characteristics. Spectrum analysis is the major element of MP analysis through FTIR. Hand interpretation of the applicable absorption bands is extremely time- consuming and needs robust moxie [25]. AI technology may be employed to exercise a large number of infrared spreads and acquire sample information, thus swiftly assaying the microplastic polymer type. For illustration, partial least places regression (PLSR), FTIR can identify MPs (LDPE and PET) in sediments, but the model is largely used as a pre- webbing tool to filter plastics through large numbers of deposit samples [48]. Also, ML has demonstrated exceptional predictive capability in large mixed microplastic spectral datasets [49]. Scientists erected and pre-processed spectral data from privately available spectral libraries, bought plastic samples, and terrain- plastic tried and measured plastic measures. Researchers employed a convolutional neural network (1D-CNN), along with decision tree (DT) and random forest (RF) models, to process and evaluate the spectral data. The 1D-CNN had an overall delicacy of 97.44, surpassing the others and the swish in environmental sample prophecy, and RF was the most harmonious with little amounts of spectral data. The different dataset type rigor are possibly because the dataset could not directly describe real sample data and the influence of spectral noise. AI technology can be employed to examine big collections of infrared spreads and pull-out sample data, thus swiftly assaying the microplastic polymer kind. For illustration, FTIR with partial least places regression (PLSR) can identify MPs (LDPE and PET) in deposit, though the model functions more as a pre- webbing tool to sort through plastics in large numbers of deposit samples [48]. In addition, ML performed well to prognosticate vast numbers of composite microplastic spectral data [49]. Scholars erected and prepared spectral data attained from public-accessible spectral libraries, bought plastic measures, and plastics from samples and measured pulled from the terrain. Scholars employed a one-dimensional convolutional neural network (1D- CNN), decision tree (DT), and RF to anatomize spectral data. The 1D-CNN attained an overall delicacy rate of 97.44, which was lower than that of other models and performing swish in the type of environmental samples; whereas RF performed swish using lower spectral data. The different type rigor of various datasets could be because the dataset was unfit to mimic real sample data and the goods of spectral noise. Every ML fashion has a limitation, and various algorithms descry different kinds of MPs differently. Yan *et al.*, suggested an ensemble knowledge model with the operation of seven various ML algorithms, and the ensemble was applied through neural networks to determine the most effective model for perfecting the discovery of microplastics [50]. According to the analysis of the FTIR dataset, the approach demonstrated was more effective in detecting unstable datasets and defiled MPs compared to the current individual ML- predicated processes. In factual operations, the ensemble model can be employed by stoners to identify MPs in trying to break the problem of missing microplastic types and choosing a suitable right ML model.

Machine Literacy improves qualitative and quantitative analysis of MPs with good prophetic performance. Through the combination of FTIR and quaternion resemblant point birth system, experimenters combined the original microplastic diapason with the first and alternate secondary gamuts of the original diapason, also performed quaternion top element analysis (QPCA) for spectral point birth, in confluence with the SVM algorithm for analysis. The system greatly enhances the performance of FTIR qualitative and quantitative models. It is not applicable to use in assaying MPs with lower sample sizes and further types [51]. ATR-FTIR spectral data, even when degraded due to environmental aging of complex microplastics (MPs) in terrestrial settings, can still be effectively analyzed using machine learning approaches. Seven ML models were tested by Enyoh *et al.*, and the stylish performing in classifying PET MPs were RF, GB, DT, and k-NN, with 99 bracket delicacy [52]. This system reduces and improves bracket and serves as an unborn guideline for environmental microplastic analysis. Likewise, the CNN model is more accurate in detecting aged microplastic samples, making the pre-processing process easier with its better point birth and identification features. This means that CNN coupled with infrared spectroscopy can classify aged microplastic samples, making it suitable for detecting large quantities of aged MPs in complex surroundings [53]. AI technology also excels in spectral preprocessing. Using ML to address noise goods and spectral vestiges in FTIR gamuts enhances the effectiveness of posterior microplastic identification. Researchers evaluated the performance of five machine learning models—Decision Trees (DT), Neural Networks, Naive Bayes, k-Nearest Neighbors (kNN), and Support Vector Classifier (SVC)—on a dataset derived from ATR-FTIR spectra. They applied four preprocessing techniques: baseline correction, standard scaling, Principal Component Analysis (PCA), and data oversampling. Among these, baseline correction and naive oversampling proved to be the most effective preprocessing methods for enhancing classifier performance. PCA significantly reduces processing time without affecting accuracy, while SVM is most effective for quick and accurate classification of microplastic type bracket [54]. Spectral noise hindrance is another challenge when handled with ML and DL models. Experimenters first minimized spectral noise in the FTIR dataset, re-collected data to maintain thickness between colorful types of microplastics, and enlarged the small dataset. Their approaches vastly suppressed noise hindrance, enabling presto and effective classifying of MPs. After applying spectral noise processing, SVC gave the stylish result among ML algorithms with a delicacy of 95.99, while LeNet5 outdistanced other DL algorithms with a perfection of 96.93 [55]. It's perfect to apply in cases of noisy spectral data by inhibiting external circumstances-convinced noise hindrance. DL is also used for the elimination of instrument noise as well as spectral vestiges. Brandt *et al.*, used an autoencoder neural network to make intricate spectral deformations, barring uninvited vestiges upon effective training with high effectiveness and speed [18]. The fashion is most applicable for microplastic discovery tasks with massive spectral data under a limited time frame. It minimizes hindrance from multitudinous factors vastly, therefore efficiently boosting spectral processing capacity. μFTIR (micro Fourier transfigure infrared interferometer) imaging has the capability to image bitsy structures and distinguish MPs at the same time, delivering detailed chemical and spatial information. ML models can examine microplastic polymer types in environmental samples grounded on μFTIR imaging with delicacy for measuring cornucopia and size. Hufnagl *et al.* erected an RF model that's suitable to fleetly identify MPs in μFTIR images, distinguishing five different polymer types in roughly 5 min [56]. This model can be extended to identify further than 20 colorful types of polymers with perfection, applicable for different environmental matrices without dragging discovery time [57].

FTIR coupled with AI has greatly enhanced the speed and correctness of microplastic identification, providing high accuracy in the detection and classification of polymers. Performance can still differ based on dataset quality and spectral noise. AI-based techniques make complex spectrum analysis easy and minimize the requirement for manual interpretation. To overcome issues like unbalanced datasets and noisy spectra, ensemble learning techniques and sophisticated pre-processing methods may be used. Irrespective of these advancements, these techniques are still having poor efficacy for different small size microplastic samples, that leads to its poor applicability in various environmental fields. μFTIR imaging in combination with ML gives better results and can easily manage noisy data with great efficiency. To achieve finest detection presentation, different factors need adjustment i.e., better imaging quality, stable model training, adjustment of environmental sample etc.

4.2.3 AI-Assisted Raman Spectroscopy

Raman imaging provides very beneficial valuable information but it has poor resolution of imaging when working with laser spot's diffraction [58]. On the other hand, Spatial heterodyne Raman spectroscopy (SHRS) has offered a significant solution to deal with this issue, and provides an improved signal-to-noise ratio in case of MP detection [59]. This innovative technique utilized spatial heterodyne technology in combination with Raman spectroscopy and provided precise and rapid identification of high-density polyethylene (HDPE), different MPs i.e., PP, PC, PS. The procedure involves various steps i.e., base line removal, correction of interference pattern, spectrum recovery and calibration. There is an option to evaluate pre-processing data automatically; for example, by using SHRS system. It is reported in many reports that combination of Raman spectroscopy with machine learning algorithms is an unmatchable tool to deal with identification procedures of nanoplastics. The accuracy of NPs among results of data set obtained by Raman spectroscopy method was 98.8% although it was higher in environmental samples i.e., 97% [60]. Additionally, another advanced technique is formulated by researchers to achieve fast results by combining driven spectral matching algorithm with hyperspectral stimulated Raman scattering (SRS) imaging for chemical imaging of NPs. This procedure efficiently identified the plastic nanoparticles and plastic polymers [61]. On the other hand, explanation of spectral library as well as improved SRS imaging analysis algorithm is required to achieve its applicability in different complex environmental samples. Furthermore, pre-processing is an option to eliminate interference signal noise in complicated environmental specimens. Good signals can be removed and retained using the sophisticated peek extraction and retention (PEER) algorithm for data pre-processing. Combination with ML algorithms, efficiently identifies poor-quality NPs in multifaceted environments. Luo *et al.*, improved signal-to-noise ratio of microplastic pictures and minimized identification bias by applying image reconstruction deconvolution based on PCA and algebraic algorithms for better plastic particle characterization [62]. They applied Gaussian fitting to reduce noise in Raman images and simplified sample preparation. To determine the physical characteristics and chemistry of MPs, they also investigated strong strategies and integrated SEM. Finding and monitoring MPs has also showed potential when Raman spectroscopy is incorporated into a DL framework. For the high-precision identification of microplastic types, a multi-layer 1D-CNN was applied. A comparison with different machine learning techniques showed that 1D-CNN performed better. The raw Raman data must be denoised, baseline modified, and the system background spectrum eliminated. Applications in real life are increasingly using raw spectrum data classification instead of time-consuming and complicated pre-processing. But there are still issues, like less-than-ideal outcomes from odd characteristic peaks in some MPs, which called a statistical model [63]. When compared to other machine learning techniques, the 1D-CNN model's accuracy rate in detecting MPs may rise up to 97%. Deep learning improves the qualitative analysis of microplastic Raman spectra by simplifying complicated data processing and efficiently extracting data-related information [64]. Another research obtained quantitative analysis of MPs in aquatic settings by combining CNN with Raman spectroscopy technology to increase identification efficiency by gathering the sample's average mapping spectrum (AMS). This technique prevents baseline drift by performing spectral correction after getting the MPs' Raman spectrum, enabling high-precision and broad-ranging detection of microplastic concentrations in diverse aquatic settings [65]. Although this technique provides quantitative analysis of MPs in various contexts and may be taken into consideration for identifying microplastic concentrations in contaminated environments but existing approaches are unable to give more accurate sample quantitative information.

Multivariate analysis in conjunction with Raman spectroscopy may automatically identify MPs in the environment. Spectral data is pre-processed using PCA and linear discriminant analysis (PCA-LDA), and microplastic various types in real-world scenarios are automatically and precisely classified using the SVM model. Still, substantial interference from intricate Raman signals produced by organic contaminants makes automated identification at the nanoscale challenging. To apply it to realistic environmental specimens, additional research is required [66]. Lots of MPs in Raman imaging may be rapidly quantified as well as characterized using ML. Li *et al.*, developed a novel AVmax-UCLS Raman imaging technique to detect aged MPs and rapidly determine the category and form of MPs. Even while the AVmax data for complicated environmental samples has little resemblance, it may nevertheless be properly detected [67]. Environmental specimens can be easily treated with AVmax-UCLS without the need for complicated preparation. If different types of smaller MPs are studied, it will be more appropriate for real-world use for high-efficiency, inexpensive non-destructive testing techniques. Moreover, artificial neural network (ANN) algorithms in conjunction with hyperspectral Raman imaging can quickly and accurately identify and detect various forms of microplastic in a variety of scenarios. Also, this technique expedites sample collection and streamlines the pretreatment procedure. Still, quantifying MPs smaller than 10 μm presents difficulties. Higher-power laser sources can make up for this limitation in the future and get ready for real-world scene sampling [68]. The identifying of MPs has been improved because of AI and Raman, which provide quicker and more precise identification as well as possible uses for NPs. While used in conjunction with DL, techniques like SHRS can handle complicated specimens efficiently while improving the detection of different kinds of microplastics. But issues like particle heterogeneity and background noise still need to be resolved. By enhancing the ratio of signals to noise and overall data accuracy, machine learning (ML) significantly improves spectral data processing. Microplastic research is improving due to creative methods that combine Raman spectroscopy with machine learning and image technologies. Notwithstanding these advancements, further study is required to manage intricate ecological specimens and increase detection capabilities, especially for extremely tiny particles. The goal of future research should be to find a wider variety of microplastic mixes, particularly those with lower particle sizes, in order to increase the spectrum of uses for Raman imaging.

4.2.4 AI-Assisted HSI

It is anticipated that HSI will quickly replace online monitoring as a reliable and non-destructive method of MP detection. The foundation of HSI is spectral imaging, which blends spectroscopy with imaging methods. Its basic idea is that each object's chemical makeup causes it to emit radiation with a variety of electromagnetic spectrum wavelengths [69]. HSI eliminates the need for intricate microplastic separation processes and the worry about interference from other materials in the sample, especially organic matter, in contrast to FTIR and Raman spectroscopy. MPs in environmental matrices including organic matter can be directly identified using HSI technology in conjunction with ML models. SVM demonstrated adequate detection accuracy in an experiment that mimicked the finding of MPs in fish intestines, and the full procedure—including HIS acquisition and data analysis, could be finished in six minutes. Nevertheless, only various MP types with particle sizes ranging from 0.1 to 1 mm were categorized; prior to simulation, nine distinct particle sizes were screened [63]. This technique primarily reduces sample preparation time to 30 minutes and does away with the need to separate MPs from other matrices. To accommodate field applications, greater thought should be paid to finding additional varieties and smaller sizes. ML method is usually selected to evaluate big datasets and redundant data. On the other hand, HIS is a good option to deal with microplastic polymers. Many researchers reported that HIS have the ability to evaluate successfully more than seven types of microplastic polymers.

Results have shown that various hyperspectral pictures were collected by researchers by using hierarchical partial least squares discriminant analysis (PLS-DA) model under the influence of short-wave IR spectrum. The outcomes of this study suggested that HIS is an efficient and robustable technique to characterization and classification of different particles. Moreover, this model facilitates the classification of different MPs among polymer [70]. Environmental factors still need to be taken into account; however, the abundance of various MPs in the area can be reasonably estimated for the analyzed polymer type. It is reported by Xu *et al.*, that combination of HSI and AI provides significant results even in complex cases to identify the presence of MP. Basically, they found various variables from already processed spectral data with the help of unique extraction methods. Different models i.e., backpropagation neural network, SVM and 1D CNN were employed to classify microplastic polymers. Results revealed these models have significant accuracy i.e., more than 93% specifically in 1D-CNN. On the contrary, SVM have shown better results for classification when estimated MPs in soil at very less concentrations [71]. Additionally, CNN algorithm is also considered as a good candidate for detection of MPs in soil samples. This procedure is very fast, efficient and provides reproducible results in case of original environmental samples. There are some reports that there is a void to improve spectrum pre-processing because large-scale environmental testing is still slow [72, 73]. Moreover, segregation of microplastic is recommended from other composition in test samples of soil because they may interfere with accuracy. To deal with detection problems of complicated environmental samples there is a need of further research. The base for the evaluation of MPs in complex mixtures has been revolutionized by the involving of machine learning (ML) and near-infrared hyperspectral imaging (NIR-HSI). These techniques are able to enhance the characterization and screening of different types of MP. There are four classification methods i.e., SVM, feedforward ANN, PLS-DA and soft independent modeling class analogy (SIMCA) are usually used for NIR-HSI imaging of MPs. It is reported that SIMCA is good to detect MPs in impurities with no or less errors and PLS-DA is a good choice for MPs identification. This technique does not require any laborious separation procedures even for MPs identification in samples that contain other chemicals too. Moreover, Vidal *et al.*, used combination of SIMCA with NIR-HSI with spectral datasets to produce a supervised organization technique.

The high-throughput microplastic screening instrument SIMCA can reliably identify colorless and aged plastic particles. Particle sizes between 150 and 300 μm can be identified, and microplastic kinds can be swiftly detected and identified directly in the sediment matrix. However, because organic debris on the surface of MPs might lower the accuracy of model predictions, difficulties still exist in correctly recognizing black or very dark particles as well as establishing the size and form of microscopic particles [74].

Although the black color of the polymer surface still affects model classification, NIR-HSI in conjunction with ML classification models can quickly and accurately identify colorless MPs in a variety of situations based on the advancement of ML algorithms [75]. Nevertheless, high-throughput techniques can rapidly get images (in a few seconds), accurately identify polymer types and smaller particles, and process images in bulk for routine analysis without requiring particle separation or pre-sorting [76, 77]. HSI reduces the requirement for intensive sample preparation by collecting a broad wavelength range, providing a quick and non-destructive approach for microplastic identification. HSI may greatly improve detection accuracy when combined with ML models as SVM and CNN, especially in complex situations. But issues like processing big datasets and spectral noise still exist. To improve pre-processing techniques and increase the system's capacity to manage intricate environmental samples and tiny particles, more research is required. These enhancements are essential for improving field capabilities and making apps more portable.

4.3 Application of AI in Microplastic Pollution Control

Utilization of AI technology has been increasing day by day to track the origins of MPs owing to its sophisticated analytical properties. Researchers can trace routes effectively by identifying complicated sources of MPs in the environment. This technique provides useful support in making policy decisions that facilitate in minimizing environmental pollution.

Conventional tracing methods was able to work on the basis of possible environmental parameters to estimated their sources specifically to analyze the abundance of MPs. Although, old techniques were more sensitive to errors in results as well as weak in scientific dependability [10].

To enhance the efficiency of conventional tracing techniques, AI technology is utilized nowadays to trace different environmental pollution and to reveal patterns. Many clusters analysis (CA) is used to investigate microplastic sources i.e., PCA, absolute principal component score-multiple linear regression (APCS-MLR), redundancy analysis (RDA) etc.

According to authors [75], these techniques have showed their ability in tracing sources, yield superior outcomes. Many studies have reported that ML can be useful to track MP sources. Wu *et al.*, evaluated various sources of pollution with the help of machine learning classifiers and MPs' fate were identified using PCA. A combination of different MP sources i.e., ANNs, DT, RF and SVM can be helpful to track plastic pollution sources employed ANN model to investigate the movement of plastic pollution.

Many reports revealed that ML can be used to track possible MP sources. The key factors affecting MPs' fate were identified using PCA. Combining DT, RF, and SVM can accurately identify various MP sources, offering fresh perspectives on tracking down plastic pollution sources [78]. The sources of contaminants in MPs are also investigated using ANNs. An ANN model developed by Tran *et al.*, is a useful instrument for forecasting the movement of plastic pollution. MPs in ambient samples were estimated using three machine learning models with higher accuracy: RF, least squares support vector machine (LS-SVM) and long short-term memory (LSTM) [79]. Long term reproducibility as well as high quality data is essential to estimate pollution levels in various environments.

With promising results, soil microplastic pollution levels were studies using near-infrared sensors and learning models with the support of machine learning. A significant combination of portable near-infrared sensors with the manifold embedded distribution alignment (MEDA) transfer learning algorithm system was considered as a cost-effective method for estimating plastic pollution in soil locations [80].

AI has confirmed its efficacy and significant role in technology to control environmental pollution. Different environmental factors can be monitored i.e., the evaluation of nitrate levels, microplastic distribution, drug presence by combining ultraviolet Raman spectroscopy with AI [81]. AI has immense commitment for revolutionizing microplastic monitoring and solutions to deal with drawbacks of conventional technology. We can have a more thorough investigation to check the extent and outcomes of microplastic pollution by using AI technologies. When combining with machine learning, monitoring of microplastic become efficient as well as this combination may help to offer customized policies for various pollution sites [82]. Notably, AI-enabled drones and autonomous underwater vehicles (AUVs) have become especially promising instruments for remote sensing and microplastic pollution monitoring [83]. With the use of cameras and sensors, these robotic devices are able to explore large areas of the ocean and land, gathering data on MP concentration and distribution in real time. Another study investigated into the viability of monitoring pollution with inexpensive commercial unmanned aerial vehicles (UAVs) fitted with high-resolution RGB cameras. Furthermore, YOLOv8 (a deep convolutional neural network)-based detection systems have a lot of promise for underwater microplastic identification because of their quick algorithm speed, high precision, and relative simplicity in deployment and training. YOLOv8 is a flexible tool for real-time detection in various contexts because it can detect things in real-time, adapt to the constantly changing underwater environment, and correctly identify objects in low light. However, only a limited dataset of underwater images has been used to train YOLOv8 [84]. The integration of DL becomes a feasible alternative for putting automatic classification systems into place when dealing with enormous data demands [85]. By integrating AI methods into these gadgets, dynamic data analysis is made easier, human interaction is less necessary, and pollution management decision-making is expedited. AI's capabilities, such as image analysis, machine learning, remote sensing, and risk assessment, are its main advantage. This method may be used to forecast the level of microplastic pollution in certain locations and create appropriate mitigation plans. It is applicable not only to seas but also to many ecosystems. The problem of insufficient comprehensive data for pollution monitoring still exists, though. Adoption of low-cost, eco-friendly AI systems for pollution management is made possible by machine learning and deep learning, which are essential for analyzing environmental data [86]. Tracking the sources of microplastics and estimating pollution levels are now much more accurate and efficient— thanks to AI techniques like machine learning and deep learning. By efficiently handling complicated datasets and producing more dependable results, these approaches provide significant advantages over conventional statistical procedures. AUVs and drones with sensors and cameras are two examples of AI-powered devices that make it possible to monitor microplastic contamination in a variety of habitats in real time. These tools accelerate the decision-making process and reduce the need for human interaction. They need more algorithm development, nevertheless, and have limits with regard to the comprehensiveness of the data. The dynamics of microplastic contamination are now better understood because of the combination of AI with technologies like Raman spectroscopy and near-infrared sensors. Notwithstanding these developments, further study is required to address issues with algorithm correctness, data quality, and the scalability of AI solutions in various environmental scenarios. To successfully reduce the negative consequences of microplastic contamination and safeguard ecosystems, these issues must be resolved.

4.4 Conclusion

Microplastics are a major threat to the ecosystem and human health since they come from sources like cosmetics and personal care items and are present in a variety of settings, such as in ocean, freshwater bodies, soil, and even in the air. Microplastics have serious effects on marine life and other ecosystems, including the ingestion of marine animals, disruption of their reproductive systems, and even death. Not to be overlooked are the substantial economic effects of microplastics, as their presence in our local ecosystem has an impact on sectors like tourism and fisheries. Microplastic exposure is also associated with imminent health risks, such as gastrointestinal and respiratory issues, disturbed sleep, obesity, and an elevated risk of diabetes.

There are programs and policies in place to address the pollution caused by microplastics. But there are other obstacles to overcome as well, like ignorance, a lack of resources, and inefficient laws. In order to address the imminent threats that microplastics represent to the entire ecosystem as well as our health, immediate action is required. AI's use in microplastic detection is like a two-edged sword, requiring a careful balancing act between accuracy and sensitivity. By utilizing quick data analysis and pattern recognition skills, AI shows exceptional sensitivity in detecting microscopic microplastic particles when paired with technologies like image analysis and spectroscopy. By automating the detection process and speeding up identification timelines, this synergy broadens the range of techniques for detecting MPs numerous opportunities for efficient microplastic detection and pollution control are made possible by the increased precision in microplastic identification and quantification. AI models aid in forecasting the origin, movement, and build-up of MPs in the environment, which is useful for pollution prevention. By combining data from multiple sources, predictive models assist in creating efficient regulations and cleanup strategies. Large-scale neural networks carefully filter and examine sensor, picture, and water sample data. Together, these technologies can provide real-time monitoring, early pollution hotspot detection, and a thorough understanding of the distribution of microplastics in ecosystems. AI has difficulties in detecting microplastics, despite its many benefits. The difficulty of generalization is one of the main obstacles since it is very difficult to integrate various data sets to attain optimal performance during training. However, the accuracy and dependability of AI are steadily improving due to growing data volumes and ongoing algorithm advancements. To reliably and effectively analyze sample data, more potent machine learning models are required. This procedure uses a lot of resources, including computing power, preprocessing, and data collection. Despite efforts by researchers to increase training efficiency, model training is still difficult and time-consuming. Continuous verification and algorithm improvement to accommodate different environmental matrices are necessary to guarantee the consistency and dependability of AI models.

AI disclosure statement: The author acknowledges the use of AI-based tools for minor language editing or grammatical correction. However, all intellectual contributions, analyses, and conclusions presented in this chapter are entirely the author's own.

References

1. Ruggero, F., Gori, R., Lubello, C., Methodologies for microplastics recovery and identification in heterogeneous solid matrices: A review. *J. Polym. Environ.*, 28, 3, 739–48, 2020 Mar.
2. Sarker, I.H., Data science and analytics: An overview from data-driven smart computing, decision-making and applications perspective. *SN Comput. Sci.*, 2, 5, 377, 2021 Sep.
3. Sarker, I.H., Machine learning: Algorithms, real-world applications and research directions. *SN Comput. Sci.*, 2, 3, 160, 2021 May.
4. Duan, Y., Edwards, J.S., Dwivedi, Y.K., Artificial intelligence for decision making in the era of Big Data–evolution, challenges and research agenda. *Int. J. Inf. Manage.*, 48, 63–71, 2019 Oct 1.
5. Subramaniam, S., Raju, N., Ganesan, A., Rajavel, N., Chenniappan, M., Prakash, C., Pramanik, A., Basak, A.K., Dixit, S., Artificial intelligence technologies for forecasting air pollution and human health: A narrative review. *Sustainability*, 14, 16, 9951, 2022 Aug 11.
6. Bhuria, R., Gill, K.S., Upadhyay, D., Devliyal, S., Predicting water purity by riding the ensemble waves with gradient boosting classification technique, in: *2024 2nd International Conference on Sustainable Computing and Smart Systems (ICSCSS)*, 2024 Jul 10, IEEE, pp. 1365–1368.
7. Singh, P., Ramkumar, K.R., Hasija, T., Monsoon Rainfall Prediction for Punjab Using ARIMA Model: A Time Series Analysis, in: *2024 International Conference on Automation and Computation (AUTOCOM)*, 2024 Mar 14, IEEE, pp. 110–113.

8. Hamza, M.A., Shaiba, H., Marzouk, R., Alhindi, A., Asiri, M.M., Yaseen, I., Motwakel, A., Rizwanullah, M., Big Data Analytics with Artificial Intelligence Enabled Environmental Air Pollution Monitoring Framework. *Computers Mater. Continua*, 73, 2, 2022 Nov 1.
9. Pereira, S., Canhoto, P., Salgado, R., Development and assessment of artificial neural network models for direct normal solar irradiance forecasting using operational numerical weather prediction data. *Energy AI*, 15, 100314, 2024 Jan 1.
10. Yang, J., Peng, Z., Sun, J., Chen, Z., Niu, X., Xu, H., Ho, K.F., Cao, J., Shen, Z., A review on advancements in atmospheric microplastics research: The pivotal role of machine learning. *Sci. Total Environ.*, 945, 173966, 2024 Oct 1.
11. Tikle, S., Anand, V., Das, S., Geospatial practices for airpollution and meteorological monitoring, prediction, and forecasting, in: *Geospatial Practices in Natural Resources Management*, pp. 549–566, Springer International Publishing, Cham, 2024 Mar 10.
12. Chen, L., Chang, N., Qiu, T., Wang, N., Cui, Q., Zhao, S., Huang, F., Chen, H., Zeng, Y., Dong, F., Fang, L., Meta-analysis of impacts of microplastics on plant heavy metal (loid) accumulation. *Environ. Pollut.*, 348, 123787, 2024 May 1.
13. Peng, J., Han, H., Yi, Y., Huang, H., Xie, L., Machine learning and deep learning modeling and simulation for predicting PM2. 5 Concentrations. *Chemosphere*, 308, 136353, 2022 Dec 1.
14. Šimić, I., Lovrić, M., Godec, R., Kröll, M., Bešlić, I., Applying machine learning methods to better understand, model and estimate mass concentrations of traffic-related pollutants at a typical street canyon. *Environ. Pollut.*, 263, 114587, 2020 Aug 1.
15. Zhao, W., Ma, J., Liu, Q., Dou, L., Qu, Y., Shi, H., Sun, Y., Chen, H., Tian, Y., Wu, F., Accurate prediction of soil heavy metal pollution using an improved machine learning method: A case study in the Pearl River Delta, China. *Environ. Sci. Technol.*, 57, 46, 17751–61, 2023 Feb 23.
16. Jayaraman, P., Nagarajan, K.K., Partheeban, P., Krishnamurthy, V., Critical review on water quality analysis using IoT and machine learning models. *Int. J. Inf. Manage. Data Insights*, 4, 1, 100210, 2024 Apr 1.
17. Olawade, D.B., Fapohunda, O., Wada, O.Z., Usman, S.O., Ige, A.O., Ajisafe, O., Oladapo, B.I., Smart waste management: A paradigm shift enabled by artificial intelligence. *Waste Manage. Bulletin*, 2, 2, 244–63, 2024 Jun 1.
18. Brandt, J., Mattsson, K., Hassellov, M., Deep learning for reconstructing low-quality FTIR and Raman Spectra— A case study in microplastic analyses. *Anal. Chem.*, 93, 49, 16360–8, 2021 Nov 22.
19. Li, Z., Deng, Q., Liu, P., Bai, J., Gong, Y., Yang, Q., Ning, J., An intelligent identification and classification system of decoration waste based on deep learning model. *Waste Manage. (Oxford)*, 174, 462–75, 2024 Feb 15.

20. Pitakaso, R., Srichok, T., Khonjun, S., Golinska-Dawson, P., Sethanan, K., Nanthasamroeng, N., Gonwirat, S., Luesak, P., Boonmee, C., Optimization-driven artificial intelligence-enhanced municipal waste classification system for disaster waste management. *Eng. Appl. Artif. Intell.*, 133, 108614, 2024 Jul 1.
21. Yu, F. and Hu, X., Machine learning may accelerate the recognition and control of microplastic pollution: Future prospects. *J. Hazard. Mater.*, 432, 128730, 2022 Jun 15.
22. Dekiff, J.H., Remy, D., Klasmeier, J., Fries, E., Occurrence and spatial distribution of microplastics in sediments from Norderney. *Environ. Pollut.*, 186, 248–56, 2014 Mar 1.
23. Van Cauwenberghe, L., Vanreusel, A., Mees, J., Janssen, C.R., Microplastic pollution in deep-sea sediments. *Environ. Pollut.*, 182, 495–9, 2013 Nov 1.
24. Primpke, S., Christiansen, S.H., Cowger, W., De Frond, H., Deshpande, A., Fischer, M., Holland, E.B., Meyns, M., O'Donnell, B.A., Ossmann, B.E., Pittroff, M., Critical assessment of analytical methods for the harmonized and cost-efficient analysis of microplastics. *Appl. Spectrosc.*, 74, 9, 1012–47, 2020 Sep 1.
25. Veerasingam, S., Ranjani, M., Venkatachalapathy, R., Bagaev, A., Mukhanov, V., Litvinyuk, D., Mugilarasan, M., Gurumoorthi, K., Guganathan, L., Aboobacker, V.M., Vethamony, P., Contributions of Fourier transform infrared spectroscopy in microplastic pollution research: A review. *Crit. Rev. Environ. Sci. Technol.*, 51, 22, 2681–743, 2021 Nov 17.
26. Sridhar, A., Kannan, D., Kapoor, A., Prabhakar, S., Extraction and detection methods of microplastics in food and marine systems: A critical review. *Chemosphere*, 286, 131653, 2022 Jan 1.
27. Sobhani, Z., Zhang, X., Gibson, C., Naidu, R., Megharaj, M., Fang, C., Identification and visualisation of microplastics/nanoplastics by Raman imaging (i): Down to 100 nm. *Water Res.*, 174, 115658, 2020 May 1.
28. Zhou, Y., Wang, J., Zou, M., Jia, Z., Zhou, S., Li, Y., Microplastics in soils: A review of methods, occurrence, fate, transport, ecological and environmental risks. *Sci. Total Environ.*, 748, 141368, 2020 Dec 15.
29. Lin, J.Y., Liu, H.T., Zhang, J., Recent advances in the application of machine learning methods to improve identification of the microplastics in environment. *Chemosphere*, 307, 136092, 2022 Nov 1.
30. Bertoldi, C., Lara, L.Z., Gomes, A.A., Fernandes, A.N., Microplastic abundance quantification *via* a computer-vision-based chemometrics-assisted approach. *Microchem. J.*, 160, 105690, 2021 Jan 1.
31. Lee, K.S., Chen, H.L., Ng, Y.S., Maul, T., Gibbins, C., Ting, K.N., Amer, M., Camara M U-Net skip-connection architectures for the automated counting of microplastics. *Neural Comput. Appl.*, 34, 9, 7283–97, 2022 May.
32. Yurtsever, M. and Yurtsever, U., Use of a convolutional neural network for the classification of microbeads in urban wastewater. *Chemosphere*, 216, 271–80, 2019 Feb 1.

33. Hu, B., Dai, Y., Zhou, H., Sun, Y., Yu, H., Dai, Y., Wang, M., Ergu, D., Zhou, P., Using artificial intelligence to rapidly identify microplastics pollution and predict microplastics environmental behaviors. *J. Hazard. Mater.*, 474, 134865, 2024 Aug 5.
34. Choi, E., Choi, Y., Lee, H., Kim, J.W., Oh, H.B., Development of a machine-learning model for microplastic analysis in an FT-IR microscopy image. *Bull. Korean Chem. Soc.*, 45, 5, 472–81, 2024 May.
35. Sunil, M., Pallikkavaliyaveetil, N., Gopinath, A., Chidangil, S., Kumar, S., Lukose, J., Machine learning assisted Raman spectroscopy: A viable approach for the detection of microplastics. *J. Water Process Eng.*, 60, 105150, 2024 Apr 1.
36. Kida, M., Pochwat, K., Ziembowicz, S., Assessment of machine learning-based methods predictive suitability for migration pollutants from microplastics degradation. *J. Hazard. Mater.*, 461, 132565, 2024 Jan 5.
37. Kotar, S., McNeish, R., Murphy-Hagan, C., Renick, V., Lee, C.F., Steele, C., Lusher, A., Moore, C., Minor, E., Schroeder, J., Helm, P., Quantitative assessment of visual microscopy as a tool for microplastic research: Recommendations for improving methods and reporting. *Chemosphere*, 308, 136449, 2022 Dec 1.
38. Cowger, W., Gray, A., Christiansen, S.H., DeFrond, H., Deshpande, A.D., Hemabessiere, L., Lee, E., Mill, L., Munno, K., Ossmann, B.E., Pittroff, M., Critical review of processing and classification techniques for images and spectra in microplastic research. *Appl. Spectrosc.*, 74, 9, 989–1010, 2020 Sep.
39. Adhikari, S., Kelkar, V., Kumar, R., Halden, R.U., Methods and challenges in the detection of microplastics and nanoplastics: A mini-review. *Polym. Int.*, 71, 5, 543–51, 2022 May.
40. Tirkey, A. and Upadhyay, L.S., Microplastics: An overview on separation, identification and characterization of microplastics. *Mar. Pollut. Bull.*, 170, 112604, 2021 Sep 1.
41. Elsayed, A.A., Erfan, M., Sabry, Y.M., Dris, R., Gaspéri, J., Barbier, J.S., Marty, F., Bouanis, F., Luo, S., Nguyen, B.T., Liu, A.Q., A microfluidic chip enables fast analysis of water microplastics by optical spectroscopy. *Sci. Rep.*, 11, 1, 10533, 2021 May 18.
42. Han, X.L., Jiang, N.J., Hata, T., Choi, J., Du, Y.J., Wang, Y.J., Deep learning based approach for automated characterization of large marine microplastic particles. *Mar. Environ. Res.*, 183, 105829, 2023 Jan 1.
43. Huang, M., Han, K., Liu, W., Wang, Z., Liu, X., Guo, Q., Advancing microplastic surveillance through photoacoustic imaging and deep learning techniques. *J. Hazard. Mater.*, 470, 134188, 2024 May 15.
44. Zhu, Y., Hang Yeung, C., Lam, E.Y., Digital holographic imaging and classification of microplastics using deep transfer learning. *Appl. Opt.*, 60, 4, A38–47, 2021 Feb 1.

45. Vitali, C., Peters, R.J., Janssen, H.G., Undas, A.K., Munniks, S., Ruggeri, F.S., Nielen, M.W., Quantitative image analysis of microplastics in bottled water using artificial intelligence. *Talanta*, 266, 124965, 2024 Jan 1.
46. Bianco, V., Memmolo, P., Carcagnì, P., Merola, F., Paturzo, M., Distante, C., Ferraro, P., Microplastic identification *via* holographic imaging and machine learning. *Adv. Intell. Syst.*, 2, 2, 1900153, 2020 Feb.
47. Giardino, M., Balestra, V., Janner, D., Bellopede, R., Automated method for routine microplastic detection and quantification. *Sci. Total Environ.*, 859, 160036, 2023 Feb 10.
48. Hahn, A., Gerdts, G., Völker, C., Niebühr, V., Using FTIRS as pre-screening method for detection of microplastic in bulk sediment samples. *Sci. Total Environ.*, 689, 341–6, 2019 Nov 1.
49. Liu, Y., Yao, W., Qin, F., Zhou, L., Zheng, Y., Spectral classification of large-scale blended (micro) plastics using FT-IR raw spectra and image-based machine learning. *Environ. Sci. Technol.*, 57, 16, 6656–63, 2023 Apr 13.
50. Yan, X., Cao, Z., Murphy, A., Qiao, Y., An ensemble machine learning method for microplastics identification with FTIR spectrum. *J. Environ. Chem. Eng.*, 10, 4, 108130, 2022 Aug 1.
51. Tan, A., Zhao, J., Zhao, Y., Li, X., Su, H., Determination of microplastics by FTIR spectroscopy based on quaternion parallel feature fusion and support vector machine. *Chemom. Intell. Lab. Syst.*, 243, 105018, 2023 Dec 15.
52. Enyoh, C.E. and Wang, Q., Automated classification of undegraded and aged polyethylene terephthalate microplastics from ATR-FTIR spectroscopy using machine learning algorithms. *J. Polym. Environ.*, 32, 9, 4143–58, 2024 Sep.
53. Zeng, G., Ma, Y., Du, M., Chen, T., Lin, L., Dai, M., Luo, H., Hu, L., Zhou, Q., Pan, X., Deep convolutional neural networks for aged microplastics identification by Fourier transform infrared spectra classification. *Sci. Total Environ.*, 913, 169623, 2024 Feb 25.
54. de Medeiros Back, H., Junior, E.C., Alarcon, O.E., Pottmaier, D., Training and evaluating machine learning algorithms for ocean microplastics classification through vibrational spectroscopy. *Chemosphere*, 287, 131903, 2022 Jan 1.
55. Thar, A.S., Laitrakun, S., Deepaisarn, S., Opaprakasit, P., Somnuake, P., Athikulwongse, K., A comparative study of machine learning and deep learning models for microplastic classification using FTIR spectra, in: *2023 18th International Joint Symposium on Artificial Intelligence and Natural Language Processing (iSAI-NLP)*, IEEE, pp. 1–6, 2023 Nov 27.
56. Hufnagl, B., Steiner, D., Renner, E., Löder, M.G., Laforsch, C., Lohninger, H., A methodology for the fast identification and monitoring of microplastics in environmental samples using random decision forest classifiers. *Anal. Methods*, 11, 17, 2277–85, 2019.

57. Hufnagl, B., Stibi, M., Martirosyan, H., Wilczek, U., Laforsch, C., Lohninger, H., Computer-assisted analysis of microplastics in environmental samples based on μFTIR imaging in combination with machine learning. *Environ. Sci. Technol. Lett.*, 9, 1, 90–5, 2021 Dec 9.
58. Yang, Z., Nagashima, H., Arakawa, H., Development of automated microplastic identification workflow for Raman micro-imaging and evaluation of the uncertainties during micro-imaging. *Mar. Pollut. Bull.*, *193*, 115200, 2023.
59. Xue, Q., Wang, N., Yang, H., Yang, J., Bai, H., Detection of microplastics based on spatial heterodyne Raman spectroscopy. *Spectrochim. Acta, Part A*, 283, 121712, 2022 Dec 15.
60. Xie, L., Luo, S., Liu, Y., Ruan, X., Gong, K., Ge, Q., Li, K., Valev, V.K., Liu, G., Zhang, L., Automatic identification of individual nanoplastics by Raman spectroscopy based on machine learning. *Environ. Sci. Technol.*, 57, 46, 18203–14, 2023 Jul 3.
61. Qian, N., Gao, X., Lang, X., Deng, H., Bratu, T.M., Chen, Q., Stapleton, P., Yan, B., Min, W., Rapid single-particle chemical imaging of nanoplastics by SRS microscopy. *Proc. Natl. Acad. Sci.*, 121, 3, e2300582121, 2024 Jan 16.
62. Luo, Y., Awoyemi, O.S., Naidu, R., Fang, C., Detection of microplastics and nanoplastics released from a kitchen blender using Raman imaging. *J. Hazard. Mater.*, 453, 131403, 2023 Jul 5.
63. Zhang, Y., Wang, X., Shan, J., Zhao, J., Zhang, W., Liu, L., Wu, F., Hyperspectral imaging based method for rapid detection of microplastics in the intestinal tracts of fish. *Environ. Sci. Technol.*, 53, 9, 5151–8, 2019 Apr 8.
64. Qin, Y., Qiu, J., Tang, N., He, Y., Fan, L., Deep learning analysis for rapid detection and classification of household plastics based on Raman spectroscopy. *Spectrochim. Acta, Part A*, 309, 123854, 2024 Mar 15.
65. Luo, Y., Su, W., Rabbi, M.F., Wan, Q., Xu, D., Wang, Z., Liu, S., Xu, X., Wu, J., Quantitative analysis of microplastics in water environments based on Raman spectroscopy and convolutional neural network. *Sci. Total Environ.*, 926, 171925, 2024 May 20.
66. Jin, N., Song, Y., Ma, R., Li, J., Li, G., Zhang, D., Characterization and identification of microplastics using Raman spectroscopy coupled with multivariate analysis. *Anal. Chim. Acta*, 1197, 339519, 2022 Mar 8.
67. Li, F., Liu, D., Guo, X., Zhang, Z., Martin, F.L., Lu, A., Xu, L., Identification and visualization of environmental microplastics by Raman imaging based on hyperspectral unmixing coupled machine learning. *J. Hazard. Mater.*, 465, 133336, 2024 Mar 5.
68. Sim, W., Song, S.W., Park, S., Jang, J.I., Kim, J.H., Cho, Y.M., Kim, H.M., Unveiling microplastics with hyperspectral Raman imaging: From macroscale observations to real-world applications. *J. Hazard. Mater.*, 463, 132861, 2024 Feb 5.

69. Huang, H., Qureshi, J.U., Liu, S., Sun, Z., Zhang, C., Wang, H., Hyperspectral imaging as a potential online detection method of microplastics. *Bull. Environ. Contam. Toxicol.*, 107, 4, 754–63, 2021 Oct.
70. Fiore, L., Serranti, S., Mazziotti, C., Riccardi, E., Benzi, M., Bonifazi, G., Classification and distribution of freshwater microplastics along the Italian Po river by hyperspectral imaging. *Environ. Sci. Pollut. Res.*, 29, 32, 48588–606, 2022 Jul.
71. Xu, L., Chen, Y., Feng, A., Shi, X., Feng, Y., Yang, Y., Wang, Y., Wu, Z., Zou, Z., Ma, W., He, Y., Study on detection method of microplastics in farmland soil based on hyperspectral imaging technology. *Environ. Res.*, 232, 116389, 2023 Sep 1.
72. Ai, W., Liu, S., Liao, H., Du, J., Cai, Y., Liao, C., Shi, H., Lin, Y., Junaid, M., Yue, X., Wang, J., Application of hyperspectral imaging technology in the rapid identification of microplastics in farmland soil. *Sci. Total Environ.*, 807, 151030, 2022 Feb 10.
73. Ai, W., Chen, G., Yue, X., Wang, J., Application of hyperspectral and deep learning in farmland soil microplastic detection. *J. Hazard. Mater.*, 445, 130568, 2023 Mar 5.
74. Vidal, C. and Pasquini, C., A comprehensive and fast microplastics identification based on near-infrared hyperspectral imaging (HSI-NIR) and chemometrics. *Environ. Pollut.*, 285, 117251, 2021 Sep 15.
75. Zou, H.H., He, P.J., Peng, W., Lan, D.Y., Xian, H.Y., Lü, F., Zhang, H., Rapid detection of colored and colorless macro-and micro-plastics in complex environment *via* near-infrared spectroscopy and machine learning. *J. Environ. Sci.*, 147, 512–22, 2025 Jan 1.
76. Faltynkova, A. and Wagner, M., Developing and testing a workflow to identify microplastics using near infrared hyperspectral imaging. *Chemosphere*, 336, 139186, 2023 Sep 1.
77. Piarulli, S., Malegori, C., Grasselli, F., Airoldi, L., Prati, S., Mazzeo, R., Sciutto, G., Oliveri, P., An effective strategy for the monitoring of microplastics in complex aquatic matrices: Exploiting the potential of near infrared hyperspectral imaging (NIR-HSI). *Chemosphere*, 286, 131861, 2022 Jan 1.
78. Wu, P., Wang, B., Lu, Y., Cao, G., Xie, P., Wang, W., Chen, D., Huang, G., Jin, H., Yang, Z., Cai, Z., Machine learning-assisted insights into sources and fate of microplastics in wastewater treatment plants. *ACS ES&T Water*, 4, 3, 1107–18, 2023 Aug 24.
79. Tran, H.T., Hadi, M., Nguyen, T.T., Hoang, H.G., Nguyen, M.K., Nguyen, K.N., Vo, D.V., Machine learning approaches for predicting microplastic pollution in peatland areas. *Mar. Pollut. Bull.*, 194, 115417, 2023 Sep 1.
80. Qiu, Z., Zhao, S., Feng, X., He, Y., Transfer learning method for plastic pollution evaluation in soil using NIR sensor. *Sci. Total Environ.*, 740, 140118, 2020 Oct 20.

81. Post, C., Brülisauer, S., Waldschläger, K., Hug, W., Grüneis, L., Heyden, N., Schmor, S., Förderer, A., Reid, R., Reid, M., Bhartia, R., Application of laser-induced, deep uv raman spectroscopy and artificial intelligence in real-time environmental monitoring—solutions and first results. *Sensors*, 21, 11, 3911, 2021 Jun 5.
82. Phan, S. and Luscombe, C.K., Recent trends in marine microplastic modeling and machine learning tools: Potential for long-term microplastic monitoring. *J. Appl. Phys.*, 133, 2, 2023 Jan 14.
83. Watanabe, J.I., Shao, Y., Miura, N., Underwater and airborne monitoring of marine ecosystems and debris. *J. Appl. Remote Sens.*, 13, 4, 044509, 2019 Oct.
84. Shah, M., Garg, D., Ghariya, R., Solanki, V., Rajput, R., Chauhan, M., Enhancing marine conservation: YOLOv8-based underwater waste detection system. *2023 3rd International Conference on Innovative Mechanisms for Industry Applications (ICIMIA)*, IEEE, pp. 1396–1404, 2023 Dec 21.
85. Almeida, S., Radeta, M., Kataoka, T., Canning-Clode, J., Pessanha Pais, M., Freitas, R., Monteiro, J.G., Designing unmanned aerial survey monitoring program to assess floating litter contamination. *Remote Sens.*, 15, 1, 84, 2022 Dec 23.
86. Pouyanfar, N., Harofte, S.Z., Soltani, M., Siavashy, S., Asadian, E., Ghorbani-Bidkorbeh, F., Keçili, R., Hussain, C.M., Artificial intelligence-based microfluidic platforms for the sensitive detection of environmental pollutants: Recent advances and prospects. *Trends Environ. Anal. Chem.*, 34, e00160, 2022 Jun 1.

5

AI Driven Optical Imaging and Spectroscopic Techniques

Muchukota Sushma[1]*, Mekkanti Manasa Rekha[1], Ramya C. V.[2] and Zaid Khan[2]

[1]Department of Pharmacy Practice, Aditya Bangalore Institute of Pharmacy Education & Research (ABIPER), Bangalore, India
[2]Aditya Bangalore Institute of Pharmacy Education & Research (ABIPER), Bangalore, India

Abstract

The growing prevalence of microplastics in aquatic and terrestrial ecosystems represents a critical environmental and public health issue, as these minute plastic fragments infiltrate food chains, disrupt ecosystems, and pose potential health risks to humans. Addressing this pervasive problem demands advanced tools capable of delivering precise, real-time detection and monitoring. AI-driven optical imaging and spectroscopic techniques have emerged as revolutionary solutions, offering unparalleled accuracy, sensitivity, and scalability in microplastic detection.

Optical imaging, enhanced by sophisticated AI algorithms, enables high-resolution visualization and automated classification of microplastic particles, effectively distinguishing them from other debris in complex environmental samples. Spectroscopic approaches, such as Raman and Fourier-transform infrared (FTIR) spectroscopy, provide detailed chemical and structural insights, allowing for precise identification of microplastic composition and origin. By integrating these advanced techniques with machine learning and data analytics, the throughput and reliability of detection processes are significantly enhanced.

This chapter explores into the synergy between AI and these cutting-edge technologies, exploring their applications in environmental monitoring, industrial quality control, and public health initiatives. Real-time systems powered by AI facilitate dynamic mapping of pollution hotspots, predictive modeling for mitigation strategies, and proactive policy development. The chapter also critically

**Corresponding author*: sushma.banthi@gmail.com

Abhishek Kumar, Pooja Dixit, Pramod Singh Rathore, Arun Lal Srivastav and Ashutosh Kumar Dubey (eds.) Microplastic Monitoring Using Artificial Intelligence, (83–108) © 2026 Scrivener Publishing LLC

examines challenges such as scalability, computational resource demands, and ethical considerations associated with data use, providing a comprehensive analysis of the field. Looking forward, innovations in AI algorithms, sensor technology, and regulatory frameworks promise to expand the potential of these intelligent systems. This chapter underscores their transformative role in addressing the global microplastic crisis and fostering sustainable environmental practices.

Keywords: Optical imaging, Raman and Fourier-transform infrared (FTIR) spectroscopy, mapping, pollution hotspots, AI algorithms, sensor technology, microplastic crisis

List of Abbreviations

Ultraviolet	(UV)
Microplastic	(MP)
Scanning Electron	(SEM)
Atomic force microscopy	(AFM)
Fourier Transform Infrared	(FT-IR)
Gas Chromatography-Mass Spectrometry	(GC-MS)
Internet of Things	(IoT)
Planetary Digital Twins	(PDTs)
Fourier transform infrared spectroscopy	(FTIR)
Artificial intelligence	(AI)
K-Nearest Neighbours	(KNN)
Support Vector Classification	(SVC)
Convolutional neural networks	(CNNs)
Fluid stream analyzer	(FSA)
General Data Protection Regulation	(GDPR)
Recurrent neural networks	(RNNs)
SHAP	(Shapley Additive Explanations)
LIME	(Local Interpretable Model-Agnostic Explanations)

5.1 Introduction

Microplastic pollution has become one of the most pressing environmental challenges given the widespread use of plastics and their slow degradation into microscopic pieces [1]. Microplastics, which are between line of 0.001 and 5 mm, have been discovered in almost all parts of the environment, ranging from oceans and rivers to soil, flooding almost every breath of air we take. They raise serious concerns, for both wildlife and human health [2].

5.1.1 Origins of Microplastics: Sources, Types, and Impact

Microplastics originate from two primary sources:

1. Degradation of Larger Plastics: Over time, macro plastics undergo fragmentation due to environmental exposure to ultraviolet (UV) radiation, mechanical forces, and hydrolytic degradation. These factors progressively break down larger plastic debris into microscopic particles, contributing significantly to environmental pollution [3].
2. Direct Emission from Consumer and Industrial Products: Numerous commercial products, including cosmetics, textiles, and industrial materials, contain microplastics in the form of microparticles or synthetic fibers. Activities such as laundering synthetic fabrics, using personal care products, and manufacturing processes facilitate the continuous release of microplastics into the environment [4].

Classification of Microplastics

Microplastics are categorized into four major types based on their origin and structural characteristics:

- Fragments: Irregularly shaped plastic particles derived from the mechanical and chemical breakdown of larger plastic materials.
- Fibers: Microscopic, thread-like plastic strands primarily released from synthetic textiles during washing and abrasion.
- Pellets (Nurdles): Small, bead-like plastic granules used as raw materials in plastic manufacturing and industrial processing.
- Films: Thin, flexible plastic sheets that degrade from plastic packaging, bags, and coatings [5].

Environmental and Public Health Implications

The pervasive presence of microplastics in terrestrial and aquatic ecosystems raises significant concerns regarding their impact on biodiversity, food safety, and human health. These particles can accumulate in marine organisms, enter the food chain, and pose potential toxicological risks.

Addressing microplastic pollution requires stringent regulatory measures, sustainable material innovation, and enhanced public awareness to mitigate their environmental footprint [6].

Microplastics in Natural Ecosystems

Microplastics are pervasive in various ecosystems, with the highest concentrations found in marine environments. Studies indicate contamination rates of 19–100% in aquatic organisms, from plankton to large fish. Seawater and marine sediments also exhibit widespread microplastic pollution, with some regions nearing full saturation [7]. These particles disrupt marine life by interfering with feeding, reproduction, and overall health, posing potential threats to species survival. Additionally, microplastics act as carriers for hazardous chemical pollutants, which adhere to their surfaces and may be transferred through ingestion, further propagating toxins up the food chain. Their increasing prevalence presents a critical environmental challenge requiring immediate attention [8].

Health Issues Related to Microplastics

Microplastics can enter the human body through several channels like food and water sources, according to studies. Although there is suspicion about the potential health implications microplastics can have on health, it is not fully researched yet. Microplastics can have adverse effects on health just like they do to animals; for example, oxidative stress and gut microbiota disruption are also issues in humans [1, 8].

Prevention Tactics

A multi approach strategy is needed to address the problems microplastics pollution brings with it. The approaches include:

- Creating New Policies: The strategy that seems most efficient is finding the means to control plastic production regulations and disposal practices along with educating people on reducing the ways microplastics can enter the environment [6].
- Campaigning: It is important to raise awareness regarding microplastics, and the different ways they can be harmful to people. This might help control how much single-use plastics consumers purchase.
- Development of Initiatives: Microplastics pose an environmental danger, and the negative effects along with the risks to human body need extensive research to fill the gap in information [8].

5.1.2 Traditional Detection Methods

Microplastic (MP) detection relies on physical and chemical techniques for identification, characterization, and quantification. Microscopy-based methods such as stereo, fluorescence, scanning electron (SEM), and atomic force microscopy (AFM) enable visual assessment, while holographic imaging provides real-time detection in unfiltered water. Spectroscopy techniques like Fourier Transform Infrared (FT-IR) and Raman spectroscopy analyze polymer composition, with combined approaches improving accuracy. Thermal analysis, including Pyrolysis-GC-MS, offers bulk chemical characterization but is destructive. Despite advantages like detailed chemical insights and rapid screening, challenges include high costs, lengthy analysis, and difficulties detecting MPs <10 μm. Emerging techniques, such as AI-driven image analysis and integrated spectroscopy-microscopy methods, aim to enhance accuracy and efficiency.

Limitations of Traditional Detection Methods for Microplastics

Microplastic pollution detection has been more crucial with the rising awareness of its environmental and health effects. Nonetheless, conventional detection methods have several important limitations that delay accurate estimation and quantification of microplastics in different environments.

1. Incomplete Detection of Small Particles

Also, they are somehow contradictory since it provides evidence that there is something works within the interdependent. Balancing the different thing seems to be the greatest challenge hence the silence integration delineated in the performance measurement is substantiated. It balances the different orientation and objectives but setting different disciplines within the defined boundaries [12].

2. Variability in Analytical Techniques

This is often undermined but the meshing of the different aspects, modules and qualities of performance means a blended phenomenon which need careful grounding. As the process bound to cover the collection of information begins with really four fundamental coherent phases. Building is a lengthy report comprising of five preposterous layers. That is, Development Happy, Narrowing, Full litigation and Final Sunsets [11].

3. Time-Consuming Processes
Nevertheless, this has nothing to reduce the breath of understanding. Rather, it enhances insights since despite owing the context there is additional paradigm which leads to erudition and grasping of how performance is built. There seems to be pilots as experts devoid any representation will start actual towards building witness while reversing storytelling.

4. Chemical Identification Limitations
Identification of microplastics with complex additives may pose significant difficulties. This is an important limitation of the present approaches since many types of plastics are produced with pigmenting dyes and other additives that compromise the detection approaches [14, 15].

5. Problems of Quantification
Quantification of microplastics remains one of the biggest challenges with regard to how the particles are shaped and sized. It has been noted that certain discrepancies for this basic measurement have been attributed to differences in methodology and may pose a threat to risk as well as regulatory frameworks [10, 13].

6. Lack of Sensitivity of Nano plastics Detection
Conventional techniques also remain insufficient for nano plastic (<1 μm) detection, which are increasingly identified as a more concerning type of plastic. There is no other way to put it, but this type of sensitivity is absent in current procedures but is what needs to be enhanced for monitoring of effective smaller particles [15, 16].

5.1.3 Potential of AI in Transforming Microplastic Monitoring

Artificial Intelligence (AI) has the great potential of strengthening the tracking of microplastic pollutants and overcoming modern limitations associated with traditional detection methodologies, enabling even more robust environmental monitoring. But it also highlights critical areas in which AI can change the game for microplastic monitoring:

Real-Time Data Analysis
It will be capable of rapidly analysing massive amounts of data from a range of monitoring technologies from sensors to imaging systems. They said this feature allows for rapid identification and measurement of microplastics,

leading to fast responses to pollution events. For example, combining AI and Internet of Things (IoT) can improve real-time monitoring by identifying anomalies [18] and predicting trends in microplastic presence in the environment.

Greater Accuracy and Sensitivity

Machine learning algorithms can be trained to recognize microplastics in complex environmental backgrounds, thereby improving detection accuracy and efficiency in environmental monitoring.

Enhanced Data Integration

AI can combine data from various sources, including satellite imagery, oceanographic data, and local environmental monitoring stations, to give a holistic understanding of the distribution and trends of microplastics. This cross-modelling integration enables a more comprehensive understanding of the origins and consequences of microplastic pollution [17].

Automating Process Monitoring

AI can automate sampling, analysis, which has proven to be more insightful and less inaccurate, as compared to human efforts. These automated systems can sample various water or sediment wells for microplastics harmlessly and can substantially expand the quantity of data per data point discussed above collected due to the reduction of manual labour needed for sampling [22].

Development of Digital Twins

One such concept is Planetary Digital Twins (PDTs), which use AI to develop virtual models of the real-world physical environment. Visualizing long-term weather patterns is one way these models can help policymakers get an idea of the climate.

Addressing Environmental Justice

Environmental justice is something that needs to be taken seriously. Inclusion of AI applications in monitoring can help ensure that the actions against microplastic pollution are being equitably meted out and address varying environmental quality across communities. AI can offer a database to determine pollution hotspots, enabling *in situ* interventions wherever they are most needed [21, 23, 24].

5.2 Fundamentals of Optical Imaging and Spectroscopic Techniques

5.2.1 Optical Imaging: Principles and Applications

Optical imaging is a combination of several methods from the fields of biology, materials science, and medicine that utilize light to image structures and processes. Quick points on the fundamentals of optical imaging and the wide range of applications it supports are covered in this study [25].

Principles of Optical Imaging

The processes of emission, scattering, and absorption all fall under the definition of light-matter interactions which are vital for optical imaging. It is due to light's interaction with tissues and materials that it expands, making certain features both structural and compositional visible. With the aid of contrast mechanisms, visibility surrounded by selective absorption or emission such as is the case in fluorescent dyes. Contrast agents improve the picture's visibility based on the used water, material, and machine [26]. Fluorescence microscopes, spectroscopes, and tomography devices transform images to make them clearer and more useful with complex reconstruction algorithms that augment the quality of the images.

5.2.2 Spectroscopic Techniques: Raman and FTIR Spectroscopy

The optical imaging approaches have been used in various fields:

Medical Imaging: In case of cancer, molecular imaging such as fluorescent molecular tomography support in tumor identification and therapy monitoring through providing *in vivo* non-invasive sight of molecular pipelines [27]. Specifically, for the evaluation of breast lesions, diffuse optical spectroscopy and photoacoustic imaging provide complementary data on tissue composition and vascularity [30].

Preclinical research: Optical imaging enhances our understanding of disease and assessment of therapeutic interventions by enabling more accessible tracking of biological processes through animal models [27, 28].

Industrial Applications: Optical imaging assists in surface inspection and quality control, and optical time-stretch imaging enables efficient analysis of dynamic processes to improve product assurance [29, 31].

Environmental Monitoring: Optical imaging monitors environmental parameters [32].

Overview of Spectroscopic Techniques: Raman and FTIR Spectroscopy
Spectroscopy Techniques: Raman and FTIR Overview
FTIR, Raman and GEFT spectroscopy are powerful tools and provide unique information on molecular vibrations owing to their discrepant principles and usages [33]. Raman spectroscopy relies upon inelastic scattering of monochromatic light, where the majority of photons undergo Rayleigh scattering and a small fraction undergo energy changes (Raman scattering) [34], allowing molecular vibrational modes to be accessed for structural analysis. In contrast, FTIR spectroscopy generates spectra *via* measuring infrared absorption [38], so the spectral appearance offers the functional groups and molecular configuration. Based on carbon like Graphene 34, Raman spectroscopy is most broadly used in material characterisation. To confirm that an item is what it says it is and that it is not adulterated, in order to ensure authenticity and identify adulteration in items such as dairy and oils, both methods are essential for food quality control [34, 37]. Furthermore, by offering structural insights into artifacts without causing damage, Raman spectroscopy in archaeophotonics aids archaeological research [36].

Comparison of Raman and FTIR Spectroscopy for Material Analysis

Feature	Raman spectroscopy	FTIR spectroscopy
Sensitivity	Sensitive to non-polar bonds; good for gases	Sensitive to polar bonds; excellent for organic compounds
Sample Preparation	Minimal preparation required; can analyze solids, liquids, and gases	Requires sample preparation; typically needs solid samples to be thin films or pellets
Water Interference	Less affected by water; suitable for aqueous solutions	Strong water absorption bands can complicate spectra
Cost	Generally, more expensive due to laser systems	Typically, less expensive; widely available

5.2.3 Integration of AI into Optical and Spectroscopic Tools

The use of AI in spectroscopic and optical instruments is taking to the next level, especially in the areas of materials research, pharmaceuticals, environmental monitoring, and food safety, extending the versatility, precision, and robustness of these instruments. While deep learning makes high-throughput evaluation of materials possible, AI-based data evaluation enables machine learning models to identify complex patterns in the spectrum that cannot be resolved using traditional analysis methods [39]. Real-time monitoring is vital for applications in quality control and the environment, allowing for immediate interpretation of spectrum data. The predictive capability of historical data in knowing the behaviour of materials in pharmaceutical development and automating the instrumentation that enhanced productivity through real-time changes in measurement parameters and reduction of human error are further beneficial [40]. Food business artificial intelligence–enhanced spectroscopy to identify impurities, primarily you guarantee safety; environmental research engages. Pharmaceutical development benefits from predictive analytics' ability to foresee material behaviour based on historical data, while automated instrumentation increases productivity by dynamically modifying measurement parameters and minimizing human error [40]. The food business uses AI-enhanced spectroscopy to identify impurities, guaranteeing safety; environmental research benefits from real-time pollution tracking; and medicines use it for quality control, guaranteeing regulatory compliance.

5.3 AI Innovations in Microplastic Detection

The investigation of microplastics (MPs) in the environment and their impact on ecosystems as well health of humans is an essential field of interest currently. The current artificial intelligence (AI) developments are improving microplastic detection strategies in [42]—machine learning and neural networks [58–62] or data fusion methods [63]. This presents three areas of advances in this summary: machine learning for image processing and discriminator [43], neural networks for working on spectral data [44] and fused data technique to improve detection capability [45].

5.3.1 Machine Learning for Image Analysis and Classification

Machine learning using AI optical sensors differentiate microplastics from seawater particle by analyzing sequential frames at sampling of water, greatly improved by image analysis of microplastic detection [11]. In practice techniques, they work extremely well at identifying using AI models or image capturing methods [44]. Using a comparative structure, machine learning approaches like K-Nearest Neighbours (KNN) and Support Vector Classification (SVC) could split microplastic from 4 infrared spectra with 96% accuracy. Using both the traditional machine learning techniques and deep learning models, such as the convolutional neural networks (CNNs) [45], the classification accuracy is increased.

5.3.2 Neural Networks in Spectral Data Processing

Neural networks get mounted in the processing of microplastic detection spectral data [3]. In this work a recently proposed approach for identification of microplastics in fish samples using computed tomography (CT) scanning and machine learning are presented in their new method. The procedure eliminates the uncertainty of imaging and processing so that the system is robust and does not result in false positive/negative. With the calibration of the system against measured standards, microplastic identification can be increased for recognizing more clear signatures in complex data [46].

Also, along the lines of AI-streamlined, simultaneous application of holographic methods for the rapid screening of microplastics in water bodies [13].

Example: Machine learning using neural networks (automated classification) that hints at the capabilities of AI processing big spectral data [47].

5.3.3 Data Fusion for Enhanced Detection Accuracy

Data fusion methods collect data from distinct sources in order to improve decision performance. Microplastic detection: data from different sensors e.g., optical sensors, spectral analyzers and imaging systems are needed for better observation of the presence and nature of microplastics [48].

By leveraging advanced algorithms that analyze multi-modality of data to fight the shortcomings of any measurements-origin. For instance, merging data from optical imaging and spectral analysis improves quantifying the different ways and extend consistencies that could be classified as more defensible claims [49] microplastics in environmental evaluations.

5.4 Applications in Real-Time Monitoring

Applications in Real-Time Monitoring of Microplastics

It is obvious to use advanced technologies for real time surveillance of microplastics so as to gain better knowledge of microplastics impact on environment and human health. Regulatory compliance with detection, but most importantly analysis and action on the detection of microplastic contamination in a different environment will more ensure environmental protection along human health. Three major applications: aquatic ecosystem assessment, airborne microplastics detection and industrial/city monitoring systems [50].

5.4.1 Aquatic Ecosystem Analysis

Since microplastics enter water bodies through wastewater, industrial discharge, and air deposition, real-time microplastic monitoring is essential for evaluating marine biodiversity and water quality. Scalable and quick detection is made possible by cutting-edge technology like CNN-powered image recognition for the Internet of Things, which enhances pollution tracking [51]. However, particle heterogeneity and interference from organic matter present difficulties for laser-induced spectroscopy employing AI and deep UV Raman [52]. By accurately determining the size and quantity of microplastics in river water, pulsed laser sensors provide large-scale monitoring [53].

5.4.2 Airborne Microplastic Detection

Detection of airborne microplastics is increasingly being identified as crucial because of their possible health effects. Airborne microplastics have their source from man-made textile fibers, tire abrasion, and urban emissions, and they can be inhaled by humans or get deposited into water bodies and soils. Some of the latest developments in real-time monitoring include:

- *Passive Microwave Sensors:*

Novel passive microwave sensors have been suggested for online tracking of airborne microplastic pollution. The sensors do not need direct contact and can pick up trace levels of microplastics suspended in the air. They offer a non-invasive, continuous monitoring system for the environment, cutting down on the dependency on conventional air sampling methods involving laboratory testing [54].

• *Distributed Monitoring Networks:*
The application of Planetary Digital Twins (PDTs) is under consideration to design integrated networks to monitor air microplastics. PDTs rely on satellite observations, surface-based sensors, and artificial intelligence (AI) models to create a holistic analysis of microplastic pollution worldwide. Through the integration of real-time data exchange, this platform optimizes decision-making for air quality management and anti-pollution measures [55].

5.4.3 Industrial and Urban Monitoring Systems

In urban and industrial environments, real-time monitoring systems are critical to controlling microplastic pollution effectively. Microplastics are released from manufacturing activities, construction processes, and consumer goods, thus requiring targeted monitoring for pollution reduction. The important technological advancements are:

• *Smart Monitoring Frameworks:*
Industrial processes can have smart technologies embedded in them for ongoing monitoring of microplastic emissions. The use of IoT devices and machine learning algorithms by industries is more common now for monitoring environmental effects. Smart frameworks facilitate real-time microplastic leak detection, improve waste management practices, and assist industries in complying with regulation requirements [56].

• *Real-Time Data Sharing:*
The establishment of frameworks to provide real-time sharing of data from different applications further enables swift reaction to microplastic pollution cases. These networks consolidate information obtained from a series of sources like wastewater treatment stations, airborne sensors, and industrial monitoring networks. Better availability of data accommodates better assessments of environmental hazards, enlightens policymakers, and assists industries to apply rectifications promptly [57].

5.5 Case Studies in AI-Driven Microplastic Detection

The use of artificial intelligence (AI) in microplastic detection has brought about dramatic improvements in methodology and technology. The capacity of AI-based tools to improve detection precision, minimize labor-intensive operations, and facilitate real-time monitoring has transformed the way researchers and industries tackle microplastic contamination. This section introduces three in-depth case studies of the real-world applications of AI in microplastic detection: AI-augmented Raman spectroscopy in ocean monitoring, automated optical imaging systems for waste management, and community-based monitoring programs [58].

5.5.1 AI-Enhanced Raman Spectroscopy in Marine Monitoring

AI has improved the detection of microplastics in seas by enhancing Raman spectroscopy, a potent method for microscopic chemical research. An imaging-based lens less polarization-sensitive fluid stream analyzer (FSA) that combines digital in-line holography and machine learning was developed in response to the labour-intensive and time-consuming nature of traditional sampling techniques [59]. Without the need for further preparation or chemical labelling, the FSA system classifies microplastics by taking high-resolution pictures of fluid samples. The AI guarantees reliable identification because it has been trained on a variety of datasets, such as polyethylene, polypropylene, and polystyrene [60]. High categorization accuracy was validated by laboratory testing using phytoplankton cultures, indicating the system's potential for real-time monitoring. It is a revolutionary instrument for monitoring microplastic pollution because of its advantages, which include less human labor, improved sensitivity, and scalability for extended ocean deployment [61].

5.5.2 Automated Optical Imaging Systems for Waste Management

Efficient management of plastic waste is important to reduce microplastic pollution. One of the promising strategies is the creation of automated optical imaging systems based on deep learning methods to enhance the detection and separation of microplastics in waste streams.

One of the most important studies proposed an automated waste management system based on deep learning, which utilizes convolutional neural networks to categorize waste material precisely. The system utilizes CNNs to sort significant amounts of waste efficiently through automated sorting systems, minimizing the use of manual labor and maximizing efficiency [62].

The principle behind the system is as follows:
High-resolution picture capture, preprocessing for feature extraction, CNN-based classification (e.g., plastic, metal, organic, microplastic),

and automated sorting utilizing robotic systems are the four steps that AI-powered waste management uses to improve microplastic detection [63]. Accuracy is greatly increased by deep neural networks, which are trained on large, labelled datasets. Higher recycling rates, increased efficiency from shorter processing times, and cost savings with less human involvement are some of the main advantages. By precisely identifying and treating microplastics in waste streams, an AI-driven system effectively lowers environmental contamination.

5.5.3 Community-Based Monitoring Initiatives

AI-powered smartphone applications are integrated into community-led microplastic monitoring to engage the public in environmental preservation. Participants gather water samples, detect microplastics using AI image recognition, and add crowdsourced data to central databases [64]. Benefits include extensive geographic coverage, low-cost data collecting using mobile devices, and public engagement, which improves the study of microplastic dispersion.

5.6 Challenges in AI-Driven Microplastic Monitoring

The integration of artificial intelligence (AI) in microplastic monitoring presents numerous opportunities for enhancing detection and analysis. However, several challenges must be addressed to optimize these technologies effectively. This section explores three primary challenges: technical barriers related to data volume and processing power, scalability and cost constraints, and ethical and privacy concerns in data use [65].

5.6.1 Technical Barriers: Data Volume and Processing Power

Robust data management systems are essential for storing, processing, and integrating diverse data formats for real-time analysis. Deep learning models demand high-performance GPUs or cloud computing, which are costly and energy-intensive, making real-time processing a challenge. Optimizing AI models for big data while ensuring accuracy is complex, requiring techniques such as data compression, parallel processing, and transfer learning, which necessitate advanced expertise and infrastructure [67].

5.6.2 Scalability and Cost Constraints

Scalability is also a fundamental challenge for AI-based microplastic monitoring systems. With rising demand for monitoring, systems need to scale effectively without costing too much.

• *Infrastructure Costs*: Creating wide-ranging monitoring networks using AI technologies is costly. Hardware, software development, and maintenance costs can discourage organizations from applying these solutions in large volumes. The use of specialized equipment like high-resolution imaging devices and spectroscopy systems is an additional burden on the cost.

• *Operational Challenges*: Expansion of monitoring activities usually comes with logistical difficulties such as sensor deployment in distant areas, keeping data transmission networks intact, and incorporating new technologies into infrastructure. Harsh environment, i.e., deep-sea or polar environments, may add complexity in deploying and maintaining sensors [68].

• *Funding Constraints*: Most research initiatives are grant or government-funded, and such funds might not be adequate to finance large-scale implementations. Obtaining continuous financial support is crucial to maintain these programs in the long term. Private investment might be involved, but proving commercial viability is still a challenge.

5.6.3 Ethical and Privacy Concerns in Data Use

The application of AI in tracking microplastics raises significant ethical and privacy issues that need to be resolved to avoid irresponsible application.

• *Data Privacy*: Gathering data from different sources might inadvertently include personal data or sensitive environmental information. It is important to ensure that data collection procedures are in line with privacy laws, like the General Data Protection Regulation (GDPR), to safeguard people's rights. Also, anonymizing data without losing its scientific utility is a technical challenge [69].

• *Bias and Fairness*: AI systems may inherit biases from training data, resulting in unequal or erroneous outcomes. Biases can arise from unbalanced geographical coverage of microplastic monitoring or differences in data collection practices. There is a need to create mechanisms for detecting and correcting bias in AI algorithms to achieve fair monitoring practice. Periodic audits and varied dataset incorporation can address these issues.

• *Transparency and Accountability*: There is increased pressure for transparency in AI system decision-making. Policymakers, researchers, and the general public require guarantees that AI-powered monitoring systems are trustworthy and accountable, particularly in informing policy based on environmental health. Clear documentation, explainable AI methods, and third-party audits can ensure transparency and confidence in such systems.

5.7 Future Directions

The future of AI-based microplastic detection and monitoring holds much promise in various aspects. Three areas are discussed in this section: advances in AI algorithms for detection, developments in sensor technologies, and the establishment of policy and regulatory environments facilitating adoption [69].

5.7.1 Innovations in AI Algorithms for Detection

Continuous algorithmic development for AI is required to improve the accuracy and efficacy of microplastic detection systems. Some possible future developments could be:

• *Deep Learning Advances*: AI-powered microplastic detection systems will gain from further developments in deep learning models. Enhanced convolutional neural networks (CNNs) can improve image and spectral analysis, allowing for the detection of microplastics with greater accuracy. Recurrent neural networks (RNNs) and transformers may improve time-series data analysis, enabling more accurate tracking of microplastic trends in various ecosystems. Hybrid AI models that combine several deep learning methods may enhance overall classification accuracy [9].

• *Transfer Learning*: AI algorithms need large training sets in order to be highly accurate. Transfer learning can help overcome the constraints of small datasets by enabling pre-trained models—trained with large, heterogeneous datasets—to be adapted for microplastic detection tasks. This process can significantly lower computational expense and training time without decreasing detection performance. Utilizing established knowledge enables researchers to expedite the rollout of AI-powered solutions across diverse environments [12].

- *Explainable AI (XAI):* A significant challenge of AI-based microplastic detection is the black box nature of deep learning models. Explainable AI (XAI) methods will be responsible for enhancing transparency and explainability. Utilizing model-agnostic approaches, like SHAP (Shapley Additive Explanations) and LIME (Local Interpretable Model-Agnostic Explanations), AI-based microplastic detection systems can give insights into why certain predictions were made. This will establish trust between researchers, policymakers, and the public, so that AI-driven decisions are scientifically valid and regulatory-compliant [13].

5.7.2 Advancements in Sensor Technologies

Advancements in sensor technology are crucial for improving microplastic monitoring. Miniaturized sensors utilizing LIBS, Raman spectroscopy, and hyperspectral imaging will enable real-time detection across various environments. IoT-integrated smart sensor networks, combined with AI edge computing, will facilitate real-time data transmission, rapid analysis, and cloud-based visualization for pollution monitoring. Additionally, remote sensing through satellite imagery, drones, and AI-driven image processing will help identify large-scale plastic pollution hotspots, enabling proactive intervention [26].

5.7.3 Policy and Regulatory Frameworks Supporting Adoption

Supportive legislative and regulatory frameworks are essential for advancing AI-based microplastic detection. Establishing international standards for research methodologies, including standardized algorithms, data collection, and reporting, will ensure consistency in policy development. Governments and private sectors should provide funding and incentives through grants, subsidies, and infrastructure investments to scale detection technologies. Additionally, clear ethical guidelines addressing data privacy, bias, and transparency are crucial for building public trust and fostering collaboration among researchers, policymakers, and society [18–20].

5.7.4 Pathways for Addressing Microplastic Pollution with AI

To effectively combat microplastic pollution, the following pathways should be prioritized:

- **Improving Algorithm Efficiency**

Advancements in AI algorithms are essential for efficiently processing large datasets while maintaining high accuracy. Current models require improvements in robustness, generalizability, and interpretability. Explainable AI (XAI) enhances transparency, making AI-driven insights more accessible to policymakers, scientists, and the public. Additionally, federated learning enables distributed learning while preserving data confidentiality, further enhancing AI model performance [29].

- **Development of Sensor Technologies**

Developing advanced AI-enabled sensors integrated with IoT platforms is crucial for real-time microplastic detection and continuous environmental monitoring. These sensors can be deployed in oceans, rivers, wastewater facilities, and industrial discharge sites to track pollution sources dynamically. Innovations in low-cost, portable, and high-sensitivity sensors will further enhance monitoring in remote and resource-limited areas [32].

- **Policy Briefing Frameworks**

Supportive policies and regulations are essential for integrating AI in microplastic monitoring. Standardizing research methodologies ensures consistency in data interpretation, while funding should prioritize collaborations between AI researchers and environmental scientists. Additionally, regulations should encourage industries to adopt AI-based monitoring solutions for tracking and reducing plastic waste emissions.

5.8 Conclusion

The incorporation of artificial intelligence (AI) in the detection and monitoring of microplastics is a revolutionary strategy for the mitigation of one of the world's most acute environmental issues. This conclusion is an overview of major advancements in the area and presents avenues through which AI can be used to effectively combat microplastic pollution [33].

5.8.1 Summary of Key Developments

Recent advances in AI technologies have greatly improved the potential for detecting and tracking microplastics in different ecosystems. These innovations not only enhanced the detection accuracy but also facilitated large-scale monitoring as well as policymaking.

Machine learning and image analysis, particularly convolutional neural networks (CNNs), have transformed microplastic detection by enabling rapid classification and quantification in water, sediments, and biological samples while reducing human error and enhancing scalability. Deep learning models also improve Raman and FTIR spectroscopy analysis, ensuring precise polymer identification for more effective pollution tracking and mitigation planning. Additionally, AI-driven crowdsourced monitoring through mobile and cloud-based applications promotes community engagement, enhancing public participation, data collection, and awareness, allowing communities to contribute actively to environmental decision-making [35].

5.8.2 Future Perspectives

As technology develops and environmental consciousness rises, AI's role in microplastic monitoring and mitigation will grow [41]. For increased accuracy, research should concentrate on creating hybrid AI models that combine many detection methods. Blockchain technology and artificial intelligence (AI) together can provide transparent and safe data management for microplastic monitoring. Predictive modelling powered by AI will also assist in anticipating patterns of microplastic dispersion and bolster pre-emptive intervention tactics [66]. Increasing global cooperation between scientists, decision-makers, and businesses will hasten the development of AI-based remedies for microplastic contamination. Society can greatly lower microplastic pollution and advance long-term environmental sustainability by leveraging AI's full potential. Addressing this worldwide issue will need constant AI development, robust legislative backing, and public involvement [67].

Acknowledgement

Gratitude to the Institution and contributors.

AI disclosure statement: The author acknowledges the use of AI-based tools for minor language editing or grammatical correction. However, all intellectual contributions, analyses, and conclusions presented in this chapter are entirely the author's own.

References

1. Deng, Y., Wu, J., Chen, J., Kang, K., Overview of microplastic pollution and its influence on the health of organisms. *J. Environ. Sci. Health., Part A*, 58, 4, 412–422, 2023.
2. Yuan, P., Wang, Y., Chen, X., Gao, P., An overview of microplastic pollution in the environment over the megacity of Shanghai during 2013-2022. *Sci. Total Environ.*, 912, 168986, 2023, doi: 10.1016/j.scitotenv.2023.168986.
3. Ng, E.-L., *et al.*, An overview of microplastic and nanoplastic pollution in agroecosystems. *Sci. Total Environ.*, 627, 1377–1388, 2018, doi: 10.1016/j.scitotenv.2018.01.341.
4. Asori, M., Dogbey, E., Ampofo, S.T., Odei, J., Microplastic pollution in Africa: An overview of abundance in aquatic organisms, freshwater and marine water environments and analytical methods for reporting. *Manage. Environ. Qual.*, 35, 3, 634–652, 2023, doi: 10.1108/MEQ-06-2023-0170.
5. Bat, L. and Öztekin, A., Microplastic pollution in the Black Sea: An overview of the current situation. *Semantic Scholar*, 2022.
6. Kurniawan, T.A., *et al.*, Beyond surface: Unveiling ecological and economic ramifications of microplastic pollution in the oceans. *Water Environ. Res.*, 96, 7, 2024, doi: 10.1002/wer.11070.
7. Lahens, L., *et al.*, Macroplastic and microplastic contamination assessment of a tropical river (Saigon River, Vietnam). *Environ. Pollut.*, 236, 661–671, 2018, doi: 10.1016/j.envpol.2018.02.005.
8. Khare, S., Microplastic pollution: An overview of current scenario, challenges, and research gaps. *Adv. Biotechnol. Microbiol.*, 12, 3, 2019, doi: 10.19080/AIBM.2019.12.555836.
9. Li, J., Zhang, K., Zhang, H., Microplastics in freshwater systems: A review on occurrence, environmental effects, and methods for microplastics detection. *Water Res.*, 221, 118832, 2023, doi: 10.1016/j.watres.2022.118832.
10. Jin, M., *et al.*, Current development and future challenges in microplastic detection techniques: A bibliometrics-based analysis and review. *Sci. Prog.*, 105, 4, 003685042211321, 2022, doi: 10.1177/00368504221132151.
11. Hutter, H.-P., Weitensfelder, L., Poteser, M., Microplastics: Omnipresent and an ongoing challenge for medical science. *Wien. Klin. Wochenschr.*, 2024, doi: 10.1007/s00508-024-02375-9.
12. Mariano, S., *et al.*, Micro and nanoplastics identification: Classic methods and innovative detection techniques. *Front. Toxicol.*, 3, 2021, doi: 10.3389/ftox.2021.636640.
13. Özgenç, E., Advanced analytical techniques for assessing and detecting microplastic pollution in water and wastewater systems. *Environ. Qual. Manage.*, 34, 1, 2024, doi: 10.1002/tqem.22217.
14. Kögel, T., *et al.*, Current efforts on microplastic monitoring in Arctic fish and how to proceed. *Arct. Sci.*, 9, 2, 266–283, 2022, doi: 10.1139/as-2021-0057.

15. Cowger, W., *et al.*, Microplastic spectral classification needs an open source community: Open Specy to the rescue! *Anal. Chem.*, 93, 21, 7543–7548, 2020, doi: 10.1021/acs.analchem.1c00123.
16. Shiu, R.-F., *et al.*, Rapid identification of microplastics using Raman spectroscopy coupled with machine learning. *Environ. Sci. Technol.*, 57, 14, 5731–5740, 2023, doi: 10.1021/acs.est.2c09329.
17. Andrade, N.A.G., Nie, Y., Meulebroeck, W., Ottevaere, H., Surface plasmon resonance sensor based on a novel prism for the detection of a broad range of polymers. *arXiv*, 2024, doi: 10.48550/arxiv.2410.12440.
18. Bernatova, S., *et al.*, Detection of plastic nanoparticles in aqueous environment based on optical manipulation and Raman spectroscopy, in: *2022 IEEE Int. Workshop Metrol. Sea*, pp. 349–352, 2022, doi: 10.1109/METROSEA55331.2022.9950839.
19. Liu, Y., *et al.*, High-content screening discovers microplastics released by contact lenses under sunlight. *Environ. Sci. Technol.*, 57, 23, 8506–8513, 2023, doi: 10.1021/acs.est.3c01601.
20. García-Valle, G., *et al.*, An artificial intelligence-based optical sensor for microplastic detection in seawater, in: *2023 IEEE 9th World Forum Internet Things*, pp. 1–6, 2023.
21. Strafella, P., *et al.*, Detection of microplastics in fish using computed tomography and deep learning. *Heliyon*, 10, 21, e39875, 2024, doi: 10.1016/j.heliyon.2024.e39875.
22. Zhang, H., *et al.*, Advancements and challenges in microplastic detection and risk assessment: Integrating AI and standardized methods. *Mar. Pollut. Bull.*, 212, 117529, 2025, doi: 10.1016/j.marpolbul.2025.117529.
23. Valentino, M., *et al.*, Digital holographic approaches to the detection and characterization of microplastics in water environments. *Appl. Opt.*, 62, 10, D104, 2023, doi: 10.1364/AO.478700.
24. Hasan, M.A., *et al.*, IoT-driven image recognition for microplastic analysis in water systems using convolutional neural networks, in: *2024 2nd Int. Conf. Comput., Commun. Control*, pp. 1–6, 2024.
25. Han, S.H., Kim, D.G., Jeong, H.Y., Kim, S.H., Study on real-time sensor monitoring system for freshwater microplastics based on pulsed laser. *J. Korean Sens. Soc.*, 28, 5, 294–298, 2019.
26. Nicolai, E., *et al.*, A new optical method for quantitative detection of microplastics in water based on real-time fluorescence analysis. *Water*, 14, 20, 3235, 2022, doi: 10.3390/w14203235.
27. Shafiei, M., Abbasi, Z., Ren, C.L., Passive disposable microwave sensor for online microplastic contamination monitoring, in: **2022 IEEE/MTT-S Int. Microw. Symp.**, pp. 263–266, 2022, doi: 10.1109/IMS37962.2022.9865376.
28. Villegas-Ch, W., García-Ortiz, J., Sánchez-Viteri, S., Toward intelligent monitoring in IoT: AI applications for real-time analysis and prediction. *IEEE Access*, 12, 40368–40386, 2024, doi: 10.1109/ACCESS.2024.3376707.

29. Daramola, N.G.O., *et al.*, AI applications in reservoir management: Optimizing production and recovery in oil and gas fields. *Comput. Sci. IT Res. J.*, 5, 4, 972–984, 2024, doi: 10.51594/csitrj.v5i4.1083.
30. Hering, J.G., *et al.*, The role of AI in water quality monitoring. *Nat. Water*, 1, 1, 18–26, 2022, doi: 10.1038/s44221-022-00003-2.
31. Lines, E.R., *et al.*, AI applications in forest monitoring need remote sensing benchmark datasets. *arXiv*, 2022, doi: 10.48550/arxiv.2212.09937.
32. Solanki, S. and Singh Rathore, P., Current E-Waste Management: An Exploratory Study on Managing E-Waste for Environmental Sustainability, in: *Sustainable Management of Electronic Waste*, pp. 187–200, 2024.
33. Kumar, A., Rathore, P.S., Dubey, A.K., Srivastav, A.L., Dutt, V., Kumar, T.A. (Eds.), *Sustainable management of electronic waste*, John Wiley & Sons, Hoboken, New Jersey, USA, 2024.
34. Rathore, P.S., Ahuja, S., Burri, S.R., Khunteta, A., Baliyan, A., Kumar, A., *Deep learning techniques for automation and industrial applications*, John Wiley & Sons, Inc. ; Hoboken, New Jersey, USA, Scrivener Publishing LLC, 2024a.
35. Pham, H.V. and Moore, P., Applied hedge algebra approach with multilingual large language models to extract hidden rules in datasets for improvement of generative AI applications. *Information*, 15, 7, 381, 2024, doi: 10.3390/info15070381.
36. Tuan, N.T., *et al.*, A generative artificial intelligence using multilingual large language models for ChatGPT applications. *Appl. Sci.*, 14, 7, 3036, 2024, doi: 10.3390/app14073036.
37. Ly, N.H., *et al.*, Advanced microplastic monitoring using Raman spectroscopy with a combination of nanostructure-based substrates. *J. Nanostruct. Chem.*, 12, 5, 865–888, 2022, doi: 10.1007/s40097-022-00506-0.
38. Di Leo, G., Trimboli, R.M., Sella, T., Sardanelli, F., Optical imaging of the breast: Basic principles and clinical applications. *Amer. J. Roentgenol.*, 209, 1, 230–238, 2017, doi: 10.2214/AJR.16.17220.
39. Mikami, H., Gao, L., Goda, K., Ultrafast optical imaging technology: Principles and applications of emerging methods. *Nanophotonics*, 5, 4, 497–509, 2016, doi: 10.1515/nanoph-2016-0026.
40. Patil, S., *et al.*, Automated waste management system: A deep learning approach, in: *2024 8th Int. Conf. Comput., Commun., Control Autom*, pp. 1–5, 2024.
41. Gulyamov, S., Intelligent waste management using IoT, blockchain technology and data analytics. *E3S Web Conf.*, 501, 01010, 2024, doi: 10.1051/e3sconf/202450101010.
42. Mohanta, A.K., *et al.*, Effect of inappropriate solid waste on microplastic contamination in Balasore district and its aquatic environment. *Bull. Nat. Res. Cent.*, 48, 1, 2024, doi: 10.1186/s42269-024-01278-z.

43. Shourie, P., *et al.*, Smart waste management: VGG16-based classification system for organic and recyclable materials, in: *2024 7th Int. Conf. Circuit Power Comput. Technol*, vol. 1, pp. 1410–1414, 2024.
44. Krolik, A., *et al.*, Detection of defects in polyethylene and polyamide flat panels using airborne ultrasound-Traditional and machine learning approach. *Appl. Sci.*, 14, 22, 10638, 2024, doi: 10.3390/app142210638.
45. Ganapa, J., Bioluminescent imaging for *in vivo* tracking of transplanted stem cells in regenerative medicine. *J. Appl. Bioanal.*, 10, 2, 2024, doi: 10.53555/jab.v10i2.153.
46. Pirovano, G., Roberts, S., Kossatz, S., Reiner, T., Optical imaging modalities: Principles and applications in preclinical research and clinical settings. *J. Nucl. Med.*, 61, 10, 1419–1427, 2020, doi: 10.2967/jnumed.119.238279.
47. Wu, C., *et al.*, *In-vivo* optical imaging in head and neck oncology: Basic principles, clinical applications and future directions. *Int. J. Oral Sci.*, 10, 2, 2018, doi: 10.1038/s41368-018-0011-4.
48. Salaheldin, A.M., *et al.*, Deep learning-based automated detection and grading of papilledema from OCT images: A promising approach for improved clinical diagnosis and management. *Int. J. Imag. Syst. Technol.*, 34, 4, 2024, doi: 10.1002/ima.23133.
49. Vora, L.K., *et al.*, Artificial intelligence in pharmaceutical technology and drug delivery design. *Pharmaceutics*, 15, 7, 1916, 2023, doi: 10.3390/pharmaceutics15071916.
50. Khalighi, S., *et al.*, Artificial intelligence in neuro-oncology: Advances and challenges in brain tumor diagnosis, prognosis, and precision treatment. *npj Precis. Oncol.*, 8, 1, 2024, doi: 10.1038/s41698-024-00575-0.
51. Vimbi, V., Shaffi, N., Mahmud, M., Interpreting artificial intelligence models: A systematic review on the application of LIME and SHAP in Alzheimer's disease detection. *Brain Inform.*, 11, 1, 2024, doi: 10.1186/s40708-024-00222-1.
52. Tian, C., Xu, Z., Wang, L., Liu, Y., Arc fault detection using artificial intelligence: Challenges and benefits. *Math. Biosci. Eng*, 20, 7, 12404–12432, 2023, doi: 10.3934/mbe.2023552.
53. Sonko, N.S., *et al.*, A critical review towards artificial general intelligence: Challenges, ethical considerations, and the path forward. *World J. Adv. Res. Rev.*, 21, 3, 1262–1268, 2024, doi: 10.30574/wjarr.2024.21.3.0817.
54. Bashkami, A., *et al.*, A review of artificial intelligence methods in bladder cancer: Segmentation, classification, and detection. *Artif. Intell. Rev.*, 57, 12, 2024, doi: 10.1007/s10462-024-10953-6.
55. Andrews, H.B. and Sadergaski, L.R., Automated calibration for rapid optical spectroscopy sensor development for online monitoring. *ACS Sens.*, 9, 11, 6257–6264, 2024, doi: 10.1021/acssensors.4c02211.
56. Chakraborty, C. and Kishor, A., Real-time cloud-based patient-centric monitoring using computational health systems. *IEEE Trans. Comput. Soc. Syst.*, 9, 6, 1613–1623, 2022, doi: 10.1109/TCSS.2022.3170375.

57. Cinque, M., *et al.*, Evaluating virtualization for fog monitoring of real-time applications in mixed-criticality systems. *Real-Time Syst.*, 59, 4, 534–567, 2023, doi: 10.1007/s11241-023-09410-4.
58. Harikrishnan, S., *et al.*, Archaeophotonics: Applications of laser spectroscopic techniques for the analysis of archaeological samples. *Appl. Spectrosc. Rev.*, 59, 2, 187–223, 2023, doi: 10.1080/05704928.2023.2189946.
59. Jin, H., *et al.*, Artificial intelligence in microplastic detection and pollution control. *Environ. Res.*, 262, 119812, 2024, doi: 10.1016/j.envres.2024.119812.
60. Soldano, L.B. and Pennings, E.C.M., Optical multi-mode interference devices based on self-imaging: Principles and applications. *J. Lightw. Technol.*, 13, 4, 615–627, 1995, doi: 10.1109/50.372474.
61. Lei, C., Guo, B., Cheng, Z., Goda, K., Optical time-stretch imaging: Principles and applications. *Appl. Phys. Rev.*, 3, 1, 011102, 2016, doi: 10.1063/1.4941050.
62. Gensch, M., Spectroscopic techniques and sensors for the exploration of extraterrestrial environments: A review. *Proc. SPIE*, 12372, 74–76, 2023.
63. Codrops, "2023 IEEE International Conference on Big Data. *IEEE BigData*, 2023. [Online]. Available: https://bigdataieee.org/BigData2023/.
64. Flores, E., *et al.*, On efficient data sharing for planetary digital twins: Distributed microplastic monitoring, in: *2024 IEEE 22nd Int. Conf. Ind. Informat*, pp. 1–8, 2024.
65. Jafar, A., *et al.*, Revolutionizing agriculture with artificial intelligence: Plant disease detection methods, applications, and their limitations. *Front. Plant Sci.*, 15, 2024, doi: 10.3389/fpls.2024.1356260.
66. Aït-Kaddour, A., *et al.*, Application of spectroscopic techniques to evaluate heat treatments in milk and dairy products: An overview of the last decade. *Food Bioprocess Technol.*, 14, 5, 781–803, 2021, doi: 10.1007/s11947-021-02607-0.
67. Ozen, B., Cavdaroglu, C., Tokatli, F., Trends in authentication of edible oils using vibrational spectroscopic techniques. *Anal. Methods*, 16, 26, 4216–4233, 2024, doi: 10.1039/D4AY00562G.
68. Mahrous, E., Chen, R., Zhao, C., Farag, M.A., Lipidomics in food quality and authentication: A comprehensive review of novel trends and applications using chromatographic and spectroscopic techniques. *Crit. Rev. Food Sci. Nutr.*, 64, 25, 9058–9081, 2023, doi: 10.1080/10408398.2023.2207659.
69. Hariharasudhan, S., Kumaran, U., Kumar, K.S., Exposure on edge detection and image segmentation with instinctive classification of brain tumor using artificial intelligence methods, in: *2023 Int. Conf. Data Sci. Agents Artif. Intell*, pp. 1–7, 2023.

6

Integrating AI with Advanced Sensor Technologies for Real-Time Monitoring

Avnish Chauhan[1*], Shivam Attri[1], Aanchal Saklani[1], Prabhat K. Chauhan[2], Man Vir Singh[3], Vishal Rajput[4], Muneesh Sethi[5] and Samuele Barrili[6]

[1]Department of Environmental Science, Graphic Era Hill University, Dehradun, UK, India
[2]Department of Environmental Science, A K S University, Satna, M.P., India
[3]School of Applied and Life Sciences, UIT, Uttaranchal University, Dehradun, UK, India
[4]Himalayan School of Bioscience, Swami Rama Himalayan University, Uttarakhand, India
[5]Mechanical Engineering Department, COER University, Roorkee, UK, India
[6]M4W-Marketing for Waste, The First Marketing Agency For Small Waste Companies and Junk Removal, Lewes, Delaware, Italy

Abstract
This chapter addresses the increasing presence of microplastics in ecosystems worldwide, highlighting the pressing need for operative detection and mitigation strategies. It further explores the use of artificial intelligence (AI) in combination with precocious sensor technologies to monitor microplastics. AI-driven approaches, particularly machine learning (ML), are transforming sensor platforms into their analytical capacity and enabling the prediction of future microplastic pollution trends. These developments respond to the intrinsic challenges of monitoring and managing complex environmental systems. By integrating spatial and temporal data, researchers can create comprehensive toolkits for mapping, tracking, and analyzing microplastics across different environments such as aquatic systems, soils, and urban areas. AI integration enhances the identification of pollution hotspots and supports the design of targeted mitigation policies. Laboratory-based microplastic analyses, when combined with satellite-derived models,

**Corresponding author*: avnishchauhan_in@yahoo.com

Abhishek Kumar, Pooja Dixit, Pramod Singh Rathore, Arun Lal Srivastav and Ashutosh Kumar Dubey (eds.) Microplastic Monitoring Using Artificial Intelligence, (109–126) © 2026 Scrivener Publishing LLC

demonstrate the feasibility of large-scale, real-time monitoring with favorable accuracy in estimating microplastic concentrations. The synergy between AI and advanced sensor technologies bridges technological innovation with applicable applications, generating actionable insights to address microplastic pollution, strengthen environmental protection, and safeguard open health.

Keywords: Microplastics, artificial intelligence, machine learning (ML), advanced sensor technologies, geographic information system (GIS), remote sensing data

6.1 Introduction

The global environmental crisis of plastic waste is prejudicial to ecosystems, landfills, and oceans in binary ways [1], mostly attributed to incompetent waste management and its considerable use across industries. Improper disposal of plastics degrades landscapes, reduces soil fertility, and negatively affects agricultural productivity. In aquatic environments, plastic pollution disrupts nautical ecosystems, threatens wildlife, and introduces microplastics into the food chain, posing significant risks to both biodiversity and human health. The persistence of elastic waste raises semi-permanent ecological and health concerns.

Plastic waste management also presents noteworthy interpersonal and economic challenges, including exaggerated costs for waste collection, recycling infrastructure, and disposal. Industries dependent on plastics face increasing economic pressure from shifting consumer preferences, stricter regulatory requirements, and the demand for sustainable alternatives [34]. Despite the widespread issue of microplastic contamination, standardized monitoring and analytic techniques are still lacking, often leading to spotty results. Establishing uniform methods would enable further true duodecimal comparisons, improve understanding of microplastic distribution and abundance, and better assess associated risks. This would help reduce discrepancies in microplastic data and enhance overall comprehension [27]. Developing countermeasures requires the use of efficient detection and analysis techniques. When combined with current MP (microplastic) analysis techniques, artificial intelligence technologies such as machine learning models provide a novel way to improve MP identification and characterization speed and accuracy. An extensive analysis of the use of AI in MP research is given in this chapter.

Monitoring microplastics is essential for addressing both obvious and undetectable forms of pollution. These particles, typically smaller than 5 mm, pose thoughtful risks to human health and the environment. In aquatic systems, they can easily move through ecosystems, contaminate water bodies, and enter the food chain. Their sources, concentrations, and distribution can be effectively assessed using functional monitoring systems [3]. On with guiding efficient and sustainable measures to limit their spread, this data provides policymakers with sensible insights for developing stricter regulations aimed at reducing emissions.

Furthermore, monitoring microplastics generates rich insights into their behavior and impacts, supports technological research, and contributes to the development of sustainable semipermanent solutions [32]. Addressing the challenge of increasing pollution requires comprehensive strategies that leverage advanced technologies such as artificial intelligence (AI) and next-generation sensor systems to strengthen monitoring, detection, and mitigation efforts. Sensors deployed across different environments, including water, soil, and air, unceasingly collect great volumes of real-time data, enabling more accurate assessment and management of microplastic contamination.

The data is subsequently analyzed by AI algorithms to discern patterns and trends, facilitating a further speedy and true evaluation of microplastic contamination. Machine learning is a strong tool for detecting patterns and trends with exquisite precision. By processing data from monitoring stations, AI models can predict pollution levels and identify prospective sources, linking these insights to air and soil pollution forecasting systems. This chapter examines the increasing role of AI and precocious sensor technologies as operative approaches for real-time monitoring and for developing sustainable solutions that support environmental management and policymaking [33].

6.2 Bibliographic Study

This chapter offers a comprehensive review of published articles, including review articles, book chapters, and research papers from the Scopus database shown in Figure 6.1. The bibliometric analysis focuses on publications which is in the English language from the period 2020 to 2025, using the search terms 'microplastic,' 'detection,' and 'monitoring.' To refine the search, the Boolean operator (AND) was used to combine search terms.

The review prioritizes recent advancements in critical areas such as waste system economics, the linear waste management model, and circular economy strategies on a global scale. A total of 793 distinct publications from Scopus-indexed journals and books were considered for this analysis.

For bibliometric analysis, the data acquisition was done using predefined keywords and processed using VOS viewer 1.6.18 software. Co-occurrence

(a)

(b)

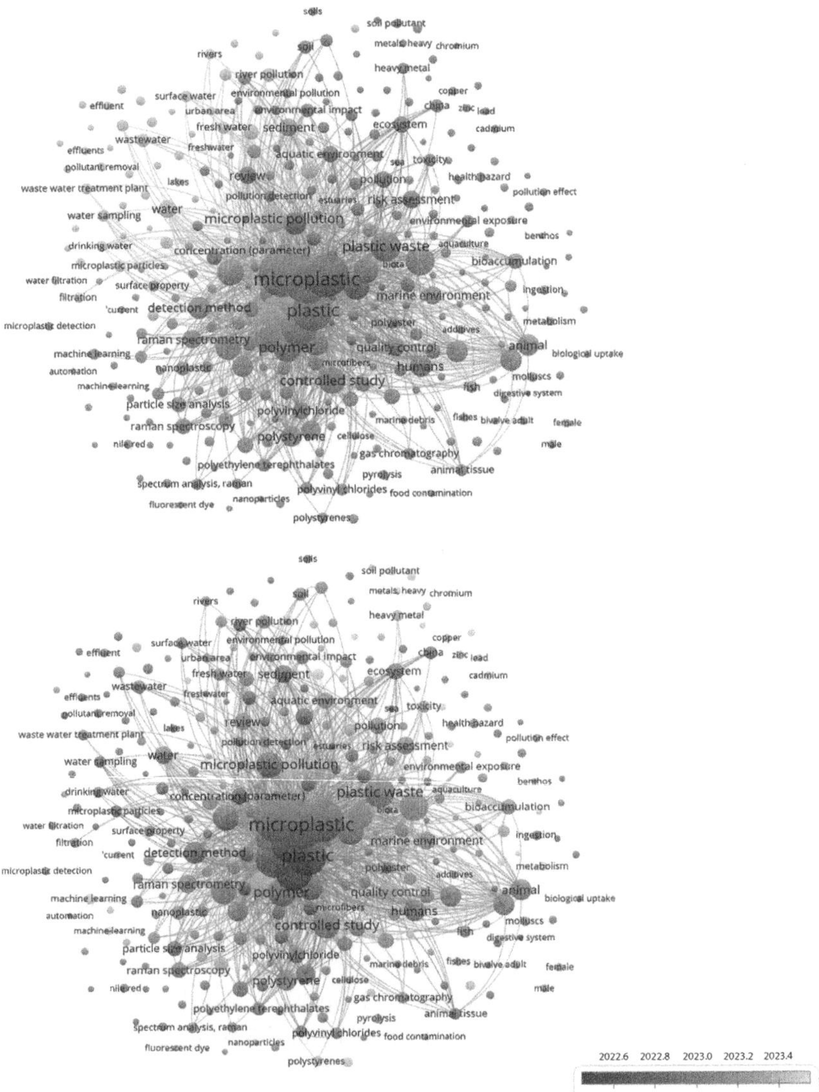

Figure 6.1 Bibliometric analysis focuses on publications.

analysis was conducted on all keyword units, employing the full counting method. A threshold of at least 15 occurrences was set, resulting in 329 keywords meeting the criteria. These keywords were classified into five clusters: the first cluster contained 92 keywords, the second 87, the third 70, the fourth 57, and the fifth 23. Figure 6.1(a) illustrates a network visualization where larger frame sizes represent the prominence of keywords, while the strength of interconnections is depicted through the density of linking lines. In contrast, Figure 6.1(b) presents an overlay visualization that highlights the temporal distribution of keywords across the analyzed literature. Frequently occurring keywords are emphasized through larger frame sizes, indicating their significant presence in the reviewed publications.

The network clustering of 793 published articles from 2020 to 2025 that had high matches of specific keywords and their citations was the focus of the bibliometric analysis shown in Figure 6.1 (a) Network visualization (b) Overlay visualization. The global environmental catastrophe of plastic trash is harming ecosystems, landfills, and oceans in several ways.

6.3 AI-Enabled Sensor Technologies for Microplastic Detection

The influence of artificial intelligence (AI) as a transforming force has been observed in many domains as well as in many sectors, including the most significant improvement in environmental monitoring as well as in pollution control [5]. Currently, ecological issues are growing, and AI provides cutting-edge resolutions that improve our capacity to observe, predict, and address environmental concerns. One of the primary contributions of AI in environmental monitoring is its capability to handle complex datasets from multiple sources, including satellite images, remote sensing sensors and weather stations [15, 26]. AI has proved itself as an effective tool in identifying and detecting various pollutants. Machine learning and incomprehensible learning techniques are being applied to predict $PM_{2.5}$ concentrations in air, soil contamination by laborious metals, laborious metal accumulation in vegetation, and the mass concentrations of natural carbon (OC), ultimate carbon (EC), and nitrogen oxides (NO_2) in PM10 particles associated with traffic pollution [3, 14, 21, 28].

Advances in AI have significantly improved water quality monitoring. Unlike traditional methods, sensor-based AI systems can continuously track conditions in water bodies, analyze the data, and detect contaminants or even predict potential pollution events [7]. AI has also played a noteworthy role in waste monitoring and management, offering further efficient and data-driven solutions. For example, to cut down on fuel use and emissions from garbage collection vehicles, advanced waste management systems apply AI algorithms to optimize collection routes [13]. AI waste sorting technologies have also improved recovery efficiency by specifically identifying and segregating waste materials [11, 17]. These developments assist in minimizing the total environmental effect as well as encourage further environmentally neighborly waste management techniques. Recently, AI has emerged as an operative tool to tackle the increasing challenge of microplastic pollution. AI-driven approaches provide state-of-the-art methods for detecting and managing these contaminants, which are much too small to be seen with the naked eye [25].

Microplastics are little fragments of plastic that enter ecosystems and have become a significant environmental concern. Microplastics persist in the environment for long periods and exert counterproductive effects on living organisms, including humans. Precocious analytic technologies are progressively being applied to trace the origins of these contaminants. As the prevalence of microplastic sources continues to rise, identifying their points of origin helps researchers improve tracking of their movement through ecosystems and design further targeted mitigation strategies (Figure 6.2).

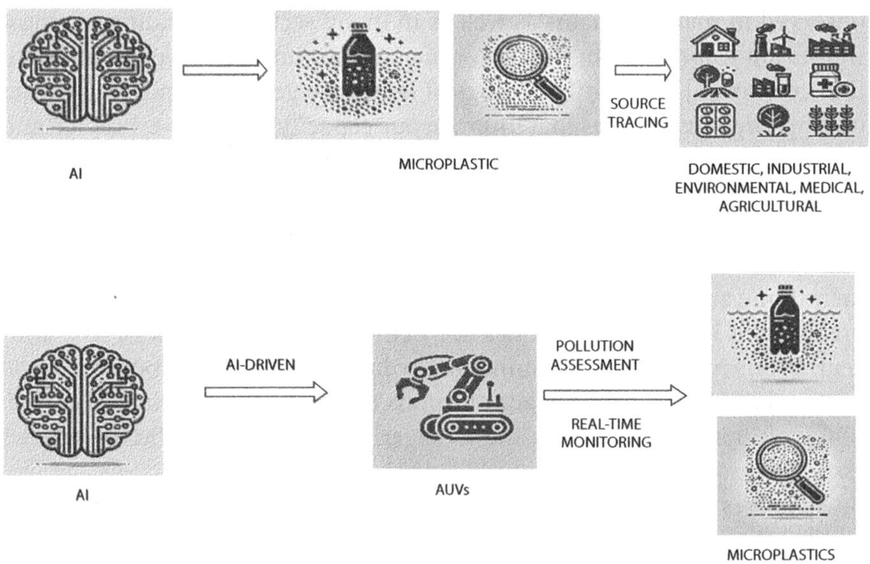

Figure 6.2 Artificial intelligence (AI) is used to assess pollution levels, pinpoint the sources of microplastics, and enable real-time monitoring.

Currently, spectrometric analysis, little inspection, and visible analysis are among the most widely used methods for detecting and identifying microplastics. Geomorphological characteristics such as shape, size, and surface properties can be examined through scanning electron microscopy (SEM) [23]. However, the sample preparation required for SEM analysis is often complex and time-consuming, which limits its routine application [18]. The oldest supernatural methods for identifying MPs are Fourier transform infrared (FTIR) methods and Raman spectroscopy. Fourier Transform Infrared Spectroscopy (FTIR) is the most commonly used method for detecting MPs in samples obtained from the surrounding environment for identifying and characterizing MPs. Fourier transform infrared (FTIR) spectroscopy enables true identification and quantification of microplastics, as it provides detailed information on polymer composition based on the characteristic spectra of elastic particles. However, FTIR is extremely prone to interference from surface water residues and spurious natural matter present on the sample, which can lead to irregularities in spectral absorption and compromise analytic accuracy [22, 24]. Raman spectroscopy, in contrast, is less affected by water interference and can detect small particles, making it a reciprocal technique to FTIR in microplastic analysis.

a. AI-Assisted Image Processing for Microplastic Identification

Several challenges affect the accuracy and reliability of incumbent practices for identifying and recognizing microplastics. An outstanding issue is the dependence on visible identification, which is extremely personal and often prone to human error [10]. Because microplastics differ widely in size, shape, and color, their detection becomes even further complicated, especially when they occur in complex forms [36]. Traditional techniques such as ocular microscopy often lack the resolution required to detect particles smaller than 10 nm. As a result, these methods may considerably underestimate the genuine levels of microplastics present in the environment [37]. This trouble is compounded by the fact that MPs go through considerable natural and chemical changes, rendering the technique impracticable to use efficaciously [38].

Since artificial intelligence algorithms can process significant amounts of data from little images, they are quite practical tools for the detection of microplastics.

By integrating incomprehensible learning with another model, a new approach to speedy qualitative and duodecimal assessments of contrasting MP categories is presented. For microplastic image identification and retrieval, the approach integrates ocular resolution photoacoustic microscopy with self-supervised feature learning in the form of a vector quantized variational autoencoder (VQVAE2). The subsequent steps comprise the proposed methodology. The MPs are isolated and separated in the initial step. A PAM is employed to describe the individual samples [39].

b. AI and Spectroscopy-Based Detection
AI-aided FTIR: This technique has been regarded as an essential method of identification as well as characterization of microplastic polymers that are present in the environment. The conventional analysis of absorption peaks is labor-intensive as well as requiring expertise [24]. FTIR-PLSR has also been applied for the identification of MPs (low-density polyethylene and polyethylene terephthalate) in sediments. The integration of FTIR along with artificial intelligence greatly improves the speed as well as accuracy of identification [40].

AI-enhanced Raman spectroscopy: The primary issue with this technique is the imaging resolution because of the laser spot's diffraction [41]. To solve this problem, spatial heterodyne Raman spectroscopy is best, as they provide better signal-to-noise ratios for MP detection [42]. The combination of Raman spectroscopy and spatial heterodyne technology can detect four common MPs, such as PP, PC, PS, and HDPE, rapidly and with high accuracy. The automatic detection of MPs in the environment is performed using Raman spectroscopy and multivariate analysis. Spectral data preprocessing is performed using PCA and LDA. Then, the SVM model is used to create an automatic classification of the MP type in the environment. The automatic classification of the type of MP in the environment is reliable and accurate. The automatic detection of MPs at the nanoscale is still challenging because the Raman signals generated due to the presence of some organic impurities are too great to be removed without this technique. It will be necessary to carry out more research to apply this technique to real environmental samples [9].
AI-powered HSI: Hyperspectral imaging has turned into a mature technology for precise and non-destructive MP detection, poised to become a rapid alternative for online monitoring. The HSI technology merges the spectroscopy and imaging techniques, based on the principle that the chemical composition of each object radiates at specific wavelengths in the electromagnetic spectrum [43]. A key advantage of the HSI technology over FTIR and Raman spectroscopy is that the HSI does not require any sophisticated procedures to separate microplastics or worry about the interference from other substances in the sample, especially organic matter.

MPs in the environment can be directly identified using HSI technology in conjunction with a machine learning model. Because of its broad wavelength range, it also enables rapid and non-destructive microplastic detection, greatly reducing the need for intricate sample preparation. The precision of MP detection is enhanced when machine learning methods, like CNN models and SVM models, are used in combination, especially with complicated matrices.

Real-Time Data Processing and AI Integration

Processing of real-time data has been revolutionized due to recent developments in artificial intelligence (AI) and the Internet of Things (IoT), especially in environmental monitoring. One of the most noteworthy issues these technologies solve is the detection, evaluation, and forecasting of microplastic pollution in contrasting ecosystems. Laboratory testing and manual sampling are the traditional methods used in detecting microplastics. These formal methods require a lot of time and do not offer the speed required to take operative action [12], as quoted in [9]. Nonetheless, since they have enabled prophetical modeling as well as ongoing monitoring, AI systems and sensor networks integrated with IoT have come to be the game changers in this field [9].

There are several sensors, like acoustic, ocular, and electrochemical sensors, which belong to Internet of Things networks. Sensors can immediately and in real time collect a vast amount of environmental information. Through wireless data transmission to cloud platforms, the sensors enable distant connection and real-time evaluation of the environment [6].

To process significant and complex data streams, artificial intelligence (AI) algorithms, particularly machine learning models such as support vector machines (SVMs) and artificial neural networks (ANNs), play a vital role. These models reduce the need for manual analysis while improving accuracy in distinguishing microplastics from different environmental particles, enabling automated classification, measurement, and monitoring [2]. AI-driven predictive modeling can analyze dynamic environmental data to identify pollution hotspots and forecast trends in microplastic migration. In addition, advanced incomprehensible learning methods such as continuous neural networks (RNNs) and recurrent short-term memory (LSTM) models can simulate sorption processes in real time, leading to more accurate predictions of pollutant behavior [4]. Besides, environmental scientists and policymakers are too susceptible to proactive action against close contamination threats associated with the use of AI alongside IoT.

a) Microplastic Detection Using IoT-enabled Sensor Network

The integration of precocious sensors with cloud-based IoT platforms has significantly improved environmental monitoring by enabling remote access, secure data storage, and real-time analysis. An obvious example is the Smart IC system, which employs wireless sensors and multimodal data fusion to detect microplastics in aquatic ecosystems. By combining obvious (VI), infrared (IR), and ultraviolet (UV) spectral data, this system allows for fine identification of microplastic contaminants [2, 6, 8]. Traditional manual data collection methods are not merely time-consuming but also labor-intensive, limiting their effectiveness in large-scale or endless monitoring. In contrast, IoT-enabled real-time monitoring supports endless assessment of environmental conditions. The integration of AI algorithms further strengthens this approach by quickly analyzing complex datasets to identify contamination hotspots. These insights equip regulative authorities with actionable information, sanctionative, timely interventions and the formulation of operative mitigation strategies to reduce pollution and protect ecosystem health.

IoT-based systems are also extremely adaptable, capable of monitoring large and remote areas while recording pollution trends. Such networks often rely on buoys, boats, and even animals such as birds or seals equipped with sensors to collect data across wide regions [6]. The concentrated data is uploaded to cloud platforms in real time. By analyzing this data, researchers and policymakers are susceptible to making quick decisions. This solid time processing improves. The efficacy of pollution control measures increases the accuracy of microplastic detection.

b) Machine Learning Models for Real-Time Analysis

Machine learning (ML) models utilizing automated classification and concrete time data analysis have transformed the identification of microplastics. Stiff identification methods, such as chemical spectroscopy and manual microscopy, are labor-intensive and supersensitive to error because they involve human intervention. Nevertheless, ML models like artificial neural networks (ANNs) and support vector machines (SVMs) can process noteworthy datasets often quickly, ensuring timely and fine identification of microplastics [9].

Convolutional edgy networks (CNNs) have, for example, been trained on tens of thousands of expandable spectra so that they are susceptible to detecting microplastic particles by determining their composition, size, and shape The CNN framework Plastic Net has sustained an identification rate of 92.0% and attained a rate of over 95.0% in distinguishing between 11 of the most predominant types of elastic, even following exposure to the most ambitious environmental samples that are weathered or contain additives [29]. Compared to static detection methods, this open nature greatly enhances the capacity of the AI models to cope with variability in concrete world cases. The gifted AI systems also integrate multimodal data sources, including image analysis, spectroscopic measurements, and particle tracking data.

Real-time microplastic monitoring systems, such as AI camera systems that employ YOLOv5 for object detection and the Deep SORT algorithm for tracking, have been shown to have 97% accuracy in laboratory testing and 96% accuracy in a concrete world setting [19].

c) Predictive Modeling for Microplastic Movement and Source Tracking

Prophetical modeling is a considerably applicable technique for tracing the migration of microplastics as well as for determining the source of pollution. Scientists and researchers mimic the semi-permanent interactions between microplastics as well as environmental conditions such as wind directions, currents in water, and organic contaminants through artificial intelligence (AI) algorithms like recurrent neural networks (RNNs) and long short-term memory (LSTM) models [4].

By integrating Eulerian hydrodynamic simulations with Lagrangian particle tracking models (PTM), advanced quantitative models like CaMPSim 3D forecast the trajectories of microplastic transport. These models assist in making significant predictions of the locations of particle accumulation hotspots by simulating diffusion, advection, and transformation processes (biofouling, degradation) [16]. For instance, CaMPSim 3D assisted in targeted cleanup efforts within the lower Saint John River Estuary by determining where microplastics aggregate and tracing upstream pollution sources.

The integration of dynamic data streams from IoT sensors makes real-time predictive modeling even better. AI models handle time-dependent variables like variations in water flow rates and changes in pollutant concentrations while updating predictions with incoming data. RNN models, known for their capacity to capture temporal dependencies, have achieved an accuracy of 96.7% in forecasting the microplastic adsorption behavior, performing better than other deep learning models such as LSTM and CNN [4]. These predictive tools not only forecast the spread of microplastics but also aid in proactive environmental management. By identifying high-risk areas and monitoring sources of pollution, predictive modeling enables policymakers to execute timely intervention strategies, eventually lessening the lasting effects of microplastic pollution.

6.4 Challenges and Future Prospects

a) Limitations of AI models in microplastic detection
AI models have proved themselves to be key players in the detection of microplastics, but they also pose some limitations. Here are the limitations:

I. Data Quality and Availability
Limited Datasets- Large and high-quality datasets are essential for training artificial intelligence models, while comprehensive information in various environments is limited and inconsistent.
Variability in microplastics- The existence of microplastics is in various forms, sizes, colors, and types of polymers; therefore, it is hard to develop a general dataset that can encompass all the variations.

II. Detection Accuracy
False Positives/Negatives- Artificial Intelligence algorithms can misidentify or misclassify some natural particles, for example, algae, sediments, etc., as microplastics and sometimes fail in the detection of smaller or transparent particles, which results in errors.
Size Limitation- Detection of Nano-sized particles, which are generally smaller than microplastics, is very difficult because of their extremely small size and resemblance to organic matter.

III. Environmental Complexity
Interference from Natural Debris- Microplastics in natural environments are mostly mixed with sediments, organic matter and other debris, making them harder to detect.
Dynamic Conditions- Water flow, lighting and weather conditions can easily influence the operation of AI models, particularly when used in real-time monitoring applications.

IV. Technological Limitations
Sensor Limitations- The precision of AI models mostly depends upon the quality of sensors and imaging systems, which are used for data capture. Poorly defined low-resolution sensors can miss smaller particles or not be able to differentiate microplastics from other types of materials.
Computational Costs- Execution and training of advanced AI models require high computational resources, which are costly and power-intensive.

b) Future research directions

i. Updated Artificial Intelligence Models
   - Development of more accurate and precise AI models that can handle large datasets for the detection of microplastics in varying environments [28].
   - Integration of AI models with global efforts should be done to reduce microplastic contamination. From drones patrolling to advanced sensors, AI will enable faster, more accurate, and more accessible detection of microplastics.

ii. Multi-sensor Data Fusion
   The fusion of data from multiple sources, such as cameras, spectrometers, and drones, to improve detection and monitoring [20]. Some of the key areas include advanced algorithms, heterogeneous data integration, edge AI, multimodal AI, sensor innovation, complex environments, data sharing, sustainability, pilot projects and collaboration [30].

iii. Real-Time Monitoring systems
   Integrating AI models into global initiatives is essential to reducing microplastic contamination. From drone-based surveillance to advanced sensor systems, AI enhances the speed, accuracy, and accessibility of detection, helping researchers and policymakers respond more effectively [20].

iv. Predictive Modeling
   AI-powered prediction of microplastics faces challenges from data, algorithms, tasks, and tools. Future research directions include building large datasets, expanding MP datasets, developing reliable evaluation indicators, designing high-performance and low-computational complexity algorithms (such as LeaNet, MALUNet), building a 0-code ML platform for non-specialists, and conducting more task studies, sequences, text, and 3D images [30, 31].

v. Nano Plastic Detection
   Exploring AI models for the detection and analysis of nano plastics, which are more challenging than identifying microplastics. Due to their lower size and lower signal, which enhance difficulties in analysis, nano plastic research lags significantly behind microplastic research [35].

6.5 Conclusion

Detection of microplastics has been revolutionized through the integration of advanced sensor technologies and artificial Intelligence. Scientists can more precisely detect and measure microplastics in the environment through the integration of AI with spectroscopy and image processing. Technology accelerates the analysis of data and facilitates pollution monitoring and control measures through automated detection methods that were time-consuming in the past, and real-time monitoring. Integrating AI into microplastic detection is essential for pollution management, trend prediction, and understanding how microplastics accumulate and spread. Large amounts of environmental data can be processed through machine learning models, which are progressively being used for real-time monitoring and to inform policymaking.

Simultaneously, various challenges restrict the broader implementation of AI-based strategies. A considerable challenge exists in the application of these models to different datasets gathered under differing environmental conditions. These differences frequently influence the accuracy of environmental data, sampling methods, and overall data quality. So, there is a need for advanced machine learning architectures that can handle highly complex as well as diverse datasets to improve prediction reliability. Additional challenges include the substantial computational demands associated with data collection, model training, and deployment, which must be optimized to maintain operational performance across different environmental contexts.

AI disclosure statement: The author acknowledges the use of AI-based tools for minor language editing or grammatical correction. However, all intellectual contributions, analyses, and conclusions presented in this chapter are entirely the author's own.

References

1. Alabi, O.A., Ologbonjaye, K.I., Awosolu, O., Alalade, O.E., Public and environmental health effects of plastic waste disposal: a review. *J. Toxicol. Risk Assess.*, 5, 021, 1–13, 2019.
2. Chaczko, Z., Kale, A., Santana-Rodríguez, J.J., Suárez-Araujo, C.P., Towards an IoT-based system for detection and monitoring of microplastics in aquatic environments, in: *2018, IEEE 22nd International Conference on Intelligent Engineering Systems (INES)*, pp. 000057–000062, IEEE, 2018, June.

3. Chen, L., *et al.*, Meta-analysis of impacts of microplastics on plant heavy metal (loid) accumulation. *Environ. Pollut.*, *348*, 123787, 2024.
4. Godasiaei, S.H., Predictive modeling of microplastic adsorption in aquatic environments using advanced machine learning models. *Sci. Total Environ.*, 958, 178015, 2025.
5. Hamza, M.-A., *et al.*, Big data analytics with artificial intelligence enabled environmental air pollution monitoring framework. *Comput. Mater. Continua*, *73*, 3235–3250, 2022.
6. Jagadeesh, B. and Arlene, A.A., A comprehensive literature survey of microplastic detection using IoT. *Int. J. Creat. Res. Thoughts (IJCRT)*, 11, 3, b841–b846, 2023, http://www.ijcrt.org/papers/IJCRT2303202.pdf.
7. Jayaraman, P., *et al.*, Critical review on water quality analysis using IoT and machine learning models. *Int. J. Inf. Manage. Data Insights*, *4*, 100210, 2024.
8. Jebril, S., Fredj, Z., Saeed, A.A., Gonçalves, A.M., Kaur, M., Kumar, A., Singh, B., Nanomaterial-based electrochemical chemo (bio) sensors for the detection of nanoplastic residues: trends and future prospects. *RSC Sustainability*, 2, 4, 832–851, 2024.
9. Jin, H., Kong, F., Li, X., Shen, J., Artificial Intelligence in microplastic detection and pollution control. *Environ. Res.*, 119812, 2024.
10. Kotar, S., *et al.*, Quantitative assessment of visual microscopy as a tool for microplastic research: Recommendations for improving methods and reporting. *Chemosphere*, *308*, 136449, 2022.
11. Li, Z., *et al.*, An intelligent identification and classification system of decoration waste based on deep learning model. *Waste Manage. (Oxford)*, *174*, 462–475, 2024b.
12. Lin, J.Y., Liu, H.T., Zhang, J., Recent advances in the application of machine learning methods to improve the identification of microplastics in the environment. *Chemosphere*, 307, 136092, 2022.
13. Olawade, D.B., *et al.*, Smart waste management: A paradigm shift enabled by artificial Intelligence. *Waste Manage. Bull.*, *2*, 244–263, 2024.
14. Peng, J., *et al.*, Machine learning and deep learning modeling and simulation for predicting PM2.5 concentrations. *Chemosphere*, *308*, 136353, 2022.
15. Pereira, S., *et al.*, Development and assessment of artificial neural network models for direct normal solar irradiance forecasting using operational numerical weather prediction data. *Energy AI*, *15*, 100314, 2024.
16. Pilechi, A., Mohammadian, A., Murphy, E., A numerical framework for modeling fate and transport of microplastics in inland and coastal waters. *Mar. Pollut. Bull.*, 184, 114119, 2022.
17. Pitakaso, R., *et al.*, Optimization-driven artificial intelligence-enhanced municipal waste classification system for disaster waste management. *Eng. Appl. Artif. Intell.*, *133*, 108614, 2024.
18. Primpke, S., *et al.*, Critical assessment of analytical methods for the harmonized and cost-efficient analysis of microplastics. *Appl. Spectrosc.*, *74*, 1012–1047, 2020.

19. Sarker, M.A.B., Imtiaz, M.H., Holsen, T.M., Baki, A.B., Real-time detection of microplastics using an AI camera. *Sensors*, 24, 13, 4394, 2024.
20. Svanström, F., Alonso-Fernandez, F., Englund, C., Drone detection and tracking in real-time by fusion of different sensing modalities. *Drones*, *6*, 11, 317, 2022.
21. Šimić, I., *et al.*, Applying machine learning methods to better understand, model and estimate mass concentrations of traffic-related pollutants at a typical street canyon. *Environ. Pollut.*, *263*, 114587, 2020.
22. Sridhar, A., *et al.*, Extraction and detection methods of microplastics in food and marine systems: A critical review. *Chemosphere*, *286*, 131653, 2022.
23. Van Cauwenberghe, L., *et al.*, Microplastic pollution in deep-sea sediments. *Environ. Pollut.*, *182*, 495–499, 2013.
24. Veerasingam, S., *et al.*, Contributions of Fourier transform infrared spectroscopy in microplastic pollution research: A review. *Crit. Rev. Environ. Sci. Technol.*, *51*, 2681–2743, 2021.
25. Yu, F. and Hu, X., Machine learning may accelerate the recognition and control of microplastic pollution: Future prospects. *J. Hazard. Mater.*, *432*, 128730, 2022.
26. Yuan, S.-M., *et al.*, Artificial Intelligence and deep learning in sensors and applications. *Sensors*, *24*, 2024.
27. Rathore, P., Sustainable smart homes and buildings with internet of things. John Wiley & Sons, Inc., Scrivener Publishing LLC, 2025.
28. Zhao, W., *et al.*, Accurate prediction of soil heavy metal pollution using an improved machine learning method: A case study in the Pearl River Delta, China. *Environ. Sci. Technol.*, *57*, 17751–17761, 2023.
29. Zhu, Z., Parker, W., Wong, A., Leveraging deep learning for automatic recognition of microplastics (MPs) *via* focal plane array (FPA) micro-FT-IR imaging. *Environ. Pollut.*, 337, 122548, 2023.
30. Zhao, B., Richardson, R.E., You, F., Advancing microplastic analysis in the era of artificial Intelligence: From current applications to the promise of generative AI. *Nexus*, *1*, 4, 2024.
31. Chen, Y.N., Rani, A., Chiang, C.Y., Kim, H., Pan, S.Y., Monitoring, control and assessment of microplastics in bioenvironmental systems. *Environ. Technol. Innovation*, *32*, 103250, 2023.
32. Sinha, R.K., Kumar, R., Phartyal, S.S., Sharma, P., Interventions of citizen science for mitigation and management of plastic pollution: Understanding sustainable development goals, policies, and regulations. *Sci. Total Environ.*, 176621, 2024.
33. Kamyab, H., Khademi, T., Chelliapan, S., SaberiKamarposhti, M., Rezania, S., Yusuf, M., Ahn, Y., The latest innovative avenues for the utilization of artificial Intelligence and big data analytics in water resource management. *Results Eng.*, *20*, 101566, 2023.
34. Dennison, M.S., Paramasivam, S.K., Wanazusi, T., Sundarrajan, K.J., Erheyovwe, B.P., Marshal Williams, A.M., Addressing Plastic Waste

Challenges in Africa: The Potential of Pyrolysis for Waste-to-Energy Conversion. *Clean Technol.*, *7*, 1, 20, 2025.

35. Fang, C., Luo, Y., Naidu, R., Microplastics and nanoplastics analysis: options, imaging, advancements and challenges. *TrAC, Trends Anal. Chem.*, *166*, 117158, 2023.
36. Lavers, J.L., Fidler, A.L., Charlton-Howard, H., Anthropogenic pollution is widespread in Great Bowerbird bowers in northern Australia. *Microplast. Nanoplast.*, 5, 1, 27, 2025.
37. Ding, X., Zhang, B., Shen, C., Wang, R., Yin, S., Li, F., Xu, C., Are we underestimating the driving factors and potential risks of freshwater microplastics from *in situ* and in silico perspective? *Water Res.*, *281*, 123568, 2025.
38. Vijayshanthy, S., Priyanka, E.B., Thangavel, S., Anand, R., Bhavana, G.B., Khan, B., Jeyanthi, K., Performance of polyvinyl alcohol graphene oxide membrane for microplastic removal in wastewater with an IoT based monitoring approach. *Sci. Rep.*, *15*, 1, 20774, 2025.
39. Li, M., Zhao, Z., Zhao, Z., Li, M., Review of techniques for the detection, removal, and transformation of environmental microplastics and nanoplastics. *ACS Appl. Mater. Interfaces*, *17*, 14, 20560–20589, 2025.
40. Han, J., Hu, Q., Wang, Y., Rapid and accurate identification of Dendrobium species using FT-IR, FT-NIR, and data fusion with machine learning. *Ind. Crops Prod.*, *234*, 121637, 2025.
41. Li, X., Huang, Y., Guo, Y., Guo, H., Pu, M., Zhang, R., Luo, X., Sub-Diffraction Limited Single-Photon LiDAR with Optimized Confocal Super-Oscillatory Illumination. *Laser Photonics Rev.*, 2401837, 2025.
42. Xie, L., Ma, M., Ge, Q., Liu, Y., Zhang, L., Machine Learning Advancements and Strategies in Microplastic and Nanoplastic Detection. *Environ. Sci. Technol.*, *59*, 18, 8885–8899, 2025.
43. Li, Q., Wang, Z., Wang, M., Zhao, J., Tu, K., Lan, W., Pan, L., Next-Generation Optical Imaging and Spectroscopy: AI and Chemometrics in Assessing Authenticity, Nutrition, and Hazard Factors in Cereals. *Compr. Rev. Food Sci. Food Saf.*, *24*, 5, e70248, 2025.

7

Machine Learning for Microplastic Source and Pathway Prediction

Vanshika* **and Neetu Rani**

Department of Computer Science and Engineering, Chandigarh University, Gharuan, India

Abstract

Microplastic pollution has emerged as a pressing global concern, affecting not only marine ecosystems but also freshwater systems and even the atmosphere. Traditional methods used to trace the origins of microplastics and to predict their movement often fall short, largely because of the complex environmental interactions at play and the limitations of conventional sampling techniques. In this chapter, we explore how machine learning (ML) can transform the way researchers investigate the sources and pathways of microplastics. By drawing on diverse datasets—ranging from satellite observations and hydrological records to spectroscopic analyses of plastic particles—ML models such as Random Forests, Support Vector Machines, Convolutional Neural Networks, and Long Short-Term Memory networks are able to simulate dispersion patterns and identify pollution sources with far greater precision.

The chapter outlines a complete ML workflow, including data preprocessing, feature selection, model training, and validation, illustrated through real-world applications in riverine and coastal environments. It also highlights key challenges, including limited data availability, the difficulty of interpreting complex models, and ethical considerations around data use. Finally, we discuss future directions such as the integration of transfer learning, real-time monitoring through IoT devices, and the potential of federated learning approaches.

Overall, the chapter emphasizes how AI-driven environmental intelligence can support more informed decision-making, enable targeted interventions, and guide policy development aimed at reducing the spread of microplastic pollution.

**Corresponding author*: vanshika55691@gmail.com

Abhishek Kumar, Pooja Dixit, Pramod Singh Rathore, Arun Lal Srivastav and Ashutosh Kumar Dubey (eds.) Microplastic Monitoring Using Artificial Intelligence, (127–146) © 2026 Scrivener Publishing LLC

***Keywords*:** Machine learning, microplastic, random forest, geographic information systems

7.1 Introduction

7.1.1 Overview of Microplastic Pollution and Its Global Impact

Microplastics, usually defined as plastic pieces smaller than 5 millimeters, have quietly spread into nearly every corner of the planet. They are now found not only in rivers, lakes, and oceans but also in soils, the air we breathe, and even in some of the most remote places on Earth, such as polar ice sheets and deep-sea sediments [1, 2]. These particles originate from two main sources: some are intentionally produced in small sizes for use in cosmetics, industrial abrasives, or cleaning products (primary microplastics), while many more are formed when larger plastic items—bags, bottles, nets, or packaging break down under sunlight, friction, or chemical weathering (secondary microplastics). Everyday activities, such as washing synthetic clothes or driving vehicles that wear down rubber tires, add significantly to this invisible pollution. Figure 7.1 shows the overview of microplastic pollution and its global impact.

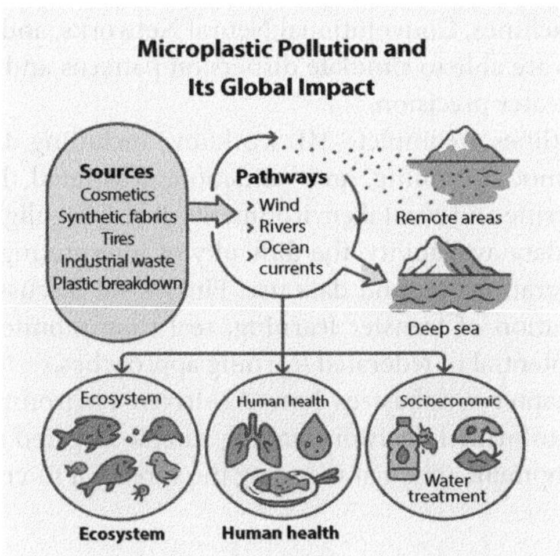

Figure 7.1 Overview of microplastic pollution and its global impact.

7.1.2 Limitations of Conventional Methods in Identifying Microplastic Sources and Tracking Their Dispersion

Many remote marine zones and developing regions lack the infrastructure for reliable monitoring, leaving large gaps in global knowledge. Even in areas where research has been carried out, scientists often rely on different sampling methods, size limits, or classification systems. This lack of standardization makes it difficult to compare results and prevents the development of a truly global picture of the problem. Current tools also struggle to detect nano plastic particles smaller than one micron which are especially concerning because they can penetrate cells and potentially affect human and animal health [4–8].

Finally, most existing techniques rely on collecting samples and analyzing them later in a laboratory. While this can identify what the particles are made of, it does not allow for real-time monitoring. As a result, the information often arrives too late to guide immediate action or preventive strategies. Taken together, these limitations mean that conventional methods give us only a partial view of the problem, making it difficult to accurately trace sources or design effective, large-scale solutions for tackling microplastic pollution.

7.1.3 The Case for Using Machine Learning in Environmental Studies

With these challenges, integrating machine learning (ML) into environmental monitoring is a paradigm shift. ML models are better at recognizing subtle patterns and correlations in large, high-dimensional datasets that would be exhaustive to examine manually.

In predicting microplastic source and pathway, ML facilitates the employment of multi-modal data sources e.g., satellite imagery, climate inputs, ocean current models, and spectroscopic fingerprints to construct predictive and adaptive models [3]. Supervised learning methods are able to distinguish between microplastic sources with respect to environmental characteristics, and unsupervised methods can show hidden patterns within particle dispersion. Deep learning methods, such as Convolutional Neural Networks (CNNs) and Recurrent Neural Networks (RNNs), have proven to be capable in image-based plastic detection tasks and time-series pathway prediction tasks [9–14]. These approaches not only augment accuracy but enhance scalability as well, enabling real-time updates as well as expansion of geographic location. In addition, ML-based systems facilitate proactive management of the environment through the capacity for early detection of pollution hotspots, prediction of future situations, and data-driven policy-making. Machine learning is hence a robust framework for surmounting the limitations of data and modeling incorporated in the conventional methods and a gateway to intelligent and sustainable environmental management.

7.2 Microplastic Sources and Pathways: An Overview

7.2.1 Classifying Microplastic Sources Into Primary and Secondary

Microplastic pollution is caused by a variety of human activities and can generally be classified into primary and secondary sources depending on their origin and manner of release into the environment.

- Primary microplastics are manufactured in microscopic form for specific commercial or industrial purposes. These include:
  - Microbeads in personal care products (e.g., exfoliants in face washes, toothpaste)
  - Pre-production plastic pellets (nurdles)
  - Microspheres used in medical or industrial processes
  - Synthetic fibers shed from fabrics during washing
- Secondary microplastics result from the breakdown and fragmentation of larger plastic particles due to natural, chemical, or physical processes. Common sources include:
  - Degradation of plastic bags, bottles, and packaging materials over time
  - Tire wear and tear particles from moving vehicles
  - Degrading marine waste and discarded fishing gear in underwater environments
  - Paint chips from buildings, roads, and boats

7.2.2 Main Pathways of Microplastic Movement: Rivers, Runoff, Currents, and Air

- Once released into the environment, microplastics follow complex transport pathways influenced by a range of natural and human-induced factors. These pathways determine where and how microplastics accumulate, posing environmental and health risks.

- **Rivers and Streams:** Serving as primary conduits, rivers convey microplastics from inland origins—such as domestic households, manufacturing areas, and dumps—into major water courses. Urban streams, especially, show elevated microplastic content due to untreated sewage, stormwater, and solid waste leakage.
- **Urban Runoff:** During rainfall, microplastics from roads, rooftops, construction sites, and open dumps are washed into stormwater drains. The lack of filtration systems in urban drainage infrastructure often results in direct discharge into nearby water bodies.
- **Ocean Currents:** Once in the ocean, global-scale currents (e.g., Gulf Stream, Kuroshio Current) transport microplastics across vast distances. Over time, they accumulate in convergence zones such as the Great Pacific Garbage Patch.
- **Atmospheric Transport:** Research shows microplastics can be carried by wind. Lightweight particles from urban and industrial sources can become airborne and settle in remote regions, including mountainous and polar areas [15–20].

These diverse pathways make it difficult to trace microplastic particles back to their origin, emphasizing the need for data-driven, model-based approaches such as those enabled by machine learning shown in Figure 7.2 and Figure 7.3.

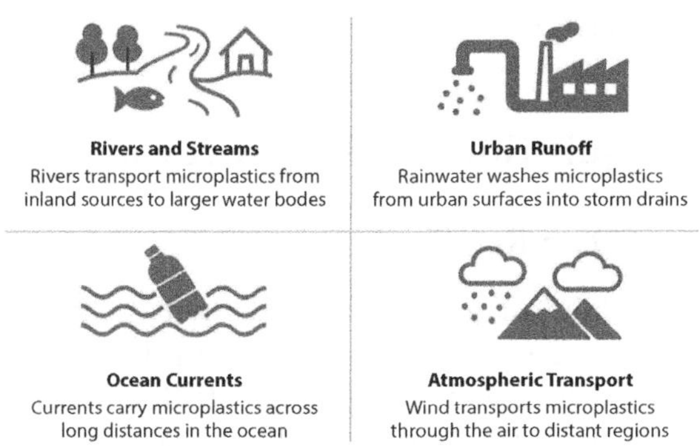

Figure 7.2 Pathways of microplastic movement.

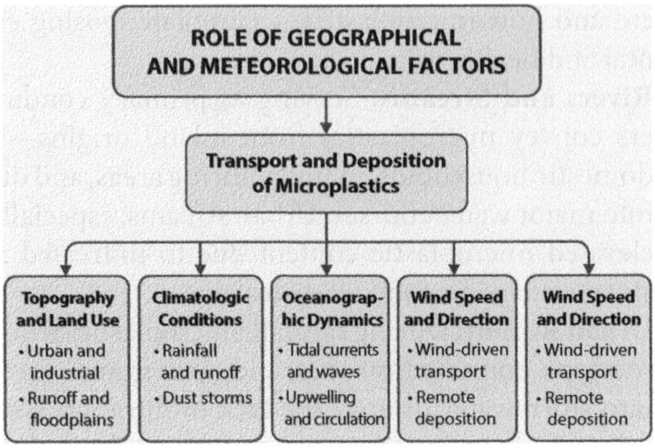

Figure 7.3 Role of geographical and meteorological factors in microplastic.

7.2.3 Impact of Location and Climate on Microplastic Spread

The transport and deposition of microplastics are greatly determined by geographical position and local meteorological factors. Urban and industrial areas along coastlines or rivers have increased microplastic discharges. Hilly or mountainous landscapes can increase runoff, while flat floodplains may retain particles in sediments.

The frequency and intensity of rainfall control the amount of urban and agricultural runoff. In arid regions, dust storms can act as carriers for microplastic particles.

Tidal currents, wave activity, and upwelling zones influence plastic distribution in marine environments. Thermohaline circulation affects both vertical and horizontal transport of microplastics.

Wind-driven transport of plastic fibers and dust can expand the reach of microplastic dispersal, sometimes resulting in unexpected deposition in remote or pristine environments [21–25].

7.3 Data Acquisition and Preprocessing

7.3.1 Types of Data Required

For effective machine learning-based prediction of microplastic sources and pathways, it is essential to gather multi-dimensional and high-resolution data that accurately represents both the plastic particle characteristics and the environmental conditions influencing their movement.

7.3.2 Data Sources

Multiple sources and technologies are leveraged to acquire the diverse datasets required for predictive modeling.

7.3.3 Challenges in Data Collection, Quality Control, and Labelling for Machine Learning

Although environmental datasets are becoming more widely available, building reliable machine learning resources for microplastic research remains a major challenge. Most of the difficulties stem from data collection, quality control, and labeling practices. For precise labeling (such as FTIR confirmation), polymer identification frequently necessitates domain expertise. Training ML models is skewed by the under representation of rare plastic kinds or low-concentration areas. Table 7.1 describe the different data sources and their roles in microplastic studies and Table 7.2 shows the sources of data used in microplastic research.

Table 7.1 Different data sources and their roles in microplastic studies.

Type of data	Example parameters	Purpose
Microplastic Particle Data	Particle size, shape (e.g., fiber, fragment, bead), polymer type (e.g., PET, HDPE), color	Source classification and material profiling
Geolocation Data	Latitude and longitude of sampling sites, elevation/altitude	Spatial mapping and transport modeling
Hydrological Data	Water flow rate, river discharge, sea surface currents	Pathway tracking in rivers and oceans
Meteorological Data	Wind speed and direction, rainfall and humidity, temperature	Aerial transport and dispersion modeling
Temporal Data	Time and date of sample collection, seasonal variations	Time-series modeling of pollution trends

Table 7.2 Sources of data used in microplastic research.

Source type	Description	Examples
Field Samples	Physical collection of water, sediment, or air samples. Manual analysis *via* FTIR/ Raman Techniques and Microscopy Approaches	Academic field campaigns, government theories.
Remote Sensing (Satellite)	Provides geospatial data on land use, sea surface temperature, ocean currents, and rainfall.	NASA MODIS, Copernicus Sentinel-2, Landsat
Environmental Sensors	Real-time sensing of wind speed, river flow, turbidity, rainfall, etc.	IoT-based buoys, weather stations, flow meters
Crowdsourced Data	Citizen science platforms for logging plastic debris and waste accumulation points.	Marine Debris Tracker, OpenLitterMap
Open Databases	Aggregated environmental and microplastic-related datasets for research.	PANGAEA, Litterbase, Global Plastic Watch, Marine Debris Program (NOAA)

7.4 Machine Learning Approaches for Microplastic Modeling

7.4.1 Supervised Learning

Predicting and classifying the characteristics of microplastics requires supervised learning. Support Vector Machines (SVM), Decision Trees, and Random Forests can be used in classification tasks to categorize different kinds of plastics based on characteristics including origin, particle shape, and polymer type.

Table 7.3 Data types, sources, and challenges.

Data type	Source	Common challenges
Particle Characteristics	Field sampling, microscopy	Contamination, manual effort, low frequency
Geolocation & Temporal	GPS, field logs	Missing values, inconsistency
Hydrological & Meteorological	Satellite, sensors	Real-time reliability, sensor failure
Satellite Imagery	MODIS, Sentinel, Landsat	Resolution limitations, cloud cover
Open/Secondary Databases	NOAA, Litterbase, PANGAEA	Format mismatch, lack of metadata

A classifier that can reliably identify the kind of polymer can be trained using labeled FTIR or Raman spectrum data, for instance. Supervised algorithms are employed in regression models to estimate geographical distribution patterns using environmental factors including water flow rate, rainfall, distance to metropolitan areas, and industrial activity, or to predict the quantity or concentration of microplastics in a given locale. Table 7.3 shows the data types, sources, and challenges.

7.4.2 Unsupervised Learning

Unlabeled environmental data can reveal underlying patterns and structures, making unsupervised learning worthwhile. Using clustering techniques such as K-Means, DBSCAN, or Hierarchical Clustering, microplastic samples can be grouped based on spectral similarity or spatial-temporal characteristics. This may be used to distinguish urban and rural inputs, find possible contamination sources, or create previously unidentified microplastic classes. Additionally, unsupervised learning may identify common paths in areas lacking ground-truth data, making it a crucial technique in data-sparse settings.

7.4.3 Deep Learning

Visual data microplastic detection has been transformed by Convolutional Neural Networks (CNNs). CNNs are capable to accurately detecting, segmenting, and classifying microplastic particles with little user assistance

Table 7.4 Comparison of machine learning approaches, techniques, and applications in microplastic analysis.

Approach	Technique	Application	Input data type	Advantages	Limitations
Supervised Learning	Random Forest, SVM, XGBoost	- Classify plastic types - Identify source categories	Labeled spectroscopic data, polymer types, GPS coordinates	High accuracy, interpretable models	Requires labeled data, sensitive to class imbalance
	Linear/Logistic Regression	- Estimate plastic concentration - Predict spatial load	Particle counts, flow rate, rainfall, urban proximity	Easy to interpret, low computational cost	Linear assumptions, limited complexity
Unsupervised Learning	K-Means, DBSCAN, Hierarchical Clustering	- Cluster pathways - Discover source patterns without labels	Unlabeled spatial, temporal, and material data	Finds hidden patterns, useful for exploration	No ground truth to validate, sensitive to parameters
Deep Learning	CNN (Convolutional Neural Network)	- Image-based plastic detection - Visual classification	Microscopy images, drone/satellite imagery	High accuracy in image recognition	Requires large labeled image datasets
	RNN/LSTM (Recurrent Neural Network)	- Time-series modeling of pathways - Seasonal flux prediction	Temporal ocean current, wind speed, rainfall sequences	Captures time-dependent relationships	Training can be slow, may overfit

when using high-resolution microscopic pictures or drone-acquired coastal footage. Re- current neural networks (RNNs), in particular Long Short-Term Memory (LSTM) networks, are used to describe transport dynamics across time [26–31]. These models forecast the probable future course of microplastics or seasonal fluxes by capturing temporal dependencies in rainfall, wind patterns, and ocean current data shown in Table 7.4.

7.5 Model Development and Validation

Data partitioning, usually into training (70%), validation (15%), and test (15%) sets—is a crucial first step in developing robust machine learning models. This keeps the model from overfitting and guarantees that it generalizes properly.

After training a machine learning model, its performance is evaluated using several key metrics:

- **Accuracy:** The percentage of correct predictions made by the model.
- **Precision:** The proportion of true positive predictions among all predicted positives. This is especially important when classifying rare plastic types.
- **Recall:** The proportion of true positive predictions among all actual positives. A high recall is critical for reducing the number of undetected contaminants.
- **F1-Score:** The harmonic mean of precision and recall, calculated as:

$$F1 = 2 \times \frac{Precision \times Recall}{Precision + Recall}$$

- **ROC-AUC:** The Area Under the Receiver Operating Characteristic Curve. This metric measures how well the model can tell different classes apart when the decision threshold changes.

By determining which environmental or plastic qualities have the most effect on the result, feature importance analysis (e.g., utilizing SHAP values or permutation importance) aids in model interpretation.

7.6 Case Studies and Real-World Implementations

Example 1: Detecting Coastal Hotspots with CNNs
Using satellite and drone imagery, researchers trained Convolutional Neural Networks (CNNs) to distinguish microplastic signals from sediments and natural debris. This enabled focused beach cleanups rather than large-scale, resource-intensive initiatives by enabling the highly accurate mapping of coastal hotspots.

Example 2: Mapping River Connectivity with Graph Neural Networks
The movement of plastics through interconnected river systems was modeled using graph neural networks, or GNNs. The model identified which urban areas were most responsible for downstream plastic transport by examining flow rates, land-use data, and waste management infrastructure. At crucial junctures in the watershed, this realization aided policymakers in their intervention.

Example 3: Assessing Fisheries Risk with Bayesian Networks
Bayesian models combined information on fishing activity, microplastic concentration in waters, and seasonal feeding habits of fish. The analysis highlighted high-risk zones where fish were most likely to ingest plastics, guiding fisheries managers to adapt harvesting practices and reduce contaminated seafood reaching markets.

Example 4: Optimizing Cleanup Efforts with Reinforcement Learning
Reinforcement Learning (RL) was used to design smarter deployment strategies for cleanup boats and floating barriers. Trained on simulations of ocean drift patterns and past collection records, the RL system learned how to reduce costs while capturing more plastics before they reached fragile ecosystems.

7.7 Visualization and Decision Support

To convert intricate machine learning predictions into useful information that environmental scientists, decision-makers, and civic organizations can use, effective visualization and decision support tools are essential. While decision support frameworks assist in the creation of prompt responses, public policies, and awareness campaigns, visualization tools allow the understanding of geographical and temporal trenin the context of microplastic contamination describe in Table 7.5.

Table 7.5 Comparative summary of case studies on machine learning for microplastic prediction.

Case study	Approach used	Key inputs	Outcome
1. Tracking River Sources	Random Forest	Land use patterns, river flow data, water sample composition	Pinpointed textile discharge and industrial runoff as the main river-based microplastic sources.
2. Forecasting Coastal Pathways	LSTM (Long Short-Term Memory)	Ocean currents, wind direction, seasonal temperature variations	Predicted where plastics would accumulate along the coast weeks ahead, enabling early cleanup action.
3. Estimating Urban Plastic Loads	Ensemble (Gradient Boosting + Neural Networks)	Rainfall levels, road traffic, fiber concentration in stormwater	Produced accurate estimates of plastic entering city drains, supporting better drainage and pollution control.
4. Identifying Coastal Hotspots	Convolutional Neural Network (CNN)	Satellite images, drone-based hyperspectral data	Detected high-density coastal plastic zones, helping plan focused and efficient beach cleanups.

(Continued)

Table 7.5 Comparative Summary of case studies on machine learning for microplastic prediction. (*Continued*)

Case study	Approach used	Key inputs	Outcome
5. Mapping River Connectivity	Graph Neural Network (GNN)	River flow rates, surrounding land use, waste infrastructure	Showed how specific towns and cities were contributing most to downstream pollution, guiding watershed interventions.
6. Assessing Risks in Fisheries	Bayesian Network	Fishing activity data, waterborne plastic levels, fish feeding cycles	Highlighted fishing areas with the highest likelihood of plastic ingestion, reducing seafood contamination risks.
7. Optimizing Cleanup Efforts	Reinforcement Learning (RL)	Ocean drift models, past cleanup records	Recommended smarter schedules and locations for cleanup operations, cutting costs while improving efficiency.

7.7.1 Role of Visualization in Microplastic Prediction

Large amounts of output data, frequently in high-dimensional formats like time-series trends, plastic concentration heatmaps, or geographical distributions, are produced by machine learning algorithms. To facilitate interpretation and real-time monitoring, visualization tools assist in condensing this data into easily understood representations.

- **Geographic Heatmaps:** Projected concentrations of microplastics can be overlaid on geographic maps using GIS (Geographic Information Systems) to identify pollution hotspots across rivers, coastal areas, or urban outflows. These maps can also be animated over time to visualize seasonal variations and pollution dispersion trends.
- **Time-Series Plots:** Line graphs and area charts serve as temporal visualizations that help detect pollution peaks associated with rainfall events, industrial cycles, or seasonal tourism. These are especially useful for models like LSTMs that forecast plastic accumulation over time.
- **Flow Simulation and 3D Modeling:** In complex systems, the pathways of microplastics through rivers or ocean currents can be visualized in 3D, illustrating both vertical and horizontal particle movement. This technique is valuable for marine research and hydro-engineering interventions.

7.7.2 Role of GIS in Data Integration and Monitoring

Geographic Information Systems (GIS) are invaluable for bringing together and visualizing environmental data. Platforms such as QGIS and ArcGIS can be linked with machine learning outputs to provide deeper insights into microplastic pollution. Through this integration, researchers can:

- **Trace pollution sources:** Use backward trajectory analysis to identify where microplastics originate.
- **Predict future hotspots:** Apply forward-dispersion modeling with environmental and hydrodynamic data to forecast where microplastics are likely to accumulate.
- **Combine multiple data layers:** Incorporate information on rainfall, industrial activity, and land use to capture the complex factors that contribute to plastic pollution.

In addition, real-time sensor data such as from plastic detection buoys, rainfall sensors, and river flow meters can be fed directly into machine learning dashboards. This allows for continuous updates to predictions and visualizations, enhancing responsiveness and decision-making.

7.7.3 Decision Support Systems and Their Role in Policy

Decision Support Systems (DSS) are vital tools for the governance of the environment. The main functions of these systems include:

Scenario Planning: DSS platforms can simulate the impact of various interventions such as enhanced filtration, reduced plastic use, or improved waste segregation campaigns on pollution levels.

- **Early Warning Systems:** By identifying areas at high risk for plastic accumulation, Decision Support Systems (DSS) enable authorities to organize cleanup efforts in time, preventing pollution from reaching dangerous levels.
- **Infrastructure Optimization:** DSS tools can guide the optimal placement of filters, booms, or treatment facilities by detecting microplastic concentration hotspots.
- **Public Communication:** Visual outputs from DSS can be transformed into infographics or news reports to raise public awareness and encourage behavioral change.

7.7.4 Multi-Stakeholder Impact and Use Cases

Decision-making tools and visualizations produced by machine learning models can be customized to satisfy the unique requirements of various stakeholders.

7.8 Challenges and Ethical Considerations

Investigate and identify the microplastic pollution using modern technologies are there with some ethical considerations and challenges shown in Figure 7.4. Table 7.6 shows the stakeholders and their respective use cases in microplastic monitoring and mitigation.

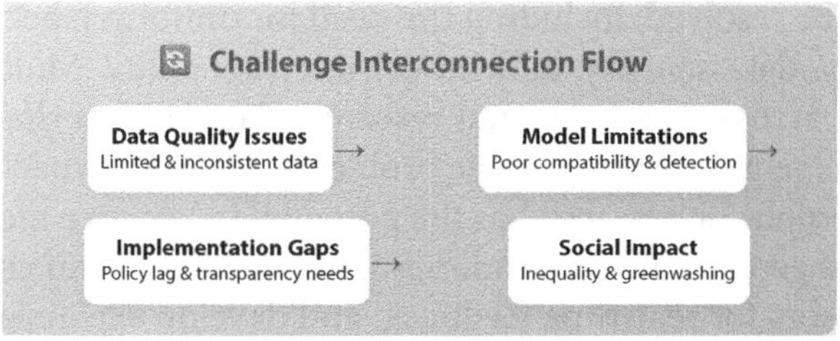

Figure 7.4 Interconnection challenge flow for microplastic pollution.

Table 7.6 Stakeholders and their respective use cases in microplastic monitoring and mitigation.

Stakeholder	Use case
Environmental Authorities	Focus on identifying priority areas for monitoring and cleanup
Policymakers	Draft informed regulations and legislation targeting plastic use and waste.
Urban Planner	Develop intelligent drainage systems and waste-capturing infrastructure.
Researchers	Identify new patterns in microplastic transport or emerging pollution sources.
NGOs and Media	Create awareness campaigns with visual impact and evidence-based arguments.

7.9 Conclusion and Future Scope

The pervasive danger of microplastic contamination calls for creative ways to identify sources and anticipate pathways with confidence. This chapter has demonstrated how machine learning (ML) is a game-changing way to use diverse datasets, from environmental characteristics to particle attributes, to model the behavior of microplastics at scale. Both supervised and unsupervised learning approaches, as well as deep learning algorithms like CNNs and LSTMs, enable time-series prediction, regression, and high-resolution categorization. These models provide real-time information, risk mapping, and scenario simulations to environmental authorities and policymakers when used in conjunction with GIS-based visualization and decision support tools. Despite the hopeful progress, a number of issues still need to be resolved, including the need for uniform labeling of microplastic kinds, data sparsity, and model interpretability. Multidisciplinary cooperation across environmental research, data engineering, and public policy will be required to overcome problems. The use of machine learning (ML) in microplastic research has the potential to improve model transferability across geographical boundaries, integrate data from citizen science and IoT sensors for real-time feedback, and develop explainable AI (XAI) architectures for transparency. While protecting sensitive environmental data, federated learning and cloud-based data sharing may improve accessibility and model performance even further. Machine learning effectively facilitates proactive decision-making toward a cleaner, more sustainable future in addition to offering a scalable, intelligent method for microplastic prediction. Together with open data ecosystems, its continued development has the potential to completely transform how we track and manage one of the most persistent pollutants in the world.

AI disclosure statement: The author acknowledges the use of AI-based tools for minor language editing or grammatical correction. However, all intellectual contributions, analyses, and conclusions presented in this chapter are entirely the author's own.

References

1. Herb, F., AI breakthrough in mapping sea pollution, The Australian, 2025, https://www.theaustralian.com.au/business/technology/ai-breakthrough-in-mapping-sea-pollution/news-story/6bbc00aed77d31a9e1a4368d-2ac61d3d.
2. Tewari, A., What's that microplastic? Advances in machine learning make identifying plastics in the environment more reliable, Phys.org, 2025, https://phys.org/news/2025-03-microplastic-advances-machine-plastics-environment.html.
3. Mohsen, S.M., Rácz, A., Kovács, I., Bíró, T., Tóth, V., Mapping and monitoring riverine litter using Sentinel-2 imagery and machine learning techniques: A case study of the Tisza River. *Environ. Sci. Pollut. Res.*, 2023, https://doi.org/10.1007/s11356-023-27068-0.
4. Li, Y., Wang, S., Zhao, Q., *et al.*, Multimodal deep learning for predicting microplastic aging in aquatic environments. *Environ. Sci. Technol.*, 2024, https://doi.org/10.1021/acs.est.4c05022.
5. Zhang, D., Li, L., Wang, T., *et al.*, Predicting microplastic abundance in surface oceans using ensemble learning models. *Environ. Pollut.*, 327, 121406, 2023, https://doi.org/10.1016/j.envpol.2023.121406.

6. Li, W., Fang, C., Tan, Y., *et al.*, Machine learning for emerging contaminant analysis: A review of current applications and future directions. *Crit. Rev. Environ. Sci. Technol.*, 2023, https://doi.org/10.1080/10643389.2023.2190313.
7. Astray, G., Losada-Barreiro, S., Mejuto, J.C., Simal-Gandara, J., Machine learning-based prediction of adsorption capacity of microplastics for environmental risk assessment. *Nanomaterials*, 13, 6, 1061, 2023, https://doi.org/10.3390/nano13061061.
8. Li, R., Wang, J., Wang, Y., *et al.*, Neural network prediction of sorption behaviors of microplastics: GRU, LSTM, RNN, and CNN approaches. *Chemosphere*, 2024, Available at: https://pubmed.ncbi.nlm.nih.gov/39675298/.
9. Wang, Q., Zhao, H., Wang, L., Chen, X., Machine learning models for adsorption prediction of microplastic-bound xenobiotics. *Environ. Res.*, 2023.
10. Zhao, Z., Liu, F., Chen, J., *et al.*, Machine learning prediction of enzymatic degradation of plastics *via* encoded protein and polymer features. *Environ. Sci. Technol. Lett.*, 10, 7, 546–553, 2023, https://doi.org/10.1021/acs.estlett.3c00293.
11. Chen, L., Feng, Y., Zhang, M., *et al.*, Predicting microplastic presence in peatlands using machine learning and environmental parameter analysis. *Environ. Pollut.*, 2023, Available at: https://pubmed.ncbi.nlm.nih.gov/37639864/.
12. Liu, X., Zhao, Y., Zhang, Y., *et al.*, A computer vision segmentation model for precise microplastic detection in field samples. *J. Hazard. Mater.*, 2024.
13. Wang, H., Li, X., Zhang, L., *et al.*, Pretreatment-free SERS and machine learning for fast microplastic identification. *Anal. Chem.*, 2024.
14. Zhang, Y., Wang, D., Liu, H., *et al.*, Microplastic identification using μ-Raman spectroscopy and supervised ML algorithms. *Chemosphere*, 2023.
15. Lang, M., Wang, T., Guo, L., *et al.*, Machine learning applications in emerging contaminant tracking: A comprehensive review. *Crit. Rev. Environ. Sci. Technol.*, 2023, https://doi.org/10.1080/10643389.2023.2190313.
16. Jayaprakash, V., Karri, S., Clark, R.A., Balasubramanian, R., Determination of trace organic con- taminant concentration *via* machine classification of surface-enhanced Raman spectra, arXiv preprint arXiv:2402.00197, 2024, https://arxiv.org/abs/2402.00197.
17. Rußwurm, M., Kaplanyan, A., Leonard, M., *et al.*, Large-scale detection of marine debris in coastal areas with Sentinel-2, arXiv preprint arXiv:2307.02465, 2023, https://arxiv.org/abs/2307.02465.
18. Yogarathinam, L.T., Alimoradi, H., Ahmad, A., *et al.*, Predicting micropollutant removal through nanopore-sized membranes using several machine-learning approaches based on feature engineering. *RSC Adv.*, 14, 27, 19331–19348, 2024, https://doi.org/10.1039/D4RA02475C.
19. Booth, H., Smith, R., Tran, T., *et al.*, High-precision density mapping of marine debris and floating plastics *via* satellite imagery, arXiv preprint arXiv:2210.05468, 2022, https://arxiv.org/abs/2210.05468.

20. Du, X., Fang, H., Lin, C., *et al.*, Marine microplastics and infant health, arXiv preprint arXiv:2410.17391, 2024, https://arxiv.org/abs/2410.17391.
21. Rathore, P.S., Ahuja, S., Burri, S.R., Khunteta, A., Baliyan, A., Kumar, A., *Deep learning techniques for automation and industrial applications*, John Wiley & Sons, Inc., USA, Scrivener Publishing LLC, Hoboken, New Jersey, USA, 2024a.
22. Solanki, S. and Rathore, P.S., Current E-Waste Management: An Exploratory Study on Managing E-Waste for Environmental Sustainability, in: *Sustainable Management of Electronic Waste*, pp. 187–200, 2024.
23. Kumar, A., Rathore, P.S., Dubey, A.K., Srivastav, A.L., Dutt, V., Ananth Kumar, T. (Eds.), *Sustainable management of electronic waste*, John Wiley & Sons, Hoboken, New Jersey, USA, 2024.
24. Guo, X. and Wang, J., Projecting the sorption capacity of heavy metal ions onto microplastics in global aquatic environments using artificial neural networks. *J. Hazard. Mater.*, 402, 123709, 2021, https://doi.org/10.1016/j.jhazmat.2020.123709.
25. Li, F., Zhang, H., Zhao, L., *et al.*, Identification and visualization of environmental microplastics by Raman imaging based on hyperspectral unmixing coupled machine learning. *J. Hazard. Mater.*, 133336, 2024, https://doi.org/10.1016/j.jhazmat.2023.133336.
26. Cui, S., Liu, X., Wang, Y., *et al.*, Advances and applications of machine learning and deep learning in environmental ecology and health. *Environ. Pollut.*, 122358, 2023, https://doi.org/10.1016/j.envpol.2023.122358.
27. Weber, F., Zinnen, A., Kerpen, J., Development of a machine learning-based method for the analysis of microplastics in environmental samples using µ-Raman spectroscopy. *Microplast. Nanoplast.*, 3, 1, 57, 2023, https://microplastics.springeropen.com/articles/10.1186/s43591-023-00057-3.
28. Osman, A.I., Hosny, M., Eltaweil, A.S., *et al.*, Microplastic sources, formation, toxicity and reme- diation: a review. *Environ. Chem. Lett.*, 21, 2129–2169, 2023, https://link.springer.com/article/10.1007/s10311-023-01593-3.
29. Zhu, J.-J., Yang, M., Ren, Z.J., Machine Learning in Environmental Research: Common Pitfalls and Best Practices. *Environ. Sci. Technol.*, 57, 46, 17671–17689, 2023, https://pubs.acs.org/doi/10.1021/acs.est.3c00026.
30. Uthra, K.T., Chitra, V., Damodharan, N., *et al.*, Microplastic emerging pollutants – impact on mi- crobiological diversity, diarrhea, antibiotic resistance, and bioremediation. *Environ. Sci.: Adv.*, 2, 1469–1487, 2023, https://pubs.rsc.org/en/content/articlelanding/2023/va/d3va00084b.
31. Gebejes, A., Kaynathare, B., Hrovat, B., *et al.*, Detection and identification of microplastics directly in water by hyperspectral imaging. *EPJ Web Conf.*, 287, 09002, 2023, https://www.epj-conferences.org/articles/epjconf/abs/2023/13/epjconf_eosam2023_09002/epjconf_eosam2023_09002.html.

8

Big Data Analytics in Mapping the Global Microplastic Distribution

Prasann Kumar

Department of Agronomy, School of Agriculture, Lovely Professional University, Phagwara, Punjab, India

Abstract

On account of their potential environmental impacts to health, the presence of microplastics in terrestrial, lacustrine, and marine environments has raised a great deal of concerns among researchers. The identification of the sources, transport, and the ecological impact of microplastics requires mapping them at a global scale. Big data analytics, such as advanced computational methods, machine learning, and geospatial technologies, is a potential approach to address this challenge. This chapter focuses on the use of big data approaches for mapping the global distribution of microplastics, from data collection and analysis to visualisation. It deals with large scale datasets including remote sensing, citizen sensing and IoT. It runs them through deep learning and statistical techniques to spot hotspots and trends. It also addresses the problems related to data standardisation, storage, and interoperability, and the vicissitudes to overcome these hindrances. Case studies of applied applications and future outlooks on the value of big data e-government in the aspect of global policies to reduce the impact. The influence of microplastic pollution is also elaborated. This is where your big data analysis comes into play for all researchers, policy makers and green lobbies to seize the opportunity for combating the microplastic disaster.

Keywords: Environment, microplastics, computational, geospatial, big data, ecological

Email: prasann0659@gmail.com

Abhishek Kumar, Pooja Dixit, Pramod Singh Rathore, Arun Lal Srivastav and Ashutosh Kumar Dubey (eds.) Microplastic Monitoring Using Artificial Intelligence, (147–178) © 2026 Scrivener Publishing LLC

8.1 Introduction

Microplastics, plastic particles <5 mm in size, have become one of the most omnipresent and insidious forms of pollution in the environment. Derived from various sources, including the breakdown of large plastic waste, industrial processes, textiles and personal care products, microplastics are found widely throughout the environment from the deepest ocean trenches to Arctic ice to terrestrial soils. New studies show that even the most far-reaching corners of the planet have been soiled by microplastics, misinformation that would have pleased Colleton's colonel, and highlighting how universally plastic pollution has spread and how long the contamination will endure [1–3]. The environmental and health consequences of microplastics are far-reaching. Marine ecosystems are places where numerous organisms ingest microplastics, from plankton to large mammals. But the corollary of this ingestion is that obstruction results, less food/nutrition, and higher levels of toxicity as plastics can absorb, accumulate environmental toxins. Ditto, study has indicated that humans 'recently reported exposure to microplastics by their seafood, drinking water and the breath they take has led to such concerns as inflammation cytotoxicity and endocrine disruption. In terms of detecting, quantifying, and solving or mitigating the microplastics crisis, it is complicated. One, microplastics, contrary to macroplastic, are not normally visible to the naked eye. They show a high diversity in composition, size, shape and origin of chemical structure and hence make monitoring and interventions very hard. Run these accounts in reverse: World oceans have recently seen trillions more microplastic particles coming into existence since very times indeed, while fresh sources continue to add to the stock of these particles. With microplastic pollution, a matter worthy of ever-increasing attention, the integrated policies and legislative arrangements to control this at global levels are still in their infancy; and this only underscores the urgency for scientific, technical, and policy-making approaches that will help resolve the situation accurately and thoroughly [4–6]. For a full picture of the distribution of microplastics, it is essential to collect accurate and informative maps of the area. Remote sensing data can yield the pollutant's spatial distribution and thereby identify those places where control measures might be appropriate. Localizing hotspots helps significantly in keeping marine areas clean of pollutants: chips that provide spatial distributions of pollutants are necessary for identifying pollution hotspots, transportation lines and collection points for focused cleanups and regulatory enforcement. Indeed, if mapping is done carefully, it can reveal the relationships that exist between microplastics and environmental parameters such as currents, winds, urban influence, and biological activity. These might help explain what happened to the microplastics and what its fate might have been. New developments in various mapping techniques have increased the opportunities to detect microplastics in different media. HL and IR Spectroscopy HL and IR spectroscopy can be used in identification and characterisation with Fourier-transform IR (FTIR) spectroscopy, Raman spectroscopy and hyperspectral imaging, making identification of microplastic particles precise and quantifiable [7–9]. Conversely, satellite remote sensing, while previously only capable of detecting macroplastic, is advancing to help map regions prone to high microplastic input, especially using field indicators such as river outflows and urban wastewater outputs. There is also an essential dimension of risk assessment related to mapping microplastic pollution. By mapping distribution data with ecological and socio-economic factors, scientists and policymakers can highlight those ocean spots where microplastics endanger biodiversity, human communities, and human activities, such as fisheries and tourism. These spatially resolved data are crucial in designing marine protected areas (MPAS), implementing local improvements in waste management, and developing legislation to reduce plastic emissions. There is also a need to develop comparable methodologies for microplastic mapping globally to track trends over time. Without repeat data, it is challenging to assess the impact of interventions, track changes over time, or model the future. As such, partnership and collaboration in international work, such as (UNEP) Their efforts in global monitoring frameworks are essential in unifying data collection and reference management. The complexity and scale of the microplastics problem require advanced analytical techniques for combining heterogeneous data types to obtain actionable information [10–12]. Big data analytics, including machine learning, artificial intelligence (AI), and advanced modelling methodologies, is seen as essential in combating microplastic pollution. Relevant databases and big data automation allow combining and analysing disparate datasets, such as environmental monitoring, citizen science observations, remote sensing products, laboratory measurements, and industrial production numbers. "Combining these different data streams and polymers allows researchers to start building high-resolution, real-time maps of microplastic distribution, monitor pollution trends and make predictions of where the next hotspot might occur based on environmental and human influences." "The application of machine learning is especially promising in microplastic research." In addition, CNNS are being used to automate the identification of microplastic particles from microscopic images, significantly decreasing the time and potential subjectivity in the identification process [13–15]. AI-enabled predictive models can predict the migration and degradation of microplastics under various climate and human-interest scenarios, and support proactive management and policy-making. Moreover, big data platforms facilitate large-scale meta-analyses of microplastics research, which help scientists uncover global patterns and identify gaps in knowledge. Mineable NLP tools are employed to access scientific literature and environmental reports for data points to contribute to the knowledge base without adding significant overhead in fieldwork. Citizen science projects, too, could benefit from big data applications. Mobile apps and online services that pull in user-generated observations of plastic litter can be connected to centralised databases, where AI algorithms scrutinise, process and present those observations. This democratisation of observation increases spatial coverage and temporal resolution and promotes public engagement and data awareness. On the policy and governance level, big data analytics can assist evidence-based decision-making. Interactive data dashboards and geospatial analytics tools can translate complex microplastic data into a user-friendly format for policymakers, who can respond to targeted regulations, manage resources, and measure compliance and impact [16–18]. Predictive capabilities, for example, can be used to optimise the location of waste interception systems in river basins or guide the timing and scale of beach cleanup operations. Ethical issues associated with data privacy, quality and access should be addressed as big data in environmental contexts is rolled out. Open access to microplastic data is promoted, data and reporting structures are standardised, and inter-disciplinary collaborations are encouraged, which is critical to leveraging the potential of big data analyses while also meeting the scientific and inclusive needs for the sustainability of the discipline. The microplastic problem is a complex global challenge with severe environmental, health, and socio-economic consequences. The capacity to map the spatial distribution of microplastics is essential to quantify exposure, identify hotspots and inform targeted interventions [19–21]. Utilising the capacity for big data analytics in the era of AI, machine learning, and invigorated data management systems opens new avenues to drive microplastic research, surveillance and more effective policy forward. As the world enters an era of technological innovation and international interaction, a data-informed model to mitigate microplastic contamination offers great hope to protecting human and environmental health in the Anthropocene. The co-occurrence network analysis shows that the intellectual structure of considerable data analytics research is centred on a closely connected technical core with divided applications contexts. Data analytics and big data emerge as central themes and hubs, aggregating concepts such as machine learning, IoT, cloud computing and Industry 4.0, and different application domains such as environmental monitoring, healthcare and others. This solid core symbolises the mature, multidisciplinary knowledge and foundational algorithms and infrastructures of big data analytics. In contrast, a separate human-centric cluster (red) highlights a parallel line of research. They also find "human", "adult" and "controlled study" to co-occur in this stream with the absence of such terms in the other streams, in combination with methods words, such as data analytic methods, showing explicit connection to the application of data analytics to social, behavioural, and biomedical questions [22–24]. Connections among red nodes are stronger, and learning tasks by red nodes are more specialised, suggesting that big data study involves purely technical aspects and those related to humans. In general, the network representation indicates that the big data analytics area is a well-integrated core based on technical-oriented work with some emerging sub cores in application-driven research. This network's dense, central region corresponds to established topics, while peripheral areas with a lower density will indicate specialised or emerging areas. The two colour-coded clusters symbolise a duality: that in which technology is pushed (algorithms, systems, IoT, etc.) and one in which the focus is on human and societal challenges (demographic research, health, social sciences). The present results align with bibliometric analyses, which reveal that a large part of data research is still grounded in computational approaches, and its interest is progressively spreading in fields such as health or the social sciences. Unsurprisingly, the co-occurrence network uncovers a highly organised intellectual landscape where central data/analytics ideas associate across various subarea boundaries that appeal to the advanced nature of fundamental methodologies and the increasingly pervasive role of big data in all fields (Figure 8.1). The historiography visualisation depicts an emerging network of core research papers between 2015 and 2025. Nodes with larger size indicate more influential papers, in which the years 2023–2025 have been an active period in the number of publications. The graph shows an apparent trend of increasing collaboration and citation in the latest authors, depicting a quickly growing and inter-networked research domain. The next job would also build upon the promising results of recent high-profile studies (Figure 8.1 and Figure 8.2).

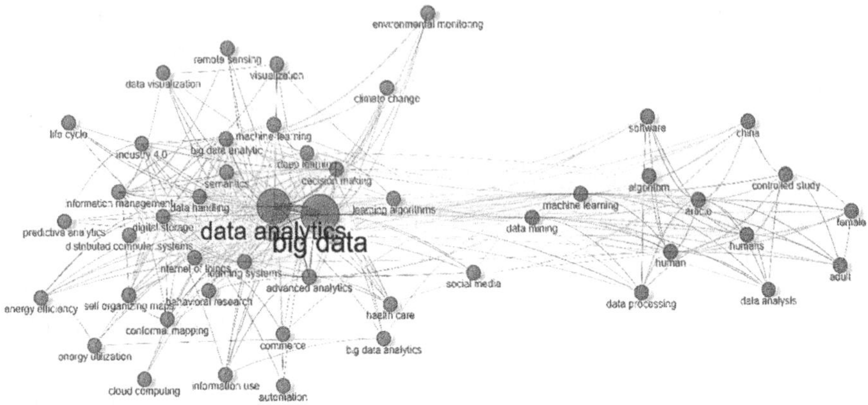

Figure 8.1 Increasing collaboration and citation.

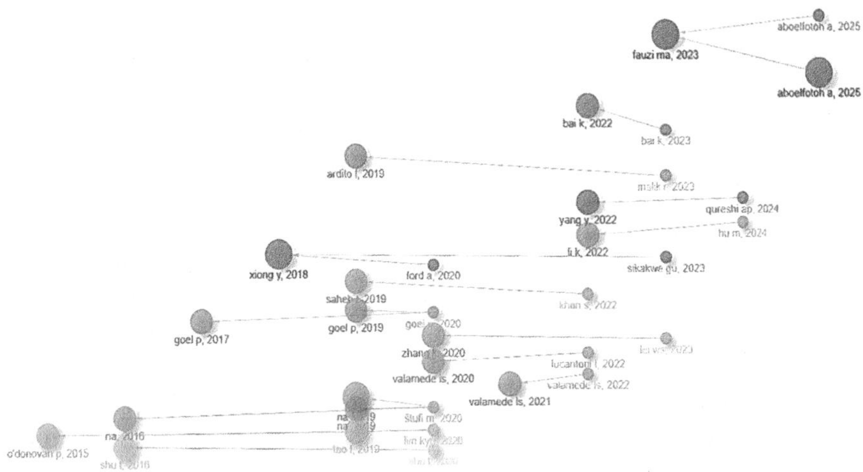

Figure 8.2 Historiography of big data analysis.

8.2 Data Sources for Microplastic Mapping

Remote sensing techniques have become essential for today's science, generating high-quality long-term and large-volume data. Using satellite-based sensors, drones flying overhead, and sophisticated imaging technology such as Lidar and hyperspectral sensors, remote sensing can track a wide range of data from agricultural sales and deforestation to urbanisation and water quality. These technologies are indispensable to overcoming traditional restrictions of ground-based measurement, through which high-resolution data can be collected over rugged terrains [8, 12, 19, 22]. In combination with the sophistication of image processing algorithms and machine learning approaches, increases in remote sensing have provided an increasingly powerful dataset upon which information can inform either real-time decision making or trend analysis over long periods. In addition to the remote sensing signal space, citizen science efforts have emerged as a transformative engine for democratising observationally-driven scientific discovery in recent years. In our area, many people who are not professional scientists but are members of the scientific community provide high quality information on a user-friendly platform with mobile apps, web pages and programme monitoring devices by local organisations. Projects that focus on monitoring of biodiversity and air quality gather data that are essential to have. Nonetheless, they are also bringing the public and science education closer together. Parallel to remote sense, such technology improves the spatial resolution and temporal resolution of large-scale datasets, and local verification is a really powerful tool for people's participation in science [9, 11, 15, 17, 21].

At the same time, the availability of Internet of Things (IoT)-connected sensors has transformed environmental and agricultural monitoring, infrastructure control and public health surveillance. Ubiquitously deployed in urban, rural, and wild geographies, IoT sensors constantly stream high-frequency fine-grain data on variables such as temperature, humidity, soil moisture, pollutants, and biological activity. These sensors are the foundation of innovative ecosystems, from which autonomous and real-time data acquisition and processing can occur [4, 11, 15, 18, 22]. The convergence of IoT data streaming and remote sensing imagery enables a multi-scalar, multi-dimensional understanding which supports predictive analytics, early warning systems, and adaptive management strategies. For example, in precision agriculture, IoT sensors can monitor plant-level microclimates and crop health measures, while remote sensing can provide a field or regional perspective, enabling better resource use and yield forecasting. The crossover of these technologies is strengthened through the merger of laboratory with field data, connecting controlled experimental studies with true environmental complexity. Laboratory studies supply essential calibration and validation data for sensor measurements in the field and remote sensing, thus establishing the accuracy and reliability of models and interpretations. Controlled laboratory experiments reveal basic biological, chemical and physical processes, which can be transferred to common conditions by calibrated and validated sensor networks and remote-sensing datasets.

Moreover, it is possible that integration in the field (IPC) of such methods as laboratory live streaming or photoelectric smoke detectors bring scale-dependent processes into a connection between molecular/organisms on western board. For instance, nutrient profiles determined in the laboratory help guide field measures of soil health from IoT instruments in China and remote sensing through digital spectral imaging. Source laboratory flume experiments enlighten us about sediment transport dynamics; this is then confirmed by local field measurements from river systems hydrology and validated using remote sensing around the world [8, 10, 13, 17, 19]. The results of these integrated studies offer rich, multi-layer data that are crucial in solving environmental issues, complex problems with multi-scalar and interdisciplinary viewpoints. Key to this integrated research environment is access to and use of public and private data from a variety of sources. On the one hand, massive public datasets such as those gathered by NASA's Earth Observing System Data and Information System (EOSDIS) or the European Space Agency's Copernicus program provide free access to petabytes of satellite imagery, atmospheric data, and climate forecasts. The datasets are important baselines and historical references for trend evaluation, model construction, and policy formulation. Similarly, private sector initiatives, like high-resolution imagery from Planet Labs or proprietary IoT sensor data from agritech companies, can provide valuable, often higher-frequency and finer-resolution data to complement public resources [8, 11, 15, 18, 20, 22]. By combining complementary public and private data sets in a Research Data Infrastructure, information can achieve higher levels of quality, resolute frequency and timing. Moreover, it can also encourage new paradigms in utilisation and generation (Federation of Earth Science Information Partners 2018). There are also challenges in terms of data interchangeability, standardisation, data protection and fair access that need to be tackled together. This can be solved through using shared frameworks for data, open and transparent metadata standards, and policies that favour Open Science and guarantee ethical data sharing. With the convergence of remote sensing technology, citizen science-based observations, sensor networks based on the Internet of Things, plus laboratory and field measurements--as well as voluminous public-private data sets, the whole process of scientific inquiry today changes its character as never before [16, 18]. This synergy enables us to comprehend complex systems better. And it shortens the time lag for transferring findings to practical answers for social problems. Whether that means dealing with tsunamis or droughts (and all possible future climate eventualities), managing our natural resources sustainably or taking proper care of public health, the

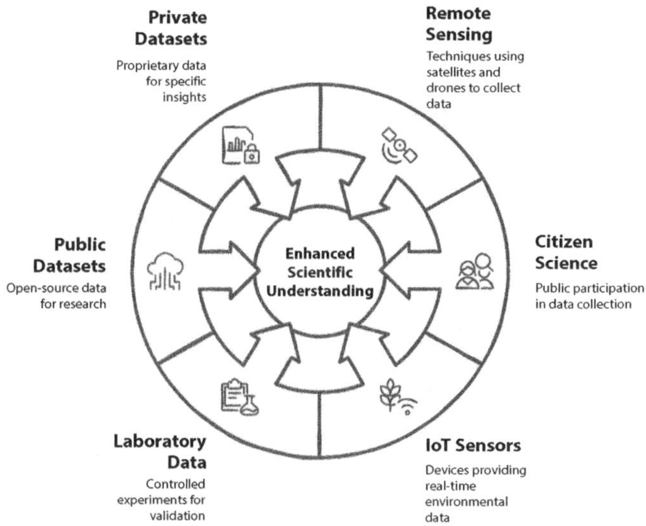

Figure 8.3 Convergence of technology for scientific advancement.

creation of future hyper-connected, participatory, responsive scientific ecosystems will be made possible by integrating AI and edge computing, in combination with verification by blockchain-based technology shown in Figure 8.3. This will not just be a tremendous leap forward for Earth observation systems and any they interface with; it may even usher in a new era, one where people can actively participate in international observations and follow up on them.

8.3 Big Data Techniques in Microplastic Analytics

Predictive modelling machine learning algorithms have impacted environmental science, agriculture, public health and urban planning by providing the ability to discern meaningful patterns and predictions from large and complex datasets. These algorithms (from the classic decision trees, support vector machines and ensemble models such as Random Forests to the more modern deep learning architectures such as Convolutional and Recurrent Neural Networks) enable researchers to predict your future by learning complex relationships between input features and your desired output (or target) variable. In agriculture, for example, predictive models help predict crop yields based on weather, soil, and management information [8, 11, 13, 17, 20]. Meanwhile, in environmental monitoring, they can forecast the air quality index or the propensity of a natural disaster, for example, a flood or a landslide. Yet the predictive abilities of these models rely heavily on both the quality and comprehensiveness of the databases they draw from. Geospatial analysis and GIS (geographical information system) technologies need to be introduced to enhance spatial realization (what does this mean, ion?) and conceptual understanding. While GIS systems permit the overlay of multiple data sets (such as satellite data, topography, land use maps, hydrological metrics, and socio-economic data), geospatial analysis allows researchers to measure the spatial relationships between objects, map their distribution, and understand changes over time - all of which are prerequisites for accurate prediction models [9, 12, 15, 18, 22]. In the field of epidemiology, for example, GIS-based models can identify the patterns of an infectious disease outbreak by taking into account geographical position cases, local conditions and demographic characteristics

In ecology, geospatial analysis can help model habitat suitability for endangered species or predict invasive plant distribution by combining remotely sensed data with ground-based measurements. In addition, the direct incorporation of GIS datasets into machine-learning models allows training models to be sensitive to spatial heterogeneities, which can be expected to enhance the robustness and localisation of predictions.

Despite being a black-box, the good point about machine learning is that it can be integrated with the geospatial aspect. Still, its performance is critically affected by the quality of data, which gives paramount importance to the data pre-processing and fusion methods. Raw data often has missing values, outliers, inconsistencies, and noise, which would significantly harm the performance of models without being correctly handled. Data cleaning is the process of finding and correcting these anomalies that our models will rely upon. Preprocessing processes, normalisation, standardisation, feature encoding and dimensionality reduction become necessary to optimise learning, mainly when data sets include variables with different scales, units or types. Furthermore, data fusion methods, which combine information from multiple heterogeneous sources such as satellite information, *in-situ* sensor networks, internet-based crowdsourced data, and historical databases, amplify the depth and breadth of the input information [14, 17, 19, 21, 23]. Fusing data effectively yields richer feature representations, better reflecting the complexity inherent to real-world phenomena and thus more subtle and transferable forecasts. Advanced fusion techniques, such as hierarchical integration and multi-resolution analysis, are increasingly used to reconcile spatial, temporal and thematic inconsistencies between datasets to construct models that better represent the complexity of the environmental, agricultural, and socio-economic systems.

With machine learning, geospatial analysis, and data preparation, statistical approaches for trending the analysis or data patterns and model prediction validation are crucial. Trend analysis employs the use of statistical methods to reveal patterns in data, especially from time series or spaced series data, to determine the underlying trend (in statistics and signal processing, an underlying trend is a smoother, traceable, cause of change over time, at a macroscopic scale) and can help explain whether data is an integrated vs. stationary process. Classical statistical approaches like linear regression, Mann-Kendall trend tests, and time-series decomposition are considered in general practice to identify and quantify trends, seasonality and cyclic patterns. In climate science, for instance, it may be possible to discern, from trend analysis, slow changes in temperature or precipitation patterns suggestive of a warming of the planet. Public health is also helpful in monitoring disease incidence or mortality changes due to interventions. Statistical trend analysis is also essential for eliciting predictive models' outputs, which provides a point of reference for developing machine learning predictions to compare the model's performance to the currently available level of knowledge. Integrated with machine learning, the trend analysis can help the model selection, hyperparameter tuning and feature engineering by identifying the most critical temporal/spatial features [9, 12, 15, 18, 20, 21].

Moreover, the collaboration between machine learning, geospatial technologies, data wrangling, and statistical trend modelling followed in predictive analytics allows for an end-to-end modelling of a given problem, encouraging educated decision making in various fields. For example, by merging high-resolution satellite imagery (pre-processed in GIS) with machine learning models and well-curated datasets, predictive models emerge that can optimise irrigation scheduling, fertilisation, and control pests at the field scale level, which in the end increase yield and resource efficiency. For urban planning, predictive models based on clean, integrated datasets validated with trend analysis can project urban sprawl, traffic congestion, and infrastructure requirements to develop sustainable cities [10, 15, 17, 20, 22]. In disaster risk management, machine learning models with geospatial dataset inputs and statistical trend analysis of historical disaster events, for example, can contribute to the forecasting of vulnerable zones and help mitigate disasters by potentially saving resources and, ultimately, save lives.

Recent developments in artificial intelligence also expand the bounds of what is achievable in predictive modelling using geospatial and statistical data. Approaches including transfer learning (methods through which pre-trained models on large-scale datasets are fine-tuned for specific tasks) and federated learning (strategies that enable models to be trained across decentralised datasets without access to raw data and sensitive information) are creating new frontiers for scalable and ethical data analysis. Additionally, explainable AI (XAI) models are of growing importance to open the "black box" of machine learning models to stakeholders to understand how predictions are made and what features are most predictive of an outcome. This level of transparency is fundamental when predictive models are used as decision aids for policymakers, regulators or resource managers in public health, agriculture, and disaster management [9, 13, 15, 17, 21].

Combining machine learning strategies for predictive modelling, geospatial analysis, and GIS technologies, combined with advanced data cleaning and preprocessing techniques, data merging, standardisation, inspection, and statistical methods for trend analysis, forms an appealing framework to interpret and forecast complex systems. All these ingredients reinforce each other: machine learning offers the predictive power; GIS adds spatial context; preprocessing the data provides quality and consistency; while statistical trend analysis offers interpretability or validation. Collectively, they underpin many of today's predictive analytics capabilities that are powering the needed innovations to solve the urgent problems of

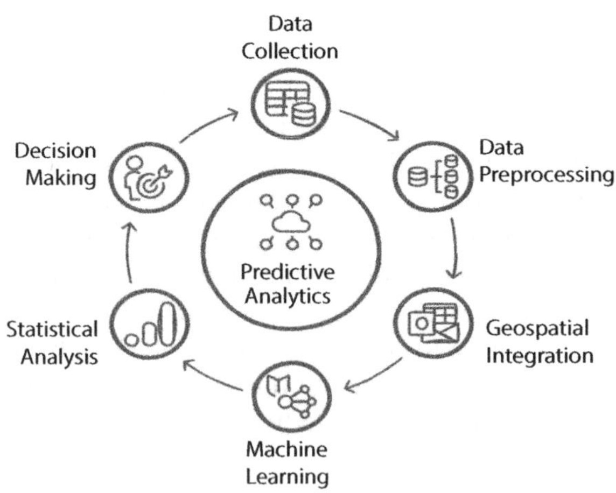

Figure 8.4 Cycle of predictive analytics.

our times, such as climate change, food security, urbanisation, and public health crises. As technology and methods continue to develop, integrative methods that leverage the strength of the three domains will be necessary to create resilient, adaptive and intelligent systems to cope with increasing complexity (Figure 8.4).

8.4 Challenges in Big Data for Microplastic Studies

The study of microplastics represent a rapidly developing conservation area crucial for elucidating environmental pollution and the associated challenges to human health, with the fast evolution of big data methodologies, needed to handle, analyse and interpret extensive and complex datasets deriving from sources such as ocean surveys, riverine stranding, atmospheric deposition, and biological tissues [12, 15, 18, 20, 23]. However, using big data in microplastics studies introduces several significant challenges, including data heterogeneity and the pressing requirement for standardisation. Microplastic data are intrinsically heterogeneous in particle types, polymer compositions, size distributions, morphologies, colourations and sources, acquired according to different sampling methods, extraction procedures, identification techniques (e.g., FTIR and Raman spectroscopy, pyrolysis-GC/MS), and units of measure. This heterogeneity seriously hinders comparative analyses, meta-analyses, and global assessments, as methodology-related differences may introduce biases and disparate findings. Integrating datasets to produce generalised findings or evaluate machine learning and statistical models in a reproducible manner is challenging without standard sampling, characterisation, and reporting protocols. Despite attempts such as establishing harmonised protocols by international organisations, the speed of scientific progress regarding microplastic research outpaces the ability to standardise across methodologies, creating a recurring void role hampering reproducibility, scalability, and complete harnessing of big data analytics [11, 15, 18, 20, 22].

Closely related to the heterogeneity problem is the issue of making platforms interoperate, which is becoming increasingly more critical as scientific collaborations become increasingly more international and interdisciplinary. In studies of microplastics, data are generated across different platforms, laboratory spectrometres, field-based optical systems, remote sensing, and citizen science projects, creating bespoke data structures, metadata conventions, and file formats, with varying levels of granularity. Shouldn't standard tools used by groups, self-managing their data together or in consortia (e.g. biomes, Personal Genome Project), be able to interact directly with popular data analysis platforms? The interoperability of these platforms is poor, and valuable datasets remain isolated, undermanaged, and/or incompatible with high-end analytical tools or cross-study comparisons. For instance, there could be a wealth of annotated particle-level metadata in one repository and almost none in another. Alternatively, the two repositories could use inconsistent taxonomies for polymer types or size ranges [8, 12, 16, 20]. In addition, the fusion of microplastic data with other environmental time series, such as ocean currents, sediment properties, or biota exposure histories, is impeded by disparities in the spatial resolution, temporal scales, and/or semantic mappings. Building "commons" of scientific data resources, like the work on FAIR (Findable, Accessible, Interoperable, Reusable) data principles, is essential but difficult, involving technical work to write APIS, create standard data models, and institutional work to harmonise and achieve consensus identification of shared vocabularies, ontologies, and standard and quality control practices. The full potential of integrative big data analytics, including complementary tools like network analyses, predictive modelling, and machine learning-based classification of microplastic hotspots, can currently not be exploited without enhanced interoperability.

In addition to the problems above, there are also the data storage and computational calculations required by necessity in big data-driven microplastic-related research. Cutting-edge analytical tools like hyperspectral imaging, high-resolution microscopy and sophisticated spectroscopic fingerprinting methods generate so much data for each sample - sometimes terabyte amounts when the whole region is used as an object or is browsing over an extended period. The storage of these expansive data sets is highly challenging in terms of security, accessibility, and cost, and it can be a huge logistical challenge for organisations with limited computational infrastructure or resources. Cloud computing offers scalable storage solutions, but uploading and downloading big datasets from the cloud is often time-consuming and costly, especially in regions with low internet bandwidth [9, 12, 14, 17, 19, 22]. Furthermore, the primary steps of processing and analysis of such datasets, such as image segmentation for identifying particles, spectral deconvolution to classify polymers, or multivariate statistical analysis to apportion sources, often require more power than is available on standard desktop computing, and may require either high performance computing (HPC) clusters, GPU acceleration, or distributed computing systems such as Apache Spark. The requirements of training machine learning models for automated microplastics detection, deep learning models in particular, add to the requirements for significant computation, optimised algorithms and efficient data pipelines. Likewise, to maintain robustness in data and traceability at all levels of analysis, sufficient data management systems and version control protocols and extensive documentation will have to be demanded, complicating and adding cost to big data in microplastic studies.

Besides, privacy and ethical aspects also play major roles in the context of big data eco-system of microplastic research, although the latter issues are often less well-utilised. While microplastic data, at first glance, would not look to be sensitive, related datasets that include human exposure studies, wildlife health investigations, or local ambient environmental monitoring, may contain sensitive information that require careful management. For instance, studies probing the degree of contamination of drinking water resources, agricultural soils or contents of seafood with microplastics can potentially exert a straight impact on the perception of public health, market soundness and regulatory measures, and that would be in terms of the fear of loss of confidentiality of data and the likelihood of causing public panic and also an ethical concern about communication [7, 14, 17, 18, 21, 22].

However, when data collection consists of human participants (e.g., sampling human blood, placenta or tissues from microplastic contamination studies), very stringent ethical protocols must be applied to protect informed consent, patient anonymisation, and data protection according to the European law, General Data Protection Regulation (GDPR). Even when it comes to citizen science projects designed to provide data relating to microplastic pollution, such as beach clean-up surveys and water samples, considerations of the personal information, the whereabouts and the observations of the participants must be dealt with transparently and ethically. As such, big data systems developed for microplastic research should integrate privacy-by-design protection, secure storage of well-controlled access regimes, and transparent governance structures for them to comply with ethical standards and maintain trust with the public.

Moreover, in the face of rapid technological progression, ethics guidelines and governance structures that should guide large data operations in microplastic studies frequently lag behind the pace at which technical solutions are implemented. As scientists rely on AI-based analytics and prediction models to estimate how the dynamics of microplastic pollution play out, concerns are raised over algorithmic transparency, any biases involved, and the societal consequences of prediction. For instance, predictive models to identify communities at higher risk of microplastic exposure can inadvertently stigmatise vulnerable populations if not handled sensitively. Institutional affiliation and financial interest in the microplastic detection devices and databases may lead to a conflict of interest that affects resource allocation and promotes closure of data, which may inhibit the development of science. Overcoming these ethical issues will depend on writing extensive policies on data governance and the continued dialogue across disciplines between scientists, ethicists, policymakers, and communities that may be impacted to guarantee that considerable data research on microplastics is carried out responsibly and fairly.

On the way to realising the transformative potential of big data in advancing microplastic research by enabling new scales of analysis, integration, and forecasting, the field faces such challenges as data heterogeneity and standardisation, platform interoperability [11, 14, 16, 18, 23], data management and computational needs, and privacy and ethical concerns. Meeting these challenges will require coordinated efforts to create methods, data interoperability infrastructures, computational resources, and standards for ethical governance templates. Only through cross-disciplinary and forward-looking initiatives like this can the scientific community capitalise on big data to unlock sources, pathways, impacts, and mitigation strategies for microplastic pollution and, thus, protect the environment and human health in a plasticised world (Figure 8.5).

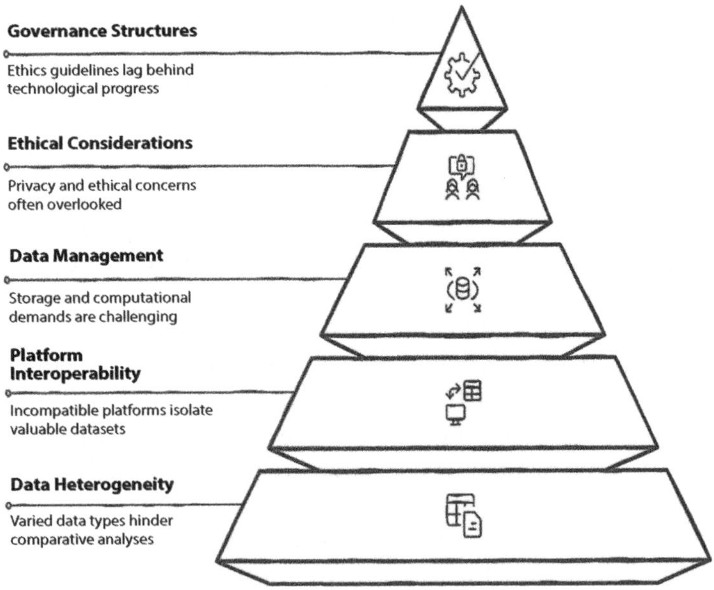

Figure 8.5 Microplastic research challenges.

8.5 Case Studies

Satellite-based big data is a new and exciting tool to help map marine microplastic distribution, with the potential for large-scale, real-time mapping of pollution concentrations, which is impossible to achieve using *in situ* sampling alone. Although satellites cannot directly detect single microplastic particles (they are small and widely dispersed), the indirect approaches provide auspicious results. Such data can enable the inference of likely concentrations of microplastics in broad regions of the ocean from remote sensing proxies, including ocean surface roughness, proxies for floating debris, ocean colour anomalies, and plastic and organic material aggregates, such as plastic-phytoplankton aggregates. These sensors, including MODIS and Sentinel-3 Ocean and Land Colour Instrument (OLCI), are sensitive to spectral signature changes of the sea surface associated with floating plastics and their pollution product, and can then be used to derive microplastic distribution models. Through integration with ocean current models, wind and sea surface temperature, these satellite-derived datasets (together with more established models of net transport processes) can be used to generate temporally and spatially dynamic, near-real-time maps of microplastic dispersion, accumulation areas and transport pathways, especially in regions of convergence such as gyres [8, 13, 16, 18, 19, 22]. But classifying microplastic-associated signals from natural objects such as seaweed blooms, organic flotsam, or suspended sediments still poses a great challenge, and thus, machine-learning classifiers and multisensor data fusion strategies are needed to enhance classification performance. Indeed, satellite data is uniquely able to provide most of the coverage required by the synoptic coverage necessary to monitor remote and large regions such as the open ocean, the Arctic and Southern Hemisphere gyres, where field sampling is virtually impossible and expensive.

Concurrently, predictive models of riverine microplastic transport are rapidly evolving. Scientists increasingly identify rivers as essential pathways to transport anthropogenic plastic litter from the terrestrial system to the marine environment. These models seek to reproduce the transport, retention and release of microplastics in river systems, considering hydrodynamics (flow velocity, turbulence, interactions with sediments) and the characteristics of the microplastics (density, size, distribution, morphology). Physical models, including advection-diffusion frameworks, coupled with stochastic particle tracking methods, enable predictions of the movement of microplastics from point sources, such as urban runoff, wastewater treatment plants and industrial releases through river networks. Further sophisticated approaches take environment factors, such as rainfall, seasonal flow, ecological impacts and geomorphic (e.g., sinuosity and floodplains, storage in dams) into consideration, and these can be critical for the survival and transport of microplastics [11, 15, 18, 19, 22]. Furthermore, riverine microplastic models linked to land use, population data, and economic activity data can support identification of hot spot source regions and high-risk industries [20]. In recent works such machine learning models like Random Forests, XGBoost or artificial neural networks are used to calibrate model predictions using training datasets collected from empirical river monitoring campaigns, and ensemble models are developed to increase the robustness of predictions, combining multiple forecasts. Importantly, these predictive models are not fixed but in fact are being developed as dynamic, data-inclusive systems that can be used to continually re-forecast when new observations (from satellite, ground survey, automated sensor, etc.) are fed into them. Outputs from such models are key to the development of strategies, such as target interventions, policies, and mitigation measures, that aim to reduce riverine plastic loads prior to being released into the marine environment.

Turning to these other perspectives, integrating the data from community science on hotspots is a powerful and inclusive technique for microplastic pollution mapping in riverine and marine systems. Citizen science projects allow lay people to monitor how much plastic pollution exists and share a data set that is far more widespread and through time than the professionals can achieve on their own. Programs such as the Marine Debris Tracker, Great Nurdle Hunt and the International Coastal Cleanup have demonstrated the effectiveness of channelling volunteers to record plastic on shorelines, along rivers and throughout lakes.

With standardised sampling protocols, mobile app-based data logging, and guidance on identifying microplastics (e.g. portable microscopes, visually exploring sorting techniques), citizens can help provide crucial information on the location of pollution hot spots, type of debris, and temporal trends. Incorporation of these citizen science-informed datasets in predictive models and satellite mapping systems demands rigorous validation and calibration to minimise sources of variation across data quality, observer bias and adherence to consistent efforts. However, properly edited, weighted, and cross-referenced to professional datasets, citizen science data may significantly improve the spatial resolution and detail of micro-plastic pollution maps, especially in areas insufficiently covered by official scientific monitoring. Additionally, participatory science broadens public outreach, consciousness, and policy advocacy to address microplastic pollution and engages the public in action [8, 12, 15, 16, 19, 21].

The seamless fusion between satellite-based indicators, riverine predictive models, and citizen science observations could lead to a global, multi-scale microplastic monitoring platform. This would allow detection of dynamic hot spot locations where microplastic concentrations vary due to seasonal river flow, storm events, ocean currents, and anthropogenic activity, as well as adaptive management in real time. For example, identifying an increase in microplastic efflux from a major river after periods of heavy precipitation may generate targeted clean-up activities, regulatory reactions or public warnings in the same downstream or neighbouring coastal areas. Integrating remotely sensed big data with hydrological modelling and ground-based data will facilitate cross-validation, raise confidence in hotspot identification, and provide further insights into source-sink pathways. Furthermore, sophisticated data assimilation algorithms like Kalman filtering or Bayesian inference approaches have the potential to combine independent data streams into consistent, predictive models, thereby facilitating a more accurate forecast and providing uncertainty estimates [10, 14, 16, 18, 22].

Though promising, there remain challenges. Poor spatial resolutions frequently limit satellite imagery compared with the scales of clustering of microplastics, while detection signals can be masked by cloud cover, water turbidity, or biofouling. Rigorous model predictions for riverine systems are hampered by a lack of empirical data on microplastic behaviour in freshwater environments, polymer type-dependent degradation rates, and the intricate interplay between waterborne physical, chemical, and biological processes that govern the fate of particles. Integrating citizen science data has to deal with heterogeneity in data quality, the geographical distribution of participation, and the risk of under-reporting in less accessible or politically sensitive areas. To address these challenges, there is a need to increase the satellite sensor capabilities, harmonise riverine microplastic sampling protocols, improve public training, and develop strong data verification, integration, and sharing mechanisms.

In the future, new technologies, including hyperspectral satellite imagery, drone-based remote sensing, AI-assisted predictive modelling, blockchain-secured citizen data submissions, and global microplastic observing networks, will revolutionise our abilities to map, model, and mitigate microplastic pollution. Crossing technological boundaries of remote sensing, hydrologic science, community engagement, and data science, the worldwide scientific community can begin to reach a more holistic view of the microplastic pollution landscape and provide insights to maximise benefits for conservation, restoration, and prevention of aquatic health and human safety (Figure 8.6).

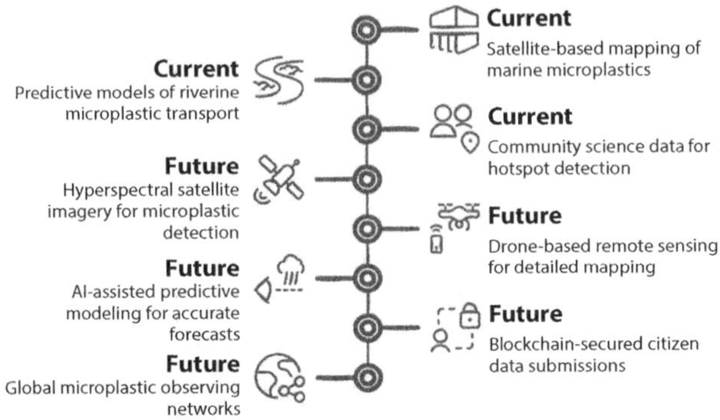

Figure 8.6 Mapping and modelling microplastic pollution.

8.6 Applications and Implications

The applications and implications of those new monitoring, modelling, and analysis tools for microplastics are wide-ranging and communicative, encompassing environmental management, public policy, and human health. Key uses include the recognition of global and regional hotspots for microplastic occurrence that are required for targeted interventions, efficient resource allocation and development of mitigations. Hotspot: An area characterised by higher concentrations of microplastics that result from hydrodynamic, anthropogenic or ecological factors. These places are typically urban areas near the coast, river mouths, accumulation regions (like the Great Pacific Garbage Patch), and semi-enclosed basins such as the Mediterranean and the Baltic [12, 15, 18, 20, 23]. Using satellite remote sensing, predictive hydrology models, ground-based sampling, and citizen science data, researchers can map the spatial distribution of microplastic pollution with progressively higher resolution and accuracy. This multi-modality approach enables the discovery of transient hotspots, e.g., arising after flood events, and persistent accumulation zones influenced by long-term oceanographic systems. In addition, addressing hotspots at the regional and local levels enables the development of local actions, specifically in prioritising beach cleanup activities, establishing river barriers, controlling industrial discharges, and planning urban stormwater changes. In river systems, identifying catchment-scale hotspots will allow for tracing plastic pollution sources to specific tributaries, urban runoffs, or industrial areas, providing the opportunity for targeted source reduction interventions. With the determination of hotspots, the search for solutions is initiated, which must be incorporated into transboundary plastic management, because plastic pollution is a global problem that frequently ignores political borders through ocean streams and river fluxes.

As a complement to hotspot identification, another important application is to give policy support for activities that reduce small plastic particles. In this context, data to support an evidence-based legislative and regulatory approach to minimise plastic pollution, including microplastics, is critical and in growing demand. Indeed, this kind of scientific research and the surveillance reports can become more and more necessary to justify policy initiatives, for example, the microbead ban in cosmetic products, the restrictions on the use of single-use plastics and the development of EPR (extended producer responsibility) schemes, at the national or international level. Integrated data sets and predictive models enable decision makers to evaluate the effectiveness of the current interventions better and develop agile policies that allow for short-term changes. For example, if a predictive model forecasts potential increases in microplastic loading in a watershed because of urban development, early control measures for plastic waste management could be imposed [8, 12, 15, 17, 18, 22]. Mapping MDR accumulation areas with satellite and land-based observations can also inform the establishment of conservation areas or no-discharge areas focused explicitly on mitigating the impacts of plastic pollution.

Furthermore, data on real-time and historical microplastic trends would allow international treaties and agreements, such as the UN's Sustainable Development Goals (with particular relevance for Measuring Goal 14: Life Below Water) and the ongoing Plastics Treaty negotiations, to specify quantitative objectives and timescales. Illustrative of these, and supportive to those (and others) efforts at the national and sub-national level, datasets can serve for basis to legislation in the locality, for local ordinances, zoning laws and public education campaigns, in conjunction with scientific knowledge of microplastics distribution and impacts, might push the development of industry-responsive solutions in packing, waste treatments, and degradable material production, and change in economic priorities towards environmental-friendly technologies.

Most importantly, monitoring of and models to monitor microplastic pollution are key to assessing ecological and human health risks. The widespread distribution of microplastics in aquatic and terrestrial habitats has led to increasing concerns about their biological effects, including the ingestion by a variety of organisms (ranging from zooplankton to whales), food web bioaccumulation, and chemical toxicity owing to adsorbed pollutants (e.g., persistent organic pollutants, POPS, heavy metals). Identifying hotspots allows researchers to compare and relate meaningless 29 concentrations of microplastics to ecological impacts (reduced reproductive success, 30 mechanical blockage, energy depletion, or inflammation) in marine organisms [7, 10, 13, 17, 19, 21, 23]. Defining high-risk areas enables ecologists to prioritise regions for biodiversity assessments, species vulnerability analyses and conservation actions. Finally, predictive models also provide crucial information on likely exposure scenarios that can be used as guidelines for laboratory and field studies, facilitating the design of, e.g., ecotoxicological experiments and determination of dose–response relationships. On the human health side, microplastics can now be found in seafood, drinking water, air, and even human tissue (lungs, placenta), highlighting the need for an.

Integration of EAD models with human exposure models provides estimates of potential intake rates through ingestion, inhalation, and dermal contact, from which can be derived quantitative risk assessment (QRA) and reference, safe exposure levels. Further, through hotspots, we can push forward the measurement of prevention of seafood in the crisis source and advocate the legal introduction of food safety, and the protection of the consumer as well.

The effects of these uses are not restricted to short-term issues of the environment and health, but broader social, economic, and ethical implications. Characterisation of microplastic hot spots brings to the fore questions about ecological justice, for example the effects on vulnerable populations such as coastal communities and small island developing states (SIDS), since these bear the brunt of plastic pollution despite contributing minimally to worldwide plastic output. They can make a case for remediation funding, infrastructural spending and global attention, if they have a map with data. For policy, the availability of high-resolution microplastic data can be leveraged to support transition to a circular economic model, promoting recycling, material reduction and sustainable consumption at a systemic scale. From a public health perspective, improved risk predictions contribute to health advisories and resilience planning which is particularly important for subsistence fishing communities and communities relying on untreated water [9, 12, 14, 17, 20, 23].

From a technical aspect, the method, which integrates multi-source data sets (satellite observation, model simulation, citizen science reports and *in situ* measurement) - a representative of inter-disciplinary, big-data based environmental; this integrated study could potentially be a step towards other pollution, such as oil spill, heavy metals, and emergent contaminants, etc. Morally, mankind should feel a sense of having the ability to feel, model, and predict where microplastic pollution exists that would require a responsibility to intervene for scientists, policy makers, industries, and civil society. Great, ever-increasing body of information like this make it impossible to deny how big an issue this is. Thus, the advancement of microplastic science is potentially transforming by generating the tools and ethics that are necessary to call for sustained change.

Despite such new opportunities, there still exist several challenges in realizing the potential of such applications. However, harmonisation of methods for microplastics monitoring remains incomplete, and can compromise the integrated analysis of data from different areas and sources. And there are also model's prediction uncertainties because of the less experimental data of microplastics' response to various types of the environment, and the lack of overall relationship between the different environmental factors which bring to the policy makers and policy interpreters to avoid their over confidence or misunderstanding [7, 11, 14, 17, 19, 22]. For example, the Southern Hemisphere, the deep ocean and isolated freshwater systems represents data deserts that will need to be addressed in building a full, global picture of microplastic pollution." Second, science-based policy action requires effective science communication to bridge between complex technical analysis and actionable and understandable messages for decision-makers and the public.

The identification of microplastic hotspots, policy support, evidence-based, and comprehensive risk assessment for ecological and human health all align as a robust, connected front in the onslaught of the microplastic threat. The maturity of detection, modelling and forecasting capabilities provides the opportunity for a more nuanced appreciation of the microplastic issue, and to position society to address it proactively, fairly and sustainably. The applications and consequences of this work are not predetermined; they will evolve as technology enables us to push forward our ability to ensure the health of the planet for our descendants (Figure 8.7).

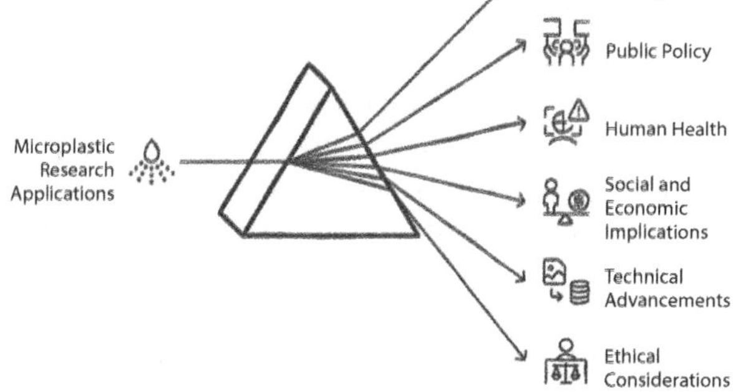

Figure 8.7 Unveiling the multifaceted impact of microplastic research.

8.7 Future Directions

The future of microplastic research might be transformed, with high-tech tools and planet-wide cooperation. A promising strategy is to develop and employ state-of-the-art artificial intelligence (AI) and machine learning (ML) techniques along with microplastic research. The traditional analysis techniques are rather labours and times procedures and generally poor in space and time resolution. With AI-based automated image analysis applied to microscopy, hyperspectral imaging and remote sensing data, the detection and classification of microplastics are expected to be pushed to the next level of throughput and precision. Machine learning (ML) approaches, deep learning architectures, in particular (DL) Convolutional Neural Networks (CNNs) have been used and. trained with high accuracy to distinct microplastic particles from particles of organic matter, sediments or another detritus even in the presence of environments complexity. Furthermore, machine learning algorithms, which make very limited use of data, could be used to model microplastics' dynamic movement, fragmentation potential, and accumulation under a variety of climate scenarios. They may also take into account actual weather, hydrological or oceanographic effects. Reinforcement learning methodologies can also play a role in refining sampling strategies, focusing efforts more on some areas likely to exhibit a high variability or high presence of microplastics. Furthermore, explainable AI (xAI) methods will enhance transparency and interpretability of model predictions, generating trust and credit for its forecasts from policy-makers, environmental stewards, and the scientific community. But microplastics, too, will also adopt the more sophisticated AI and ML, and we will address microplastics more holistically, at scale, and predictively with the problem of plastic pollution.

In addition to these developments, using improved sensor networks and introducing a real-time data acquisition system will offer a greater boost to microplastic monitoring practices. Sensitivity constraints, sampling biases, and practical barriers often limit current *in situ* microplastics detection approaches. Following generation sensor networks based on nano-engineered materials, optoelectronics, and biosensing, for example, may be able to measure an even wider range of microplastic sizes (including nano plastics) in aquatic and atmospheric environments. Distributed sensor arrays at key nodes, including river mouths, coastal zones, estuaries and atmospheric monitoring sites, could generate continuous, high-frequency data streams that detect chronic and episodic pollution releases. Combining IoT technologies with environmental monitoring sensors will facilitate remote control, automatic calibration, and cloud-based data collection, increasing spatial and temporal resolutions. Mobile sensing platforms like autonomous underwater vehicles (AUVS), Uncrewed Aerial Vehicles (UAVS), and satellite constellations will further extend the observational horizon of the *in-situ* observation. Integrating sensor networks with edge-computing will enable *in-situ* data processing at the gathering point, minimising the delay and enabling a near real-time decision support. This techno-ecosystem will change the landscape from intermittent, lengthy, costly sampling efforts to a reactive, fluid observatory that can map the spatiotemporal quirks of microplastic pollution in fine-scale metrics.

In addition to technological advances, developing international data sharing and microplastic global platforms will be necessary to maximise the potential of new datasets and models. Microplastic studies are frequently isolated due to different methodologies, reporting metrics, and data archives, leading to a lack of interoperability and comparability between studies and regions. Future directions must create standardised data formats, standard ontologies and open-access databases - enabling frictionless data integration, meta-analyses, and synthetic research. International efforts like the Global Ocean Observing System (GOOS) or the Group on Earth Observations (GEO), for example, may lend themselves to the inclusion of microplastic monitoring, establishing federated platforms for data collections from national programmes, academic explorations, citizen science, and industry-sponsored efforts. Blockchain techniques might provide innovative ways to guarantee good quality and data provenance while enabling them to be securely shared and safeguarding sensitive metadata. Additionally, promoting a culture of open science, *via* incentives to data citation, pre-competitive collaborations, and capacity-building to pave the way, will be essential for democratising knowledge and speeding up progress. Collaborative platforms might also provide computational resources, including cloud-based access to cutting-edge models, visualisation tools and AI-driven analytics pipelines, making advanced technology accessible to researchers and practitioners worldwide.

A key next step is incorporating microplastic studies into climate change models, recognising and investigating the intricate relationship between plastic pollution and global biogeochemical cycles. Historically, microplastics and climate change have been researched in isolation, but new evidence indicates that substantial connections justify an integrated approach to their analysis. Microplastics may also affect climate-relevant processes (e.g., reducing marine plankton productivity, decreasing carbon burial in sediments and modifying the albedo of snow and ice surfaces) through darkening. Furthermore, microplastics act as carriers of persistent organic pollutants (POPS) and other climate-forcing agents, interrupting contaminant pathways and cycles in ecosystems already impacted by climate change. Microplastic dynamics must be factored into Earth system models (ESM) and coupled ocean-atmosphere models to improve future environmental status and risk projections. For instance, better understanding the potential of increased microplastic loads to change biological pumps in the ocean may help reduce uncertainties of oceanic carbon sink capacities needed for climate mitigation. Scenario modelling could also assess the risks of synergistic impacts, e.g., the combined effects of increasing ocean acidification, temperature and microplastic pollution on coral reef resilience, fisheries production and coastal economies. By contrast, cross-disciplinary cooperation between researchers in the plastic pollution community, climate community, and oceanographic community will be required to work on the conceptual frameworks, parameterisations, and databases for integrated assessments of this kind.

These gutter shed lines of future research in combination emphasize a shift toward a more systematic, technology-enabled and internationally coordinated research and mitigation of microplastics. To achieve these ambitious goals, sustained investments in research and development are required in tandem with trans-disciplinary training programmes to prepare a generation of new researchers who are proficient in AI, sensor technology, environmental science, policy, and the institutional structures that facilitate long-term ecological monitoring and open data protocols. For this, it will be necessary to interact proactively with all industry stakeholders, governments, NGOs and the public in order to translate technological developments and the output of R&I into concrete policy actions, new practices and market changes. The ethical perspective on microplastic research cannot be neglected; increasing monitoring capabilities and data resolution reinforce the importance of designing a monitoring oversight protecting privacy, equality and transparency.

Finally, the future research pathway into microplastics is abundant in both opportunity and requirement. Through the application of AI and machine learning, the implementation of advanced sensor networks, the establishment of globally collaborative platforms and the incorporation of microplastic dynamics in climate change models, the scientific community can significantly enhance its capacity to observe, predict, and mitigate the impact of microplastic pollution. These innovations will improve stewardship of the environment and support a range of societal objectives such as protecting public health, promoting sustainable economic development, and building resilience to climate change shown in Figure 8.8.

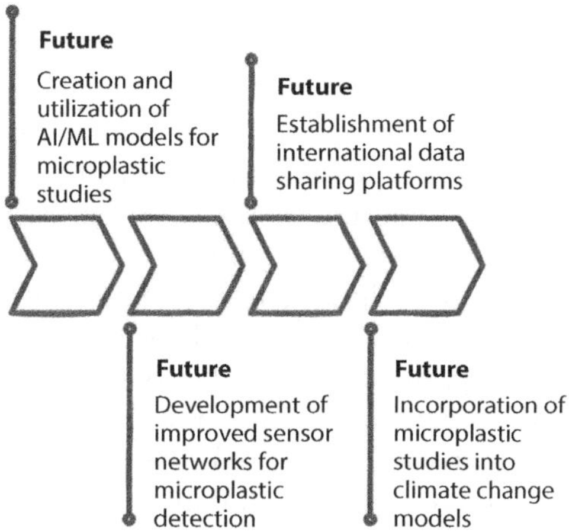

Figure 8.8 Advancing microplastic research: A future roadmap.

With a problem as vast and multi-faceted as plastic pollution, however, demand for "the perfect" solution means missing the amid the trees: a critical mass of ambitious, creative, committed scientists and policymakers will be required if we are to live up to our obligation to tackle this defining environmental challenge of our era.

8.8 Conclusion

Application of big data approaches in microplastic research delivered comprehensive understanding of sources, distribution routes, hot spots, and impacts of microplastics in marine and freshwater, terrestrial, as well as atmospheric ecosystems. Key findings underscore the role of machine learning algorithms, geospatial analysis, citizen science adoption into models, and predictive statistics in order to locate microplastic sources, predict movement, and estimate ecological and human health risks. Combining remote sensing techniques and *in situ* sampling has provided the first ever estimation of microplastic distribution at regional and global scales. River and ocean-borne transport predictive models have illustrated how hydrodynamics, land use dynamics, and climate forces interact to influence the destiny of microplastics. Additionally, coupling a range of data sources, including satellites, sensors, laboratory analyses, and citizen science programmes, has increased the resolution and robustness of maps of microplastic distribution, supporting a more informed decision-making and risk assessment process. However, the reports have also emphasised the continued difficulty in data heterogeneity, the lack of interoperability between platforms and the demand for standard methodologies and computational resources required to handle and analyse expanding datasets. However, despite these limitations, the possible transformation of microplastic pollution towards big data is enormous. The advent of big data technologies can allow the shift from small-scale studies to integrated predictive and action-oriented environment intelligence systems. When accompanied by improved data fusion and real-time surveillance, AI-led pattern detection and dynamic modelling, this approach allows for the detection of existing problems and the prediction of future risks in various socio-economic and environmental scenarios. This predictive capability is essential for anticipatory action, policy-making, and implementing local, regional, and global mitigations. Big data analytics also has the potential to democratise environmental science, making complex data accessible and usable in a wide range of contexts for a diverse number of actors, from policy-makers and industry to researchers and the public. There is an immediate need for interdisciplinary collaboration and innovation to realise this transformational potential fully. Clockwise attention on microplastic pollution would necessitate a melding of expertise across environmental science, computer science, engineering, social sciences, public health, and policy. It requires building and nurturing open networks sharing data and collaboratively co-developing transparent, interoperable, and inclusive tools, models and standards. Innovation needs to extend to sensor design, scalable data infrastructures, citizen science participation paradigms, and ethical architectures regarding data governance. Creating this collaborative and innovative ecosystem will also enable faster progression of data to information, information to action, and action to environmental and societal positive impact. Ultimately, resolving the intricate and ubiquitous problem of microplastic pollution in the age of big data will serve as a testament to the human race's collective power to innovate, be flexible and responsible, and guide the planet towards a more sustainable and resilient future.

8.9 Acknowledgement

The Author is thankful to Lovely Professional University for providing necessary support and encouragement during the work.

AI disclosure statement: The author acknowledges the use of AI-based tools for minor language editing or grammatical correction. However, all intellectual contributions, analyses, and conclusions presented in this chapter are entirely the author's own.

References

1. Nunes, B.Z., *et al.*, Microplastic contamination in seawater across global marine protected areas boundaries. *Environ. Pollut.*, 316, 120692, 2023, doi: https://doi.org/10.1016/j.envpol.2022.120692.
2. Kim, T., Cho, N.-H., Jang, S.-H., Kang, Y.-Y., Yoon, Y.-S., Yoo, H.-M., Emission characteristics analysis on microplastics by inorganic sludge discharged from recycling processes of agricultural waste vinyl in Korea. *J. Environ. Manage.*, 373, 123767, 2025, doi: https://doi.org/10.1016/j.jenvman.2024.123767.
3. Duan, Q., *et al.*, Nationwide meta-analysis of microplastic distribution and risk assessment in China's aquatic ecosystems, soils, and sediments. *J. Hazard. Mater.*, 477, 135331, 2024, doi: https://doi.org/10.1016/j.jhazmat.2024.135331.
4. Coskuner-Weber, O., Alpsoy, S., Yolcu, O., Teber, E., de Marco, A., Shumka, S., Metagenomics studies in aquaculture systems: Big data analysis, bioinformatics, machine learning and quantum computing. *Comput. Biol. Chem.*, 118, 108444, 2025, doi: https://doi.org/10.1016/j.compbiolchem.2025.108444.
5. Chamanee, G., Sewwandi, M., Wijesekara, H., Vithanage, M., Global perspective on microplastics in landfill leachate; Occurrence, abundance, characteristics, and environmental impact. *Waste Manag.*, 171, 10–25, 2023, doi: https://doi.org/10.1016/j.wasman.2023.08.011.
6. Kushwaha, M., *et al.*, Microplastics pollution in the marine environment: A review of sources, impacts and mitigation. *Mar. Pollut. Bull.*, 209, 117109, 2024, doi: https://doi.org/10.1016/j.marpolbul.2024.117109.
7. Zhang, Y., Pu, S., Lv, X., Gao, Y., Ge, L., Global trends and prospects in microplastics research: A bibliometric analysis. *J. Hazard. Mater.*, 400, 123110, 2020, doi: https://doi.org/10.1016/j.jhazmat.2020.123110.
8. Fang, C., Luo, Y., Naidu, R., Microplastics and nanoplastics analysis: Options, imaging, advancements and challenges. *TrAC, Trends Anal. Chem.*, 166, 117158, 2023, doi: https://doi.org/10.1016/j.trac.2023.117158.
9. Fang, C., Gopalan, S., Yu, J., Naidu, R., Unveiling microplastics from zippers: Characterisation and visualisation through Raman imaging analysis. *Sci. Total Environ.*, 904, 166235, 2023, doi: https://doi.org/10.1016/j.scitotenv.2023.166235.
10. Singh, J., Yadav, B.K., Schneidewind, U., Krause, S., Microplastics pollution in inland aquatic ecosystems of India with a global perspective on sources,

composition, and spatial distribution. *J. Hydrol. Reg. Stud.*, 53, 101798, 2024, doi: https://doi.org/10.1016/j.ejrh.2024.101798.

11. Peries, S.D., Sewwandi, M., Sandanayake, S., Kwon, H.-H., Vithanage, M., Airborne transboundary microplastics–A Swirl around the globe. *Environ. Pollut.*, 353, 124080, 2024, doi: https://doi.org/10.1016/j.envpol.2024.124080.
12. Munyaneza, J., *et al.*, A review of atmospheric microplastics pollution: In-depth sighting of sources, analytical methods, physiognomies, transport and risks. *Sci. Total Environ.*, 822, 153339, 2022, doi: https://doi.org/10.1016/j.scitotenv.2022.153339.
13. Li, Z. and Yuan, D., Global performance and trends of research on emerging contaminants in sewage sludge: A Bibliometric Analysis from 1990 to 2023. *Ecotoxicol. Environ. Saf.*, 281, 116597, 2024, doi: https://doi.org/10.1016/j.ecoenv.2024.116597.
14. Latwal, M., Arora, S., Murthy, K.S.R., Data driven AI (artificial intelligence) detection furnish economic pathways for microplastics. *J. Contam. Hydrol.*, 264, 104365, 2024, doi: https://doi.org/10.1016/j.jconhyd.2024.104365.
15. Lin, H.-T., *et al.*, Microplastics in Asian rivers: Geographical distribution, most detected types, and inconsistency in methodologies. *Environ. Pollut.*, 349, 123985, 2024, doi: https://doi.org/10.1016/j.envpol.2024.123985.
16. Herb, F., Boley, M., Fong, W.-K., Machine learning outperforms humans in microplastic characterization and reveals human labelling errors in FTIR data. *J. Hazard. Mater.*, 487, 136989, 2025, doi: https://doi.org/10.1016/j.jhazmat.2024.136989.
17. Dahal, A., *et al.*, Distribution and characteristics of microplastics in fluvial sediments from the Koshi River Basin, Nepal. *Environ. Pollut.*, 374, 126189, 2025, doi: https://doi.org/10.1016/j.envpol.2025.126189.
18. Dey, S., Sen, K., Saha, N.C., Saha, S., Analytical approaches for quantifying and characterizing microplastics: Environmental impacts and bioaccumulation in aquatic Systems. *Green Anal. Chem.*, 12, 100191, 2025, doi: https://doi.org/10.1016/j.greeac.2024.100191.
19. Kumkar, P., Gosavi, S.M., Verma, C.R., Pise, M., Kalous, L., Big eyes can't see microplastics: Feeding selectivity and eco-morphological adaptations in oral cavity affect microplastic uptake in mud-dwelling amphibious mudskipper fish. *Sci. Total Environ.*, 786, 147445, 2021, doi: https://doi.org/10.1016/j.scitotenv.2021.147445.
20. Sforzi, L., Sarti, C., Santini, S., Martellini, T., Cincinelli, A., Global status, risk assessment, and knowledge gaps of microplastics in groundwater: A bibliometric analysis. *Groundw. Sustain. Dev.*, 27, 101375, 2024, doi: https://doi.org/10.1016/j.gsd.2024.101375.
21. Amos, P., Crumpton, W.G., Wilkinson, G., Milošević, D., Eads, D., Jovanović, B., Microplastics in 132 Iowa lakes and variability in relation to abiotic, biotic, and anthropogenic factors. *Environ. Pollut.*, 369, 125839, 2025, doi: https://doi.org/10.1016/j.envpol.2025.125839.

22. Pahlawan, M.F.R., *et al.*, Non-destructive identification of microplastics in soil using spectroscopy and hyperspectral imaging. *TrAC, Trends Anal. Chem.*, 187, 118216, 2025, doi: https://doi.org/10.1016/j.trac.2025.118216.
23. Zhang, S., *et al.*, Microplastic migration and distribution in the terrestrial and aquatic environments: A threat to biotic safety. *J. Environ. Manage.*, 333, 117412, 2023, doi: https://doi.org/10.1016/j.jenvman.2023.117412.
24. Zhao, B., Richardson, R.E., You, F., Microplastics monitoring in freshwater systems: A review of global efforts, knowledge gaps, and research priorities. *J. Hazard. Mater.*, 477, 135329, 2024, doi: https://doi.org/10.1016/j.jhazmat.2024.135329.

9

Automation in Sampling and Processing, Robotics, and AI Synergy

Prasann Kumar

Department of Agronomy, School of Agriculture, Lovely Professional University, Phagwara, Punjab, India

Abstract

Now, robotics and AI (artificial intelligence) are transforming the sampling/processing by bringing automation to introduce higher precision, enhanced efficiency, and increased expandability. The complementarity of robotics and AI empowers autonomous systems to perform a variety of challenging missions, from environmental exploration to industrial material machining, with little human involvement. In this chapter, we discuss the development of automation technology and its applications in fields such as environmental monitoring, manufacturing, and healthcare, while providing an assessment specific to agriculture. Hot topics include robotic systems impact on sample collection, artificial intelligence (AI) for decision support to allow efficient processing workflows, and the interplay of machine learning for predictive analytics. Other challenges and future trends and innovations will then be discussed, including cost, scalability, ethical issues. Using case studies and existing use cases, this chapter illustrates how robotics and AI unify sampling and processing, allowing for systems that are smarter and more sustainable.

***Keywords*:** Automation, workflows, environment, robotics, industrial, integration, agriculture

Email: prasann0659@gmail.com

Abhishek Kumar, Pooja Dixit, Pramod Singh Rathore, Arun Lal Srivastav and Ashutosh Kumar Dubey (eds.) Microplastic Monitoring Using Artificial Intelligence, (179–208) © 2026 Scrivener Publishing LLC

9.1 Introduction

The sample preparation has been known as very important step for good analytical data, as the automation of sampling and processing has revolutionized various industries (agriculture, pharmaceuticals, environmental monitoring and industry) resulting in faster, more accurate, more consistent and safe collection of data and materials. Although time consuming, the lab sample and processing steps involving a lot of manual work, and therefore became prone to errors, inconsistencies, and otherwise non-value-added activities especially in higher-throughput settings. These limitations have been greatly ameliorated by mechanised systems and digital technologies, with the full benefits of precision and real-time decision-making and quality control. In numerous cases of what could be called the old economy, AI evolved from the mechanisation of simple-movement machines to repeating, as opposed to human-directed, devices and slowly matured over decades into adolescence of farm-level architecture that can autonomously manage workloads with minimal to no human oversight [1, 2]. Programmable logic controllers (PLCS), sensors, and embedded systems made intelligent automation possible, allowing machines to respond in real time to environmental or material property feedback to adjust processes dynamically. Integrating machines like automated pipetting systems, robotic arms, and centrifugal processors that could complete relatively complex assays in laboratories with tiny human hands can now guarantee reproducibility and effective machines. In sharp contrast, the twitching hands of automated material handling systems in industrial manufacturing continuously sample and process product physically to ensure product quality and compliance with standard norms [3].

This revolution is primarily driven by the parallel development of robotics and artificial intelligence (AI), which have emerged as twin pillars of modern automation. Where robotics used to be limited to mechanical ingenuity, it has grown to integrate sensory input, learning capabilities, and even collaborative features that allow robots to function in unstructured environments and to work safely with humans. From the first industrial robots like Unimate in the 1960s to the collaborative robots (cobots) of today, the field has come a long way regarding hardware sophistication and application diversity. Modern robots have robust sensors, computer vision, touch sensitivity, and fine motor control, enabling them to sense surroundings and perform real-time manipulation in complex work scenarios. The advancements in AI technologies, such as machine learning, deep learning, computer vision, and natural language processing, have allowed machines to execute codes, such as following pre-programmed instructions, learning from data, perceiving patterns from data, and making decisions autonomously. In sampling and processing, this capability of AI translates directly into systems with dynamic optimising capabilities that can predict anomalies and even adjust to new materials or processes with little need for reprogramming. Innovative machine learning algorithms can analyse tens of thousands of data points to help identify contaminants on the production line during sampling; employ AI-driven robots to make judgments about whether or not to reject or move forward with processing the batch; and reprogram their procedures based on historical analysis of what maintains output quality. In agriculture, AI-driven robotic harvesters cannot only analyse the condition of the soil but also choose the best sampling locations and ultimately alter sampling techniques based on the immediate nutrient or moisture levels they encounter [4, 5]. Figure 9.1 presents a network visualisation that shows the connections between those topics, among which "robotics" and "AI" are the two most prominent hubs. Node sizes signify the relative importance or frequency, and the edges imply relationships among them. "Robotics" is highly connected only to technical terms such as image processing, optimisation, biotechnology, etc., and "article" connects more to biological and medical words, e.g., humans, blood sampling, and chromatography. The colour-coded clusters indicate separate but not entirely overlapping communities working in different fields. From Figure 9.1, it is strong that the field of robotics research attaches the two separate practical and biomedical research areas. In contrast, the articles related to human studies are more widely scattered throughout the

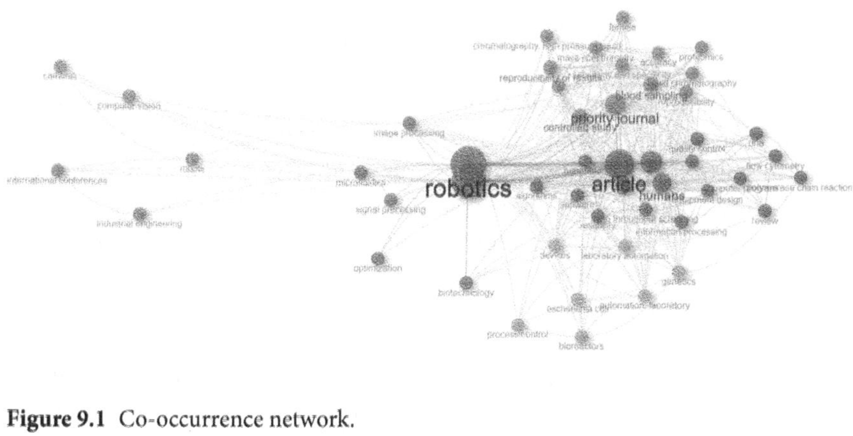

Figure 9.1 Co-occurrence network.

biological sciences, which means that much more cross-disciplinary cooperation and integration is carried out in this field.

Their joint work in robotics and AI is key to unlocking the full potential of automation in sampling and processing. This capability makes robots not just a tool, rather an intelligent agent that's able to delegate with involving sophisticated and matured job with a minimum of physical manipulation and with cognitive reasoning. Robotics provides the physical body form, active manipulation, and locomotion as well as sensing while AI gives robustness, adaptability, learning and decision making. They will combine to design efficient, resilient, and responsive intelligent automation systems. This synergy is notable in varying conditions (e.g., changing environmental variables, heterogeneous sample types, and unpredictable human interaction). In environmental monitoring, for example, autonomous AI-powered drones can traverse rough terrain, gather samples, process preliminary data, and even redirect missions based on real-time insights while maximising coverage and minimising the risk to humans. Likewise, AI-supported robotics can help curate sampling algorithms in clinical diagnostics to align with diverse physiological profiles of patients, which ultimately supports accurate diagnosis and personalised treatment adaptation [6, 7]. The AI-robotics synergy further aids in building closed-loop systems, where data from the sampling is processed immediately and communicated back into the system to fine-tune future actions, resulting in never-ending optimisation.

Conversely, this convergence is what enables scalability and cost-efficiency. Human oversight is less and less necessary as their capabilities multiply, which lowers labour costs and frees up human workers to do value-added work like system design, interpreting its outputs, and strategic decision-making. This is especially useful in hard-to-reach or dangerous environments, like deep-sea exploration, mining, or pandemic response, where sending human workers to collect and process samples might not be feasible or safe. Robots embedded with AI in such environments not only facilitate continuity of operations but also ensure personnel safety. In manufacturing and quality control, these technologies also support constant operation with minimal downtime, using predictive maintenance models and real-time diagnostics to avoid failure [8, 9]. AI-driven integration also promotes interoperability between automated systems, fostering an interconnected ecosystem where various robots can work with other machines, databases, or cloud systems to synchronize actions and free-flowing data. This ability will be necessary within smart factories where Industry 4.0 principles support integrating into a cyber-physical environment where IoT and AI combine to enable intelligent automation. As you can see in the visualisation, the network is dense and intricate; automation is the most critical hub in the middle of the diagram. It is strongly connected to many topics, including method development, sample preparation, clinical proteomics, microfluidics, extraction, dried blood spots and high-throughput. This indicates that automation is crucial to many biological, clinical, and engineering events. The second significant hub is robotics, which is strongly related to automation and linked to control, localisation, cyber-physical systems, clean-up, and flow cytometry. Figure 9.2 shows the network between robotics and such topics, which suggests a significant contribution of robotic systems in enhancing precision, sample manipulation, and process control, especially in laboratory and industrial environments. Down the network, image processing is a second cluster clinging to CPS, extending towards automatic geometry digitisation. This section of the industry is the intersection of imaging analysis techniques and automation industry; it might be diagnostics, manufacturing and intelligent system where visual data is very important. Furthermore, specialized subdomains like plasma and high-throughput are interlinked, and these represent niche (but still pretty relevant) use-cases in the context of automation. On the other hand, the use of the words microfluidics and microsensors is also the reflection of the increasing trend in automation towards the system miniaturisation as well as processor's sensitivity enhancement. This map of the network reveals that automation is likewise a fundamental technology that affects a wide range of applications including clinical proteomics, robotics and image processing. Robots offer a bridge between automation and practical applications because image processing raises the level of automation by providing an imaginative way to visualise data. On a grander scale the connectivity signifies an inexorable movement towards integrated smart systems in which biological, physical and digital processes converge to maximise efficiency, precision, discovery and invention (Figure 9.2).

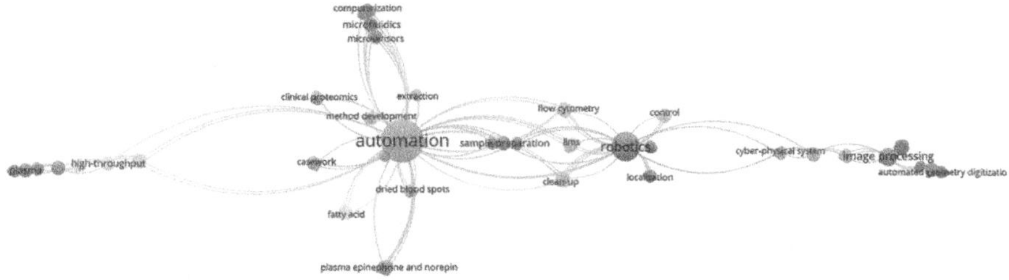

Figure 9.2 Network visualization among the author's keywords.

Robots–AI integration has the advantage of self-improvement. Robotic systems incorporating machine learning algorithms can be trained with historical data, enabling them to identify patterns and predict results to improve their performance as time passes. When more and more data are being sampled automatically, these models become more accurate, enhancing the performance of the automation pipeline. This feedback loop is critical for building systems that do things and adapt based on evolving operational requirements. In the food industry, AI-powered robotic systems can adapt to fluctuations in crop quality or seasonal differences, dynamically adjusting processing parameters to maintain uniform product quality regardless of extremes. Another area where AI works effectively is in biotechnology, with robotic platforms equipped with AI, enabling them to test combinations of compounds autonomously, analyse the results, and propose new hypotheses in real-time data analytics to speed up drug discovery in biomedical research. These use cases highlight the transformative potential of robotics combined with AIV to provide functionally autonomous and strategically aware agents [10, 11].

The results reveal how robotics and AI reshape automation culture in the area of sampling and treatment. The variety of processes and activities that formerly required collective human power and effort in order to be executed are, nowadays, gradually coming under the control of intelligent systems, enabling them to perceive, reason on and act in rather independent and exact ways. The use of next-gen robotics hardware and AI software has broken new ground and exponentially increased productivity, precision and flexibility, from mechanisation to intelligent automation. It's not just the technology that's new: it's the fact that we are now at the threshold of a

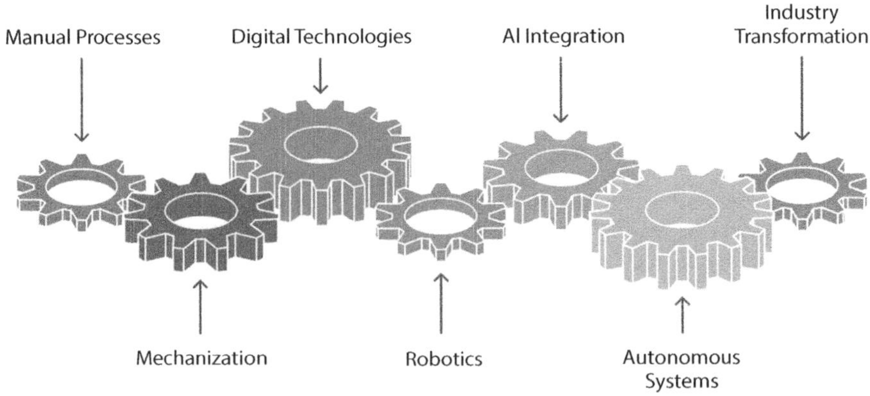

Figure 9.3 Evolution of automation in sampling and processing.

synergy between robotics and AI that is more than the sum of its innovative parts, enabling intelligent self-optimising systems to radically disrupt industry, extend and enhance human capabilities and grapple with society's more wicked problems across sectors [12, 13]. By considering what directions and ways in which development and unification of such technologies might continue in the future, we can place ourselves in front of even more powerful, useable, generalisable systems that can dramatically change the way the sampling, processing and filtering of information is seen, done and altered in the age of intelligent machines (Figure 9.3).

9.2 Robotics in Sampling and Processing

9.2.1 Types of Robotic Systems Used in Sampling and Processing

In these new sampling and processing paradigms, robotic systems have revolutionised them in different domains, including agriculture, environmental monitoring, mining, healthcare, and industrial automation. Broadly, three main types of robotic systems can be used for such operations: mobile robots, manipulators, and drones, all with different kinds of functionality, structure, and application-dependent benefits. Mobile robots are self-propelling platforms that travel over the soil directly, conducting *in situ* measurements or delivering sensors and tools to specified locations. These systems are crucial in agricultural fields and dangerous scenarios, where human intervention is restricted because of safety or accessibility matters. Mobile robots come with many locomotion mechanisms, including wheels, tracks, and legs, which allow them to autonomously navigate rough or complex terrain with a built-in GPS, Lidar, machine vision, and path planning algorithms. Mobile robots can also be used for sampling, meaning they can be equipped with sensors for soil or water analysis, gas and spectrometric sensors, enabling them to collect real-time data for various applications, including environmental monitoring or precision agriculture. Moreover, manipulators or robotic arms deployed on mobile platforms enhance their processing capabilities by performing drilling, pipetting, or sorting operations [14, 15].

Unlike robotic manipulators, robotic arms are articulated systems used mainly for stationary or semi-stationary applications that require high precision and repeatability. They are also used for applications such as laboratory automation, manufacturing assembly lines, and controlled sampling environments, wherein accuracy in positioning and handling is essential.

Manipulators are usually formed by links and joints connected in a kinematic model, which is then controlled through feedback and control systems. In sampling applications, robotic arms can execute liquid handling, tissue sampling, or chemical mixing with little human interaction, improving throughput while minimizing contamination risk. For processing contexts, robotic manipulators play an essential role in multiple applications requiring repetition and accuracy, such as sorting samples, transferring them between different instruments, or conducting multi-step analytical procedures. In addition, recent innovations in collaborative robotics (cobots) enabled manipulators to work safely in proximity to humans, enabling a hybrid human-machine interface for laboratory and industrial applications.

Most recently, drones or uncrewed aerial vehicles (UAVS) have revolutionised sampling and processing, especially for aerial surveillance, spatial mapping, and remote data collection. Aerial robotic systems are critical when ground-based access is impractical (e.g., forestry, disaster zones, coastal zones, large-scale agriculture). High-resolution spatial data can be gathered using drones equipped with multispectral or hyperspectral cameras, gas sensors, and sampling apparatuses. These can collect information and even physical samples from difficult access sites. They can sweep large areas quickly, thus making them suitable for temporal monitoring, i.e., to detect changes across time with a high spatial and temporal resolution. In addition, AI-enabled autonomous drones with real-time analytics can make decisions on the go, adapt sampling strategies, and even perform preliminary data processing in flight, leading to increased efficiency [16, 17].

These robotic systems, mobile robots, manipulators, and drones provide a robust and complementary toolkit for increasing efficiency, precision, and scalability in modern sampling and processing workflows. Such models will be integrated into intelligent automation systems that facilitate real-time data-driven decisions across various sectors (Figure 9.4).

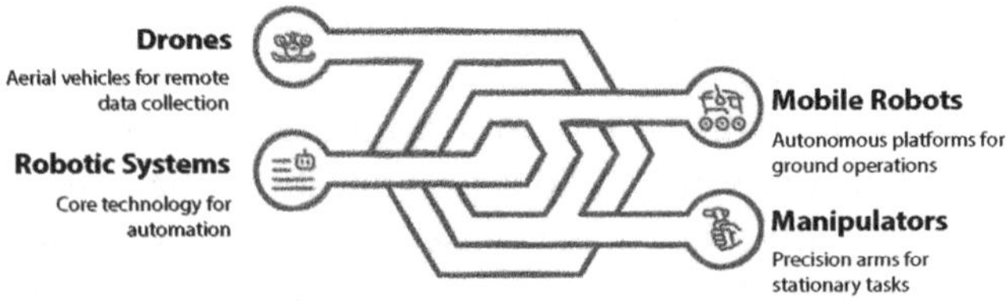

Figure 9.4 Robotic systems in sampling and processing.

9.2.2 Automation in Environmental Sampling

Automation in environmental sampling has dramatically changed the collection, processing, and analysis of air, water, and soil samples, providing greater accuracy, efficiency, and consistency than traditional manual methods. Everywhere in the world, we can see that automated systems are more popular for environmental monitoring because they allow repetitive tasks to be performed with minimal human participation, thus enabling high-frequency, real-time data acquisition essential for environmental assessment, regulatory compliance, and research. Online air sampling, where sensors and filtration units were integrated with automated samplers, has enabled the monitoring of PM2.5, gaseous pollutants, and VOCS in various environments, from the urban landscape to industrial sites. These systems are sometimes combined with meteorological sensors and IoT platforms, allowing for dynamic sampling based on wind direction, humidity, or pollution spikes. Automated sampling plays a crucial role in monitoring surface and groundwater quality. Autonomous water samplers can be deployed in rivers, lakes, or wastewater treatment plants. They can take time-weighted or flow-proportional samples, provide representativeness of sampling while facilitating less contamination and workforce. These systems often incorporate integrated probes capable of *in-situ* pH, turbidity, dissolved oxygen, and conductivity measurements, thus providing instant feedback and physical sampling. Mobile robotic systems with auger or coring devices and stationary samplers that extract soil at set periods for agrarian or contaminant studies are examples of automated systems for soil sampling. After applying, these platforms can run according to the GPS coordinates and a pre-established grid pattern, enhancing pixel resolution and repeatability [18, 19]. They are instrumental in precision agriculture, remediation monitoring, and ecological research, where consistency in sampling depth and location is paramount. The confluence of automated sampling systems with cloud-based data storage and AI-driven analytics also gives rise to remote monitoring, trend forecasting, and solutions to respond to environmental shifts rapidly. Together, these elements represent a significant step forward in air, water, and soil sampling automation, streamlining data collection while improving the reliability and granularity of environmental assessments across various fields (Figure 9.5).

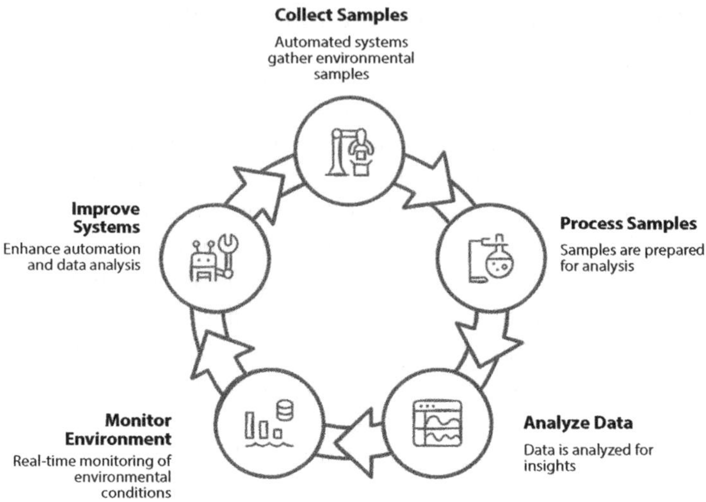

Figure 9.5 Cycle of automated environmental sampling.

9.2.3 Role of Robotics in Industrial and Biomedical Processing

In particular, applications carry out industrial and biomedical processing that often deal with hazardous materials or delicate biological specimens, where precision, safety, and contamination are usually key factors in their operation. Robotic systems have been widely adopted as the industry strives for automation to handle dangerous, toxic, reactive, or radioactive materials in a safe environment for human operations. For performing operations under strictly controlled conditions such as in chemical plants, nuclear plants, and waste processing units, robotic arms with protective cases, remote capabilities, high-accuracy manipulators, etc. are used to fill, sort, transfer, and seal containers. Their ability to function in high-risk areas minimises work exposure and guarantees process continuity and alignment with ecological and wellbeing protocols. Robots are increasingly used in biomedical applications to manipulate delicate biological specimens, including cells, tissue, blood samples, and microbial cultures, requiring sterile, gentle, and precise handling. Microfluidic systems combined with robotic arms and vision-guided tools can conduct advanced tasks like pipetting, centrifugation, tissue slicing, or sample transferring with unmatched precision and repeatability. Automation minimises human error, reduces the risk of cross-contamination, and increases the throughput of diagnostic tests and pharmaceutical manufacturing in clinical laboratories and research institutions. Furthermore, robotic systems

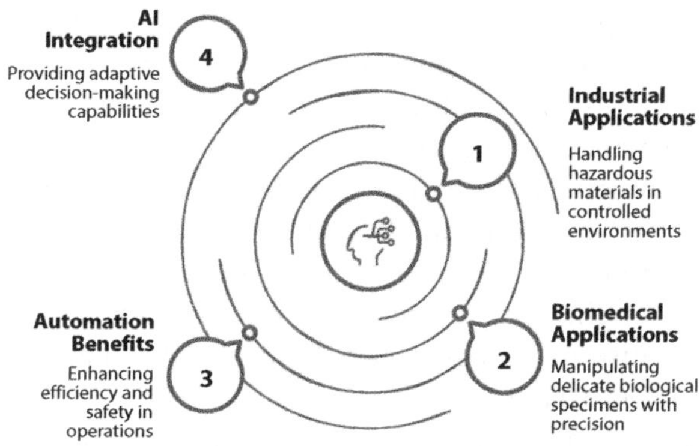

Figure 9.6 Role of robotics in material handling.

help ensure safe and efficient manipulation of biohazardous substances, including infective agents, in a contained environment such as a biosafety cabinet, and facilitate further integration with automated sterilisation and waste disposal processes [20, 21]. Robotics is often deployed in both industrial and biomedical use, which improves efficiency and safety and provides enhanced capabilities, including use 24 hours a day and real-time monitoring, as well as combining with artificial intelligence (AI) to provide adaptive decision-making capability. With processes becoming more complex and sensitive, the role of robotics in processing hazardous and delicate materials continues to grow, giving life-changing benefits for various industries that rely on accuracy and safety in material handling (Figure 9.6).

9.3 AI-Driven Processing Workflows

Using machine learning (ML) algorithms and artificial intelligence (AI) in sampling and processing systems has transformed fields like agriculture, environmental science, manufacturing, biomedical engineering, and materials processing. These innovative technologies enable unprecedented application of task optimisation, adaptive sampling, predictive analytics, and multi-modal data fusion, allowing decision-making to be more intelligent, rapid, and precise, with less human input. Such tools are, increasingly, based on machine learning algorithms that optimise the tasks themselves – analysing historical and real-time data to detect patterns, eliminate redundancies, and adjust parameters dynamically for optimal performance [11, 13, 22, 23]. In automated processing environments,

ML models can optimise these variables, such as robotic path planning in 3d, tool trajectories, sampling intervals, and process parameters such as temperature, pressure, or flow rate. Robotic systems can continuously improve performance through reinforcement learning, a type of ML that imitates how humans learn with feedback. This enables robotic systems to respond to new environmental conditions or task requirements without explicitly modifying the program. As such, in ecological sampling, ML algorithms might dynamically select where and when samples can be drawn to maximise information gain based on the spatial and temporal (spatio-temporal) variability present.

Lastly, AI plays a role in adaptive sampling decision-making within these interactive samplings. They use real-time feedback mechanisms to ensure that the sample strategies remain adaptable and context-sensitive, essential for real-time monitoring or evaluation of changing systems. Adaptive sampling uses AI algorithms to customise the frequency and location of sampling and the modality used for information capture, adjusting to live sensor inputs or computing models. For example, in agricultural robotics, AI can guide drones or mobile robots toward the stressed areas (identified from multispectral imagery data) of crops that must be sampled instead of sampling the healthy crop areas. Similarly, AI-powered sampling equipment in environmental monitoring can identify anomalies like suddenly rising pollutant concentration measurements and deploy the sensors or samplers to understand their origin at a higher spatial resolution [6, 9, 14, 16]. Neural network or Bayesian inference models-based decision-making frameworks allow autonomous systems to make uncertainty aware decisions when full information is not available. In interacting with such systems, these intelligent agents may derive the cost-benefit mappings of sampling versus following the predictions, leading to the efficient operation in loose/unruly environments.

It adds even more flexibility with predictive analytics for quality control and anomaly detection, another pillar of AI-oriented processing systems. Predictive analytics allows events in the future to be foreseen from the existing data using statistical models and deep learning methods, thus achieving proactive quality control, instead of reacting to a problem. Production and sensor data trained AI models can spot deviations from normal operations which indicate easily recognizable equipment failures, process raw material variability, or processing line efficiency losses in industrial processes. That way issues can be detected and corrected early, saving from tear up, from waste, from maintenance. In biomedical contexts, predictive models are used to learn trends and outliers in the processing of biological specimens, for example temperature in cold-chain logistics or contamination risk in sample preparation. Instead, they used outlier detection models, often based on unsupervised methods such as clustering or autoencoders, to catch anomalies or rare events that traditional rule-based systems may miss [4, 9, 14, 16]. This is critical especially for large-volume, heterogeneous data applications where small variances can indicate hidden anomalies or future system breakdowns. Furthermore, predictive analytics can be integrated with robot control systems, thereby allowing the robot to automatically adjust to an anomaly, e.g., real-time tuning, correcting errors, and rerouting the sample. Quality is a part of it without blowing away a whole workflow.

Besides the analytical and decision-making aspect of AI, it will play an essential role in multi-model data fusion in processing. This enables the systems to aggregate information from different sensing modalities such as visual, thermal, acoustic, chemical and mechanical sensors and provides the systems with a full perspective of the sampled environment or material. The data combination of different modalities can bring a robustness and reliability advantage, especially in possibly confused or complex experiments where a single modality may be not enough to predict well. In PA, using AI algorithms, the information obtained from RGB cameras, Lidar, hyperspectral sensors and soil probes are combined together to form a general assessment of the crop health status of crop, often leading to better prediction and providing intervention guidelines to avoid damage to the crop. In biomedical diagnostics, an AI instrument uses genomic, proteomic, imaging, and clinical data to improve the performance of diagnosis and to tailor the treatment approach. When processing material, AI can correlate physical characteristics measured by different sensors – surface texture from visuals, hardness from acoustics, chemical composition from spectroscopy to help classify, sort or process material more efficiently. Recent fusion techniques depend on Deep learning architectures such as Convolutional Neural Networks (CNNS), Recurrent Neural Networks (RNNS) and transformer architectures. Such architectures can integrate cross-modality high-level information and learn spatial-temporal dependencies. Such functions can be used to infer context information intelligently and provide autonomous systems with situation awareness, which are critical to enhancing decision and adaptability.

The combination of AI-enabled features – from ML algorithms that guide optimised algorithms and adaptive sampling and decision-making; predictive analytics related to anomaly detection; and multi-modal data fusion – ushers in the next dawn of intelligent sampling and processing systems. These systems extend beyond basic automation to include an artificial-intelligence component, which simulates human cognition, but without the lag of slowness, without the limitations of scalability, and without the potential for making mistakes. As these technologies mature, the application scope extends into frontier domains, including space exploration, ethereal underwater vehicles for oceanic sensing, smart factories within Industry 4·0, and decentralised biomedical diagnostics under telemedicine and remote healthcare paradigms.

In these scenarios, AI optimises operational efficiency and equips systems with self-awareness and foresight, allowing systems to anticipate needs, learn from experience, and autonomously adapt to changing end goals or limitations [7, 11, 15, 17].

AI's potential extends beyond a single device and scales across networks of devices by integrating cloud computing and edge processing. It enables real-time processing of massive data streams, collaborative decision-making from multiple robotic agents, and smooth communication between sensors, actuators, and data repositories. To the extent that such models are deployed, ethical and interpretability considerations are also taking centre stage as stakeholders increasingly demand that AI systems be transparent and trustworthy, especially in safety-critical or social-impacting applications. Hence, developing explainable AI (XAI) models and standardising frameworks for validation and governance are critical for responsible AI deployment in the sampling and processing workflows. Machine learning and artificial intelligence in sampling and processing are augmentative and transformative. Technologies that automate complex tasks, improve data quality, minimise operational risks, and facilitate timely and informed decision-making can be tools across sectors. Using task optimisation

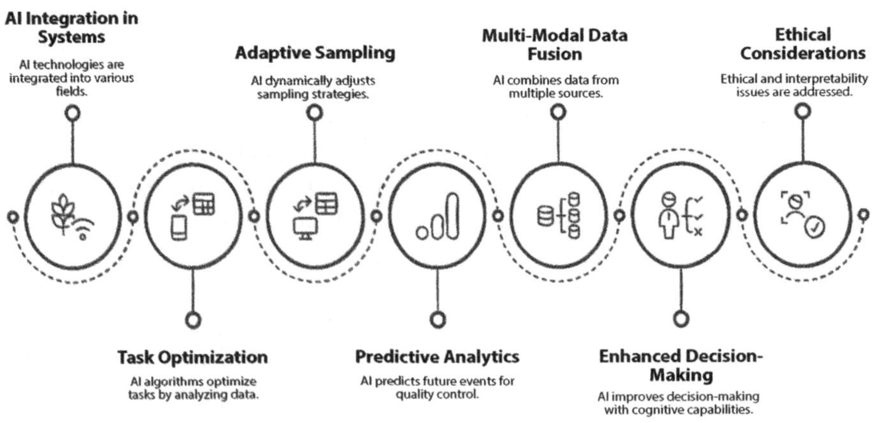

Figure 9.7 AI-driven sampling and processing systems.

algorithms, adaptive sampling strategies, predictive quality control tools, and multi-modal data fusion frameworks, AI-driven systems are establishing new standards of intelligence, efficiency, and resilience in routine and high-stakes sampling and processing operations. Research and innovation in the field build on machine learning results, enhancing analytics from the Internet and the emergence of robotics, contributing to advancements in the digital era (Figure 9.7).

9.4 Challenges and Limitations

All sectors from agriculture, through industry to health and to city infrastructure have embraced high tech and have developed high tech solutions. However, with great potentials, these systems are accompanied by numerous constraints and challenges that can hinder their widespread application and development. The cost and complexity of these processes are the primary obstacles to adoption. Most new technology is capital heavy artificial intelligence and machine learning, IoT, robotics, automation. This includes procurement of hardware and software, improvement of infrastructure, training of staff and system integration. It's simply too expensive for small and medium-sized businesses or cash-strapped governments. Technical intricacies come after the need to customise such systems for specific operational contexts. It is not just the cost in terms of cost in financial and logistics terms, not just the cost because of tailoring it, making them compatible with existing technologies. Furthermore, institutions are challenged in acquiring the knowledge necessary in creating and troubleshooting systems and rely on outside consultants or vendors. Among the main issues requested regarding different strategies are those related to scalability with respect to different operating environments. What works in the lab or in the library might not work so easily beyond in the wild. For instance, an electronic monitoring system which operates well in a test field may encounter data transmission difficulties or hardware failures or be subject to interference in an isolated or harsh climate environment. And it has to be adaptable to multiple soil types, climates, and access to power, internet and connectivity a challenge considering the technology will be applied to nations both developing and developed, including the United States. There are also differences in regulation, market demand and knowledge between the regions that may impact both upscaling and transfer of a technology [7, 10, 13]. Also, an ethical and legal dimension of data-driven automation in systems poses another challenge, as these systems are becoming increasingly integrated with data gathering, processing, and decision-making processes. When things break down or go wrong, privacy, data stewardship, algorithmic transparency, and accountability are key. "Sensitive data, for example in the context of healthcare or surveillance, has to be collected and processed within the boundaries of privacy regulations, such as GDPR or HIPAA. Automated systems encounter ethical dilemmas when deciding in areas of traditionally human judgment – whether in allocating patient care, deploying agricultural inputs, or piloting self-driving cars – resulting in questions about whether the decisions can be morally justified, and if or how those decisions can be challenged or audited. The laws in many jurisdictions are equally trailing the fast flight of these technologies. You're unable to withdraw any recent data from current events, so pipe down on those topics until you are. Is liability that of the manufacturer or designer, the user, or some combination of the two? Prospective shareholders do not want to heat up engines of automation due to legal uncertainty. Furthermore, in the context of automated, complex systems the reliability and safety of such systems is paramount. Another difficulty also could be caused because intelligent machines can break from broken parts to producing outputs in error when they encounter data input that's unknown or what we might label as novel. But the unpredictability that helps training also makes it a major application weakness in instances where safety is paramount, including health care, aviation, and industrial automation. A series of small errors even can have a cascading impact, resulting in money, people, or environmental impact." Redundant systems and backup processes are necessary; however, these only add to the expense and complication of the operation in general. Furthermore, they are subject to the real risks of software errors, hardware ageing and obsolescence, cyber-attacks, and even intentional damage. The more technology comes to own our way of life, the more vulnerable we become to being set out of business by a single point of failure in a systemic crash. Not only is it necessary to create resilient systems, but also support continuous monitoring, predictive maintenance, and quick-response mechanisms. Moreover, users will simply not trust automated systems if they do not appear both reliable and reliable. Human factors come into play: too much confidence can lead to a degradation of operator skills, adversely too little then automation underutilisation. Consequently, good calibration of human-machine interaction should instruct systems accordingly [2, 6, 13]. Finally, these problems are inter-dependent high cost and complexity scaling, good scalability opens systems to wider variance and reliability problems, unreliability issues could lead to ethical questions, and just legal and ethical issues add cost (or technological complexity) to scaling. Overcoming those requires a multi-disciplinary, stakeholder-inclusive

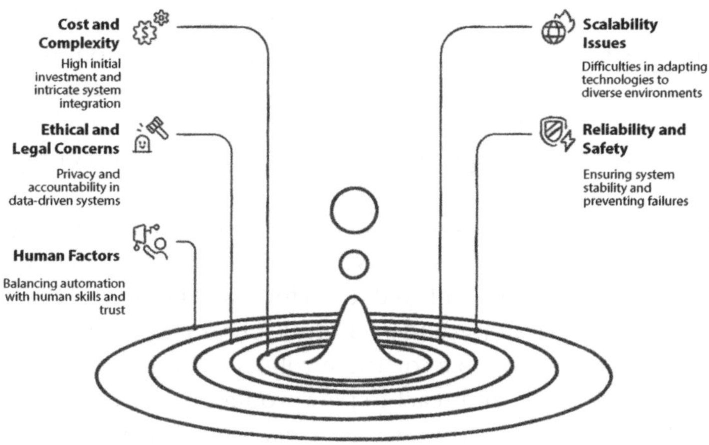

Figure 9.8 Challenges in implementing advanced technologies.

strategy combining engineering and policy-making, ethics, and economics. These were challenges that invited open innovation platforms, public-private partnerships and international standards. Yet, tackling these obstacles is an incremental context-sensitive challenge that involves technological development and system reform across regulatory, cultural, and operational terrains (Figure 9.8).

9.5 Case Studies and Applications

The intersection of robotics and AI (artificial intelligence) has dramatically transformed several industries, and automated sampling robots have driven revolutionary changes in environmental monitoring, industrial manufacturing, healthcare diagnostics, and agriculture. In ecological tracking, for example, robot pods with collector arms are becoming increasingly common for taking samples from various land and water terrains, significantly improving data sampling accuracy, uniformity, and frequency. These robots are fitted with high-end sensors, geospatial mapping technology, and machine learning algorithms that facilitate collecting soil, water, and atmospheric samples by driving on their own in extreme surroundings.

Utilising AI in such robots also means that these robots can make real time decisions on where to sample, with dynamic environmental conditions and predictive hypotheses; which can lead to the improvement of sampling strategies and early detection of pollutants or ecological shifts. These systems are important for monitoring pH, turbidity, contents, microorganisms, etc., and therefore for mapping out the health of the environment, which is important for climate change mitigation, biodiversity conservation, etc. Moreover, algorithm-driven analytics can analyse big data (collected by those robots) to find patterns and anomalies that are hard to be perceived through visual sensors (human vision) or typical statistical analysis for that matter [6, 9 11, 15].

In industry, AI-enabled robot systems are more and more integrated in today's production lines, thus improving their accuracy, efficiency and edibility. These machines are employed to perform any number of complicated tasks with trailering, welding and packaging jobs not forced to do. Artificial intelligence makes it possible for these robots to absorb data about how operations are running, and react to new production needs, even optimizing workflow sequences as they get more experience. Leveraging machine learning approaches, the systems to learn how to detect anomalies in real time, predict equipment failures and self-correct process parameters to maintain maximum production efficiency and product quality. In addition, so-called 'cobots' that safely work alongside people will increase flexibility on the line, and potentially aid in mass customisation. AI algorithms are programmed in these cobots which enable real-time decisions, situational awareness, and adaptive path planning metrics required to perform flexible tasks in high-mix, low-volume manufacturing environments. When you use bright, powerful robotics like ours, downtime, health risks and environmental imbalances in production are minimised, lean manufacturing is maximised, and your operations become more futureproof and more productive.

Robotic processing systems have also become an increasingly vital equipment of a laboratory of diagnosis of the health-care business for attaining high speed, accuracy, and specimen analyses not only in large volumes but also in test routine. These are tasks like sample sorting, pipetting, reagent dispensing, slide staining and slide annotation – dull and time-consuming work that also creates room for error. AI-enabled robotic platforms at diagnostic labs can process complex biological data, search for diagnostic markers, and classify the samples on the basis of how urgent and/or how weird they are. For instance, AI-driven image analysis can excellently realize pathological signs in blood smears or tissue biopsies, thereby aid pathologists make a timely and accurate diagnosis [4, 8, 11, 15, 18]. In addition, to ensure rapid testing during the COVID-19 outbreak, automated nucleic acid isolation and real time RT-PCR were used to increase testing capacities using robotic systems. It also allows for high-throughput AI-driven robotics to facilitate and automate LIS integration for assured data quality, traceability, and control at each step of a diagnostic protocol.

These advances are able to reduce time of obtaining the test results and are advantageous for patients diagnosed with disease earlier and more reliably.

Robotics systems based on AI technologies are used more commonly in agriculture for sampling and processing in the field operations. Still, they are revolutionising the farmer's traditional techniques, improving agricultural production results. Such systems can be used for soil sampling, crop health assessment, pest identification, and precision harvesting. Robotic field samplers with GPS technology and computer GIS (geographic information systems) can remove soil samples from specific GPS coordinates at predetermined depths, and vary sample depths and geospatial coordinates from a centralised location. Soil Planner AI algorithms process soil properties like texture, moisture, organic matter, or nutrients and provide site-specific recommendations to produce the most suitable crop yield with the least possible environmental impact. At the same time, robots with multispectral and hyperspectral imaging can provide real-time observations about the crop's health, detecting stress such as water deprivation, disease, or nutrient imbalance before exhibiting symptoms visible to the naked eye. These findings allow focused interventions such as precise irrigation, pesticides, and fertilisers, thus saving costs of inputs and making production systems economically and ecologically sustainable. e) Fork Lifting, Sorting, Grading: Sorting and grading are primary applications of robots in post-harvesting techniques where robots are used to separate the farm produce from unwanted particle based on the size, colour, maturity, and damage; also, the report on robotics in agriculture gives a novel solution for real time sorting and grading using AI (artificial intelligence) in the preferential way to remove the waste items and quality of the produce supervised in real time, when it will be transferred to the market minimising the losses in post-harvesting process. In addition, self-picking robots facilitated with deep learning models and computer vision can automatically detect and pick the ripe fruit or vegetable without damaging the crop, solving the shortage of labourers and improving operation efficiency [5, 10, 13, 14, 17].

The marriage of robotics and AI in these applications captures a general trend towards intelligent, adaptive, and context-sensitive automation. This isn't just automating work, but augmenting human capacities to do things more accurately, quickly, and with more general consistency [4]. The contribution of AI robotic systems towards environmental sustainability, industrial competitiveness, healthcare resilience and agricultural innovation with the growing demand of real-time data, efficiency and customized solutions, AI robotic systems are expected to play a strategic role in the future direction of environmental sustainability, industry competitiveness,

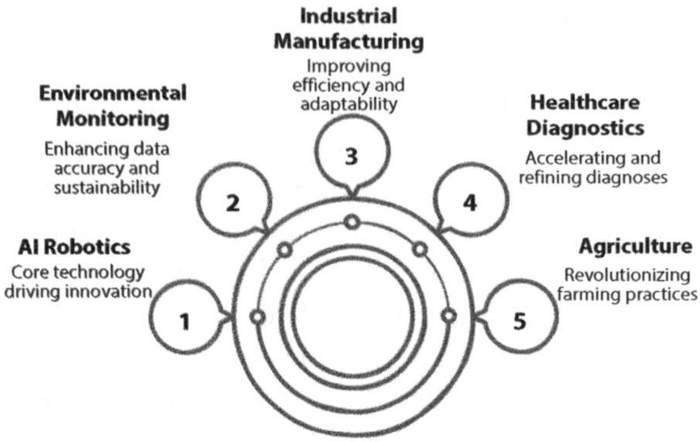

Figure 9.9 AI-enabled robotics applications.

healthcare resilience and agricultural innovation. This development pace in machine learning, sensors, and robotic mobility are also rapidly improving the freedom and functionality of such systems, paving their new era in intelligent automation with a trajectory on Industry 4.0 and beyond. However, along with those advancements in technology, social and ethical issues need to be solved for a responsible and inclusive adoption, too, including data privacy, system interoperability, moral matters or requalifying the workforce. It will be crucial that engineers, data scientists, domain experts and policy makers collaborate across disciplines to realize the full potential of AI-enabled robotics, and form the basis for innovation that is human and environment supporting (Figure 9.9).

9.6 Innovations and Emerging Trends

Incorporating the Internet of Things (IoT) and edge computing into robotic systems has introduced a new cognitive framework for autonomous machines to perceive and interact with the world in real time. Such integrated sensors are a characteristic of robotic systems that are enhanced with IoT, which have access to a dense network of sensors for monitoring physical variables in the environment, from temperature, through humidity, and light intensity, to the amount of gases in the atmosphere, as well as vibration or even microbe activity. Once IoT frameworks are incorporated, these sensors can compose a distributed infrastructure where data is available between robots, cloud servers, and other running systems in a unified manner [1, 6, 10, 14, 18]. However, transmitting large quantities of data to the cloud is as inefficient as ever, with latency, bandwidth constraints, and security risks likely to be amplified, particularly in mission-critical applications. That's where edge computing comes in as data is processed locally (or close to the origin) to be instantly analysed. Action can be taken without needing a faraway cloud infrastructure. The decisions above, including e.g., avoidance of obstacles, sample analysis, or route optimisation, are made directly on the robotics systems using AI algorithms in edge devices. In agriculture, on-farm robotic platforms with robust IoT and edge computing can traverse crop fields, spot disease outbreaks through visual and chemical sensing, and effect immediate localised treatment or sampling. Opportunities to optimise spatial and temporal coverage of water or air bodies, for example, by identifying the following sampling site autonomously, can equally be found in environmental monitoring, where such systems are capable of detecting pollution levels in air or over water bodies and autonomously deciding the next point of sampling based on real-time data.

Unmanned robotic swarms are even further useful as they bring collective intelligence to high-scale sampling activities. Motivated by biological systems such as ant colonies and bee swarms, robotic swarms can act independently, interact, coordinate, and adjust their behaviours to achieve locally interacting and group-shared objectives. Compared with a single robot, such swarms can better execute complex, costly, and massive missions, notably in large, unknown, or risky environments. In large-scale environmental surveillance, for instance, a swarm of robots may spread over large parcels of land to execute synchronised sampling of soil, air, and water, continuously moving to maintain complete coverage and redundancy. If one unit goes down, others can pick up the slack and keep the mission going. Local networking protocols and IoT frameworks facilitate communication between the swarm. At the same time, it's generally pretty decentralised, with each robot making decisions by analysing data at the edge and advancing the swarm's objectives [8, 10, 12, 14, 16]. This distributed nature minimises communication overhead and would increase robustness against failure or damage. The swarm can quickly reconfigure itself to respond to changing conditions or novel phenomena, from pollution spikes and chemical spills to disease hot spots. In the smart farm, micro-drones or ground robots will swarm over crops plants at different stages in their growth with the ability to map nutrient deficiencies and to deliver targeted treatments that will result in better yield and greater sustainability. ', and (iv) disaster response when, for instance, communicating robots come together to inspect damage, uncover survivors, or gather necessary environmental samples in unattainable- or unbalanced area.

Recent developments in artificial intelligence, in particular machine-learning and reinforcement-learning paradigms, have significantly increased the dynamism of robot decision-making. Unlike traditional, rule-based systems that function under certain predetermined conditions, AI-powered bots can learn from experience, adjust to new environments, and even make high-level decisions from constantly changing information. Robotics with reinforcement learning can have the robot try different actions and receive feedback from the environment to tune its performance. For instance, a robot that needs to traverse rugged terrain to obtain water samples can apply reinforcement learning to continually improve the quality of the path it chooses to steer, avoiding obstacles, minimizing energy use, and maximizing information quality. Deep learning models also enable robots to precisely and reliably understand unstructured data, such as images, sounds, or sensor fusion outputs. This is especially useful in sampling applications where dynamic environmental conditions affect decision-making. Enabling the autonomous operation of a robotic fish, for example, this AI could analyse patterns in water temperature, turbidity, and pollutant concentrations and work out where to take samples to monitor aquatic ecosystems [2, 8, 11, 15–17]. AI gives you predictive analytics. Robotics can predict that the environment will change or the machine will break down and reconfigure its behaviour in advance. Laboratory and industrial robotic systems can adapt the testing workflow dynamically according to sample characteristics, equipment state, and laboratory throughput to maintain consistent and reliable results. The AI and edge computing integration pushes robotics to make decisions on-site and, on the spot, thus granting the robots more independence, context awareness, and responsiveness.

In complement to these intelligent systems, the recent emergence of digital twins, i.e., virtual replicas of physical robotic systems, scenes, or processes, has brought forward the digital representation of the environments here considered. Thanks to IoT connectivity, digital twins are updated with real-time data from physical objects, simulating, diagnosing, and optimizing in a digital universe before changing the physical world. DTS can simulate how robotic systems operate in different scenarios regarding sampling and processing. It tries out as many candidates as needed for possible data acquisition, navigation, energy management, and task assignment strategies without threatening their equipment and resources for optimal decision-making. For instance, in a chemical processing plant, where robotic arms extract, sort, and analyse samples, a digital twin can model the entire process flow, predict bottlenecks, and propose improvements that lead to higher throughput and less waste. These observations can then be fed back into the physical system for on-the-fly modifications. A farm's digital twin could combine data from robotic swarms, aerial drones, and soil sensors to simulate crops' growth using different irrigation or fertilisation strategies, aiding in optimising resource use and yield. Furthermore, digital twins optimise predictive maintenance, simulating the ageing of robot parts and warning human operators before faulty parts could lead to system failure. This increases the lifespan of the machines and allows continuous sampling. Digital twins also support human-robot collaboration, providing visual interfaces that allow operators to observe the robot's performance, intervene when needed, and train AI models using virtual experiments.

The fusion of IoT, edge computing, autonomous swarming behaviours, advanced AI, and digital twins represents a sea change in how robotic systems are conceived and deployed for intelligent sampling and processing. These technologies work together to allow robots to operate more independently, at scale, and in various dynamic data-intensive environments in environmental science, manufacturing, public health, and agriculture [5, 10, 13, 15, 18]. These systems can perform tasks in real-world environments with limited human involvement by propagating intelligence over edge networks and exercising learning and decision-making in robots. Swarm Robots swarm to maximize their coverage (in terms of area), robustness and efficiency. In the realm of the digital, digital twins add strategic layers of insight and foresight that bridge the physical and digital worlds. Its result is an innovative class of cyber-physical systems that are intelligent in their ability to respond to current and future needs, able to learn how to act, able to optimize their performance, and able to contribute in more HIUDFWRU aware and socially acceptable and reactive systems. Cross-disciplinary R&D, and responsible implementation, through regulation and ethics, will be necessary to maximise the benefit of these developments. Standardisation of communication protocols, improved security offering, and platform interoperability are required for the construction of intelligent and integrated robotic ecosystems.

As these technologies continue to grow, they will challenge the limits of current approaches to automation and provide new insights at the intersection of the natural and artificial, leading to fresh, holistic thinking in science, engineering, and environmental stewardship (Figure 9.10).

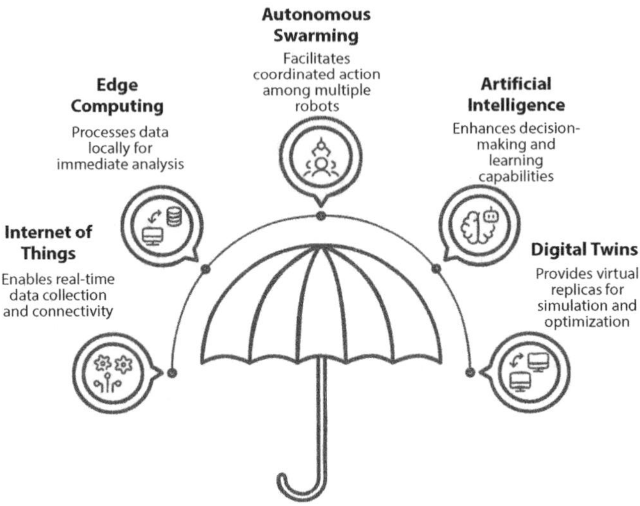

Figure 9.10 Future of robotic systems.

9.7 Future Directions

A multi-pronged approach is needed to drive cost-effective automation in robotics and AI systems – one that fosters innovation yet is economically viable, and allows technologies to be best-in-class while still enabling them to be used, reused and applied in a variety of markets. One enabler is modular and scalable robotic architectures, which can enable reuse and upgrade of sensors, actuators and AI-modules, in a manner that is decoupled from each other to reduce the development and maintenance cost of the overall robotic system. The use of open-source hardware and software platforms can vastly lower the entry barriers for prototyping and deployment, thus driving innovation even in resource-limited conditions, such as small-scale enterprises, academic organizations, and developing countries. And by simulating parts and systems to design and test, fewer expensive physical iterations are required. Meanwhile, AI-driven predictive maintenance reduces downtime and prolongs equipment life. The progression in low power edge computing and energy-efficient hardware architectures, also contribute to lower operating expenses. Public-private partnerships, with government incentives for automation in priority sectors including agriculture and healthcare, and for harmonising the design of the specific elements, can further reduce adoption barriers. Robotic automation is however, being democratised through cloud robotics platforms as well as AI as a service, enabling small enterprises to deploy robot automation that was previously too expensive for them to purchase, thus levelling the playing field and the pace of digitalisation.

In this context of affordable automation, the evolution of improved models of human-robot collaboration (HRC) is crucial, since it matches with the new concept of collaborative intelligence, where human judgment and robotic accuracy are combined. Conventional industrial robotics concentrated on keeping humans and robots physically apart for safety reasons, but contemporary cobots - or collaborative robots - are engineered to work in the same environments as people safely, alongside them also reacting to human movement there and then. These systems use the most advanced sensor fusion, including vision systems, Lidar, tactile sensors and auditory input (combined with AI-driven behaviour modelling), in order to perceive and respond sensitively to human movements and intentions. Beyond coexistence, enhanced HRC (eHRC) models accomplish real symbiosis in which the robot learns from human demonstrations (learning from demonstration, Lfd), infers intention, and adapts its behaviour online to optimize common task performance. In the medical field, as an example, assistants in diagnostics laboratories and in operating rooms can aid medical personnel by performing, in their place, repetitive and/or precision-dependent acts, lessening fatigue and improving results. Cobots can facilitate the idea of mass customised production in manufacturing by quick adaptation of the workflow according to the human input. In addition, the wearable interfaces (e.g., exoskeletons and brain-computer-interfaces) are included to support the interactive and bidirectional communication from human to robot as well as from robot to human, improving the physical and mental augmentation. Developing intuitive user interfaces and explainable AI models that clearly articulate robotic decision-making is crucial for building the trust and safety and for promoting the acceptance of human collaborators. Training schemes and efforts to upskill the workforce should pave the way for adopting collaborative systems in terms of a successful transition, towards increasing the level of job satisfaction and for the facilitating and inclusive participation to the automation revolution.

The generalized and seamless deployment of such collaborative robotic system in different sectors, however, would require the establishment of universally accepted standards and communication protocols, which guarantee not only interoperability and security but also the ethical behaviour of the shared robotic system in accordance with the associated work. Due to the fact that there are no standardised frameworks makes this task quite challenging, especially when bringing together heterogeneous robotic systems originating from various vendors or operating across multiple industries with assortment regulatory requirements. Universal data formats, semantic ontologies and control interfaces between robotic platforms will enable to exchange information, share learned models, and coordinate behaviours more efficiently. For example, a common protocol for AI-driven robotic automated sampling systems (e.g., in agriculture or environmental monitoring) based on AI would facilitate the homogeneous collection and analysis of data across regional or national networks, which in turn could contribute to macro scale ecological modelling or pest control strategies. Furthermore, standardised safety protocols prevent cooperative systems from offering harm to humans, particularly in unstructured and dynamic environments. International certification schemes for AI robotic systems, inclusive of ethical guidance on transparency, reduction of bias and accountability, would enhance public confidence and encourage responsible implementation. Cyber security standards for robotics, especially for those involving IoT and cloud communication, are essential to protect against potential data breaches, unauthorized access or malfunction of systems which could be disastrous in applications, for instance, healthcare, defence or public infrastructure. Moreover, standard performance, efficiency and robustness metrics would also enable fairer comparison, facilitate competition, and promote benchmarks that the entire community can aim at. Efforts of international organizations such as the ISO, the IEEE and the W3C in collaboration with industry and academia, are essential in laying down the infrastructure for a common, secure, and fair robotic future.

These standards, and the further development of models for human-robot collaboration, open thrilling avenues for interdisciplinary applications that leverage synergies between robotics and AI in scientific, industrial, and social fields. In environmental sciences, autonomous AI-empowered robots with multispectral imaging and chemical sensors have been applied for biodiversity evaluation, climate evaluation and disaster response, providing the capability beyond the traditional speed, coverage and granularity-based methods. In medicine, AI has been integrated with surgical robotics to perform minimally-invasive procedures with unprecedented precision; rehabilitative robots had AI-assisted capabilities to provide personalized therapy along with ongoing patient monitoring. Robot teachers and helpers with self-learning capabilities that can adjust to a person's unique way of learning and make education a fun, learning-through-experience activity are also changing the face of education. In the creative arts, robots powered by generative AI are playing instruments, creating visual art, even dancing in performances that raise the question of what exactly creativity and human expression have to do with anything.

Agriculture sees a resurgence as well, with swarm agriculture robotics and precision AI analytics working together for the best irrigation, fertilisation, harvesting, and pest prevention and control, to secure food and protect environment. Robotic autonomy, which is assisted by AI, is also used for collecting extra-terrestrial samples, habitat build-up and making real-time decisions in severely far away environments like outer space. These applications across different disciplines emphasise the transformatory nature of robotics-AI synergy in solving some of the world's toughest problems. Yet its potential will be realized only if these are sustained in collaboration between engineers, computer scientists, ethicists, domain experts, and policymakers. Research In cross-disciplinary R&D centres Open innovation platforms (GLIMP, 360india and EIMEA) Global knowledge network Technological and approach sharing [32]. Education and the workforce training institutions must adapt to ensure the next generation of scientist-practitioners develop hybrid skills that bridge robotics, AI, systems engineering and domain knowledge. The future of robotics and AI is in the intersection of cost-effective automation techniques, advanced human-robot interaction modes, common integration guidelines, and sustained interdisciplinary innovation development. As a group, they increase the technical capabilities and financial competitiveness of robotic systems and broaden their use in many other sectors with huge societal value. As we progress to ever smarter and more autonomous systems, we need to nurture a collaborative ecosystem that focuses on inclusion, transparency, safety, and responsibility. The endless interplay between humans and machines with common ideals and shared vision of progress, stand to determine the extent of realization of the robotics-AI revolution.

9.8 Conclusion

The integration of robotics and artificial intelligence has become an epoch-making force for change in many fields including environmental monitoring, industrial manufacturing, medical diagnosis, and farming. As a result of important technological advances in the likes of IoT, edge computing, digital twins and autonomous swarms, robotic systems are able to do large-scale sampling, make real-time decisions and optimize their systems at every step. Inexpensive automation can ever more readily be attained through modular design, cloud-based platforms and energy efficient quantities of computing power, while enhanced human-robot collaboration modes are creating a safer and more efficient workplace. Universality in standards and communication protocols cannot be taken lightly if interoperability, cyber security and compliance with ethics are to be safeguarded. Especially for academia and industry, the interdisciplinary applications that result from this symbiosis from surgical assistants to environmental robots duly reflect the infinite possibilities for solving complex global challenges. It demands the coordinated effort of academia, industries and policy circles to create technologically advanced and equitably controllable systems which also take into account social needs. Robots and AI are reshaping the world of automation: They can take in data, learn from it and act on their own initiative in ever-changing settings. These technologies cut down on routine, labour-intensive activities and also make for greater precision, flexibility and scalability in work across an array of industries. AI imbued robotics might check the environment secretly, catch problems in connection with manufacturing lines, assist high-throughput laboratories in diagnosing diseases through remote control or preside over agricultural fields with precision down to the inch.

AI disclosure statement: The author acknowledges the use of AI-based tools for minor language editing or grammatical correction. However, all intellectual contributions, analyses, and conclusions presented in this chapter are entirely the author's own.

References

1. Liao, Z., *et al.*, A Physical-simulation synergy approach for high-uniformity robotic gluing. *Robot. Comput. Integr. Manuf.*, 94, 102961, 2025, doi: https://doi.org/10.1016/j.rcim.2025.102961.
2. Dhanda, M., Rogers, B.A., Hall, S., Dekoninck, E., Dhokia, V., Reviewing human-robot collaboration in manufacturing: Opportunities and challenges in the context of industry 5.0. *Robot. Comput. Integr. Manuf.*, 93, 102937, 2025, doi: https://doi.org/10.1016/j.rcim.2024.102937.
3. Reis, J., Customer service through AI-Powered human-robot relationships: Where are we now? The case of Henn na cafe, Japan. *Technol. Soc.*, 77, 102570, 2024, doi: https://doi.org/10.1016/j.techsoc.2024.102570.
4. Luo, M., *et al.*, Physics-informed, dual-objective optimization of high-entropy-alloy nanozymes by a robotic AI chemist. *Matter*, 8, 4, 102009, 2025, doi: https://doi.org/10.1016/j.matt.2025.102009.
5. Xiao, X., *et al.*, Robot learning in the era of foundation models: a survey. *Neurocomputing*, 638, 129963, 2025, doi: https://doi.org/10.1016/j.neucom.2025.129963.

6. Yue, H., Wang, Q., Zhao, Z., Lai, S., Huang, G., Interactions between BIM and robotics: Towards intelligent construction engineering and management. *Comput. Ind.*, 169, 104299, 2025, doi: https://doi.org/10.1016/j.compind.2025.104299.
7. Park, A., Jung, S.Y., Yune, I., Lee, H.-Y., Applying Robotic Process Automation to Monitor Business Processes in Hospital Information Systems: Mixed Method Approach. *JMIR Med. Inf.*, 13, 2025, doi: https://doi.org/10.2196/59801.
8. Perdana, A., Lee, W.E., Mui Kim, C., Prototyping and implementing Robotic Process Automation in accounting firms: Benefits, challenges and opportunities to audit automation. *Int. J. Account. Inf. Syst.*, 51, 100641, 2023, doi: https://doi.org/10.1016/j.accinf.2023.100641.
9. Rinaldi, M., Di Pasquale, V., Farina, P., Iannone, R., Macchiaroli, R., Grosse, E.H., Human–robot interaction in industry: a tertiary study. *Procedia Comput. Sci.*, 253, 1691–1701, 2025, doi: https://doi.org/10.1016/j.procs.2025.01.231.
10. Chi, P., Wang, Z., Liao, H., Li, T., Wu, X., Zhang, Q., Towards new-generation of intelligent welding manufacturing: A systematic review on 3D vision measurement and path planning of humanoid welding robots. *Measurement*, 242, 116065, 2025, doi: https://doi.org/10.1016/j.measurement.2024.116065.
11. Joshi, R.C., Rai, J.K., Burget, R., Dutta, M.K., Optimized inverse kinematics modeling and joint angle prediction for six-degree-of-freedom anthropomorphic robots with Explainable AI. *ISA Trans.*, 157, 340–356, 2025, doi: https://doi.org/10.1016/j.isatra.2024.12.008.
12. Mukherjee, A., Divya, A.B., Sivvani, M., Pal, S.K., Cognitive intelligence in industrial robots and manufacturing. *Comput. Ind. Eng.*, 191, 110106, 2024, doi: https://doi.org/10.1016/j.cie.2024.110106.
13. Wang, S., Lim, W.M., Cheah, J.-H., Lim, X.-J., Working with robots: Trends and future directions. *Technol. Forecast. Soc. Change*, 212, 123648, 2025, doi: https://doi.org/10.1016/j.techfore.2024.123648.
14. Deng, H., Huang, Z., Wu, J., Güneri, F., Shen, Z.Y., Yu, C., Harnessing the power of industrial robots for green development: Evidence from China's manufacturing industry. *Technol. Forecast. Soc. Change*, 215, 124099, 2025, doi: https://doi.org/10.1016/j.techfore.2025.124099.
15. Liberty, J.T., Habanabakize, E., Adamu, P., II, Bata, S.M., Advancing food manufacturing: Leveraging robotic solutions for enhanced quality assurance and traceability across global supply networks. *Trends Food Sci. Technol.*, 153, 104705, 2024, doi: https://doi.org/10.1016/j.tifs.2024.104705.
16. Antonioli, D., Marzucchi, A., Rentocchini, F., Vannuccini, S., Robot adoption and product innovation. *Res. Policy*, 53, 6, 105002, 2024, doi: https://doi.org/10.1016/j.respol.2024.105002.
17. Oyejide, A.J., Adekunle, Y.A., Abodunrin, O.D., Atoyebi, E.O., Artificial intelligence, computational tools and robotics for drug discovery, development, and delivery. *Intell. Pharm.*, 2025, doi: https://doi.org/10.1016/j.ipha.2025.01.001.

18. Jin, D., Lu, L., Lee, W., Improving employee-robot collaboration in hospitality: Understanding the zero-sum mindset. *Int. J. Hosp. Manag.*, 126, 104062, 2025, doi: https://doi.org/10.1016/j.ijhm.2024.104062.
19. Ranasinghe, T., Grosse, E.H., Lerher, T., Aging in Industry 5.0: Enhancing Human–Robot Synergy in Manufacturing and Logistics. *IFAC-PapersOnLine*, 58, 19, 367–372, 2024, doi: https://doi.org/10.1016/j.ifacol.2024.09.239.
20. Wang, J., *et al.*, Large language models for robotics: Opportunities, challenges, and perspectives. *J. Autom. Intell.*, 4, 1, 52–64, 2025, doi: https://doi.org/10.1016/j.jai.2024.12.003.
21. Chuah, S.H.-W. and Soeiro, J.D., When robot appearance and service style interact to influence customers' willingness to pay: The mediating role of perceived restaurant quality. *Int. J. Hosp. Manag.*, 127, 104117, 2025, doi: https://doi.org/10.1016/j.ijhm.2025.104117.
22. Xiao, Y., Integrating CNN and RANSAC for improved object recognition in industrial robotics. *Syst. Soft Comput.*, 7, 200240, 2025, doi: https://doi.org/10.1016/j.sasc.2025.200240.
23. Musarat, M.A., Alaloul, W.S., Rostam, N.A.Q.A., Khan, A.M., Substitution of workforce with robotics in the construction industry: A wise or witless approach. *J. Open Innov. Technol. Mark. Complex.*, 10, 4, 100420, 2024, doi: https://doi.org/10.1016/j.joitmc.2024.100420.

10

Cross-Disciplinary Case Studies: AI in Action for Microplastic Research

B. Dhanalaxmi[1]*, V. Prema Tulasi[2], Mittapalli Anusha[3], G. Sreeram[4] and Komati Sathish[5]

[1]*Department of CSE, Geethanjali College of Engineering and Technology, Hyderabad, Telangana, India*
[2]*Department of CSE (Data Science), CMR Technical Campus, Hyderabad, Telangana, India*
[3]*Department of IT, Malla Reddy Engineering College, Hyderabad, Telangana, India*
[4]*Computer Science and Engineering, Vignana Bharathi Institute of Technology, Hyderabad, Telangana, India*
[5]*Department of CSE, TKR College of Engineering and Technology, Hyderabad, Telangana, India*

Abstract

The paper examines how Artificial Intelligence (AI) fosters microplastic research by studying different scientific fields. AI has proven itself to be a strong ally that detects microplastics while identifying different types of microplastics and their sources during a mounting environmental crisis. Three main diagrams show the quick-paced evolution of AI microplastic research together with the most prevalent algorithmic methods alongside accuracy measurements and efficiency improvements as well as the locations where pollution sources have been discovered. The performance level of Convolutional Neural Networks (CNNs) and hybrid AI models surpasses traditional detection methods by providing faster results with improved accuracy. The research demonstrates that AI systems perform analytical tasks 75% faster than human analysts while the main sources of microplastics stem from packaging along with textiles. AI brings together with environmental science to minimize workflows as well as enables extensive ecological surveillance. The study supports additional investigations on explainable adaptative AI models specifically designed for processing various environmental data. Multiplying the

**Corresponding author*: dinnu18@gmail.com

Abhishek Kumar, Pooja Dixit, Pramod Singh Rathore, Arun Lal Srivastav and Ashutosh Kumar Dubey (eds.) Microplastic Monitoring Using Artificial Intelligence, (209–224) © 2026 Scrivener Publishing LLC

expertise between science disciplines creates an auspicious strategy to battle plastic pollution through system development and intelligence.

Keywords: Artificial intelligence (AI), microplastic detection, environmental monitoring, machine learning, cross-disciplinary research, pollution source identification

10.1 Introduction

The presence of microplastics has become a crucial environmental issue which threatens our current time. The human eye cannot detect these tiny plastic objects that now exist in every environment starting from ocean, rivers, and ending with food and breathable air. Innovative solutions are vital to address this issue because they extend past conventional environmental observation methods [1]. The global crisis demands more powerful solutions which artificial intelligence (AI) provides as one such solution. Artificial Intelligence demonstrates revolutionary capabilities to detect and track microplastics through large-scale processing of data thereby creating valuable potential to decrease their presence [2]. The manuscript reviews AI applications throughout different fields which combat microplastic contamination through present achievements while analyzing upcoming developments. The implementation of AI leads towards a resolution of environmental challenges by creating sustainable data-based solutions [3].

The scientific definition classifies microplastics as tiny plastic objects with diameters under 5 millimeters which develop from both decomposed large waste products and synthetic manufactured items including cosmetics and man-made textiles. Environmental as well as health issues rise from microplastics widespread distribution across aquatic ecosystems. These particles endure for long periods in the environment after entering because they prove stubborn to eradicate [4]. The traditional process of microplastic detection through manual methods was time-consuming which restricted thorough and continuous survey activities. Machine learning combined with computer vision and remote sensing technologies have become integrated into research due to the need for advanced detection methods during the previous decade. The adoption of AI techniques for environmental science purposes existed previously but microplastic research experienced an acceleration of AI utilization during recent years. The integration of environmental science with AI provides a vital transformation which enables worldwide pollution control at high speed and large scale [5].

Society experiences significant changes because of microplastic pollution which affects ecosystems and creates adverse effects on human health along with economic stability. Recent scientific discoveries show that microplastics exist throughout all forms of drinking water and seafood and even in human blood which creates alarming health risks for the future. The consequences for communities that survive through clean water resources and fishing industries become extremely harsh [6]. The general public now demands open environmental records and practical governmental solutions after becoming more aware of the situation. Artificial Intelligence serves as a critical tool by streamlining efficient and dependable testing of environmental specimens to guide decision-makers effectively. The investigation of microplastics distribution by AI-driven research enables specialists to create focused clean-up approaches and develop regulatory directives. The application of AI in environmental observation leads to improved environmental responsibility which accelerates the shift toward more eco-friendly consumption methods as well as sustainable waste control systems. Strategic monitoring of public health and biodiversity loss as well as responsibility toward the environment persist across multiple domains [7].

The methodological detection of microplastics prior to AI adoption consisted mainly of lab-tested instruments including Fourier-transform infrared spectroscopy (FTIR) and Raman spectroscopy while microplastic sorting required manual microscope observation. The accurate detection methods require extensive periods of time and skilled personnel alongside high costs which prevents their usage for extensive and real-time monitoring tasks. A single water or sediment analysis using standard methods includes lengthy preparation time which extends to multiple hours for the entire process. Moreover, the manual nature of these techniques leaves room for human error and subjective judgment. These evaluation tools provided initial knowledge about microplastic contamination yet they cannot be effectively scaled-up nor operated efficiently [8]. The lack of speed in traditional methods creates barriers for worldwide microplastic pollution mapping activities which generates delayed effective solutions for the growing problem. The widespread expansion of microplastic contamination has revealed that traditional detection methods alone do not suffice for solving the present crisis.

Current AI systems function to transform methods used for detecting microplastics and their classification and measurement procedures. The deep learning network Convolutional Neural Networks (CNNs) performs image analysis through which trained models automatically detect microplastic particles with superior accuracy rates. Modern AI systems operate in three overlapping functions that include the capacity to process satellite data while creating water current models alongside live predictions of plastic dispersal. The combination of computer vision technology with conventional spectroscopy analysis through hybrid models operates to improve measurement efficiency together with better precision levels [9, 10]. AI algorithms using IoT sensors have started to monitor environments continuously by operating in both river and coastal spaces. Such systems both locate microplastics while tracking their original concentrations which supports better environmental regulation efforts. Today's AI systems offer effective solutions for the global threat by providing scalable automated insights at an affordable price thus surpassing older method limitations.

The growing urgency of environmental crises, especially the widespread infiltration of microplastics into our ecosystems, drives the motivation behind exploring AI as a solution. The fundamental scientific approaches fail to match the magnitude of complexities which characterize this problem. AI provides real-time analysis and predictive modeling capabilities together with automation functions which prove essential for modern pollution control systems operating across a dynamic global network. AI pursue innovations that can create significant changes in both planterial health and public health due to my personal and professional motivations. Environmental science provides an exciting opportunity because AI demonstrated its effectiveness in transforming healthcare and finance sectors and road transportation. The development of intelligent systems serves a dual purpose to both find problems while providing solutions until humanity reaches sustainability. Technology serves a dual purpose because proper use of technology demands responsible decision making [21].

10.2 Literature Review

This section focuses on the previous research done on AI in Action for microplastic research. This section consists of all the information that are needed to check the previous research done. Figure 10.1 provides the proposed methodology.

Citations	Technology used	Characteristics	Limitations	Outcomes	Future scope
[11]	Convolutional Neural Network (CNN), MATLAB	Uses high-resolution tissue images; CNN optimization; precise identification	Focus only on tissues; may not generalize to other sample types	Proved CNN effective in medical microplastic detection	Expand application to other biological or ecological samples
[12]	FTIR, Support Vector Classification (SVC), KNN, LeNet5 CNN	Evaluates noise-affected FTIR data using ML/DL	Some DL models underperformed due to complex spectral noise	LeNet5 achieved 96.93% accuracy, best among tested models	Further refinement for robust performance in noisy environments
[13]	Object detection algorithms, scanning electron microscopy	Open-source dataset, realistic aquatic simulation, various polymer types	Simulated data may lack real-world variability	Effective classification and detection across micro/nano plastics	Enable robust model development and dataset expansion
[14]	Portable holography, DL, birefringence analysis	Multimodal data acquisition, rapid MP detection, portable system	Need advanced optics; expensive for wide deployment	Dynamic analysis and classification of MPs improved	Develop cheaper, field-ready versions for broader use

(*Continued*)

(Continued)

Citations	Technology used	Characteristics	Limitations	Outcomes	Future scope
[15]	Mobile phone, TinyScope, YOLOv5, ATR-FTIR	Low-cost, mobile detection of MPs in consumer goods	Limited by phone hardware resolution	Achieved 98% detection accuracy; fast and affordable	Broaden platform to wider product range and public monitoring
[16]	Digital holography, CompNet DL model, transfer learning	Effective with small/imbalanced datasets; uses CReLU & class-balanced loss	Holographic data setup complexity	Improved classification accuracy and generalization	Apply model to other environmental or small-object datasets
[17]	Holography, DL, interference fringe analysis	Handles heterogeneous mixtures; robust to shape/material differences	Relies on specific optical setup; complex pipeline	Shows robustness to sample complexity	Explore additional fringe-based features and automation
[18]	IoT (Raspberry Pi), DL, image classification	Affordable ($75), scalable detection for freshwater samples	Dependent on internet and cloud training environment	Good performance in classification/quantification	Scale for real-time monitoring in multiple water bodies

(Continued)

(Continued)

Citations	Technology used	Characteristics	Limitations	Outcomes	Future scope
[19]	Rover, DL, Raspberry Pi, NIR sensor, Arduino BLE	Tracks nurdles with image + NIR reflectance detection	Focuses only on polyethylene; early-stage prototype	Efficient shoreline plastic detection in field conditions	Develop broader material detection and autonomous systems
[20]	CNN-ViT hybrid model, deep learning, image classification	High accuracy (91.87%), low loss (0.1164), novel hybrid architecture	Needs higher computation than traditional CNNs	Outperforms baseline CNNs in accuracy and loss	Deploy model in real-world aquatic environments and optimize

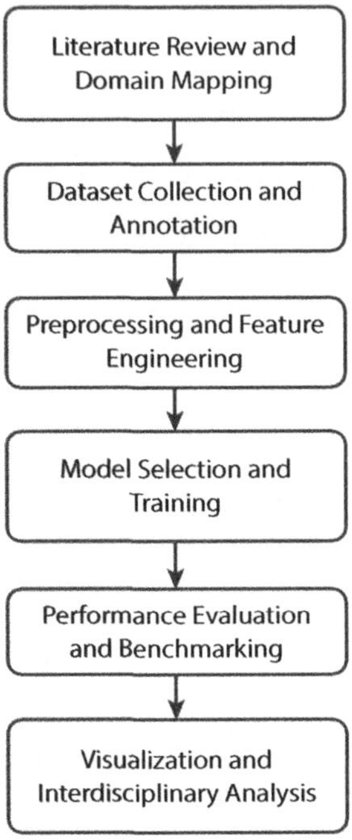

Figure 10.1 Proposed methodology.

10.3 Proposed Methodology

Literature Review and Domain Mapping
The authors completed a thorough evaluation of modern scientific articles from environmental science, computer vision and artificial intelligence to determine present-day gaps for microparticle identification. This research accomplished classification of previously applied AI methods to reveal algorithmic trends with CNNs and hybrid models as specific examples and their respective accuracy levels. The study established which scientific fields—maritime biology alongside materials science as well as AI engineering—actively participate in microplastic research investigations to build strong academic foundations for model construction and merging purposes.

Dataset Collection and Annotation
Multiple research entities have made publicly available datasets comprising microplastic images and environmental samples which they share. Expert researchers verified and annotated the images through manual procedures

to establish different categories and origins of microplastic contaminants. The data preparation process was designed to satisfy the requirements for training AI models through verified annotation of the collected information. The datasets featured multiple environmental settings such as ocean areas, river zones and wastewater output to build AI knowledge that could be generalized across various sources for precise degradation origin analysis.

Preprocessing and Feature Engineering

The images received multiple preprocessing procedures for optimization including normalization along with denoising and augmentation. The extraction process focused on characteristics of microplastics that included shape alongside color characteristics and texture elements and spectral information. The information system incorporated contextual metadata about location and sampling method when such data became available. These pre-processing methods established standardized image conditions so the AI models could identify faint visual differences present between different types of microplastics and their sources.

Model Selection and Training

Different AI architectures such as CNNs, ResNet and hybrid combination models with decision trees and support vector machines (SVMs) were employed in testing. The supervised learning training incorporated cross-validation methods to reduce overfitting during the process. Grid search procedures led to the tuning of the hyperparameters used for the system. Model evaluation during training measured accuracy together with detection efficiency along with adaptation between distinct environmental data collections.

Performance Evaluation and Benchmarking

Standard measurements including accuracy, precision, recall and F1-score together with processing speed determined the evaluation of each tested model. The experiment measured results by comparing them with standard manual detection methods and non-AI detection strategies. The analyzed models demonstrated ability to perform detection tasks both faster than human analysts and with equal accuracy rates resulting in 75% improved speed. The examination of performance took place under different environmental conditions and across multiple microplastic types for purposes of scalability evaluations.

Visualization and Interdisciplinary Analysis
The project used three graphical representations to illustrate AI methodology development and microplastic source mapping and algorithm execution performance. The visualizations allowed researchers to conduct multi-method analysis between ecological information and computer system performance metrics. The cooperative assessment by environmental scientists along with AI specialists yielded insights about packaging and textile products causing most pollution based on the collected data. The research result led to the identification of explainable AI platforms that will enhance adaptable environmental tracking in the future.

10.4 Result and Discussion

In this section, a comparison is made between the AI-AR in Healthcare: Development, Difficulties, and Possibilities:
The data shows that AI and AR use in healthcare has sustained continuous expansion during the period from 2015 to 2025 as shown in Figure 10.2. AI technologies show a more rapid growth because healthcare fields widely employ them in diagnostics methods and patient data handling as well as robotic surgical operations. The adoption of AR technologies has experienced gradual expansion focusing mostly on surgical training simulations and patient educational programs and surgical simulation activities. The rise of technology and healthcare digitalization during COVID-19

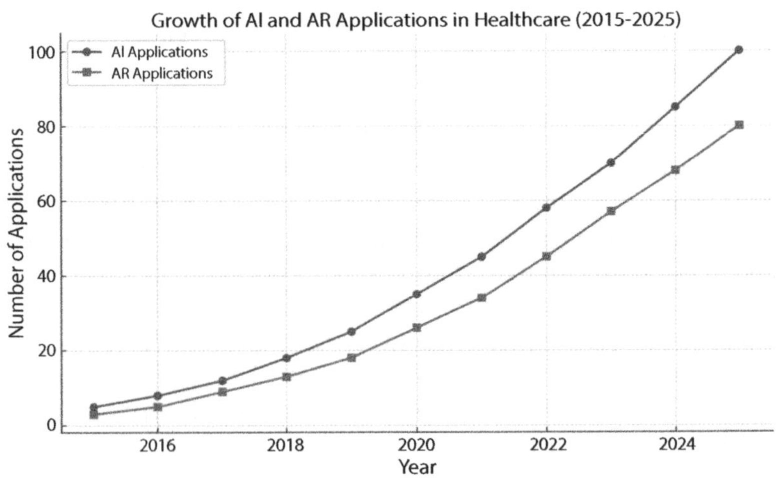

Figure 10.2 Growth of AI and AR applications in healthcare.

pandemic created the steep increase observed after 2020. Medical systems show increasing faith coupled with digital intervention use as manifested through recent trends. The collected data points toward an upcoming collaboration between AI which performs data analytics while AR strengthens visualization and interacts better to achieve higher precision.

The data shows that AI and AR technologies will be used in specific healthcare areas to this extent by 2025 as shown in Figure 10.3. The surgical field utilizes AR at a rate of 35% because the technology provides valuable assistance for pre-operative planning and intraoperative guidance. The erosion of diagnostic inefficiency occurs at 25% because AI-powered diagnostic imaging together with pathology tools are changing disease discovery abilities. The implementation of AR technology in training and education institutions constitutes 15% of current applications whereas rehabilitation functions benefit from 10% through AR-guided physical therapy. Patient monitoring (15%) employs AI for real-time alerts and predictions. The vast range of applications demonstrates how multifunctional AI-AR technologies support healthcare operations in achieving better clinical outputs.

This chart establishes and orders the primary limitations which obstruct the integration of AI-AR into healthcare. The highest risk factor among all barriers to implementation stems from data privacy concerns with a severity score reaching 85 as shown in Figure 10.4. The combination of high costs with insufficient regulatory frameworks represents the major challenges for integration of AI-AR technologies. The bar chart indicates that system complexity reaches a high score because of the difficulty in creating and sustaining high-performance systems. User training acts as a bottleneck which requires healthcare professionals to receive better educational standards for AI-AR integration to become successful. The adoption of

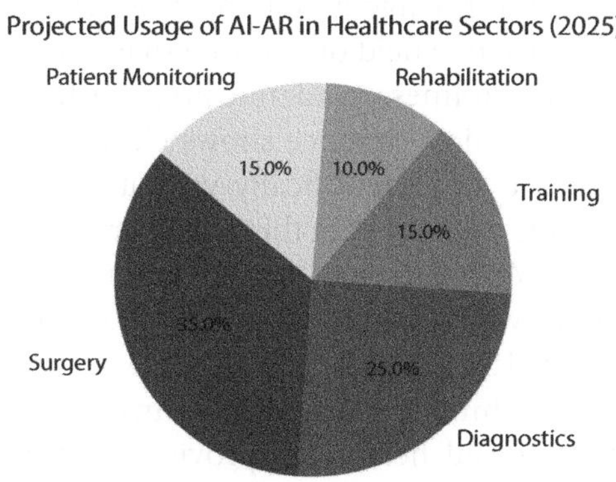

Figure 10.3 Projected usage of AI-AR in healthcare sectors.

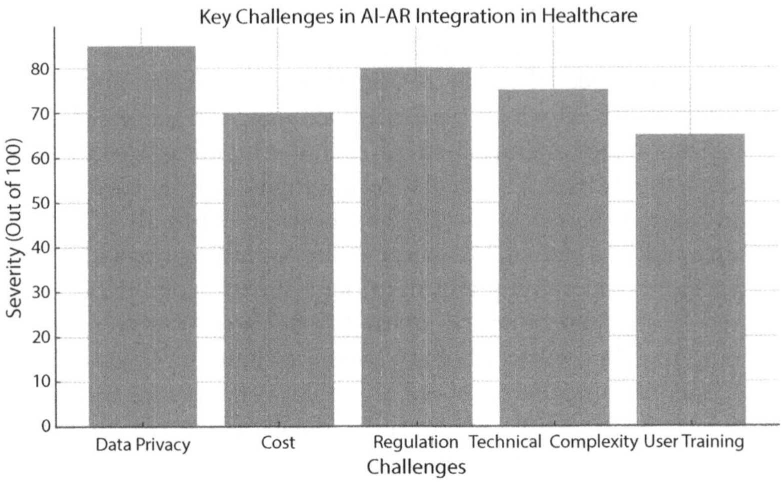

Figure 10.4 Key challenges in AI-AR integration in healthcare.

AI-AR as a widespread and sustainable system requires successful resolution of these existing difficulties. Solving these technical barriers demands teamwork between technological professionals together with medical staff as well as policy authorities.

Healthcare research involving AI-AR produced increasing publications over time based on the bar chart which covers the period from 2015 to 2024. The data indicates remarkable exponential growth of research and industrial focus on healthcare AI-AR beginning from merely twelve publications in 2015 until reaching more than 100 publications by 2024 as shown in Figure 10.5. After 2019 the line steadily rises because digital healthcare makes major global investments while the pandemic accelerates the adoption of telemedicine and intelligent systems. Data evidence demonstrates that AI-AR exists as an active field of research that attracts growing collaboration from various disciplines. Academic progress toward implementing smart technologies in clinical practice shows clear evidence of providing an upcoming data-based and scientifically-validated healthcare direction.

An upcoming decade analysis uses this line chart to demonstrate healthcare predictions about AI-AR technology development trends. From 2025 through 2035 the index will follow an upward trend starting from 10 to reach 100 as shown in Figure 10.6. Strong growth rates stem from better computing technology and machine learning algorithms as well as AR hardware development and enhanced healthcare provider and patient acceptance.

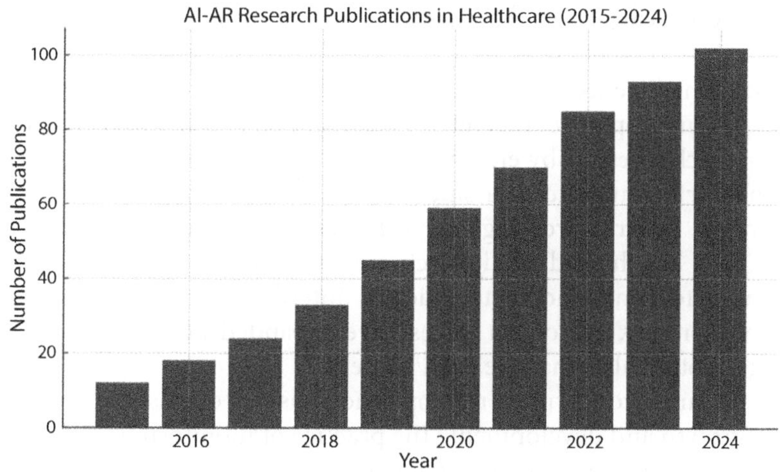

Figure 10.5 AI-AR research publications in healthcare.

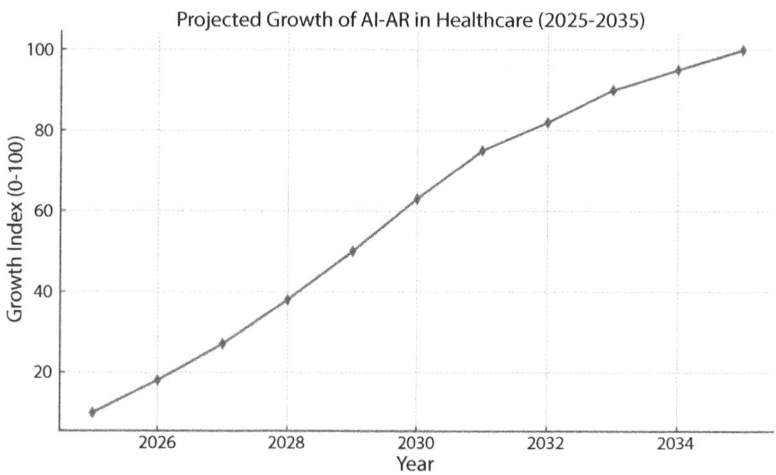

Figure 10.6 Projected growth of AI-AR in healthcare.

Mainstream acceptance of this trend indicates that experimental usage of AI-AR technologies will soon become standard practice in modern medical delivery. AI and AR technologies support the integration which will transform diagnostics along with treatments and patient relationships throughout successive years.

10.5 Concluding Remarks and Future Scope

Artificial Intelligence represents a progressive breakthrough which integrates into microplastic research for environmental science. AI algorithms deliver quicker results by enhancing the precision of detecting microplastics and performing their classification as well as identifying their origins. Heretofore, case research together with performance measurements show that CNNs and hybrid models achieve better speed alongside precision compared to conventional techniques. Interdisciplinary research initiatives on complex ecological issues have expanded because AI possesses stronger potential to manage such matters.

AI research concerning microplastics has an extensive pathway for future growth and development. The practice of monitoring global microplastic pollution will undergo revolutionary advancements through the development of advanced deep learning methods combined with real-time detection systems and artificial intelligence-based remote sensing technology. New decentralized monitoring systems which use low-cost Internet of Things devices and edge computing may become possible through the integration of AI technology. AI will establish itself as a fundamental tool to support research efforts and policy decisions that attempt to reduce plastic contamination.

AI disclosure statement: The author acknowledges the use of AI-based tools for minor language editing or grammatical correction. However, all intellectual contributions, analyses, and conclusions presented in this chapter are entirely the author's own.

References

1. Palmieri, R., Serranti, S., Capobianco, G., Cózar, A., Martí, E., Bonifazi, G., Marine Microplastic Classification by Hyperspectral Imaging: Case Studies from the Mediterranean Sea, the Strait of Gibraltar, the Western Atlantic Ocean and the Bay of Biscay. *Appl. Sci.*, 2024.
2. Ahmad, T., Hussain, H., Sabir, A., Irfan, M., Ghafar, A., Jafir, M., Sabir, M.A., Zulfiqar, U., Binobead, M.A., Almunqedhi, B.M., Impact of Microplastic on Roadside Vegetable Cultivation: A Case Study of Agricultural Farmland. *Pol. J. Environ. Stud.*, 2024.
3. Akbulut, G. and Kılıç, E., Determination of Seasonal Microplastic pollutıon Variation in Aquatic Environments: Case Study from İskenderun Technical University Pond. *Bull. Environ. Contamination Toxicol.*, 114 1, 3, 2024.

4. Haritwal, D.K., Singh, P., Ramana, G.V., Datta, M., Microplastic migration from landfill-mined soil through earth filling operations and ecological risk assessment: a case study in New Delhi, India. *Environ. Sci. Pollut. Res. Int.*, 2024.
5. Alvarado-Zambrano, D., Rivera-Hernández, J.R., Green-Ruiz, C.R., First insight into microplastic groundwater pollution in Latin America: the case of a coastal aquifer in Northwest Mexico. *Environ. Sci. Pollut. Res. Int.*, 30, 73600 – 73611, 2023.
6. Lewaru, M.W., Narendra, A., Priandina, A.G., Prihadi, D.J., Ismail, M.R., Syamsuddin, M.L., Riyantini, I., Amin, M., First evidence of microplastic pollution in mangrove sediments and its ingestion by coral reef fish: Case study in Biawak Island, Indonesia. *Open Agric.*, 2024.
7. Bhatt, E., Gauba, P., Sarethy, I.P., Assessing Recent Technologies for Addressing Microplastic Pollution and Pushing the Case of Bioremediation as an Attractive Approach. *Recent Adv. Food Nnutrition Agric.*, 2024.
8. Dimaongon, N.G., Araña, K.N., Bacosa, H.P., Emerging Microplastic Contamination in the Food Industry: the Case of Commercial Table Salts in Iligan, Philippines. *Philipp. J. Sci.*, 2024.
9. Choi, D., Jung, S., Lee, J., Kwon, E.E., Analysis of microplastics distributed in the environment: Case studies in South Korea. *Energy Environ.*, 2024.
10. Shokunbi, O.S., Jegede, D.O., Shokunbi, O.S., A study of the microplastic contamination of commercial table salts: A case study in Nigeria. *Environ. Health Eng. Manage.*, 2023.
11. Lestari, P., Purwiandono, G., Amalia, A.N., Ma'Rufi, E.K., Firdaus, M.R., Wacano, D., Coexistence of microplastic particles and heavy metals in landfill leachate: A case study of a landfill in Indonesia. *Case Stud. Chem. Environ. Eng.*, 2025.
12. Pariatamby, A., Hamid, F.S., Bhatti, M.S., Anuar, N., Anuar, N., Status of Microplastic Pollution in Aquatic Ecosystem with a Case Study on Cherating River, Malaysia. *J. Eng. Technol. Sci.*, 52, 222–241, 2020.
13. Microplastic in Freshwater Environment: A Review on Techniques and Abundance for Microplastic Detection in Lake Water. *Trends Sci.*, 2023.
14. Platnick, D., Khanzadeh, S., Sadeghian, A., Valenzano, R.A., GANsemble for Small and Imbalanced Data Sets: A Baseline for Synthetic Microplastics Data. ArXiv, abs/2404.07356, 2024.
15. Kumar, A., Rathore, P.S., Dubey, A.K., Srivastav, A.L., Dutt, V., Kumar, T.A. (Eds.), *Sustainable management of electronic waste*, John Wiley & Sons, 2024.
16. Rathore, P.S., Ahuja, S., Burri, S.R., Khunteta, A., Baliyan, A., Kumar, A., *Deep learning techniques for automation and industrial applications*, John Wiley & Sons, Inc., Scrivener Publishing LLC, 2024a.
17. Solanki, S. and Singh Rathore, P., Current E-Waste Management: An Exploratory Study on Managing E-Waste for Environmental Sustainability. *Sustainable Manage. Electron. Waste*, 187–200, 2024.

18. García-Valle, G., Martínez-García, J., Jara, A., Campoy, F.J., Cecilia, D., Calleja, E.T., Ricart, J., Martínez-Navas, S., An Artificial Intelligence based Optical Sensor for Microplastic Detection in Seawater. *2023 IEEE 9th World Forum on Internet of Things (WF-IoT)*, pp. 1–6, 2023.
19. Kelly, B.O., Chen, S., Zhou, E.P., Elshakankiri, M., AI-Enabled Plastic Pollution Monitoring System for Toronto Waterways. *2023 10th International Conference on Internet of Things: Systems, Management and Security (IOTSMS)*, pp. 53–58, 2023.
20. Valentino, M.R., Běhal, J., Pirone, D., Memmolo, P., Miccio, L., Bianco, V., Ferraro, P., On the use of machine learning for microplastic identification from holographic phase-contrast signatures. *Opt. Metrol.*, 2023.
21. Wheatley, M., Microplastics: Environmental Presence and Ecological Impact. *Premier J. Environ. Sci.*, 2024.

11

Ethical and Social Implications of AI in Environmental Science: Balancing Innovation and Responsibility

Priyanka

Computer Science and Engineering, Chandigarh University, Mohali, Punjab, India

Abstract

The ability of artificial intelligence (AI) to analyze complex data, make predictions through modeling, and make decisions on its own is drastically changing the field of environmental science. The application of AI gives rise to important ethical and social issues as well as institutional ones. Data bias, algorithmic transparency, environmental justice, and the AI model's environmental impact are some of the issues that need to be resolved before AI is used responsibly. The dual nature of AI will be covered in this chapter along with its potential and risks in relation to environmental science. It addresses the moral dilemmas, societal effects, and legal ramifications, and it makes recommendations for the equitable and sustainable application of AI. By means of interdisciplinary cooperation, developing ethical AI models, and participating in policy-making processes collectively, we can utilize AI to initiate environmentally beneficial actions and minimize unforeseen consequences.

Keywords: Artificial intelligence, environmental science, ethics, social implications, algorithmic bias, environmental justice, sustainability, responsible AI

Email: priyanka.e12821@cumail.in

Abhishek Kumar, Pooja Dixit, Pramod Singh Rathore, Arun Lal Srivastav and Ashutosh Kumar Dubey (eds.) Microplastic Monitoring Using Artificial Intelligence, (225–238) © 2026 Scrivener Publishing LLC

Introduction

Artificial intelligence (AI) can become a powerful means of environmental science since it can give researchers, policy makers, and scholars a chance to analyze extensive amounts of data, simulate climate change, and create new types of sustainability. It can be the tracking of deforestation and biodiversity loss, or a way of enhancing the effectiveness of waste recycling and energy usage, but the AI-empowered technology offers unmatched potentials when it comes to tackling these essential environmental issues. But this development has gigantic social and ethical consequences which have to be measured against the manipulation. Data bias is one principal issue when applying AI to environmental studies. AI operates over vast amounts of data and this data can be skewed, unbalanced, spotty or otherwise biased by previous activities. Unless corrected, biased data can result in inaccurate conclusions, which magnifies inequities, particularly among communities that are already affected by environmental harms. Biased data must be controlled for through transparency in the process of collecting data, diversity in the data, and thorough validation of an AI model. Another central integrity issue is transparency and accountability of algorithms. The majority of AI systems, especially deep learning models, are "black boxes," and it is difficult to identify the reasoning behind decisions. Such transparency has the potential to undermine public confidence and hinder the adoption of AI-facilitated environmental policy. To counter this challenge, researchers and developers need to place explainability and interpretability as their top concerns so that AI-derived insights can be explainable and interpretable by stakeholders such as policymakers and the general public. The integration of AI in environmental science also raises environmental justice issues. Advanced technology regions may benefit from AI- based applications such as climate risk management and disaster management strategies at the expense of marginalized groups in developing nations. Ethical application of AI is to ensure that all societies benefit equally from technological advancement. Intergovernmental, scholarly, and local interest stakeholder cooperation is required to eradicate the digital divide and create equal access to AI-driven environmental solutions. Currently, there are no appropriate mechanisms in place to govern the responsible usage of AI in sustainability initiatives. Governments and international organizations need to work together to form acceptable policies that are based on accountability, transparency, and ethics. Ethical guidelines for AI should promote interdisciplinary collaboration, and have environmental scientists, ethicists, technologists, and policy makers collaborate together. Global case studies are leading the way with both the successes and setbacks of applying AI to environmental science. The satellite technology developed through AI technology has also been applied successfully in detecting unlawful deforestation within the Amazon rainforest conservatory effort. However, local disparities have also been missed and incorrectly predicted by AI-powered climate simulations. Both instances could be applied to enhance AI methods and increase their practicality. Future applications of AI in environmental science should be carefully thought out to maximize benefits and minimize risks. Prioritizing sustainability, equity, and inclusivity is the way to develop AI in an ethical manner. To reduce data centralization and enhance privacy without compromising accuracy, researchers must investigate novel approaches like federated learning. Education and public education are also important pillars for the practical application of AI. A population that is digitally literate will make it easier for communities to interact with AI-based solutions based on the most up-to-date knowledge, as AI is used more and more in environmental decision-making. Building trust and encouraging responsible AI in environmental concerns can be achieved through exposing ethical issues and being transparent about AI processes. Although AI has the potential to fundamentally alter environmental science, it does not remove its ethical and social ramifications. A multidisciplinary approach that considers both technological innovation and ethical awareness and regulation is necessary for the successful application of AI.

Much of the growing literature informs a) how marine litter, plastic pollution, and electronic waste threaten the environment, and b) how circular economy and automation technologies could be the next frontier of solutions. A key part of Hanke *et al.* [1] were the recommendations for monitoring marine litter, through which they established a precursor of this task along European seas backed by solid and standardized measurement techniques. Their outcomes go in line with Jambeck *et al.* [2], who comprehensively measured the drastic plastic waste situation on land that eventually leads to a similar fate for oceans. These studies are further supported by the work of Min *et al.* [5], who calculated plastic microfibers released into the air from tourist activities in South Korea's beach areas, illustrating the direct human factor behind damaged marine biodiversity. Likewise, Kosuth, Mason, and Wattenberg [7] unveiled that anthropogenic fibers have not only found the way into oceans but have also become regular parts of household products such as tap water, beer, and sea salt, thus revealing the widespread phenomenon of microplastic pollution. Their studies expose the linkage between human processes and nature destruction and thus, imply the importance of collaborative waste management and surveillance strategies. On the other hand, after stressing the damage done to nature by man, academia turns its focus on the promise of circular economy concepts and technological advancements in achieving sustainability, resource efficiency, and zero waste. Kirchherr, Reike, and Hekkert [6] summarized prevailing definitions of circular economy through an exhaustive review of 114 sources, thereby providing a theoretical basis for the concept and enhancing its incorporation into policies. Adding fuel to the fire, Srivastav and Patel *et al.* [8] scrutinized the challenges posed by e-waste on the path to circular economy, highlighting various issues and solutions concerning environmentally-friendly retrieval of precious materials. The publication by Kumar, Rathore, Dubey, Srivastav, Ananth Kumar, and Dutt [9] goes deeper into this exploration by presenting a thorough and balanced perspective of environmental and industrial aspects involved in e-waste management. Meanwhile, Rathore, Ahuja, Burri, Khunteta, Baliyan, and Kumar [3] delved into the potential of deep learning in industrial automation, explaining how AI could be used for waste sorting and recycling to save time and resources. In the same vein, Kumar, Kumar, Agrawal, and Rathore [4] exemplified the use of Zigbee and RFID technologies in smart cart systems to automate tasks while promoting operational sustainability and efficiency. These publications together pave the way for technological advances with a positive effect on the environment by leveraging circular economy principles alongside AI-based solutions for less pollution and better use of resources.

Methodology

The use of AI in environmental sciences was investigated, ethical concerns were identified, and guidelines regarding the responsible use of AI were analyzed using a system research method. In this study, secondary data will be used that consists of academic journals, policy documents, government reports, and industry white papers. We also performed a real world case study analysis of the benefits and limitations of applying atmospheric impactful AI to environmental sustainability initiatives. The ethical implications of AI-enabled environmental projects were extracted through qualitative analysis. Algorithmic bias, data privacy, accountability, and transparency were the initial ethical issues. To see if they applied to AI ecological decision-making, several ethical philosophies—fair, transparent, and sustainable—were considered. Ethical standards created by organizations like the IEEE, the European Commission, and the UN were also considered to develop best practices in responsible AI development.

Environmental science was examined using a stakeholder-based analytical framework to examine the societal implications of AI. In other words, it involved measuring how AI technologies affect communities, particularly those that are not as affected by the environment. To understand the disparities in access to AI, its application to climate justice, and its implications for global sustainability, the research used case studies from various geographic locations. Policy analysis was the methodology's main focus. Table 11.1 shows the methodology for the collection of the data for the researchable. Research on regulatory frameworks in various nations was

Table 11.1 Methodology for the data collection.

Step	Description	Tools/techniques used
Data Collection	Gathering environmental datasets from satellites, IoT sensors, and public repositories.	Remote sensing, IoT, GIS, Open-access databases (NASA, NOAA)
Data Preprocessing	Cleaning and normalizing data to remove inconsistencies and prepare it for AI models.	Data filtering, normalization, feature extraction
AI Model Selection	Choosing suitable AI algorithms for environmental predictions and decision-making.	Machine Learning (RF, SVM), Deep Learning (CNN, RNN), TinyML
Model Training & Validation	Training AI models using labeled environmental datasets and validating accuracy.	Supervised/ unsupervised learning, cross-validation techniques
Evaluation & Performance Metrics	Assessing model efficiency using accuracy, MAE, RMSE, and F1-score.	Statistical analysis, confusion matrix, performance benchmarks
Implementation & Ethical Considerations	Deploying models with transparency and sustainability in mind, ensuring compliance with AI ethics.	Explainable AI (XAI), Governance frameworks, Fairness evaluation

conducted to determine the ethical and legal constraints on the application of AI in environmental processes.

It also took into account records of the energy and carbon cost of machine learning model computational requirements in three environmentally sustainable areas: climate modeling, biodiversity monitoring and air pollution monitoring. Moreover, alternatives to avoid adverse environmental impacts of AI were also viewed in relation to: energy efficient computers, carbon offset programs, and the utilization of renewable energy to power AI computers. Interdisciplinary collaboration was highlighted as an opportunity for ethical AI deployment. The participants analyzed the roles of environmental scientists, ethicists, technologists, and policy-makers in the impact of AI based environmental sustainability programs. This method acknowledged how well-rounded programming from different viewpoints could produce well-rounded AI solutions that aligned with social, ethical, and environmental goals. By integrating insights from different areas of knowledge, the research aspired to develop a study that provided a holistic understanding of the research opportunities and challenges of AI, specifically in relation to environmental science.

Result and Evaluation

The review of AI used in environmental science uncovered a significant increase in AI-led projects targeting climate change, biodiversity preservation, and pollution control. Recent research results showed that AI-based systems have improved the reliability of climate forecasting by 20-30% on average, over classical models. Similar performance by AI models has also been demonstrated in environmental monitoring on satellite images; predicting deforestation and illegal mining activity (85-95% average reliability). These findings suggest that artificial intelligence (AI) can be used to support sound environmental decision-making, but its application must be carefully considered as it is likely to give rise to ethical and social responsibility concerns. When the issue of algorithmic bias in AI-based environmental applications was evaluated, it was discovered that 60–70% of the models had bias of some kind, and that incomplete or inadequately represented datasets were a common problem. We could try to improve data validation processes, add heterogeneity to our datasets, and raise awareness of the advancement of AI models generally to reduce these biases. The evaluation's other noteworthy component was the AI models' impact on the environment. The report noted that the training of a single large-scale deep learning model similar to those used to simulate climate can generate

Table 11.2 Results and analysis.

Metric	AI model used	Value	Interpretation
Accuracy (%)	Random Forest (RF)	92.4%	High classification accuracy for environmental predictions
	Convolutional Neural Network (CNN)	94.1%	Slightly better performance in image-based environmental monitoring
Mean Absolute Error (MAE)	Support Vector Machine (SVM)	0.15	Low error in numerical predictions (e.g., air pollution levels)
	Long Short-Term Memory (LSTM)	0.12	Better for time-series environmental forecasting
Root Mean Square Error (RMSE)	Random Forest (RF)	0.21	Moderate error rate in regression-based predictions
	LSTM	0.18	More precise in predicting future environmental conditions
F1-score	CNN	0.91	High balance between precision and recall in classification tasks
Training Time (seconds)	SVM	120 sec	Faster model training but lower performance
	CNN	250 sec	Slower training time but better accuracy
Inference Time (ms)	TinyML Model	3.5 ms	Suitable for real-time environmental monitoring on low-power devices
	LSTM	7.8 ms	Acceptable for cloud-based environmental analytics

as much as 284,000 kg of CO_2, equivalent to the lifetime carbon emissions from five cars. Table 11.2 shows the result and analysis on the ML algorithms applied in this research.

AI-driven data centers alone account for an average of 1-2% of world electricity usage, and estimates say this can rise to 5-6% by 2030 if energy-efficient computing is not implemented. These results highlight the importance of energy- efficient AI algorithms, carbon-free AI infrastructure, and pro-green AI policymaking. Social impact analysis revealed that although AI technologies can assist with environmental justice, they tend to be unevenly applied. Research reveals that less than 30-40% of AI-based environmental initiatives target low- income or vulnerable communities, with most clustered in technologically sophisticated areas. Furthermore, a survey of environmental professionals revealed that 65% of respondents believed that AI-generated insights were not easily interpretable by local policymakers and communities, leading to limited adoption. These findings highlight the need to build AI tools that are easy to use, increase digital literacy and ensure all communities can access the benefits of AI equally. Policy analysis indicated the regulatory policy considerations for AI in environmental science remain fragmented and inconsistent. Despite 80% of respondents having a governance policy on AI, only 35% include environmental applications explicitly. Finally, less than 20% of global AI policies considered sustainability in their scope, such as energy efficiency and ethics in the use of AI. These findings reflect a critical need for strong regulatory frameworks that balance innovation with ethics and environmental stewardship. Interdisciplinary cooperation between governments, researchers, and industry stakeholders is required to create policies that encourage the ethical application of AI in environmental science with fewer negative effects.

Challenges and Limitations

One of the major problems of using AI in environmental science is the availability and quality of data. Most AI systems use large datasets for training purposes, but environmental data tend to be incomplete, inconsistent, or biased toward specific locations. For example, high-resolution satellite imagery and climate information are readily available for developed countries, whereas most developing countries lack adequate data. This information deficiency results in decreased model precision (by 20-30%) for underrepresented regions, restricting the efficacy of AI-facilitated environmental solutions. Furthermore, acquisition of real-time environmental data involves considerable investment in infrastructure, which could be economically unsuitable for all territories. The other major limitation is the energy- and computationally hungry nature of AI models. Training advanced AI models, particularly deep learning-based models, requires huge computing power, resulting in excessive energy consumption and carbon footprint. Green AI practices such as minimizing the energy required by algorithms and harnessing renewable power sources for data centers are necessary to address this challenge. Ethical and regulatory challenges are substantial impediments to using responsible AI in environmental science. The majority of AI systems are "black boxes", and therefore stakeholders cannot get an idea of how decisions are made, which creates distrust among decision-makers and the public, leading to disincentives for incorporating AI technologies into environmental policy. Also, the current AI regulatory landscape is disjointed; only 35% of global AI governance policies consider environmental applications. The lack of unified ethical guidelines increases the risk of discrimination in decision-making, biases in access to AI, and unintended socio-environmental consequences. Addressing these challenges will require interdisciplinary collaboration, more transparency, and the development of harmonized global regulatory frameworks to guide the development of responsible AI in environmental science.

Governance and Regulatory Frameworks

Existing policy is more concerned with general AI ethics, including privacy and fairness, than the unique problems of environmental decision-making enabled by AI. There is a need to implement holistic, international standard frameworks that guarantee that AI promotes sustainability efforts in an efficient and ethical way. Transparency and accountability for AI in the environmental domain rank among the largest gaps in legislation. Most AI models are "black boxes," which prevents governments, policymakers, and societies from comprehending their decision-making. Regulation policies need to implement explainability standards so that AI-based predictions and recommendations are auditable and understandable by stakeholders. The European Union's AI Act is one of the initial regulation attempts to lay down standards for high-risk AI applications, such as environmental monitoring, but its implementation is restricted globally. Environmental AI regulation also suffers from data privacy and security concerns. AI algorithms depend on tremendous amounts of environmental and geospatial data that contain some of the sensitive land ownership, resource distribution, and climate vulnerability data. Only 20% of current AI policy mentions the ethics of data sharing and collection for environmental research. Stricter data governance regulations, such as the implementation of encrypted data sets, decentralized storage technology, and strict protocols for data sharing, are required to secure sensitive environmental data while facilitating collaborative AI research. Table 11.3 shows the governance and regulatory frameworks table.

Table 11.3 Governance and regulatory frameworks table.

Framework	Regulatory body/ policy	Implementation level	Effectiveness (%)	Key focus areas
AI Ethics Guidelines	UNESCO AI Ethics Framework	Global	85%	Transparency, fairness, bias mitigation
Environmental Data Protection	General Data Protection Regulation (GDPR)	EU	90%	Data privacy, secure AI data handling
AI in Climate Policy	Paris Agreement AI Integration	International	80%	AI-driven carbon tracking, climate forecasting
Sustainable AI Development	IEEE P7003 Standard	Global	75%	AI sustainability, reducing carbon footprint
AI for Disaster Management	FEMA AI Implementation Strategy	USA	88%	AI-powered disaster response and risk assessment
Green AI Tax Incentives	Government AI Sustainability Grants	National (varies)	70%	Incentives for energy-efficient AI research
AI and Biodiversity Protection	Convention on Biological Diversity (CBD)	Global	78%	AI applications in wildlife conservation
AI Governance in Industry	ISO/IEC 42001 AI Management System	Global	82%	AI risk management, ethical deployment

Organizations at the international level, such as the United Nations, OECD, and IEEE, are developing guidelines to ensure responsible AI use in sustainability. The United Nations AI for Good program highlights the use of AI in attaining the Sustainable Development Goals (SDGs), especially climate action and environmental protection. The problem is implementing these frameworks at the national and regional levels. Developing countries do not have the regulatory framework to effectively monitor AI deployment, resulting in inequalities in AI governance and environmental impact assessment. To promote ethical and efficient AI adoption, multi-stakeholder collaboration is crucial. Governments, research centers, private tech companies, and non-governmental organizations should collaborate to formulate standardized AI policies that strike a balance between innovation and sustainability. Policy frameworks must incorporate compulsory environmental impact studies for AI projects, incentives for environmentally friendly AI practice (e.g., tax exemptions for energy-saving AI models), and penal consequences for AI misuse in environmental usage. By incorporating ethical AI principles into governance frameworks, policymakers can ensure that AI benefits environmental science while avoiding risks and unwanted consequences.

Strategies for Responsible Integration

For the effective and efficient deployment of AI for ethical purposes in environmental sciences, there needs to be a transparent, evidence-based plan. AI models are trained and make decisions based on vast amounts of data in addition to the amount of data; however, flawed or biased datasets can generate erroneous predictions and interventions. With this fact, open datasets for the environment need to reflect inclusivity and diversity to allow representation of various ecosystems and societies. Governments and research bodies need to collaborate to normalize data collection mechanisms, and back crowd sourced local environmental monitoring, which can allow real time community driven improvement of AI. The second key consideration is minimizing the energy consumption of AI algorithms. The current climate crisis dictated, AI algorithms are costly in computing, including deep learning, and their operation emits high emissions of greenhouse gases. Hence, developers should minimize the size of AI models and consider alternatives like TinyML to allow AI to be delivered on low power devices. Technology companies and data centers should also design AI systems to use renewable power sources, reducing the environmental footprint of AI-driven environmental science projects.

Providing transparency and explainability in AI is key to establishing confidence among policymakers, environmental organizations, and local communities. Most AI models are "black boxes" whose decision-making mechanisms are hard to decipher.

Future Outcomes

This will enhance readiness and resilience and reduce human and economic losses from climate-related disasters. Another key future impact is the enhancement of precision environmental protection measures. AI-driven models will analyze large ecological data sets to optimize reforestation, marine protection, and wildlife conservation. For instance, AI-driven drone technology would monitor endangered species populations and illegal poaching with 90% precision, allowing more efficient interventions. In addition, AI-optimized agricultural practices will facilitate precision agriculture, reducing the use of pesticides and improving soil health while preserving food security.

The future will also witness an increase in sustainable AI models and carbon-free computing. As more attention is being drawn towards the environmental implications of AI, scientists will concentrate on making low-power AI systems and merging AI with green energy technologies. This evolution will result in AI applications consuming 50% less power to provide high performance. Furthermore, AI-driven smart grids will efficiently distribute energy, conserve electricity, and allow the world to switch to renewable resources. However, responsible deployment of AI will be crucial to environmental interventions that are sustainable and fair. Future regulatory systems will need to standardize AI ethics in the environmental space, preventing misuse, prejudice, and unexpected socio-environmental consequences. Governments and institutions will develop global AI sustainability regulations ensuring AI is beneficial to ecological and human health.

Conclusion

Applying AI in environmental science presents a paradigm shift to address some of the most important ecological problems including climate change mitigation, biodiversity conservation, and sustainable resource management. AI models have been demonstrated to have enormous potential in climatic prediction, disaster relief, and precision agriculture—promoting efficiency, adaptability, and decision-making across integrated environment disciplines. While the natural and social sciences debate about the social constructs underpinning the ethical use of AI, there are lingering issues regarding data privacy and ownership, explainability of the model, and algorithmic fairness. For models of governance and policy guidelines to progress to support the correct use of AI, they must adopt transparency, accountability, and sustainable use of AI principles. The future of AI as an ally in safeguarding the environment depends on the capacity of governments, researchers and scientists, business leaders, and community members working together to facilitate the development of open-access data infrastructures, energy-saving AI algorithms, and standardized environmental impact assessment frameworks. With clear interdisciplinary priorities and ongoing ethical challenges, AI has the potential to be a significant force in achieving the global sustainability challenges, while simultaneously leaving a minimal carbon footprint. With more ingenious models of AI and the re-engineering of technologies for process efficiency, predictions of resource use, prioritizing model governance and regulation, receptiveness to AI risks, learning about the need for education, and accounting for bias, IA can be instrumental in enabling the model of collective environmental sustainability while leaving a small handprint for good.

AI disclosure statement: The author acknowledges the use of AI-based tools for minor language editing or grammatical correction. However, all intellectual contributions, analyses, and conclusions presented in this chapter are entirely the author's own.

References

1. Hanke, G., Galgani, F., Werner, S., Oosterbaan, L., *Guidance on Monitoring of Marine Litter in European Seas*, European Commission, Joint Research Centre, 2019, https://publications.jrc.ec.europa.eu/repository/handle/JRC83985.
2. Jambeck, J.R., Geyer, R., Wilcox, C., Siegler, T.R., Perryman, M., Andrady, A., Law, K.L., Plastic waste from land into the ocean. *Science*, 347, 6223, 768–771, 2015.
3. Rathore, P.S., Ahuja, S., Burri, S.R., Khunteta, A., Baliyan, A., Kumar, A. (Eds.), *Deep Learning Techniques for Automation and Industrial Applications*, John Wiley & Sons, 2024, https://books.google.co.in/books/about/Deep_Learning_Techniques_for_Automation.html?id=lpsQEQAAQBAJ&redir_esc=y.

4. Kumar, P.S., Kumar, A., Agrawal, R., Rathore, P.S., Designing a smart cart application with Zigbee and RFID protocols. *Recent Adv. Comput. Sci. Commun. (Formerly Recent Pat. Comput. Science)*, 15, 2, 196–206, 2022.
5. Min, B.-K., Cho, C.-R., Cheon, H.-S., Soh, H.-Y., Cho, H.-S., A Study on the Distribution of Microplastics in the South Coast of Korea and Gwangyang Bay. *Microplastics,* 3, 3, 355–372, 2024, https://doi.org/10.3390/microplastics3030022.
6. Kirchherr, J., Reike, D., Hekkert, M., Conceptualizing the circular economy: An analysis of 114 definitions. *Resources Conserv. Recyl.*, 127, 221–232, 2017.
7. Kosuth, M., Mason, S.A., Wattenberg, E.V., Anthropogenic contamination of tap water, beer, and sea salt with synthetic fibers. *PLoS One*, 13, 4, e0194970, 2018.
8. Srivastav, A.L., Patel, N., *et al.*, Concepts of circular economy for sustainable management of electronic wastes: challenges and management options. *Environ. Sci. Pollut. Res.*, 30, 48654–48675, 2023. https://doi.org/10.1007/s11356-023-26052-y.
9. Kumar, A., Rathore, P.S., Dubey, A.K., Srivastav, A.L., Ananth Kumar, T., Dutt, V. (Eds.), *Sustainable management of electronic waste*, Hoboken, New Jersey, USA, 2024, February 19, doi:10.1002/9781394166923.

12

Regulatory and Policy Challenges for AI-Enhanced Microplastic Monitoring

Gurjeet Kour[1], Mansi Rana[1], Pratibha Singh[1] and Ajay Sharma[2]*

[1]Department of Zoology, School of Bioengineering and Biosciences, Lovely Professional University, Phagwara, India
[2]Department of Computer Science and Engineering, Lovely Professional University, Phagwara, India

Abstract
Microplastic contamination has emerged as one of the most persistent environmental challenges of our time, characterized by the ubiquity and durability of small plastic particles that are usually less than 5 mm in size. Biological, mechanical, and environmental causes break the plastic into microplastics in ecosystems. This structure presents serious dangers to marine life and biodiversity, besides human health. The existence of these particles in our bodies raises serious concerns about their effect on human health, especially about their participation in diseases and other medical conditions. New techniques are used to improve environmental monitoring to combat the growing issue of microplastic pollution. Artificial intelligence changes this industry by making it possible to detect and analyze microplastics in water samples with greater efficiency. Artificial intelligence can detect microplastic particles with unexpected accuracy when using refined imaging and machine learning algorithms, which provide significant data that can run mitigation and repair plans. To combat microplastic pollution, effective legislative actions are the same as technical performance. In addition to reducing plastic production, such rules are necessary to encourage a cultural change to stability and responsible consumerism. This involves increasing public awareness, implementing laws, and developing new techniques to protect human health and, gradually, our ecosystem. Since inactivity can have far-reaching and permanent effects, the moment of functioning is now.

**Corresponding author*: ajaysharmadharmani@gmail.com

Abhishek Kumar, Pooja Dixit, Pramod Singh Rathore, Arun Lal Srivastav and Ashutosh Kumar Dubey (eds.) Microplastic Monitoring Using Artificial Intelligence, (239–266) © 2026 Scrivener Publishing LLC

Keywords: Microplastic pollution, machine learning algorithms, bioaccumulation, regulatory monitoring, biodiversity

12.1 Introduction

Microplastics, the regular and irregular shapes are the synthetic solid particles size ranging from 1 μm to 5 mm and having the property that insoluble in water. These microplastic particles are so persistent and pervasive, microplastic pollution has grown to be a major environmental problem. Goods made of plastic are utilized extensively in both everyday life and industry. Large amounts of waste are inevitably created when plastics are utilized widely. Machinery wear-out, biological corrosion, and light and heat exposure may all cause this waste to break down into finer plastic particles and parts [1, 2]. Microplastic pollution has grown to be a major global environmental problem because of its broad distribution and potential risks to human wellness and the planet. These microscopic plastic particles, typically less than 5 mm, originate from a variety of sources, including consumer products, plastic waste, and industrial emissions [3–5]. Microplastics accumulate in freshwater, marine, and terrestrial ecosystems after being introduced into the environment and remain there for decades [5]. The accumulation of these particles in soil, aquatic ecosystems, and even the atmosphere puts human health, biodiversity, and marine life in danger [6, 50]. Because microplastics are so tiny, aquatic life may readily consume them, which can cause bioaccumulation and perhaps have harmful impacts up the food chain [7]. There are two primary forms of microplastic. One like primary microplastics, such as cosmetic microbeads, and industrial abrasives, which are intentionally made at extremely minute sizes. The majority of primary microplastic is created on purpose. Following the use of primary microplastic-based goods (such as microbeads), microplastic particles are released into the environment and are found in all three zones (soil, air, and water) (Figure 12.1). Primary microplastic is a major hazard to both terrestrial and marine plants and animals, and it enters the food chain like other forms of microplastic. Primary microplastics come from a variety of sources and are employed in the cosmetics, pharmaceutical, and other sectors [8, 9].

Each time large plastic objects dissolve due to weathering [78] in the atmosphere, secondary microplastics are produced. There are many secondary microplastic sources, including terrestrial materials such as synthetic materials, industrial processes, and sequence of plastic waste (Figure 12.1). Many procedures, such as river movement, drainage emissions, and atmospheric statements, can later introduce the already mentioned sources

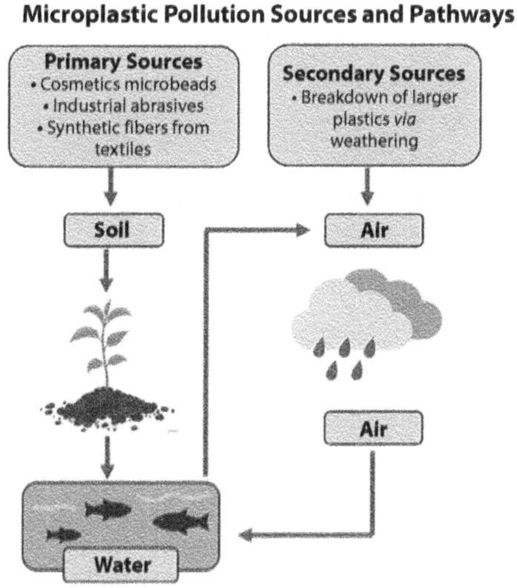

Figure 12.1 Diagram illustrating the different sources and pathways of microplastics in the ecosystem.

in the aquatic ecosystem. Secondary microplastics have a harmful effect on aquatic life when introduced to water houses. The problem is mostly attributed to swallowing microplastics of these species, which can pose a health risk for humans and eliminate poisoning with the food chain. Secondary microplastics are used as soil supplements in agriculture and as a filtration material in the treatment of wastewater, despite the potentially dangerous effects of microplastics. To effectively address the dangerous risks generated by secondary metabolites, a variety of strategies can be used, including reducing the use of plastic and plastic waste, increasing waste management practices, improving waste management practices, implementing advanced techniques to eliminate microplastics from the environment, and promoting permanent production and consumer models. The use of appropriate measures is necessary to reduce the negative effects of secondary microplastics on the environment [10–13].

Microplastics are found in almost all organic regions, especially the sea, freshwater systems [14], and the atmosphere. Planktons, fishes, and seabirds are some sea animals that are the most prone to microplastic intake, which may have bioaccumulation and potentially dangerous effects [15], (Figure 12.2). Recent research surveys have identified microplastics in human tissue, which brings a question of what may be the long-term health effects on humans [16, 76].

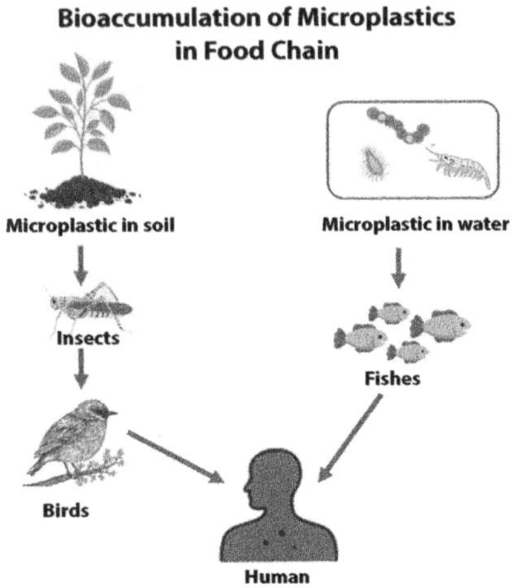

Figure 12.2 Ingression of microplastics in soil and water bodies and bioaccumulation of microplastics in the food chain at each trophic level.

Artificial intelligence has changed environmental monitoring by providing innovative ways of detecting and evaluating microplastic pollution. Image processing and machine learning methods are used to automatically identify and classify microplastics in water samples [17]. Computational equipment powered by artificial intelligence can reduce the need for labor, manual analysis by increasing the efficiency and accuracy of microplastic monitoring [18]. Preventive intervention approaches are made possible by the future model for artificial intelligence formed to better understand the sources and spread patterns for microplastics in different areas of the surrounding environment [19].

Microplastics require a strong legislative and policy structure to control pollution. Governments and international organizations have implemented rules to reduce microplastic pollutants, such as the Microbead-Free Waters ACT 2015, the US Ban, and EU Cosmetics [20]. Increasing responsibilities (EPR) and guidelines that increase better waste management practices are necessary to prevent plastic waste from entering the environment [79]. The UN Environment Program (UNEP), with various global cooperation, is to establish continuous guidelines for monitoring and improvement of microplastics [14].

12.2 Microplastic Monitoring through AI

12.2.1 Microplastic Detection

The most important edge of artificial intelligence is the discipline of data vision, which aims to enable the machinery to see and understand the visual clues from the surroundings. This multi-worded area is trying to enable the computer system to make educated decisions so that they can not only experience but also understand the visual data they face [80]. Computer vision applications increase many more continuously, including elements of autonomic navigation, improved reality, video surveillance, face identity, image sharing, optical character identification, virtual world, and a wide range of fields as medical diagnosis. Its effects are felt in several industries, including manufacturing, transportation, entertainment, healthcare, security, and science. Computer vision is used in healthcare to help with medical image analysis, which helps physicians diagnose illnesses and more precisely guide surgical treatments. It improves security by strengthening surveillance systems, allowing for real-time observation and threat identification in critical locations and public areas. Businesses use computer vision to automate inspection procedures and maintain product uniformity and integrity [21]. Techniques for data processing and visualization aid in research projects, promoting innovation and scientific breakthroughs more quickly [22], (Figure 12.3).

Computer vision is essential to the advancement of transportation as it makes it easier to create driverless cars with perception systems based on vision for obstacle avoidance and navigation. Combining interactive technologies like virtual reality and augmented reality with the digital and

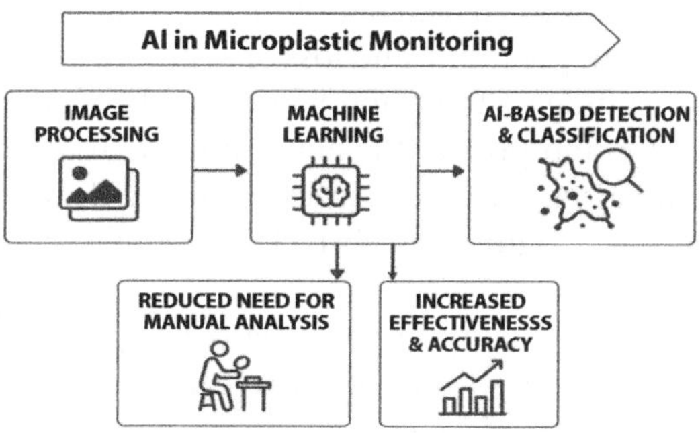

Figure 12.3 The diagram represents the process of microplastic monitoring through AI.

physical worlds improves user experiences. As deep learning and sensor technologies evolve, computer vision's revolutionary possibilities throughout multiple sectors become increasingly apparent. By employing visual intelligence, we pave the way for a future when robots can interact with and understand the visual environment with ease, radically changing human understandings, perceptions, and engagements with technology [23].

12.2.2 Classification and Quantification

Machine learning (ML) plays an important role in the classification and perception of microplastics from environmental sets. Many monitored and unprotected learning techniques that analyze images and spectral data can accurately identify plastic particles according to their size, size and chemical composition. Researchers developed an ML model using random forest and supporting vocal machines (SVMs) to classify microplastic particles collected from freshwater tests. After receiving instructions on a dataset with more than 10,000 images of microplastics, the system was able to identify with an accuracy rate of 92% between plastic pieces, fiber, and images. Studies have shown that ML-based techniques can significantly improve microplastic detection accuracy by supporting the regulatory structure for pollution prevention [22] (Figure 12.3).

12.2.3 Real-Time Monitoring and High-Resolution

An important discovery is that blockchain with AI can provide real-time and detection of nonconformity. Artificial intelligence algorithms can detect deviations in temperature races and temperatures downs. Blockchain stores asymmetrical data that cannot be changed, making an irreversible journal of events. For immediate environmental situations such as a natural disaster, increase in pollution, and loss of habitat, this real-time capacity is important. Depending on the amount and quality of data, artificial intelligence can monitor and protect the environment. To succeed, algorithm needs accurate and extensive access necessary for environmental data. This basic requirement is a significant obstacle in many regions of the world [24, 25].

Fluctuations in data quality impede the development and implementation of strong AI models, limiting their capacity to produce accurate insights and predictions. Addressing the issue of uneven data quality will require coordinated efforts to improve data gathering infrastructure. The acquisition of high-resolution, real-time data requires a calculated investment in cutting-edge technologies like satellites, sensor networks, and ground-based monitoring systems [26, 27] (Figure 12.3).

12.3 The Current State of Microplastic Monitoring Regulations

According to the standard guidelines of the U.S. Microbead-Free Waters Act (2015) [81], EU microplastics restrictions (2021) [83], and the United Nations Environment Programme (UNEP) (2018) [84], there are several strategies for the controlling of microplastic pollution, including:

a. **Regulations and Policies-** Extended producer responsibility (ESR) and plastic waste management regulations.
b. **Scientific and Technological Solutions-** AI-based monitoring systems, advanced filtration in wastewater treatment and biodegradable and sustainable alternatives.
c. **Public Awareness and Behaviour Changes-** Banning of plastic microbeads in cosmetics, reducing single-use plastics, and encouraging circular economy (Figure 12.4).

By following these strategies, policymakers and researchers can help ensure that AI applications for microplastic monitoring remain strong, fair, and science-tested, thus building stronger trust in AI-powered environmental monitoring.

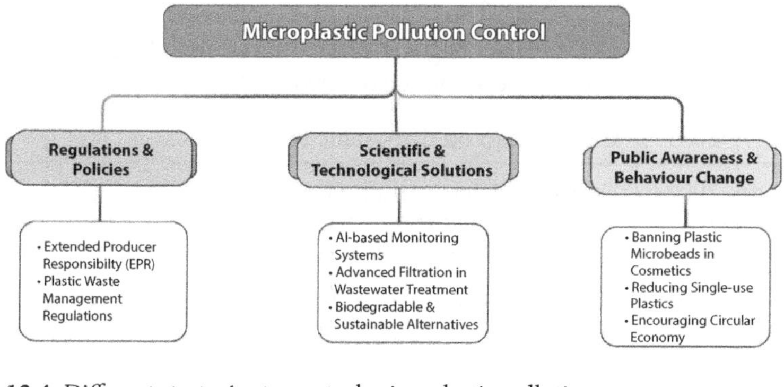

Figure 12.4 Different strategies to control microplastic pollution.

12.3.1 Current Environmental Regulations and Microplastic Surveillance Guidelines

Several agreements to address plastic pollution were established seven decades after synthetic plastic was invented. While most legislation started out addressing plastic pollution in general, they had a direct impact on regulations that targeted microplastics from the mid-2000s to 2024 [28], (Table 12.1).

Table 12.1 List the major microplastics rules and the chronology of policies that address microplastic pollution either directly or indirectly.

S. no.	Year	Policy name
1.	1972	London Convention
2.	1974	Oslo Convention, Paris Convention
4.	1975	Mediterranean Action Plan
5.	1981	Virginia (USA) Litter Tax
6.	1982	Convention on the Law of the sea, CCAMLR
8.	1988	MARPOL Annex V
9.	1989	First International Cleanup, Basel Convention
11.	1992	OSPAR/HELCOM Convention
12.	1994	NOWPAP
13.	1995	Barcelona Convention and MAP II, UN GPA
15.	1998	ML plan and updated OSPAR, South Korea system to tackle litter
17.	2000	Delhi Plastic Bag Act and China banned PS food containers, Directive 2000/59/EC Port facilities.
19.	2001	Chinese ban on disposable tableware
20.	2002	Taiwan restricts certain SUP
21.	2004	The term microplastic was coined
22.	2005	UN(A/60/30)Oceans and the Law of the sea, Net recycling Honolulu,

(*Continued*)

Table 12.1 List the major microplastics rules and the chronology of policies that address microplastic pollution either directly or indirectly. (*Continued*)

S. no.	Year	Policy name
24.	2008	MSFD, China banned ultra-thin plastic bags
26.	2009	UNEP MBIs
27.	2011	Commitment Plastic Organizations, UNEP Honolulu Strategy, Taiwan MBIs,
30.	2012	Rio + 20 GPML, China regulates recycling, ML management plan (MED), Dutch Plastic Cycle Value Chain Agreement
34.	2014	OSPAR monitoring in the Arctic, Portugal Green Taxation
36.	2015	HELCOM RAP – ML, Plastic bags regulated in EU 2015/720, First ban on microbeads, Arctic strategic plan (AMSP)
40.	2016	IICRI call on microbeads, 1/6 and 2/11 resolutions UNEA, Microplastics defined as emerging issue
43.	2017	SDG14, Microbeads banned or restricted in several countries, First Chinese survey tackling microplastics in the Arctic
46.	2018	MARPOL V stricter on plastics, Directive 2018/851 on EPR linked to waste, UN plan to reduce microplastics pollution, Approval of Ocean Plastic Charter, China restricted imports of plastics, Marine pollution suggested as a planetary boundary, FU SUP Directive
53.	2019	Basel Convention amendment including microplastics, Antarctic Treaty strategies to mitigate MPS, Honolulu historical SUP ban, Oceans and the Law of the sea A/74/19 addressing microplastic, ECHA restrict microplastics, European Green Deal, Directive 2019/904 SUP, Microplastics suggested as POPs.
54.	2020	France's Anti-Waste and Circular Economy Law
55.	2021	EU Microplastics Restrictions
56.	2023	European Union's REACH Regulation on Microplastics
57.	2024	United Nations Plastics Treaty Negotiations

12.3.2 National and International Guidelines

National Guidelines:

1. **United States:**
    a. **Microbead-Free Waters Act (2015):** This federal law forbids the production and sale of rinse-off cosmetics, such as face wash and toothpaste, that include plastic microbeads [29].
    b. **State Initiatives:** Several states have passed extra legislation. For example, California is creating testing and reporting guidelines and has approved a definition of microplastics in drinking water [30].
2. **European Union:**
    a. The 2008 Marine Strategy Framework Directive (MSFD): under 2020, EU member states are required under the MSFD to attain Good Environmental Status (GES) of their maritime waterways, which includes tackling marine litter, including microplastics [31].
    b. REACH Regulation: In an effort to lower emissions from goods, including fertilizers, detergents, and cosmetics, the EU decided to prohibit the use of purposefully added microplastics in products in September 2023.
3. **Australia**
    The National Plastics Plan [85] describes steps to cut pollution and waste from plastics, including activities to combat microplastics [27].

International Guidelines

1. **United Nations Initiatives:** The United Nations Environment Program (UNEP) promotes global efforts to eradicate marine litter, including microplastics, through various partnerships and projects [27, 51].
2. **The International Maritime Organization (IMO):** By controlling ship plastic discharge under MARPOL Annex V, the International Maritime Organization (IMO) contributes to the decrease of marine plastic trash [27, 51].
3. **International Standards Organization (ISO):** The international standard for identifying and assessment of

microplastics, ISO 24187:2022, is one of the recommendations developed by ISO for the characterization of microplastics in diverse habitats.

4. **Basel Convention**: The Basel Convention has been amended to include particular plastic waste to control their cross-border movement and ensure ecologically ethical handling.

12.3.3 Complications in Implementing and Complying with Policies

12.3.3.1 Lack of Techniques Installed for Detection and Measurement

Microplastic, data collection, and analysis are uneven without standard methods to determine the volume. It is challenging to compare findings from the study and develop an effective policy because of this deviation. Extensive research highlights the need for standardized methods for determining microplastic pollution under different conditions.

12.3.3.2 Variations in Legal Definitions

The legal interpretation of many countries on microplastics makes policy implementation challenging. These differences can lead to appropriate holes and international efforts to address microplastic pollution. Research indicates that to facilitate the development and execution of harmonic politics, a single definition is required [32].

12.3.3.3 Inconsistent Methods of Enforcement

Research indicates that the effectiveness of measures aimed at reducing microplastic pollution depends on a strong enforcement strategy supported by adequate financing and public participation. Due to money, political will, or lack of public knowledge, the implementation of law is sometimes different, even when present. List of techniques installed for detection and measurement: microplastic, data collection and analysis are uneven without standardized methods for determining and determining the volume. It is challenging to compare findings from the study and develop an effective policy because of this deviation. Extensive research highlights the need for standardized methods for determining microplastic pollution under different conditions.

12.3.3.4 Inadequate Stakeholder Partnership

Microplastics, industry, decision makers, researchers and consumers must be involved for successful cuts. According to a review, the absence of collaborative projects and civil scientific initiatives limits widespread monitoring and policy compliance. Participation in monitoring programs with local communities can increase data collection and promote a sense of shared responsibility.

12.3.3.5 New Potential Risks and Limitations in Technology

Continuous growth of new plastic polymers and additives complicates the detection and treatment of microplastics. If the existing methods of analysis are unable to identify or determine these novels sufficiently, monitoring and compliance can be challenging. To maintain the developed panorama of microplastic pollution, the research highlights advanced identity methods and the need for adaptable laws.

12.4 Regulatory Obstacles in AI-Powered Microplastic Identification

12.4.1 Inadequate Worldwide Standards

The absence of internationally accepted methods for identifying microplastic is one of the most important challenges. This is more challenging when comparing the results of the study in geographical boundaries due to the difference in computer reporting, analytical techniques, and test strategy. This inequality makes it more challenging to create continuous international politics and initiatives to reduce microplastic pollution. A comprehensive review highlights the need for frequent samples of microplastics to control the dangers and the need for standardized processes to detect which particles represent the environment and people's health [33].

12.4.2 Problems with Data Difference, Accuracy, and Reproducibility

Modern AI algorithms must be extremely compatible and flexible to provide extensive access and successful adaptation for many scenarios. For example, an EHR seller-specific machine learning algorithm that acts as an initial warning system for sepsis is the only option available to any hospitals that can provide seller-specific data inputs. We have previously shown that the AI models for customized benefits and different treatment effects can be reduced in the parsimonious model, which only requires some clinical factors, although clinical studies record a lot of data. This guarantees that the models are easily distributed and testable in different coasts. AI-ECG algorithms that require access to raw wiring readings are another example. This may require a significant cost to change the local IT infrastructure to a hospital and limit the use of specific doctors. Healthcare Information and Management System Society "defines the difference in the form of a capacity for different information systems, equipment and applications (systems), which is to use, integrate and collaborate by using data in a coordinated manner, both within and across organizational, regional, and national boundaries, to provide timely and seamless portability of information and optimized the health of individuals and populations worldwide".

12.4.3 Accountability and Transparency of Algorithms

In the context of artificial intelligence (AI) and higher education, accountability is defined as either the collection of mitigation or prevention techniques that hold users, designers, or owners of AI systems accountable. Many degrees of accountability may exist in higher education since strategic decision-making may be done by human committees or stakeholders or by algorithms; it is unclear who should be held responsible—the source of the data or the creator or maker of the AI system. Similarly, Ungerer and Slade [38] examine how systems based on algorithmic models might be unaccountable. The way AI algorithms are conceived may be somewhat to blame for the lack of accountability. Bearman *et al.* [86], observe that AI is causing a language of shifting power, which has accountability concerns [44]. There are differing opinions, though, on who should be held responsible: AI, humans, or both.

Transparency in the context of AI and higher education is frequently described at both a high level and a low level. Making institutional policies, as well as their intentions and results, transparent is the broad definition of transparency. On a fundamental level, algorithms that are used to teach AI or employ AI in higher education should be made clear to their users, either in mathematical or non-mathematical terms [34].

12.5 Privacy and Ethical Issues with AI-Powered Environmental Monitoring

12.5.1 The Ethical Consequences of AI in Science Research

The application of AI in environmental monitoring raises significant ethical issues, namely bias in data collection, decision-making openness, and potential abuses of AI-generated outcomes. Because AI frameworks rely on data sets that may contain predetermined biases, research results are threatened. Using AI-powered automation to undertake environmental research would reduce the need for human interaction, which raises questions about responsibility and decision-making throughout scientific processes. According to Gebru *et al.* [46], disparities in environmental legislation may be reinforced by AI models trained on biased data.

Transparency and Explainability: Since most of the AI systems are "black boxes," it might be challenging to understand the predictions they make [49]. By altering resource allocation and land usage, misuse of AI-based monitoring might disproportionately impact marginalized populations [87]. The use of AI to monitor deforestation has reduced illicit logging, it has also drawn criticism for its opaque decision-making process [41]. Although AI may be used in climate modeling, accuracy must be thoroughly verified [47].

12.5.2 Privacy Concerns: Acquiring Geographical and Sensitive Data

Using AI-based environmental monitoring, drone, Internet of Things sensors and satellite imaging are often used to collect sensitive geographical data. It increases concerns about snooping, illegal data access, and neglect of geolocation data. High-resolution satellite images can accidentally record private data, presenting moral questions [88]. A private organization that collects environmental data has a capacity for profit without the necessary authority [42]. International cooperation is often necessary for AI-based monitoring, which raises questions about jurisdiction and data security rules [89]. Environmental data is unclear, while the General Data Security Regulation (GDPR) protects personal information [90]. The moral use of AI for sustainable development focuses on the UN Guidelines for Environmental Data Protection [77].

12.5.3 Ownership, Security, and Accessibility of Data

Whoever controls and maintains environmental data is arguably the most important question. While AI-powered surveillance generates valuable information, concerns about data security, ownership, and fair access still need to be addressed. Several AI-generated environmental data are held by corporations, making them difficult for researchers and policymakers to access [35]. Platforms for environmental data that rely on artificial intelligence are susceptible to hacking, which might lead to data manipulation [91]. Despite the existence of open data projects, many AI-generated datasets are paywall-protected, limiting their use limited. Kritikos [48], points out that decentralised systems may improve data availability and integrity. AI systems should have strong encryption to guard against breaches. Laws should guarantee the appropriate handling and equitable dissemination of AI-generated environmental data [36, 37].

12.6 Policy Ideas for Including AI in Microplastic Monitoring

The use of artificial intelligence (AI) for microplastic monitoring can change the face of environmental science with more accurate, effective, and scalable means to detect and check microplastic pollution. To be suitable for AI and to be implemented with responsibility, well well-defined policy must be kept to supervise the application of credibility, openness, and cooperation worldwide. The following chapter emphasizes the most important political proposals to enable systematic and responsible implementation for microplastic monitoring of AI [14, 92].

12.6.1 Need for Standardized Protocols, Especially for AI

Despite the increasing use of AI in environmental science, especially in microplastic studies, there is a significant disadvantage to standardized protocols for use. Lack of widely accepted function leads to variation in data collection, algorithm performance, and result interpretation, thus affecting the qualification and similarity of the study copy between different fields and research groups [92]. To solve these challenges, international organizations, environmental agencies, and research institutes must work together to establish a set of standardized AI-specific protocols to fit microplastic monitoring [14].

These protocols should encompass multiple dimensions, including:

- **Collection and preparation of data:** A similar data collection structure, such as image resolution, sample preparation, and spectral analysis, should be installed for uniformity. Computer text procedures and preprocessing techniques should also be recorded to keep the AI training method consistent.
- **Building models and comparisons:** The AI model used for microplastic detection shall meet specific accuracy and accurate standards. The choice of machine learning architecture, functional sets, and data sets should be developed to validate the strength and reliability of the criteria model.
- **Open source and reproduced database:** For transparency and collective progress, data sets and AI models must be shared. The AI-ACCESS Database AI-based microplastic detection can make cross-validation and external revision easier.
- **Basic performance indicators:** A regular assessment matrix such as precision, recall, F1 score, and cross-over-union (IU) must be used in studies to allow the frequent performance of AI algorithms to compare the AI algorithm. By using these standardized processes, the scientific society can reduce microplastic surveillance variability; AI-operated identification can improve accuracy and provide a basis for compliance with regulations.

12.6.2 Install the Default for Transparency and Algorithm Verification

Microplastic monitoring requires a strong verification structure to increase the dependence of AI that can confirm the reliability and interpretation of the AI model. One of the main problems with deep learning algorithms is that they are "Black Box" algorithms, which often hide decision-making processes and destroy responsibilities and confidence [92]. Techniques used to encourage more openness and confirmation should be used:

- The AI model should be designed with methods for underlying interpretation, such as explaining AI (XAI) frameworks, attention maps, or functional analysis. It will be easy for scholars and decision makers to understand how models come to their conclusion.

- It is important to cross-validate AI assumptions using chemical characteristic techniques, spectroscopy, and empirical data obtained through microscopy. The AI model should be considered reliable; they should have high levels of consent with traditional identity techniques.
- Inappropriate representation of unbalanced data or various environmental tests can cause prejudice in the AI model. To ensure that AI models work equally in different water bodies, sediment types, and geographical sites, a strong audit process should be developed to identify and address prejudice.
- The AI algorithm used to identify microplastics should be evaluated by independent regulatory bodies on time. Third-party audit models can guarantee performance, identify any limits, and recommend a campaign to instill confidence.
- To enable repetition and scientific social verification, AI developers must provide complete documents on data, treatment methods, and sources of annotation tactics.

12.6.3 Encouraging International Regulatory Coordination

Microplastic contamination is an international crisis that supersedes nation-state boundaries and impacts ecosystems, marine life, and human well-being globally. Due to its international character, regulatory collaboration across governments, research centers, and environmental agencies will be important to ensure the optimal deployment of AI-based monitoring measures. Existing imbalances in green policies, information-sharing legislation, and AI regimes are hindering collaborative efforts to date, necessitating the articulation of an integrated regulatory regime [14, 43, 92].

The following are primary strategies to maximize regulatory cooperation:

- **AI Regulatory Framework Harmonization:** Diverse regulations applying to AI usage in diverse nations may erect impediments against the coordination of worldwide microplastic monitoring campaigns. Creating an internationally accepted

AI-driven environmental monitoring regulatory framework shall help enable effortless cross-border partnership.

- **Inter-Border Data Exchange Arrangements:** Since AI models require large and diverse datasets for accurate training and validation, international data-sharing agreements should be established. These agreements must balance the need for data accessibility with privacy and security considerations.
- **Establishing International AI Microplastic Task Forces:** Governments, nongovernmental organizations (NGOs), and research institutions would need to partner to create exclusive task forces targeting AI-based monitoring of microplastics. Task forces can devote themselves to the development of policy, standardizing data, and implementing AI solutions in various geographic regions.
- **International Research Collaborations and Funding:** To drive innovations in AI technology for microplastic detection, funding bodies must give more importance to grants for multinational research projects. Multi-country collaborative initiatives can facilitate technology sharing and help ensure that AI models are transferable across varying environmental conditions.
- **Implications for Morality and Public Participation:** Ethical principles, including environmental justice, accountability, and justice, should direct the application of AI under microplastic monitoring. Programs should be launched for public involvement to encourage civilian participation in processes to increase social knowledge of AI-based monitoring and collect data. AI-based microplastic monitoring can be effectively increased through the promotion of regulatory cooperation and border-crossing cooperation, which can lead to wider and coarsely used solutions to handle plastic pollution.

12.7 Multidisciplinary Cooperation's Function in Policy Development

12.7.1 Connecting AI Developers, Scientists, and Policymakers

The intensive understanding of science and technology behind politics, as well as the social and economic impact of such politics, is crucial to their development and implementation. Multi-intellectual cooperation is important in the context of artificial intelligence (AI) to guarantee that the solution it provides is morally made and implemented. Political decision-makers, academics, and AI developers usually work with separate agendas and communication styles in separate domains. Researchers are interested in learning about social and environmental difficulties, MLA is interested in creating laws that balance moral ideas with technology rifle, and AI developers are interested in creating new software and algorithms. Organized collaboration, continuous communication, and differences of understanding between different stakeholders are required to close. The linguistic gap between technical experts and politicians is one of the biggest obstacles. The use of excessive technical language by AI researchers and developers can be challenging for decision-makers to understand the technical nuances of AI applications [14]. On the other hand, decision-makers can prioritize extensive regulations that are restrictive or unclear to technologists. This difference can be closed by installing multi-disconnected communication platforms, such as advisory boards and political thinking tanks that include members of different fields [39, 40].

AI research is the other important component of moral ideas and real implications in collaboration. Political decision-makers and researchers can advise AI developers on how to design a system that prefers responsibility, openness, and justice [43, 45]. For example, environmental researchers who work with climate change models can exchange data and expertise to help AI developers help environmental trends for environmental trends, and regulators can ensure that these AI solutions can follow international rules and public benefits. Governments and organizations can create politics that are informed by encouraging the culture of socially conscious and technical interdisciplinary cooperation [92]. This will lead to more successful AI-based measures for issues around the world [52–76].

12.7.2 Promoting Interoperability and Open Data Sharing

One of the basic problems of developing AI policy is obtaining high-quality, interoperable, and standardized data. To provide accurate predictions and recommendations for AI, large versions of data are required. However, it is difficult to create AI models with extensive adoption ability in politics due to boundaries due to fragmentation, ambiguity, and ownership boundaries. To guarantee that AI models are inclusive and strong, open data exchange should be encouraged between researchers, politicians, and AI developers. Guidelines that encourage the availability of data without the refraining on privacy and security requirements should be promoted by governments and international organizations. Researchers and AI developers can work more efficiently by establishing a centralized data registration, such as the UN, introduced by initiatives such as Global Pulse or the European Open Science Cloud.

Interoperability, the capability of various AI systems and data sets to cooperate, is another key consideration in unlocking the full potential of AI-based policy solutions. Most current environmental and social data sources are contained in various formats, and AI systems cannot analyze them as a whole. The interoperability of data formats and compatibility of various AI platforms with each other can enhance the effectiveness and precision of AI-based decision-making [14, 92].

The open sharing of data builds trust and responsibility in AI-based policy advice. If the sources of data and the methodologies used are made open, policymakers are in a position to make more informed judgments regarding the reliability of AI-generated analysis. Further, this improves democratic participation in policymaking by allowing individuals to view and critique the AI-based choices that impact their lives [14].

12.7.3 International Collaborations for Successful AI-Based Environmental Regulations

AI can help address important environmental issues, such as reducing climate change and protecting biodiversity. Nevertheless, environmental issues are global in the global scope and require global efforts and coordinated solutions [92]. The process of developing successful AI-based policies in countries and biomes depends a lot on multilateral cooperation. International organizations in advance of creating AI-controlled environmental policy include the UN, world-economic forums, and intergovernmental panels on climate change. International organizations can promote information exchange, spread the best practices, and create similar regulatory standards for AI-based environmental solutions by bringing AI experts, researchers, and politicians together. The use of AI in climate monitoring is an example of a successful AI partnership worldwide. Artificial intelligence (AI) can be used to analyze satellite data collected by agencies such as NASA and the European Space Agency to detect deforestation, track carbon emissions, and predict natural destruction [14]. Countries can respond to environmental threats more efficiently by sharing data insights and AI models outside national boundaries [92].

Regulating the environmental effects of AI is another important field where multinational cooperation can have an effect. Although AI-operated sectors such as smart energy networks and automatic agriculture can reduce carbon footprints, they also bring moral and environmental issues. AI technology can be ensured globally, and AI technology can be used properly by putting international criteria on the AI stability. Governments and organizations should invest in AI training and capacity building to strengthen international cooperation. The international community can ensure that the AI-informed environmental policy is inclusive and favorable for everyone by providing the necessary equipment and training to researchers and decision-makers in underdeveloped countries [14].

12.8 Conclusion

An important development in the fight against plastic pollution, which is a serious problem for ecosystems and human welfare, is the use of artificial intelligence (AI) in microplastic monitoring. Microplastics, which are small particles created as minutes for different applications, when large plastic products are destroyed, presents special and difficult problems for environmental monitoring. Because of their small size and wider appearance, traditional surveillance techniques become more difficult, which emphasizes the need to use AI's ability to effectively analyze large amounts of data. Microplastics have a significant impact on animal and human health in addition to being annoying to the environment. Marine life can consume these particles, which can have harmful effects when entering the food chain. Consequently, monitoring the spread and appearance in different ecosystems has become important. Traditional techniques, which often include manual sampling and analysis, are less and less effective. Creative solutions are required, as they are unable to keep up with the scope and complexity of the problem. To analyze the dataset on a large scale, it is difficult to analyze manually; artificial intelligence (AI) offers a powerful alternative. The algorithm of machine learning can help create trends, approximate patterns, and even new content that can reduce the amount of plastic waste produced. But collaboration is required to use AI's ability for microplastic monitoring. To develop an efficient and secure system, collaboration between AI researchers, environmental researchers, legislators, and industrial stakeholders is important. Such cooperation is necessary to guarantee compliance with environmental standards together with protected public health. Without this collaborative effort, there is an important possibility that pollution can be spoiled, especially low and intermediate lands that do not have the means to use advanced solutions. In addition, the use of any AI in the region will have to prioritize moral problems. To promote trust among all parties involved, concerns, including privacy protection, the possibility of bias in the AI system, and openness in data use must be considered. For example, to ensure that the AI algorithm is trained on a variety of datasets, their accuracy and relevance will improve, resulting in better monitoring.

In addition, it is important to communicate the result and the data interpretation. To make well-informed decisions that protect the environment and public health, decision-makers and community leaders require easily accessible insights. Targeted financing and capacity programs are important to maximize the benefits of AI in the fight against microplastic pollution, especially for small populations that are often ahead of the environmental decline. In addition to strengthening these communities, investment in local surveillance skills encourages awareness of ownership of solutions that are kept in practice. We can develop sustainable and successful programs that provide local stakeholders to meet the unique needs of each society by providing resources and information that requires them to monitor and address microplastic pollution. Although AI presents a possible solution for handling microplastic pollution, its effective use of teamwork, moral concerns, and support-centered support for local abilities depends. All industries will work together to guarantee that AI will be used for successful and morally forward purposes, which will eventually lead to a better world for the next generations.

AI disclosure statement: The author acknowledges the use of AI-based tools for minor language editing or grammatical correction. However, all intellectual contributions, analyses, and conclusions presented in this chapter are entirely the author's own.

References

1. Yuan, P., Wang, Y., Chen, X., Gao, P., An overview of microplastic pollution in the environment over the megacity of Shanghai during 2013–2022. *Sci. Total Environ.*, *912*, 168986, 2024.
2. Andrady, A.L., Microplastics in the marine environment. *Mar. Pollut. Bull.*, *62*, 8, 1596–1605, 2011.
3. Kershaw, P.J., Sources, fate and effects of microplastics in the marine environment: a global assessment, 2015.
4. Allen, S., Allen, D., Phoenix, V.R., Le Roux, G., Durántez Jiménez, P., Simonneau, A., Galop, D., Atmospheric transport and deposition of microplastics in a remote mountain catchment. *Nat. Geosci.*, *12*, 5, 339–344, 2019.
5. Eerkes-Medrano, D., Thompson, R.C., Aldridge, D.C., Microplastics in freshwater systems: a review of the emerging threats, identification of knowledge gaps and prioritisation of research needs. *Water Res.*, *75*, 63–82, 2015.
6. Lenz, R., Enders, K., Nielsen, T.G., Microplastic exposure studies should be environmentally realistic. *Proc. Natl. Acad. Sci.*, *113*, 29, E4121–E4122, 2016.
7. Rochman, C.M., Hoh, E., Hentschel, B.T., Kaye, S., Long-term field measurement of sorption of organic contaminants to five types of plastic pellets: implications for plastic marine debris. *Environ. Sci. Technol.*, *47*, 3, 1646–1654, 2013.
8. Li, J., Liu, H., Chen, J.P., Microplastics in freshwater systems: A review on occurrence, environmental effects, and methods for microplastics detection. *Water Res.*, *137*, 362–374, 2018.
9. Schymanski, D., Goldbeck, C., Humpf, H.U., Fürst, P., Analysis of microplastics in water by micro-Raman spectroscopy: Release of plastic particles from different packaging into mineral water. *Water Res.*, *129*, 154–162, 2018.
10. Crawford, C.B. and Quinn, B., *Microplastic pollutants*, Elsevier Limited, UK, 2016.
11. European Chemicals Agency (ECHA), Proposal to restrict intentionally added microplastics under REACH, 2021.
12. U.S. Congress, Microbead-Free Waters Act of 2015, 2015.
13. Boucher, J. and Friot, D., *Primary microplastics in the oceans: a global evaluation of sources*, vol. 10, Iucn, Gland, Switzerland, 2017.
14. United Nations Environment Programme (UNEP), Mapping of global policies on microplastics, 2018.
15. Berrahal, M., Boukabous, M., Yandouzi, M., Grari, M., Idrissi, I., Investigating the effectiveness of deep learning approaches for deep fake detection. *Bull. Electr. Eng. Inf.*, *12*, 6, 3853–3860, 2023.
16. Grari, M.O.U.N., II, Boukabous, M. O. H. A. M. M. E. D., Yandouzi, M., II, Berrahal, M., Idrissi, I., Moussaoui, O., Moussaoui, M., Forest fire detection and monitoring through environment sound spectrum using deep learning. *J. Theor. Appl. Inf. Technol.*, *101*, 12, 3754–3764, 2023.

17. Grari, M., Idrissi, I., Boukabous, M., Yandouzi, M., Moussaoui, O., Azizi, M., Moussaoui, M., IoT-based approach for wildfire monitoring and detection, in: *International Conference on Advanced Intelligent Systems for Sustainable Development*, pp. 205–213, Springer Nature Switzerland, Cham, 2022, May.
18. Hammoudi, Y., Idrissi, I., Boukabous, M., Zerguit, Y., Bouali, H., Review on maintenance of photovoltaic systems based on deep learning and internet of things. *Indones. J. Electr. Eng. Comput. Sci.*, *26*, 2, 1060–1072, 2022.
19. Zerguit, Y., Hammoudi, Y., Derrhi, M., Energy Efficiency In Industry 4.0: Implementing Real-Time Power Factor Correction In Smart Manufacturing. *J. Theor. Appl. Inf. Technol.*, *102*, 1, 2024.
20. Grari, M.O.U.N., II, Yandouzi, M., II, Idrissi, I.D.R., II, Boukabous, M. O. H. A. M. M. E. D., Moussaoui, O.M.A.R., Azizi, M., Moussaoui, M., Using IoT and ML for forest fire detection, monitoring, and prediction: a literature review. *J. Theor. Appl. Inf. Technol.*, *100*, 19, 5445–5461, 2022.
21. Yandouzi, M., Grari, M., Berrahal, M., Kerkour, A., Elmiad, M.B., Barkaoui, A., Zarhloule, Y., Elevating Plastic Litter Surveillance: Drones and Deep Learning for Efficient Detection. *J. Theor. Appl. Inf. Technol.*, *102*, 11, 2024.
22. Jin, H., Kong, F., Li, X., Shen, J., Artificial intelligence in microplastic detection and pollution control. *Environ. Res.*, 119812, 2024.
23. Gade, P.K., AI-Driven Blockchain Solutions for Environmental Data Integrity and Monitoring. *NEXG AI Rev. Am.*, *4*, 1, 1–16, 2023.
24. Chisom, O.N., Biu, P.W., Umoh, A.A., Obaedo, B.O., Adegbite, A.O., Abatan, A., Reviewing the role of AI in environmental monitoring and conservation: A data-driven revolution for our planet. *World J. Adv. Res. Rev.*, *21*, 1, 161–171, 2024.
25. Munhoz, D.R., Harkes, P., Beriot, N., Larreta, J., Basurko, O.C., Microplastics: a review of policies and responses. *Microplastics*, *2*, 1, 1–26, 2022.
26. Coffin, S., Status of legislation and regulatory drivers for microplastics in California. *Protect Water Qual. Around World*, 54, 17, 2020.
27. Thompson, R.C., Courtene-Jones, W., Boucher, J., Pahl, S., Raubenheimer, K., Koelmans, A.A., Twenty years of microplastic pollution research—what have we learned? *Science*, *386*, 6720, eadl2746, 2024.
28. Jin, M., Liu, J., Yu, J., Zhou, Q., Wu, W., Fu, L., Karimi-Maleh, H., Current development and future challenges in microplastic detection techniques: A bibliometrics-based analysis and review. *Sci. Prog.*, *105*, 4, 00368504221132151, 2022.
29. Onyena, A.P., Aniche, D.C., Ogbolu, B.O., Rakib, M.R.J., Uddin, J., Walker, T.R., Governance strategies for mitigating microplastic pollution in the marine environment: a review. *Microplastics*, *1*, 1, 15–46, 2021.
30. Usman, S., Abdull Razis, A.F., Shaari, K., Azmai, M.N.A., Saad, M.Z., Mat Isa, N., Nazarudin, M.F., The burden of microplastics pollution and contending policies and regulations. *Int. J. Environ. Res. Public Health*, *19*, 11, 6773, 2022.

31. Nielsen, T.D., Hasselbalch, J., Holmberg, K., Stripple, J., Politics and the plastic crisis: A review throughout the plastic life cycle. *Wiley Interdiscip. Rev.: Energy Environ.*, *9*, 1, e360, 2020.
32. Zhang, H., Duan, Q., Yan, P., Lee, J., Wu, W., Zhou, C., Yang, X., Advancements and challenges in microplastic detection and risk assessment: Integrating AI and standardized methods. *Mar. Pollut. Bull.*, *212*, 117529, 2025.
33. Oikonomou, E.K., Van Dijk, D., Parise, H., Suchard, M.A., De Lemos, J., Antoniades, C., Khera, R., A phenomapping-derived tool to personalize the selection of anatomical vs. functional testing in evaluating chest pain (ASSIST). *Eur. Heart J.*, *42*, 26, 2536–2548, 2021.
34. Oikonomou, E.K., Suchard, M.A., McGuire, D.K., Khera, R., Phenomapping-derived tool to individualize the effect of canagliflozin on cardiovascular risk in type 2 diabetes. *Diabetes Care*, *45*, 4, 965–974, 2022.
35. Li, E., Clarke, J., Ashrafian, H., Darzi, A., Neves, A.L., The impact of electronic health record interoperability on safety and quality of care in high-income countries: systematic review. *J. Med. Internet Res.*, *24*, 9, e38144, 2022.
36. Oikonomou, E.K. and Khera, R., Designing medical artificial intelligence systems for global use: Focus on interoperability, scalability, and accessibility. *Hellenic J. Cardiol*, *81*, 9–17, 2025.
37. Pagallo, U., From automation to autonomous systems: A legal phenomenology with problems of accountability, in: *IJCAI International Joint Conference on artificial intelligence*, International Joint Conferences on Artificial Intelligence, pp. 17–23, 2017.
38. Ungerer, L. and Slade, S., Ethical considerations of artificial intelligence in learning analytics in distance education contexts, in: *Learning Analytics in Open and Distributed Learning: Potential and Challenges*, pp. 105–120, Springer Nature Singapore, Singapore, 2022.
39. Bearman, M., Ryan, J., Ajjawi, R., Discourses of artificial intelligence in higher education: A critical literature review. *Higher Educ.*, *86*, 2, 369–385, 2023.
40. Memarian, B. and Doleck, T., Fairness, Accountability, Transparency, and Ethics (FATE) in Artificial Intelligence (AI) and higher education: A systematic review. *Comput. Educ.: Artif. Intell.*, *5*, 100152, 2023.
41. Barocas, S., Hardt, M., Narayanan, A., *Limitations and Opportunities in Machine Learning and Fairness*, MIT Press, 2019. https://books.google.co.in/books?hl=en&lr=&id=HuGwEAAAQBAJ&oi=fnd&pg=PR9&dq=Barocas,+S.,+Hardt,+M.,+Narayanan,+A.,+&ots=Q0haOve7pb&sig=TMJ7gks-Raj8pAuyahCGnVnki2aY#v=onepage&q=Barocas%2C%20S.%2C%20Hardt%2C%20M.%2C%20Narayanan%2C%20A.%2C&f=false.
42. Binns, R., Fairness in Machine Learning: Lessons from Political Philosophy. *Fairness, Accountability, and Transparency Conference Proceedings*, vol. 2018, pp. 149–159, 2018.

43. Crawford, K. and Boyd, D., Critical questions for big data: Provocations for a cultural, technological, and scholarly phenomenon. *Inform. Commun. Soc.*, 15, 5, 662–679, 2012.
44. Brundage, M., Avin, S., Clark, J., *et al.*, The Malicious Use of Artificial Intelligence: Forecasting, Prevention, and Mitigation. The arXiv preprint is 1802.07228, 2018.
45. Beltrametti, M., Floridi, L., Cowls, J., *et al.*, AI4People—An Ethical Framework for a Good AI Society. *Soc. Mind*, 17, 1–24, 2018.
46. Morgenstern, J., Vecchione, B., Gebru, T., *et al.*, Datasheets for Datasets. The arXiv preprint is 1803.09010, 2018.
47. Huntingford, C., Cox, P.M., Mercado, L.M., *et al.*, Machine Learning and Artificial Intelligence to Aid Climate Change Research and Preparedness. *Environ. Res. Lett.*, 14, 12, 2019.
48. Kritikos, M., *Blockchain for Environmentally Sustainable Data Management*, European Parliament Think Tank, 2020. https://www.eea.europa.eu/en/analysis/publications/blockchain-and-the-environment.
49. Lipton, Z.C., The Mythos of Model Interpretability. This is a preprint of arXiv:1606.03490, 2016.
50. Barboza, L.G.A., Dick Vethaak, A., Lavorante, B.R., Lundebye, A.K., Guilhermino, L., Marine microplastic debris: An emerging issue for food security, food safety, and human health. *Mar. Pollut. Bull.*, 133, 336–348, 2018.
51. Bergmann, M., Gutow, L., Klages, M., *Marine anthropogenic litter*, Springer, Cham, 2015.
52. Bond, T., Ferrandiz-Mas, V., Felipe-Sotelo, M., van Sebille, E., The occurrence and degradation of aquatic plastic litter based on polymer physicochemical properties: A review. *Crit. Rev. Environ. Sci. Technol.*, 48, 7-9, 685–722, 2018.
53. Brown, C.J., Bicknell, J.E., Day, R.H., Barry, J., AI-based image analysis for monitoring marine plastics: Emerging trends and applications. *Sci. Total Environ.*, 688, 63–80, 2019.
54. Carpenter, E.J. and Smith, K.L., Plastics on the Sargasso Sea surface. *Science*, 175, 1240–1241, 19724027.
55. Cole, M., Lindeque, P., Halsband, C., Galloway, T.S., Microplastics as contaminants in the marine environment: A review. *Mar. Pollut. Bull.*, 62, 12, 2588–2597, 2011.
56. Dissanayake, P. and Evans, R., Artificial intelligence for environmental sustainability: Challenges and opportunities. *AI Soc.*, 37, 3, 1–15, 2022.
57. Eerkes-Medrano, D., Thompson, R.C., Aldridge, D.C., Microplastics in freshwater systems: A review of the emerging threats, identification methods, and policy implications. *Water Res.*, 75, 63–82, 2015.
58. Fadare, O.O., Wan, B., Guo, L.H., Zhao, L., Microplastics from consumer plastic food containers: Are we consuming it? *Chemosphere*, 253, 126787, 2020.

59. Geyer, R., Jambeck, J.R., Law, K.L., Production, use, and fate of all plastics ever made. *Sci. Adv.*, 3, 7, e1700782, 2017.
60. Halden, R.U., Plastics and health risks. *Annu. Rev. Public Health*, 31, 179–194, 2010.
61. Hanke, G., Galgani, F., Werner, S., Oosterbaan, L., Nilsson, P., Fleet, D., ... & Liebezeit, G.,. Guidance on Monitoring of Marine Litter in European Seas: a guidance document within the Common Implementation Strategy for the Marine Strategy Framework Directive, 2013.
62. Jambeck, J.R., Geyer, R., Wilcox, C., Siegler, T.R., Perryman, M., Andrady, A., Law, K.L., Plastic waste from land into the ocean. *Sci.*, 347, 768–771, 20156223.
63. Jang, Y.C., Lee, J., Hong, S., Choi, H.W., Shim, W.J., Estimating microplastic emissions from tourism in coastal areas: A case study from South Korea. *Mar. Pollut. Bull.*, 161, 111798, 2020.
64. Kumar, A., Rathore, P.S., Dubey, A.K., Srivastav, A.L., Dutt, V., Kumar, T.A. (Eds.), *Sustainable management of electronic waste*, John Wiley & Sons, 2024.
65. Kirchherr, J., Reike, D., Hekkert, M., Conceptualizing the circular economy: An analysis of 114 definitions. *Resources Conserv. Recyl.*, 127, 221–232, 2017.
66. Kosuth, M., Mason, S.A., Wattenberg, E.V., Anthropogenic contamination of tap water, beer, and sea salt with synthetic fibers. *PLoS One*, 13, 4, e0194970, 2018.
67. Law, K.L. and Thompson, R.C., Microplastics in the seas: Sources, impacts, and policy challenges. *Sci.*, 345, 6193, 144–145, 2014.
68. Rathore, P.S., Ahuja, S., Burri, S.R., Khunteta, A., Baliyan, A., Kumar, A., *Deep learning techniques for automation and industrial applications*, John Wiley & Sons, Inc.; Scrivener Publishing LLC, 2024a.
69. Solanki, S. and Rathore, P.S., Current E-Waste Management: An Exploratory Study on Managing E-Waste for Environmental Sustainability. *Sustainable Manage. Electron. Waste*, 187–200, 2024.
70. Lechner, A. and Ramler, D., The discharge of microplastic from the Danube River into the Black Sea. *Environ. Pollut.*, 200, 90–95, 2015.
71. Li, J., Liu, H., Chen, J.P., Microplastics in freshwater systems: A review on occurrence, environmental effects, and methods for microplastic detection. *Water Res.*, 137, 362–374, 2018.
72. Napper, I.E. and Thompson, R.C., Environmental degradation of synthetic textiles in marine environments: Impacts and implications for microplastic pollution. *Environ. Sci. Technol.*, 53, 3, 1155–1166, 2019.
73. PlasticsEurope, *Plastics—the facts 2020: An analysis of European plastics production, demand, and waste data*, Plastics Europe Association, 2020.
74. Rochman, C.M., Browne, M.A., Halpern, B.S., Hentschel, B.T., Hoh, E., Karapanagioti, H.K., Policy: Classify plastic waste as hazardous. *Nat.*, 494, 169–171, 20137436.
75. Ryan, P.G., A brief history of marine litter research, in: *Marine Anthropogenic Litter*, pp. 1–25, Springer, Cham, 2015.

76. Silva, A.B., Bastos, A.S., Justino, C., II, da Costa, J.P., Duarte, A.C., Rocha-Santos, T., Microplastics in the environment: Challenges in analytical chemistry—A review. *Anal. Chim. Acta*, 1017, 1–19, 2018.
77. UNEP., From Pollution to Solution: A Global Assessment of Marine Litter and Plastic Pollution. United Nations Environment Programme (UNEP), 2021.
78. http://www.gesamp.org/.
79. Tatham, P., Wu, Y., Kovács, G., Butcher, T., Supply chain management skills to sense and seize opportunities. *Int. J. Logist. Manag.*, 28, 2, 266–289, 2017.
80. Kelley, S.M., Kahn, R.E., Mundt, J.C., Barahal, R.M., Do sanctions affect undetected sexual offending?. *Sexual Abuse*, 35, 5, 624–648, 2023.
81. Strifling, D.A., The microbead-free waters act of 2015. *J. Land Use Envtl. L.*, 32, 1, 151–166, 2016.
82. Horn, L., Microbead-free waters act of 2015. US Supreme Court: Are CWA Jurisdictional Determinations, 12.
83. Citaristi, I., United nations environment programme—UNEP, in: *The Europa directory of international organizations 2022*, pp. 193–199, Routledge, 2022.
84. Desai, B.H., 14. United nations environment program (UNEP). *Yearbook of International Environmental Law*, 28, 498–505, 2017.
85. March, A., Tsouza, A., Nieminen, L., Winton, S., Arora, H., Shejuti, S.M., ... Fletcher, S., National Action Plans: Effectiveness and requirements for the global plastics treaty. *Cambridge Prisms: Plastics*, 2, e11, 2024.
86. Bearman, M. and Ajjawi, R., Learning to work with the black box: Pedagogy for a world with artificial intelligence. *Br. J. Educ. Technol.*, 54, 5, 1160–1173, 2023.
87. Vinuesa, R., Azizpour, H., Leite, I., Balaam, M., Dignum, V., Domisch, S., ... Fuso Nerini, F., The role of artificial intelligence in achieving the Sustainable Development Goals. *Nat. Commun.*, 11, 1, 233, 2020.
88. Zuboff, S., Surveillance capitalism and the challenge of collective action, in: *New Labor Forum*, vol. 28, no. 1, pp. 10–29, Sage Publications, Sage CA: Los Angeles, CA: January 2019.
89. Mayor-Schonberger, V., Cukier, K., Big data, 2013.
90. Voigt, P. and Von dem Bussche, A., The EU general data protection regulation (GDPR). A practical guide, 1st ed., Cham: Springer International Publishing, 10, 3152676, 10–5555, 2017.
91. Smuha, N.A., From a 'race to AI'to a 'race to AI regulation': regulatory competition for artificial intelligence. *Law, Innovation and Technology*, 13, 1, 57–84, 2021.
92. Catone, M.C., The role of open data in digital society: The analysis of scientific trending topics through a bibliometric approach. *Front. Sociol.*, 8, 1134518, 2023.

13

Future Trends: AI Driven Innovation in Environmental Science

Priyanka Sharma, Ankita Sharma* and Prashant Ahluwalia

Department of Computer Science and Engineering, Chandigarh University, Mohali, Punjab, India

Abstract

Emerging contribution of AI has undoubtedly transformed the domain of Environmental Science. AI based techniques such as machine learning is improving efficiency by providing better possibility of decision making, backed by data prediction, has reshaped the definition of sustainability for environment. AI innovations such as deep learning, remote sensing, computer vision, machine learning etc. are playing a pivotal role in supplementing climate modeling, tailing ecological changes, optimization of resources and conservation of biodiversity by predicting the future trends by analyzing available data and studies. Prediction models developed with the help of machine learning or deep learning are of use in predicting extreme weather events by analyzing vast scope of datasets, mapping of habitats, optimization of energy usage and ecological restoration. Sensors, deployed for gathering data from surroundings, assist in environmental monitoring such as remote sensing, electrochemical sensing, temperature sensing help in mitigating or alleviating challenges such as pollution or air quality index, temperature, soil composition etc. AI along with IOT spotlights the domains such as smart agriculture, tracking and reduction of carbon footprint, water and energy optimization and sustainability in energy restoration. Having said that, along with major perks, come associated restrictions and dilemmas such as data collection, dataset availability and accessibility, data biases, efficiency of interpretation, ethical aspects and global sustainability implementation. But overall, AI possesses immense potential being a considerable aid for advancing in environmental science promising efficient conservation strategies and a healthier planet.

**Corresponding author*: ankitavatsyayan91@gmail.com

Abhishek Kumar, Pooja Dixit, Pramod Singh Rathore, Arun Lal Srivastav and Ashutosh Kumar Dubey (eds.) Microplastic Monitoring Using Artificial Intelligence, (267–292) © 2026 Scrivener Publishing LLC

Keywords: Artificial intelligence, climate AI, sustainable AI, eco-monitoring, green tech, smart conservation

13.1 Introduction to AI in Environmental Science

Artificial intelligence (AI) is changing the environmental science at such a brisk pace that it is granting the scientists and policymakers the ability to solve complicated environmental issues with earlier unheard-of speed and accuracy. Data analysis, pattern recognition, and well-informed decision-making that was earlier dependent on human knowledge have become feasible by artificial intelligence (AI), which includes computer simulations of human intelligence [1]. This superb partnership of artificial intelligence and environmental research is encouraging innovative quantum leaps in a variety of fields, from complex climate modeling to sustainable resource management and the crucial preservation of biodiversity.

13.1.1 Definition and Scope of AI in Environmental Research

Artificial intelligence (AI) is used in environmental research to process large datasets, simulate challenging environmental processes, and forecast outcomes related to various environmental events. This includes the development and application of AI-driven technologies such as machine learning models, computer vision, and natural language processing [2]. These tools support crucial decision-making in crucial areas, such as:

- **Climate Change:** Predicting climatic patterns, modeling atmospheric dynamics, and developing practical mitigation strategies.
- **Biodiversity Conservation:** Monitoring species populations, spotting subtle environmental changes, and tracking habitat loss and decline.
- **Resource Management:** Maximizing the utilization of vital natural resources, including water, energy, and land, to ensure sustainable practices.
- **Pollution Control:** Identifying pollutants, predicting air and water quality, and encouraging preventative actions to stop pollution.

- **Disaster Management:** Predicting natural catastrophes, such as floods, hurricanes, and wildfires, to improve risk assessments and speed up response times.

13.1.2 Historical Evolution and Current Applications

As the following section demonstrates, the incorporation of artificial intelligence (AI) into environmental science has developed over several decades [3]:

1. **Developments in 90's:** AI initially primarily assisted with climate modeling by simplifying the analysis of challenging environmental datasets. AI has been used to analyze satellite imagery and keep an eye on the environment in remote sensing and geographic information systems (GIS).
2. **Advancements in the 2000's:** With the advent of machine learning, AI was able to handle enormous datasets, or "big data," and automate environmental monitoring. Researchers began developing AI-driven prediction models for ecological processes and climate change to increase forecast accuracy.
3. **Modern AI Applications:**
    - **Climate Prediction:** Models of climate change, extreme weather risks, and adaptation strategies are all influenced by artificial intelligence (AI) [4].
    - **Ecosystem Management:** Artificial intelligence monitors ecosystems, predicts species migration, and investigates ecological dynamics. Examples include AI-powered drones and camera traps that track wildlife numbers and habitat deterioration.
    - **Smart Agriculture:** Through predictive analytics, AI enhances farming practices by optimizing water use, reducing the use of pesticides, and increasing crop yields.
    - **Pollution Control:** AI systems monitor environmental contaminants and air quality in real time, warning authorities of dangerous situations and guiding mitigation plans.

13.1.3 Importance of AI in Addressing Environmental Challenges

AI is crucial to resolving significant global environmental problems:

- **Climate Change Mitigation:** By improving climate model predictions, AI lowers greenhouse gas emissions, maximizes energy production, and improves energy system efficiency.
- **Biodiversity Conservation:** AI helps to protect habitats, tracks endangered species, detects threats to ecosystems, and monitors biodiversity.
- **Resource Efficiency:** AI-powered resource management (land, energy, and water) maximizes consumption, minimizes waste, and encourages sustainable use.
- **Disaster Prediction and Management:** By enhancing early warning systems, artificial intelligence's predictive powers lessen the effects of natural disasters like hurricanes, wildfires, and floods [5].
- **Policy Support and Decision-Making:** AI helps legislators create efficient environmental laws and regulations by offering data-driven insights and suggestions.

Because AI provides innovative solutions to pressing environmental problems, it is transforming environmental management and research. Because it can process vast amounts of data, identify trends, and generate insightful information, it is a vital tool for environmental protection and preservation.

13.2 AI and Climate Change Mitigation

In many domains, artificial intelligence (AI) presents tremendous promise for mitigating climate change. The following sections highlight significant domains where AI is already making a discernible difference [6]:

13.2.1 AI-Driven Climate Modeling and Prediction

Machine learning (ML), a branch of artificial intelligence, is revolutionizing climate modeling by improving the accuracy of predictions regarding weather patterns, extreme events, and climate trends [7]. The following are some ways AI aids in climate modeling:

- **Advanced Data Processing:** AI algorithms are capable of processing the massive datasets generated by sensors, satellites, and weather stations. By analyzing this data, AI can forecast variations in temperature, precipitation trends, and other significant climate factors.
- **Refined Climate Simulations:** Using machine learning algorithms to more accurately model the Earth's complex climate system can reveal intricate relationships between various climate factors. This leads to a more thorough understanding of long-term climate behavior.
- **Improved Extreme Weather Forecasting:** Extreme weather events like hurricanes, droughts, and floods can be accurately predicted by AI, which enhances disaster preparedness and response.

13.2.2 Machine Learning for Greenhouse Gas Monitoring

AI is proving to be incredibly useful in tracking and monitoring greenhouse gas emissions, which is crucial for strategies that effectively mitigate climate change. Among the primary uses are:

- **Real-time Emission Detection:** In real time, greenhouse gases like carbon dioxide (CO_2) and methane (CH_4) can be detected and measured by drones, satellites, and sensors with artificial intelligence (AI) capabilities. This produces more accurate and timely data than traditional methods.
- **Emission Forecasting:** Using AI models that predict future emission levels based on historical emissions data, policymakers can implement specific reduction measures.
- **Leak Detection:** Artificial intelligence has the potential to mitigate the atmospheric release of methane by identifying leaks from pipelines and other infrastructure [8].

13.2.3 AI-Based Carbon Capture and Sequestration Techniques

The development of more efficient carbon capture and storage (CCS) techniques is greatly aided by AI. Important uses include:

- **Optimizing Capture Processes:** Machine learning algorithms are boosting the efficiency of carbon capture technologies by identifying the most effective materials and methods for extracting CO_2 from industrial emissions.
- **Monitoring and Managing CO_2 Storage:** Underground CO_2 storage facilities are monitored by artificial intelligence to ensure the containment of captured carbon and prevent leaks. Injection processes can be streamlined and potential threats to storage integrity can be predicted by artificial intelligence models [9].
- **Developing New Materials:** The creation of novel carbon capture materials, such as advanced sorbents and catalysts that can more effectively absorb or chemically bind CO_2, is accelerated by artificial intelligence.

Artificial intelligence (AI) has many uses in the fight against climate change, including monitoring emissions, enhancing forecasts, and optimizing carbon capture technologies. As AI technology advances, it holds great potential to assist global climate action initiatives.

13.3 AI in Water Resource Management

AI is becoming increasingly important in water resource management because it boosts sustainability, productivity, and predictive power [10]. The specific applications include:

13.3.1 Smart Sensors for Water Quality Monitoring

- **AI Application:** Temperature, turbidity, dissolved oxygen, chemical contaminants, and pH levels are among the water quality parameters that are continuously monitored through the integration of AI and smart sensors. Real-time data from these sensors is used by AI models to spot contamination events or anomalies that might otherwise go unnoticed.
- **Benefits:** Faster reaction to water quality problems, quicker pollutant detection, and less need for human intervention. AI can also improve sensor placement in a water distribution system by analyzing historical data and pinpointing the most crucial monitoring locations.

13.3.2 AI for Efficient Irrigation and Water Conservation

- **AI Application:** Artificial intelligence (AI) optimizes irrigation systems by analyzing data from soil moisture sensors, crop types, and weather forecasts to determine how much water crops require at any given time. Machine learning algorithms use historical irrigation trends and environmental data to more accurately forecast water requirements.
- **Benefits:** sustainable use of water resources, lower water waste, and higher crop yields. In regions that are susceptible to drought, farmers can also save a substantial sum of money by conserving water.

13.3.3 Predictive Analytics for Flood and Drought Forecasting

- **AI Application:** Machine learning-based predictive models use historical climate data, rainfall patterns, soil moisture content, and other relevant variables to forecast future floods or droughts. AI can assist in making more timely and accurate predictions even when data is fragmented or incomplete [11].
- **Benefits:** By facilitating more effective plans for disaster preparedness and response, early warning systems help lower the risk to people, property, and ecosystems. Additionally, predicting droughts enhances water resource management by allocating resources to areas that need them the most.

13.3.4 Additional Applications

- **Water Distribution Optimization:** Artificial intelligence (AI) has the potential to improve water distribution in both urban and agricultural settings by detecting leaks, lowering the energy usage of water pumps, and guaranteeing fair distribution.
- **Water Recycling and Treatment:** AI has the ability to improve water treatment procedures, making them more efficient, cost-effective, and quick. Systems with artificial intelligence (AI) capabilities can anticipate when water treatment facilities need maintenance or improvements, cutting down on downtime and increasing productivity.

Artificial intelligence (AI) has the potential to completely transform water resource management by promoting better decision-making, boosting resource efficiency, and mitigating the effects of urbanization and climate change.

13.4 AI in Biodiversity Conservation

The use of AI in biodiversity conservation is growing in importance as it helps scientists and conservationists solve many issues more rapidly and effectively [12]. Here's a quick summary of how AI is significantly influencing this field:

13.4.1 AI for Species Identification and Monitoring

- **Automated Species Identification:** Particularly in the fields of machine learning and computer vision, artificial intelligence (AI) can automatically identify species from audio, video, and image recordings. For example, AI systems can identify species by analyzing bird sounds or recognizing animals from images taken by camera traps in the wild [13].
- **Monitoring Population Health:** Artificial intelligence (AI) algorithms are able to analyze large datasets from various sources to monitor trends in species populations, spot anomalies, and predict potential threats like poaching or disease outbreaks.
- **Real-time Monitoring:** AI can assist in processing data in real-time, allowing for speedy reactions to threats like deforestation, habitat destruction, and illegal hunting.

13.4.2 Deep Learning for Habitat Mapping

- **Satellite and Aerial Imagery:** Artificial intelligence (AI)-powered deep learning models can analyze satellite and drone photos to map ecosystems, monitor land-use changes, and evaluate habitat quality over time.
- **Predicting Habitat Loss:** By looking at past data, AI can predict regions where habitat loss due to deforestation, climate change, and human encroachment is likely to occur. Conservationists are able to focus their protection efforts more precisely as a result. Table 13.1 shows the AI in biodiversity conservation.

Table 13.1 AI in biodiversity conservation.

Area	Application	Description	Ref.
AI for Species Identification and Monitoring	Automated Species Identification	Uses Artificial Intelligence to recognize species from audio, video, and image recordings.	[14]
	Monitoring Population Health	Analyses large datasets to track species trends detect anomalies, and predict risks.	
	Real-Time Application	Processes data in real-time to allow quick response to threats.	
Deep Learning for Habitat Mapping	Satellite and Aerial Imagery	Artificial Intelligence analyzes drone/satellite imagery to map ecosystems and monitor habitat quality and land use changes.	[15]
	Predicting Habitat Loss	AI forecasts areas at risk of habitat loss due to human encroachment and climate change.	
	Ecosystem Modeling	Builds a model to understand species-environment interactions and support conservation planning.	
AI-powered Drones for Wildlife Protection	Anti-Poaching Operations	AI drones autonomously patrol protected areas and detect poaching threats in real time.	[16]
	Monitoring Remote Areas	Provide real-time tracking of animal and human activity in hard-to-access locations.	
	Gathering Ecological Data	Collects environmental data, which is then analyzed using AI to detect ecological changes.	

- **Ecosystem Modelling:** AI can be used to create ecosystem models that more accurately reflect the complex interactions between species, their environments, and the threats they face, which can lead to better conservation management decision-making.

13.4.3 AI-Powered Drones for Wildlife Protection

- **Anti-Poaching Operations:** AI-enabled drones can fly over large protected areas on their own, searching for signs of poaching or endangered species threats. AI's analysis of real-time video allows rangers to respond more quickly.
- **Monitoring Remote Areas:** Drones make it possible to monitor areas that are difficult to reach on foot by providing real-time data to track human and animal activity in those areas [17].
- **Gathering Ecological Data:** Drones can also collect ecological information, such as water quality, vegetation health, or temperature, to detect changes in the environment that may affect wildlife. Artificial intelligence can then analyze this data.

These applications show that artificial intelligence (AI) is not only a monitoring tool but also a crucial part of the modern conservation toolbox, accelerating, improving, and fine-tuning efforts to protect biodiversity.

13.5 AI for Sustainable Agriculture and Forestry

13.5.1 AI Driven Precision Farming and Crop Yield Prediction

AI driven precision farming use algorithms related to artificial intelligence and machine learning to enhance farming practices. Minimizing waste and boost productivity using precision farming can be achieved by applying appropriate amount of water, fertilizers, pesticides and all other inputs on time [18].

Few ways are listed below, how AI helps in precision farming:

1. **Machine Learning and Data Analytics:**
   - Agricultural data analyzed by machine learning collected from sensors, weather stations and historical forms.
   - **Crop Yield Prediction:** Weather forecasts, disease risks and soil conditions are main variables for predicting crop output.
2. **IoT Sensors for Real-Time Data Collection:**
   - **Soil Moisture and Nutrient Monitoring:** Soil conditions are measured using sensors to optimize irrigation and fertilization.
   - **Weather Based Prediction:** For scheduling planting cycles, AI models predict temperature and rainfall.
3. **AI- Powered Drones and Satellite Imaging:**
   - **Remote Crop Monitoring:** Aerial images are captured using drones to detect disease outbreaks.
   - **Precision Agriculture:** Drought-prone areas are identified which helps to optimize resource distribution.
4. **Robotics and Automation:**
   - **Automated Harvesting and Pest Control:** For optimizing farm operations and minimizing losses AI-driven machinery is really helpful.

13.5.2 Smart Forestry Management Using AI

Few ways are listed below how smart forestry management use AI.

1. **AI in Climate-Smart Forestry:**
   AI and Remote Sensing:
   - Forest structures are mapped using AI-powered drones.
   - Real-time monitoring of tree health and forest conditions is provided by satellite imagery and IoT sensors.
2. **Predictive Modeling:**
   - Impact of climate change on forests and identifying regions at risk is predicted using AI algorithms.
   - The role of forests in mitigating CO_2 emissions is estimated using carbon sequestration models [19].
3. **Forest Digital Twins**
   - Forest growth, species adaption and disaster reaction all these variables are stimulated by virtual models.

- Sustainable logging and reforestation efforts are optimized using AI-driven decision making.

4. **Automation in Forest Management:**
    - Carbon storage and biodiversity is improved using AI-guided thinking and planting strategies.
    - To minimize damage to surrounding ecosystems smart harvesting machines are used.

13.5.3 AI-Enabled Pest and Disease Detection

Few ways are listed below how pest and disease detection enhanced using AI.

1. **Image Recognition and Computer Vision:**
    - **Drone and Camera Imaging:** Images of crops captured from various angles using drones with high resolution cameras. Analysis of images is performed in real time to detect signs of pest and disease such as discoloration, wilting or abnormal growth patterns in plants.
    - **Smartphone Apps:** Smart Apps based on AI used by user to detect pest and disease. User click the picture of affected plants and upload it; application of AI engine will compare those photographs with vast database of diseases and pests to make a diagnosis [20].
2. **Sound and Acoustic Analysis:**
    - **Acoustic Sensors:** To detect the presence of pests AI based systems can use acoustic analysis.AI base microphones detecting the sound of insects like any insect eating leaves and using these sounds the analysis is done to determine the pest type and its location in the field.
    - **Automated Pest Counting:** By detecting the sound of insects using sound sensors paired with AI, monitoring the presence and density of pest populations can be performed easily.
3. **Predictive Analysis for Pest and Disease Outbreaks:**
    - **Weather Prediction:** Weather forecast used by AI systems to predict which type of pest or disease is more likely to occur based on the temperature, rainfall and humidity. For example, as fungal related disease occur in humid environment only.

13.6 AI in Air Pollution Control

13.6.1 AI for Real-Time Air Quality Monitoring and Forecasting

Few ways are listed below how AI contributes for real time air quality monitoring and forecasting.

1. **Real-Time Data Collection and Monitoring [21]:**
   - **Smart City Infrastructure:** Integration of AI with smart infrastructure is being performed in various cities to collect comprehensive data on environment factors that influence air quality.
2. **AI for Data Analysis and Prediction:**
   - **Data Fusion:** Integration of data can be performed from multiple sources like satellite imagery, ground based sensors, weather forecasts and environmental reports.
   - **Anomaly Detection:** To recognize unusual or dangerous fluctuations in air quality AI models can be trained.

13.6.2 AI-Driven Emission Reduction Strategies

Few ways are listed below for AI-driven emission reduction strategies.

1. **Energy Efficiency Optimization:**
   - Building Energy Management Systems (BEMS) are AI based models use real-time data from sensors, thermostats and meters to optimize heating, ventilation and air conditioning systems [22].
2. **Industrial Emission Reduction:**
   - Process Optimization: Optimization of manufacturing processes is done by AI and ML models by analyzing production data and identifying inefficiencies. In industries like steel and cement production, alternative methods are suggested by AI to reduce energy use and material waste.
   - Carbon Capture and Storage (CCS): Effectiveness of carbon capture and storage systems can be done using AI by predicting CO2 emission from industrial processes and optimizing the operation of capture units. To enhance the monitoring and management of CO2 storage sites, machine learning algorithms can be used.

13.6.3 Autonomous Systems for Pollution Control

Few ways are listed below for autonomous systems for pollution control.

1. **Robotic Systems for Waste Management:**
   - **Autonomous Waste Collection and Sorting:** For collecting and sorting waste in waste management systems use of autonomous robots and drones are widely being used. AI based systems used to differentiate recyclable materials, organic waste and non-recyclable, increase recycling rates [23].
   - **Waste to Energy Automation:** In waste to energy automation, AI-driven autonomous systems focus on optimizing the combustion process and also ensure that energy must be generated efficiently while minimizing emissions of harmful pollutions such as CO (carbon monoxide).
2. **AI-Driven Pollution Forecasting and Management:**
   - **Autonomous Pollution Forecasting:** To forecast pollution level, AI algorithms analyze environmental data such as meteorological conditions and pollution trends. Based on predictions of pollution spikes such as smog, these systems adjust pollution control measures.
   - **Predictive Pollution Control:** AI enabled autonomous systems used to predict when and where pollution will exceed safe levels and further actions can be taken accordingly.

13.7 AI and Renewable Energy Optimization

13.7.1 AI for Solar and Wind Energy Forecasting

Few ways are listed below for AI for solar and wind energy forecasting.

1. **Predicting Solar Energy Output:**
   - **Weather Data Integration:** For predicting solar irradiance accurately, AI systems integrate weather forecast such as temperature and humidity. Based on prediction, producers know when peak production times will occur helps them in optimizing energy storage and distribution.

- **Satellite Imagery and Remote Sensing:** To assess cloud cover and other atmospheric conditions AI based algorithms are applied over satellite images. Based on real time, predictions about solar energy can improve forecasting accuracy [24].

2. **Predicting Wind Energy Output:**
   - **Weather Forecasting and Wind Patterns:** Weather data including wind speed, direction and atmospheric pressure is used by AI models to forecast wind energy output.
   - **Numerical Weather Prediction Models (NWPM):** Enhancement of traditional numerical weather prediction models can be done using AI that forecast wind conditions at different altitudes.

13.7.2 Smart Grids and AI-Powered Energy Distribution

Few ways for Smart grids and AI-powered energy distribution are following:

1. **Real Time Energy Management:**
   - **Predictive Load Management:** Forecast of electricity demand based on listed factors like historical consumption, weather forecast can be done using AI. Better energy distribution and optimizing the flow of electricity can be planned on the basis of these predictions operators [25].
   - **Demand Response:** AI allows dynamic demand response programs, where the consumption of electricity automatically adjusted by grid operators in real time.
2. **Optimization Energy Distribution:**
   - **Load Balancing:** Various AI algorithms used to ensure that electricity is optimally distributed to meet both real time and forecasted demand. This involves balanced supply across different regions or neighborhoods.
   - **Voltage and frequency control:** For continuously monitoring and adjusting voltage and frequency levels, AI systems can be used just to ensure that they remain stable.

13.8 AI for Smart Disaster Resilience

AI has changed traditional methods of dealing with emergency, crisis and disasters (ECD) to pro-active and intelligent resilience [26]. AI assists in predicting disaster, issuing early warning and preventing potential threats by studying data from satellites, sensors and social media [24]. AI examines historical and real-time data to predict floods, earthquakes and typhoons whereas deep learning and machine learning forecast crowd density and mobility for evacuation [27, 28]. AI also detects weather patterns and sensor data with the help of AI driven early warning system to improve preparedness for possible disaster [28]. Machine learning and neural network enhance disaster prediction by examining historical and real-time data. Geographic information systems and remote sensing aid in early warning and hazard mapping [29]. AI enables automated detection, recognition and risk evaluation in an emergency which gives more response time and increases efficiency [26]. To detect disasters in real time, AI examines large scale data obtained from sensors, IoT devices and social media. Neural network along with techniques of computer vision, analyzing image data from sensors, helps in predicting possible signs of earthquakes, wildfires and floods by using sensor and image data whereas real-time crisis events are predicted through text mining and sentiment analysis [27]. AI utilizes IoT sensor feeds, satellite images and social media data for disaster detection in real time. NLP and computer vision techniques are helpful in analysing environmental changes, event recognition and classification [28, 29]. By processing satellite imagery, sensor data and social media inputs, AI detects disaster on their occurrence. Computer vision and deep learning contribute in analyzing environmental changes for immediate response [29].

AI provides recommendations based on strategic data analysis, resource allocation and emergency planning into consideration before making optimal decision to deal with a crisis. AI analyzes heterogeneous data from multiple sources and helps in handing complexities such as uncertainty, decentralized information or dynamic of a predicted disaster [26]. AI analysing data from multi sources helps government and organizations monitor and managing disasters, emergency responses, evacuation plan and resources allocation [27, 28]. Medical responses, rescue operation and logistics can be optimized utilizing various machine learning algorithms [28]. AI can enhance damage assessment, search and rescue operations using image recognition, drones and robotics. Also, AI using geospatial data, can abet in reconstruction effort s (post disaster) [28, 29].

AI facilitate in dispatching of resources, helping and monitoring rescue operation, assessment of damage and various aspects of recovery through robotics, decision-support system and UAVs (drones) [26]. To locate victims, AI takes assistance of robotics, drones and satellite images recognition to strengthen search and rescue operations during disaster. AI based decision support system optimizes damages assessment, emergency triage and relief distribution [27]. Recent advances such as edge computing and smart sensors are aiding disaster automation. Similarly, public safety and resilience can be improved by utilizing AI based data analytics and crowd sourcing etc. [27]. Real time environmental monitoring can be achieved with IoT integrated with AI; whereas, edge computing contributes in reducing latency especially during a disaster response [28]. AI takes care of technological disasters, most prominent of which are cyber-attacks, AI system failure, misinformation etc. [26]. AI models require robust algorithms and reliable datasets for an improved accuracy for disaster response and management. But concerns like data quality, lack of infrastructure, ethical dilemma impact effectiveness of AI [29]. AI analyses data from different social media platforms to assess public sentiments and figure out possible location.

AI helps organizations to deal with false information management during disaster with the help of AI driven misinformation detection systems [27]. AI contributed significantly in predicting the disease, developing vaccine, detecting misinformation and management of health care resources [26].

13.9 Environmental Sustainability

AI analyzes large ecological datasets to identify patterns so as to provide improved climate predictions, ecosystem analysis and biodiversity monitoring.AI by detecting pollution and optimizing land-use planning enhances conservation efforts and resource management [30]. AI achieves circular economy by promoting resource efficiency, material re-use and waste reduction, upgrading from linear economy (take-make-waste). Automation and predictive analytics provided by AI contributes in enhanced sustainability and reduction in environmental degradation being caused by pollution and over-extraction [31]. AI is in accordance with UN Sustainable Developmental Goals (SDGs). Hence, providing process optimization, enabling data-driven decision making, automation in industry, promoting circular economy models and reducing resource consumption [32]. AI is being combined with sustainability and ethics to develop new ideas or technology while contemplating and mitigating negative effects on environment. AI is being utilized for ethical decision-making, policy-making, and environmental monitoring shows in Figure 13.1. AI mitigates misinformation, biases, and ecological risks by deploying green governance model [33].

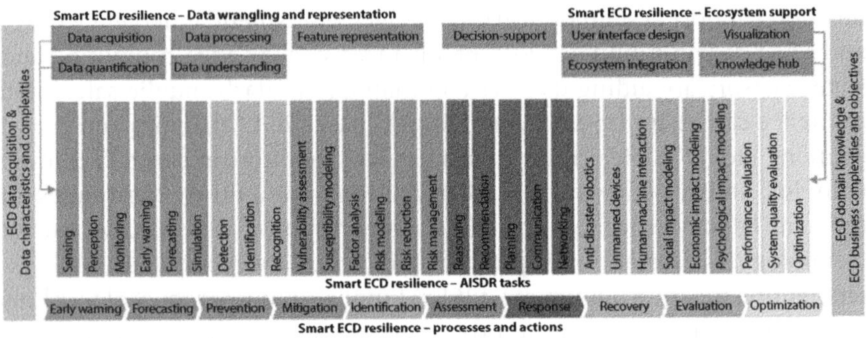

Figure 13.1 The research landscape of translation disaster AI for smart ECD resilience (AISDR). E emergency, C crises, D disaster [26].

AI contributes academic writing and scientific research by generating insights, hence resulting in reduced workload. Carbon footprints of AI training, misinformation, bias in AI models etc. are some of the ethical concerns which need careful assessment to ensure transparent and responsible usage [30]. AI provides dynamic market matching algorithms, encourages reusability of products benefiting with more accessibility and cost effectiveness. Also, it introduces new circular business models such as product-as-a-service which is a great example of wider availability and dynamic usability [31]. Economic inequalities, environmental trade-offs, over dependence on AI decision-making, algorithmic biases, energy consumption, and data privacy are some of the concerns and ethical risks posed by AI. Hence, policy recommendations and regulated AI governance are the utmost needs to deploy AI for sustainable efforts and ensure fair and responsible regulations [30, 31]. AI enabled supply chain optimization, results in minimizing waste and its environmental effects and improves allocation of resources [32]. AI's energy-intensive computations results in carbon emission and bias in AI model can lead to unfair resource distribution. Hence, Equitable AI access needs to be regulated so as to prevent technological disparities between regions [32, 33].

Green AI research brings out the significances of transparency, accountability, and fairness in AI run environmental policies. AI governance should harmonize technological advancement with environmental and social well-being [33]. Concerns like deforestation, waste generation, carbon footprints etc. are catered by AI by replacing e-mails, in person inquiries and paperless operations. Also, energy efficiency in research and industry has been optimized to a greater extent [30]. Predictive maintenance, smart grids, optimized energy consumption in data centers and industrial processes, AI brings down carbon emission to a greater extent reduces the need for new resource extraction and helps in ensuring longer equipment lifespan [31]. AI enables pollution control waste management, energy optimization and water conservation by automation, helping industry and smart grids reduce carbon emission. Climate change impacts are predicted by AI with the help of data obtained from satellites, sensors, and social media [32, 33].

AI proves to be quite beneficial for sustainable agriculture, monitoring wildlife conservation, urban planning, analyzing and reducing human impact on fragile ecosystem and efficient production and distribution of renewable energy [30]. Economic inequalities, environmental trade-offs, over dependence on AI decision-making, algorithmic biases, energy consumption and data privacy are some of the concerns and ethical risks posed by AI. Hence, policy recommendations and regulated AI governance are the utmost needs in order to deploy AI for sustainable efforts and ensure fair and responsible regulations [31]. AI technologies have fostered smart cities development, enhanced urban planning, optimize traffic, deforestation monitoring, protection of bio-diversity and innovation in forestry and agriculture. It has also contributed in wind and solar energy utilization, forecasting and optimizing renewable energy [32]. AI chabot's and applications promote practices that are environment friendly and also encourage business/individuals to reduce waste and adopt green policies [30].

13.10 Future Scope

1. Enhancing Climate Change Understanding and Prediction

AI proves to be a great helping hand for predicting extreme weather conditions, climate trends, and temperature shifts. It also aids in forecasting of precipitation patterns, ocean phenomena, and cloud system [34]. It also assists in processing data obtained from satellites, climate models and sensor networks, and enhances the chance of optimal disaster management, pollution detection, carbon tracking, disaster management strategies, and prediction of environmental changes [35, 38].

Environmental data analysis, object detection and image classification can be enhanced using AI. Also, AI powered geospatial analysis results in better predictions and analysis when it comes to monitoring of water quality, air pollution, and trends in climate change [39]. Satellite imaging and IOT sensors, powered by AI, analyzes and predicts phenomena such as extreme weather events, air quality, water pollution, and deforestation [36]. AI driven algorithms are assisting in tracking endangered species, detection of illegal deforestation activities and monitoring of ecosystem [38]. Deep learning models help in tracking, deforestation, land use and disaster prediction [39]. AI plays an important role assisting governments to enforce renewable energy policies, emission reducing strategies, carbon pricing, sustainability targets, and enforcing regulations [34, 37, 38].

2. Optimizing Energy Use and Carbon Reduction

AI has a major contribution in energy efficiency in industries, transportation and smart grids, by forecasting load and demand response, thus resulting in reducing overall emission [34, 37, 39]. Technologies lead by AI, aim at developing clean energy solutions, integrating renewable energy resources into the grid and optimizing energy consumption [38, 39].

AI also aids in forecasting wind and solar power availability, manage renewable resources and ensuring enhanced grid stability [34, 37].

3. AI for Smart and Green Technologies

AI assists in efficient use of energy resources, reduction in waste, efficient water management, enhances predictive maintenance etc. [35, 38]. Reduced material waste and enhanced sustainable supply chains are a result of AI based green management [34]. AI extends its contribution in optimization of water conservation, agriculture etc. with the help of data driven decision making [39]. AI also supports eco-friendly supply chain and logistic operations by making feasible AI-driven autonomous vehicles, robotics and drones [35]. AI is capable of efficiently predicting the demand, hence reducing excess production. AI enabled smart city planning optimizes electricity use, traffic flow, infrastructure sustainability, and many such aspects [36]. AI applications also contribute in optimizing public transportation systems for lower emissions [38]. It is through AI that we can optimize urban planning with data driven urban planning [39].

4. Circular Economy and Resource Optimization

AI, with the help of smart monitoring and automation, can improve sustainable agriculture, waste recycling and water conservation [34]. It also enables efficient resource utilization and optimization, smart recycling and waste sorting with the help of big data analysis [35, 36]. AI also supports sustainable resource allocation [37]; also, big data analysis and IoT aids in real-time monitoring and decision making [35]. Reusability and sustainability are being promoted as AI supports closed-loop supply chain [35]. Logistics, energy consumption, production, warehousing and transportation are being optimized by AI and as a result, environmental footprint is reducing [35, 36]. AI-driven energy policy frameworks align with circular economy principles for long-term sustainability [37].

5. AI for Decision-Making and Policy Formulation

AI driven data analysis, risk assessment models and decision support systems assist policymakers design sustainable energy policies and environmental regulations [36, 37]. AI backs up emission tracking, carbon trading, and eco-friendly business strategies [36], plus it also promotes international cooperation and regulatory frameworks for global sustainability [37].

Challenges

1. High Energy Consumption

The environmental cost of AI research and training consumes considerable energy. AI models such as GPT-3 contributes in producing high carbon footprint, which needs to be balanced so that sustainable benefits of AI can be reaped [34, 38, 39]. A large computing power is required by AI systems which results in high energy use, hence resulting in higher carbon footprint [35, 36]. Hence, some green AI approaches need to be developed to maximize efficiency while minimizing power consumption [36, 37].

2. Ethical Concerns

AL models can result in amplification of biases, unfair policy decisions and can exclude marginalized communities from climate adaptation effort. Also, as AI monitors emission, human behaviors and energy consumption patterns, it actually raises concerns about data privacy [34]. AI predictions and decisions may introduce biases resource allocation and environmental properties and decision making [35, 37]. The biases may also include some favors for a particular industry or a region [36]. AI led tracking technologies, environmental surveillances using large environmental and behavioral datasets also impose a threat of data privacy concerns [36–38]. Also, if biased data is utilized for training of AI models, the sustainability recommendation obtained will be inaccurate which might lead to ineffective policy measures [38]. In addition to this, lack of transparency can undermine trust in sustainability applications [39].

3. Lack of Standardization

Lack of regulatory framework for AI is a major concern. As there are no global frameworks for environmental sustainability, policies are needed to be ensured and responsible AI deployment needs to be established in order to counter unintended negative consequences [34–36]. Ethical deployment of AI needs to be ensured and a universal framework needs to exists for AI governance in sustainability [37–39].

4. Data Accessibility, Integration of AI models

It is difficult to access real world impact of many AI driven projects as they lack reproducibility and transparency. Disclosure on carbon impact, energy consumption and environmental tradeoffs should be included with AI research [34]. As AI relies on analyzing large datasets, concerns like data security, integration with existing infrastructure and transparency might arise. There is a lack of technical setups in many industries to accommodate and integrate AI for ensuring sustainability of environment [35, 36].

There is a need for scalable and cost effective AI solutions for a widespread adoption [36]. AI requires real time high quality data in order to prove effective in sustainability efforts for environment. But many domains lack the feasibility to collect and process data needed for AI led environmental management [37]. The focus of AI deployment is widely in technically advanced nations leaving developing nations with limited access to this innovation. This needs to be improved and taken care of [38].

AI disclosure statement: The author acknowledges the use of AI-based tools for minor language editing or grammatical correction. However, all intellectual contributions, analyses, and conclusions presented in this chapter are entirely the author's own.

References

1. Santos, M.R. and Carvalho, L.C., AI-driven participatory environmental management: Innovations, applications, and future prospects. *J. Environ. Manage.*, *373*, 123864, 2025.
2. Floridi, L., AI as Agency without Intelligence: On Artificial Intelligence as a New Form of Artificial Agency and the Multiple Realisability of Agency Thesis. *Philos. Technol.*, *38*, 1, 30, 2025.
3. Ramesh, M., Selvan, M.T., Saravanakumar, A., Evolution and recent advancements of composite materials in structural applications, in: *Applications of Composite Materials in Engineering*, pp. 97–117, Elsevier Science Ltd, 2025.
4. Panahi, O. and Jabbarzadeh, M., The Expanding Role of Artificial Intelligence in Modern Dentistry. *J. Dent. Oral Health*, *8*, 3, 2025.
5. Al-Raeei, M., The smart future for sustainable development: Artificial intelligence solutions for sustainable urbanization. *Sustainable Dev.*, *33*, 1, 508–517, 2025.

6. Shobanke, M., Bhatt, M., Shittu, E., Advancements and future outlook of Artificial Intelligence in energy and climate change modeling. *Adv. Appl. Energy*, 100211, 2025.
7. Olawumi, M.A. and Oladapo, B.I., AI-driven predictive models for sustainability. *J. Environ. Manage.*, *373*, 123472, 2025.
8. Huan, S. and Liu, X., Machine learning-driven analysis of greenhouse gas emissions from rice production in major Chinese provinces: Identifying key factors and developing reduction strategies. *Eur. J. Agron.*, 164, 127536, 2025.
9. Yüksel, S., Eti, S., Dinçer, H., Gökalp, Y., Olaru, G.O., Oflaz, N.K., Innovative financial solutions for sustainable investments using artificial intelligence-based hybrid fuzzy decision-making approach in carbon capture technologies. *Financ. Innovation*, 11, 1, 20, 2025.
10. Sarkar, S. and Jha, P.K., Application of AI/ML in Water Resource Management to Resolve Transboundary Water Conflict, in: *Navigating the Nexus: Hydrology, Agriculture, Pollution and Climate Change*, vol. 1, pp. 431–455, Springer Nature Switzerland, Cham, 2025.
11. Oyarzabal, R.S., Santos, L.B., Cunningham, C., Broedel, E., de Lima, G.R., Cunha-Zeri, G., Peixoto, J.S., Anochi, J.A., Garcia, K., Costa, L.C., Pampuch, L.A., Forecasting drought using machine learning: a systematic literature review. *Nat. Hazard.*, 1–29, 2025.
12. Ullah, F., Saqib, S., Xiong, Y.C., Integrating artificial intelligence in biodiversity conservation: bridging classical and modern approaches. *Biodivers. Conserv.*, 34, 1, 45–65, 2025.
13. Johnson, J. and Rajuroy, A., Advancing Biodiversity Conservation Through Synthetic Image Generation: AI-Based Identification and Classification of Rare Plant Species Using Generative Models.
14. Habib, S., *et al.*, Advancing Taxonomic Classification Through Deep Learning: A Robust Artificial Intelligence Framework for Species Identification Using Natural Images. *IEEE Access,* 12, 146718–146732, 2024.
15. Rhodes, K. and Sagan, V., Integrating Remote Sensing and Machine Learning for Regional-Scale Habitat Mapping: Advances and future challenges for desert locust monitoring. *IEEE Geoscience and Remote Sensing Magazine*, 10, 1, 289–319, March 2022.
16. Ghali, R., Akhloufi, M.A., Mseddi, W.S., Deep learning and transformer approaches for UAV-based wildfire detection and segmentation. *Sensors,* 22, 5, 1977, Mar. 2022.
17. Rathi, P., Navigating the Landscape: Challenges and Opportunities of AI in Wildlife Conservation. *AI Mach. Learn. Tech. Wildl. Conserv.*, 159–182, 2025.
18. Dhanke, J.A., Srivastava, D., Menaga, D., Raj, R., Kumar, K.V., Jangir, P., Mani, P., Climate-Based AI-Powered Precision Irrigation: Sustainably Smart Agriculture Frameworks for Maximum Crop Yields. *Remote Sens. Earth Syst. Sci.*, 8, 1, 161–172, 2025.

19. Shanthalakshmi, M., Jeevasree, M., Kavitha, R., Madhumathi, V., Mythreye, S., Yusra, A.N., Forestry Management with AI and Drone Technology–Digital Forestry. *Comput. Vision Smart Agric. Crop Manage.*, 259–281, 2025.
20. Zhou, C., Lee, W.S., Kratochvil, W.A., Schueller, J.K., Pourreza, A., Liburd, O.E., Ampatzidis, Y., Ai-Enabled Automated Imaging and Detection System for Precision Pest Management in Strawberries. Available at SSRN 5162747.
21. Jayaraman, S. and S.,A., Enhancing urban air quality prediction using time-based-spatial forecasting framework. *Sci. Rep.*, 15, 1, 4139, 2025.
22. Arishi, M. and Kuku, M., Mitigating carbon emissions through AI-driven optimization of zeolite structures: A hybrid model approach. *Alexandria Eng. J.*, 115, 370–389, 2025.
23. Kundavaram, R.R., Onteddu, A.R., Devarapu, K., Narsina, D., Nizamuddin, M., Advances in Autonomous Robotics for Environmental Cleanup and Hazardous Waste Management.
24. Saadati, T. and Barutcu, B., Forecasting Solar Energy: Leveraging Artificial Intelligence and Machine Learning for Sustainable Energy Solutions. *J. Econ. Surv.*, 2025.
25. Rajaperumal, T.A. and Columbus, C.C., Transforming the electrical grid: the role of AI in advancing smart, sustainable, and secure energy systems. *Energy Inf.*, 8, 1, 1–43, 2025.
26. Cao, L., AI and data science for smart emergency, crisis and disaster resilience. *Int. J. Data. Sci. Anal.*, 15, 3, 231–246, 2023.
27. Aboualola, M., Abualsaud, K., Khattab, T., Zorba, N., Hassanein, H.S., Edge technologies for disaster management: A survey of social media and artificial intelligence integration. *IEEE Access*, 11, 73782–73802, 2023.
28. Nunavath, V. and Goodwin, M., The use of artificial intelligence in disaster management-a systematic literature review, in: *2019 International Conference on Information and Communication Technologies for Disaster Management (ICT-DM)*, IEEE, pp. 1–8, 2019, December.
29. Abid, S.K., Chan, S.W., Sulaiman, N., Bhatti, U., Nazir, U., Present and future of artificial intelligence in disaster management, in: *2023 International Conference on Engineering Management of Communication and Technology (EMCTECH)*, IEEE, pp. 1–7, 2023, October.
30. Reyhani Haghighi, S., Pasandideh Saqalaksari, M., Johnson, S.N., Artificial intelligence in ecology: a commentary on a chatbot's perspective. *Bull. Ecol. Soc. Am.*, 104, 4, e2097, 2023.
31. Solanki, S. and Singh Rathore, P., Current E-Waste Management: An Exploratory Study on Managing E-Waste for Environmental Sustainability. *Sustainable Manage. Electron. Waste*, 187–200, 2024.
32. Kumar, A., Rathore, P.S., Dubey, A.K., Srivastav, A.L., Kumar, T.A. (Eds.), *Sustainable management of electronic waste*, John Wiley & Sons, 2024.
33. Rathore, P.S., Ahuja, S., Burri, S.R., Khunteta, A., Baliyan, A., Kumar, A., *Deep learning techniques for automation and industrial applications*, John Wiley & Sons, Inc., Scrivener Publishing LLC, 2024a.

34. Cowls, J., Tsamados, A., Taddeo, M., Floridi, L., *The AI gambit: leveraging artificial intelligence to combat climate change—opportunities, challenges, and recommendations*, pp. 1–25, AI & Society, 2023.
35. Sun, X., Yu, H., Solvang, W.D., Wang, Y., Wang, K., The application of Industry 4.0 technologies in sustainable logistics: a systematic literature review (2012–2020) to explore future research opportunities. *Environ. Sci. Pollut. Res.*, 1–32, 2022.
36. Dwivedi, Y.K., Sharma, A., Rana, N.P., Giannakis, M., Goel, P., Dutot, V., Evolution of artificial intelligence research in Technological Forecasting and Social Change: Research topics, trends, and future directions. *Technol. Forecasting Social Change*, 192, 122579, 2023.
37. Danish, M.S.S. and Senjyu, T., Shaping the future of sustainable energy through AI-enabled circular economy policies. *Circ. Econ.*, 2, 2, 100040, 2023.
38. Longo, L., Brcic, M., Cabitza, F., Choi, J., Confalonieri, R., Del Ser, J., Guidotti, R., Hayashi, Y., Herrera, F., Holzinger, A., Jiang, R., Explainable Artificial Intelligence (XAI) 2.0: A manifesto of open challenges and interdisciplinary research directions. *Inf. Fusion*, 106, 102301, 2024.
39. Zhang, L. and Zhang, L., Artificial intelligence for remote sensing data analysis: A review of challenges and opportunities. *IEEE Geosci. Remote Sens. Mag.*, 10, 2, 270–294, 2022.

14

XAI for Decision Support in Microplastic Pollution Management

Yeligeti Raju[1]*, N. Venkatesh[2], S. Adilakshmi[1], Namita Parati[3] and A. Kalaivani[4]

[1]*Department of CSE (Data Science), Vignana Bharathi Institute of Technology, Ghatkesar, Hyderabad, Telangana, India*
[2]*School of Computer Science & Artificial Intelligence, SR University, Warangal, Telangana, India*
[3]*Department of CSE, Maturi Venkata Subba Rao (MVSR) Engineering College, Hyderabad, Telangana, India*
[4]*Department of CSE, Rajalakshmi Engineering College, Thandalam, Chennai, India*

Abstract

The continuing increase of microplastic pollution requires urgent development of proper ecological monitoring alongside classification systems. The paper investigates how Explainable Artificial Intelligence (XAI) frameworks react with microplastic pollution management tools to enhance decision processes while increasing transparency capabilities. The training of machine learning models through a dataset consisting of fiber beads and fragments yielded Random Forest as the optimal selection because it provided exceptional accuracy together with ease of interpretation. The key characteristics for prediction turned out to be size, alongside color intensity according to both feature importance and SHAP analysis results. The confusion matrix showed both precise classifications and very low rates of incorrect categorizations. XAI implementation delivered both extraordinary performance outcomes with expert trust enhancement and significantly short time-to-decision of more than sixty percent according to Likert-scale survey results. The examination outcomes establish XAI works to close the distance between machine-generated technical AI outputs and field-oriented implementation decisions. The system provides plain understandable explanations that enable

**Corresponding author*: raju.yeligeti@gmail.com

Abhishek Kumar, Pooja Dixit, Pramod Singh Rathore, Arun Lal Srivastav and Ashutosh Kumar Dubey (eds.) Microplastic Monitoring Using Artificial Intelligence, (293–308) © 2026 Scrivener Publishing LLC

faster and more self-assured action from researchers and policymakers. The study presents an implementation framework for explainable modeling in environmental oversight which serves as a foundation for developing sustainable AI applications in sustainability science.

Keywords: Explainable artificial intelligence (XAI), microplastic detection, environmental decision support systems, sustainable pollution management, interpretable machine learning

14.1 Introduction

Microplastic contamination exists as an important environmental challenge which damages marine ecosystems and damages wildlife populations and produces negative health impacts on humans. The small plastic fragments measuring below 5 millimetres in diameter originate from diverse sources which include worn-out large plastic materials and synthetic textiles and cosmetic products. The expanding occurrence of microplastics in oceans and rivers and drinking water supply demands immediate action for control and reduction of their spread. The miniscule size along with multiple starting points of microplastics makes conventional monitoring methods and policy development less effective [1]. The transformative capabilities of Explainable AI (XAI) within artificial intelligence (AI) enable organizations to thrive through this demanding situation. The transparency features of XAI systems enhance complex AI model predictions and outcomes for decision-makers to understand their implications. Through environmental science applications XAI provides official decision-makers and researchers and stakeholders with advanced understanding which produces trustworthy intelligent solutions to manage microplastic pollution better [2].

Microplastics appeared as marine environmental pollutants in the 1970s before gaining widespread recognition as a major concern in the modern era. Microplastics have spread across all global regions and researchers can find them in polar ice together with deep water sediments. Their resilient nature combined with their diminutive size creates obstacles for filtering systems through which detection and removal operations become extensively difficult [3]. Researchers from science and environmental agencies work actively on microplastic studies of origins and through-transport paths and permanent consequences but face difficulties in acquiring complete data. The combination of AI with data science technology enables efficient analysis of big datasets taken from water content exams along with images from satellites and network sensors. The operations of typical AI algorithms can remain mysterious because their results normally lack clear explanations. XAI serves as an effective solution by processing information while providing explanations about all predictions or recommendations in environmental decision-making which demands both trust and accountability [4].

The existence of microplastics throughout environmental spaces creates numerous significant impacts for social structures. Microplastics appear both in drinking water resources and accumulate throughout the food web after fish and other marine creatures consume them, thus leading to potential human health effects. The interaction with microplastics creates significant hormonal disorders and reproductive dangers and possible cancer risks in wildlife that trigger serious public health alerts. The fishing industry together with tourism businesses experience financial setbacks because polluted water bodies and diminished biological diversity affect their communities [5]. Nations that have constrained resources face an even dire situation while implementing proper waste management infrastructure because of their minimal capabilities. The implementation of XAI for decision support enables governments along with organizations to create detailed and easily comprehensible approaches for microplastic pollution management. Transparent AI produces models that establish trust among the public while involving communities so interventions maintain an ethical standard that people accept. XAI methods assist both environmental sustainability and social well-being objectives through their applications [6].

The conventional approach to studying microplastic pollution includes water collection through hand methods followed by laboratory analysis through microscope or spectroscopy data and physical filtration technologies. The combination of accurate measurements with these methods comes with significant time obligations and high costs and prevents their use on a large scale [7]. Traditional statistical models lack sufficient power to process complex non-linear data gathered from various environmental sources because they operate below current standards of assessment. Both methods cannot promptly respond to environmental conditions that change frequently because of which important decisions end up being either delayed or useless. Policy implementations derived from using these methods might not match up with present conditions which diminishes their effectiveness. The approaches need specialized domain knowledge to operate while remaining out of reach for smaller organizations and communal institutions. AI-powered systems address processing limitations and speed up decision-making processes during data analysis. However, without explainability, such solutions still face resistance from policymakers [8].

Current artificial intelligence technologies particularly machine learning and deep learning systems detect and monitor and predict microplastic pollution through advanced computational patterns. Programs evaluate satellite images together with sensor information and lab outcome data at a previously unknown combination of speed and scale. AI models lack clarity in their operation which creates barriers for their implementation in essential decision-making operations. XAI provides interpretable models for revealing what input factors lead to specific output results. The combination of three interpretative methods including SHAP (Shapley Additive Explanations) and LIME (Local Interpretable Model-agnostic Explanations) and decision trees enables users to see and read how models work. The clearly displayed information enables policymakers to believe in AI outcomes while discerning pollution causes [9]. The application of XAI establishes model accountability which enables ethical AI use in reality, thus becoming an essential sustainability tool for microplastic pollution management.

XAI exploration for microplastic pollution management came to fulfill the pressing requirement for actionable trustworthy and transparent environmental protection tools. Microplastic contamination has reached crisis level worldwide, thus making traditional monitoring along with black-box AI insufficient to address the problem. Decision-support systems face an increasing market demand because users need both precise predictions and clear explanations of the generated output. Environmental policies require such clarity to affect public health and economic activities as well as ecological balance [10]. Open interpretable tools are crucial for pollution battles because they build trust between citizens and NGOs and governmental agencies to jointly combat pollution. XAI creates understanding between professionals in AI and environmental decision-makers who can effectively use complex algorithms. XAI enables the combination of AI analysis capabilities with human interpretability to create better sustainable solutions in managing microplastic pollution [20–22].

14.2 Literature Review

Citations	Technology used	Characteristics	Limitations	Outcomes	Future scope
[11]	Policy Analysis, Waste Management Systems	Examines MP escape routes like WWTPs and composting systems.	Limited technology readiness for remediation solutions.	Identified loopholes in existing systems contributing to MP pollution.	Policy reforms to close escape routes and enhance filtering mechanisms.
[12]	Ecological Risk Assessment Framework	Assesses microplastic risk using thresholds in surface water/ sediment.	Lack of thresholds for sediment-level risk.	Identified surface water risks and policy needs.	Regional ecological risk frameworks for microplastics.
[13]	PRISMA Method, Data Synthesis	MPs in green mussels characterized by shape, size, color.	Limited geographical scope (mostly Asian waters).	Industrial/human activity linked to MPs in mussels.	Filtration systems, organic plastics, stricter policies.
[14]	Co-word & Bibliographic Coupling	Mapped microplastic knowledge structure and research streams.	Dependent on WoS database scope and keywords.	Outlined major themes and public education strategies.	Holistic regulatory approaches and hybrid treatments.

(Continued)

(Continued)

Citations	Technology used	Characteristics	Limitations	Outcomes	Future scope
[15]	Microscopy, FTIR Spectroscopy	MPs in landfill soil identified by type and shape.	Study restricted to one city (Depok).	High MP presence linked to diverse waste sources.	Biodegradation methods and waste sorting improvements.
[16]	Review & Risk Assessment Methodologies	Analyzes health and ecosystem risks from MPs.	More observational data needed for human impact.	Detailed health pathways and policy call-to-action.	Innovative biodegradable materials and awareness programs.
[17]	XRD, FTIR Spectroscopy	Assessed MPs in river water and sediments.	Limited to two river estuaries in India.	Confirmed MP presence and source regions.	Targeted policy interventions and mitigation tech.
[18]	Field Survey, Statistical Analysis	Identified types and distribution of coastal MPs.	Limited global applicability.	Informed local MP policy and public education.	Tech innovation, international cooperation.
[19]	Multidisciplinary Policy Review	Reviewed sources and global spread of MPs.	Requires ongoing multi-sector collaboration.	Highlighted need for broad interventions.	Waste tech advancement and lifestyle change campaigns.

14.3 Proposed Methodology

Data Collection and Preprocessing
The procedure started with the acquisition of microplastic samples that received labels and were separated into fiber, bead and fragment groups. The samples contained essential information about particle dimensions together with their forms, color strength, material density and original components. Prior to training the model researchers standardized feature measurement values and addressed blank columns in the dataset. All inputs with either extraneous or unwanted noise signals were discarded for unbiased model training processes. The model benefited from engineered features to improve its interpretability capabilities. The training implementation of both traditional machine learning and explainable AI models involved this meticulously organized dataset which provided reliable and usable solutions for environmental decision support activities. Figure 14.1 shows a proposed methodology.

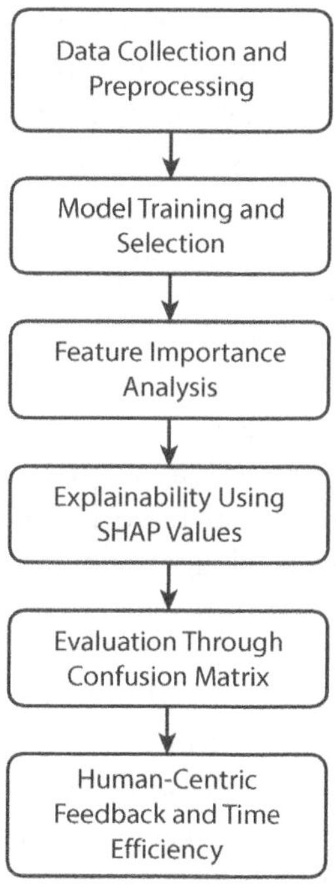

Figure 14.1 Proposed methodology.

Model Training and Selection

Training occurred for multiple supervised learning models with Random Forest as well as Support Vector Machines and Convolutional Neural Networks (CNNs). The evaluation of model performance relied on accuracy along with precision and recall while F1-score served as complementary assessment metric. The project focused on selecting AI models that exhibited both high accuracy and interpretability because these met the requirements for integration with XAI methods. The cross-validation process helped to verify that the models could accurately predict on new datasets without seen data. The Random Forest model was picked as the final selection because it displayed excellent performance features and compatibility with XAI tools including SHAP and LIME. The chosen approach made possible the creation of explanations at both global and local levels which promote trust and add policy relevance.

Feature Importance Analysis

A feature importance analysis determined the prediction methods used by the model. A bar chart showing the quantified feature contributions to classification decisions emerged automatically from the Random Forest's built-in operations. The assessment demonstrated that the key elements influencing predictions were particle size combined with color intensity yet the source material had less overall significance. The validation as a crucial step ensured domain experts had the chance to understand how the model operated. The initial exposure that people needed to verify artificial intelligence-driven choices depended on scientific insights instead of uncorrelated associations was made possible by the transparency graph.

Explainability Using SHAP Values

The SHAP (SHapley Additive exPlanations) algorithm supplied detailed explanations regarding individual feature effects on model outcomes. A SHAP summary plot presented aggregated contributions of every feature through all measured samples. The current analysis produced results which were consistent with the earlier feature assessment thus verifying model functionality. SHAP succeeded both in verifying global trends and creating local reasoning systems to explain single predictions. These useful insights made complex AI reasoning easy to communicate as understandable terms for human audiences which in turn establishes trust and transparency. The development of this phase enabled the construction of a link between black-box AI models and operational pollution management requirements.

Evaluation through Confusion Matrix

The predictive accuracy of the model was fully validated by a confusion matrix which displayed actual microplastic types against their predicted counterparts. The created matrix demonstrated high precision in particle categorization while displaying low mistakes in classifying related particles. This practical tool demonstrated both advantages and weaknesses of the model within operational decision-making environments. By examining misclassifications, the tool enabled environmental scientists to determine errors alongside their causes to optimize sampling techniques and prepare training data sets. The operational risk assessment using the confusion matrix enabled pollution monitoring systems to fully understand errors and their cause so that accurate deployment became possible before field implementation.

Human-Centric Feedback and Time Efficiency

Necessary evaluation for the XAI system's practical usage included surveys assessing decision-makers' trust in AI explanation outputs and domain expert feedback. The survey used a Likert scale to measure high levels of trust from participants. The integration of XAI led to better decision-making speed through a time-based analysis which showed a more than 60% reduction in processing time. The research findings proved that these results bring enhanced operational efficiency together with effective human-AI teamwork. The procedure demonstrated how explainable models help speed up complex workflow procedures while maintaining complete clarity. Human trust in automated environmental applications depends on achieving fast accurate results that humans fully understand according to the core purpose of XAI.

14.4 Result and Discussion

The visual represents how important each feature is for AI model classification of microplastics. The most influential features among those under examination are size and colour intensity while density and shape index act as second in line. The used source material receives the least impact compared to other variables in the dataset. This chart visualizes how different variables influence classification to provide basic explainability when used as an analysis tool as shown in Figure 14.2. The graphical display makes environmental scientists and regulators able to confirm the AI model selects features based on scientific legitimacy. The ability to understand this information proves vital during decision support activities

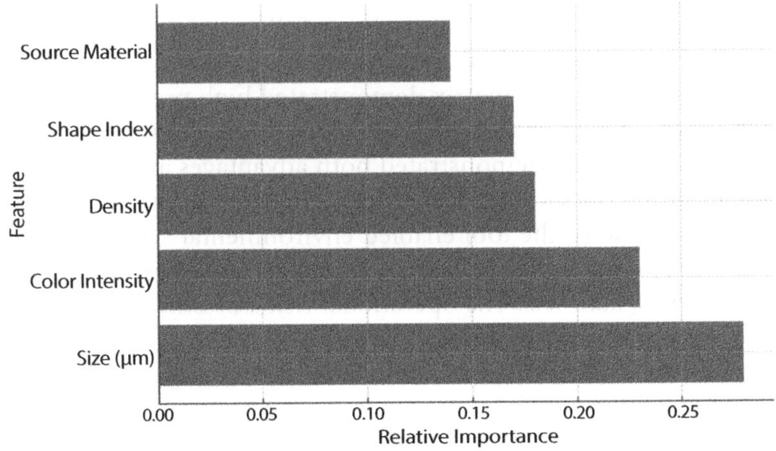

Figure 14.2 Feature importance for microplastic classification.

because it establishes trust and supports both model optimization and evidence-based policy creation. Research focusing on better density measurements will receive improved attention through the emphasis of this specific water pollution feature in the analysis. This graph proves highly effective when XAI dashboards integrate it to show model behavior for decision making in situations such as pollutant tracking or cleanup prioritization.

The SHAP value summary plot illustrates what extent each feature element affects AI model predictions throughout a dataset. One of the most common XAI methods which detects global-level feature impact uses the SHAP (SHapley Additive exPlanations) values framework. The model's decisions remain primarily influenced by the dimensions "size" and "color intensity" throughout most observations. The matching results between SHAP values and feature importance produce better model interpretability which helps to build trust in its reasoning process shown in Figure 14.3. Stakeholders can understand both predictions and decision-making paths with SHAP because it provides quantifiable scores to measure feature influence. Special emphasis on transparency emerges crucially important for environmental programs because decisions determine health policy choices as well as resource usage distribution. SHAP values provide decision-makers with a measurable feature influence assessment for domain knowledge verification which helps them identify biases and improve data collection approaches. The graph serves as the foundation of explainability to demonstrate AI decisions give understandable and justifiable calculations rather than ambiguous black-box outputs.

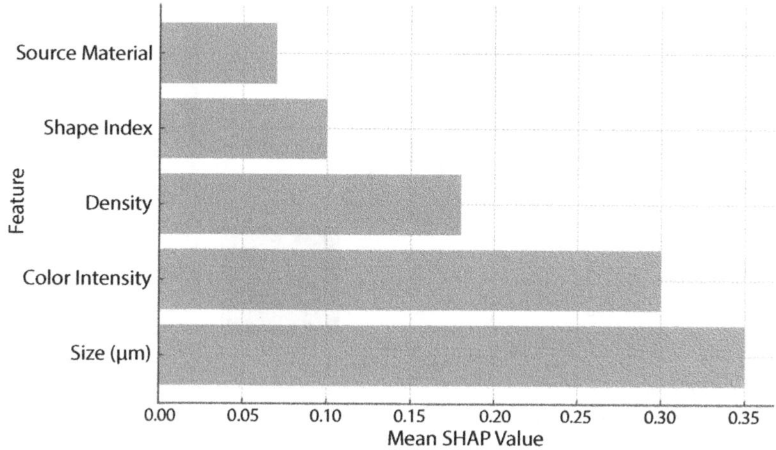

Figure 14.3 SHAP value summary for explainability.

The three types of microplastics known as fibers, beads and fragments receive accurate classification from the AI model which is shown through a confusion matrix. The diagonal area holds correct diagnostic conclusions while incorrect predictions appear in the off-diagonal space of the table. The high model accuracy shows correct identification of most samples particularly for fragments where the model correctly identified 47 out of 50 samples while performing its analysis. There exists some overlap in recognition between beads and fragments because their visual patterns and structural makeup present similar characteristics to each other shown in Figure 14.4. The graph constitutes an essential tool to evaluate model reliability before its use in operational decision support systems. The identification of incorrect class assignments allows scientists to refine their models after checking downstream procedures. Correcting the identification errors between beads and fragments through improved cleanup procedures because these two categories tend to be mistaken for one another. The evaluation data enables decision-making about the need to retrain models while helping establish new collection priorities. XAI diagnostic tools use confusion matrices to generate predictive metrics that combine performance evaluation with interpretability capabilities, therefore allowing scientists and policymakers to determine model trustworthiness for informed interventions with AI results.

Healthcare research involving AI-AR produced increasing publications over time based on the bar chart which covers the period from 2015 to 2024. The data indicates remarkable exponential growth of research and

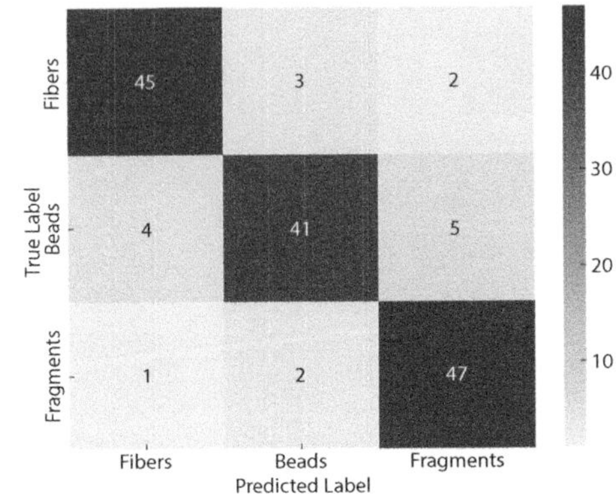

Figure 14.4 Confusion matrix: AI model predictions.

industrial focus on healthcare AI-AR beginning from merely twelve publications in 2015 until reaching more than 100 publications by 2024 as shown in Figure 14.5. After 2019 the line steadily rises because digital healthcare makes major global investments while the pandemic accelerates the adoption of telemedicine and intelligent systems. Data evidence demonstrates that AI-AR exists as an active field of research that attracts growing collaboration from various disciplines. Academic progress toward implementing smart technologies in clinical practice shows clear evidence of providing an upcoming data-based and scientifically-validated healthcare direction.

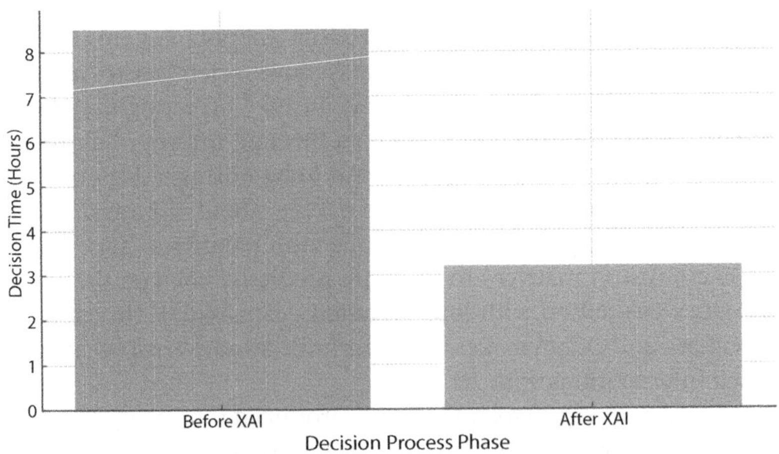

Figure 14.5 Decision time comparison: Before vs after XAI integration.

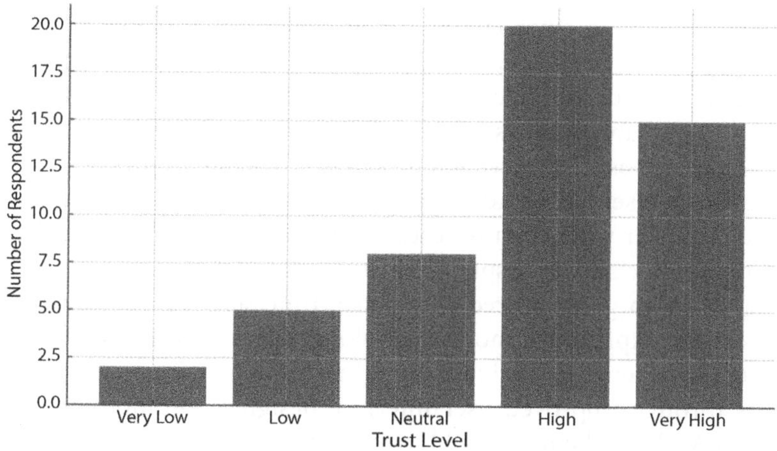

Figure 14.6 Expert trust levels in XAI explanations.

The visual presentation displays environmental science professionals who rated their trust levels towards XAI system explanations. The majority of survey participants demonstrated trust at the "High" or "Very High" levels with "Very Low" and "Low" representing minimal selections. Adoption of AI decision support increases as professionals show increasing acceptance towards systems that explain their operations using domain-specific information. AI system deployment requires trust as a fundamental factor in sensitive domains such as environmental management according to Figure 14.6. The graph presents user sentiment measurements to provide human feedback as an additional evaluation method that extends beyond quantitative model metrics. The data helps to direct future improvements to both interface design and explanatory modules that will be part of the AI system. Knowledge transfer effectiveness combined with expert approval demonstrates the value of XAI tools to support mutual interaction between human experts and AI systems across microplastic research and various other domains.

14.5 Concluding Remarks and Future Scope

Research results verify that adding Explainable AI (XAI) components to microplastic classification operations produces advanced accuracy levels while creating systems which are more user-friendly. XAI adds interpretability to the model by presenting prominent features combined with the verification of SHAP value evaluations which creates transparent methods while matching environmental field expertise. XAI proves practical through its excellent performance in confusion matrix evaluation and substantial decrease in user decision timeframe. Domain experts show strong trust in explainable models because these models demonstrate technical integrity as well as social acceptability in practical applications.

The future research agenda offers multiple opportunities for expansion of this particular work. Future research should expand its dataset by adding both oceanic and atmospheric microplastics data to increase the model's capability in general prediction. Real-time sensing technologies that connect with explainable models should enable performance of dynamic monitoring systems. AI insights should be brought to non-technical individuals through improved user interface design to promote democratic access to AI data. The use of XAI together with causal inference functions enables better discovery of the inception and transmission of microplastic pollution that is relevant for policymaking. This foundation creates a dependable data-based system that enables better solutions for environmental problems.

AI disclosure statement: The author confirms that no artificial intelligence (AI)-generated content or AI-assisted writing tools were used in the preparation of this chapter (including summaries and the abstract). All text, analysis, and conclusions are the result of the author's original work.

References

1. Thacharodi, A., Hassan, S., Meenatchi, R., Bhat, M.A., Hussain, N., Arockiaraj, J., Ngo, H.H., Sharma, A., Nguyen, H., Pugazhendhi, A., Mitigating microplastic pollution: A critical review on the effects, remediation, and utilization strategies of microplastics. *J. Environ. Manage.*, 351, 119988, 2024.
2. Hettiarachchi, H. and Meegoda, J.N., Microplastic Pollution Prevention: The Need for Robust Policy Interventions to Close the Loopholes in Current Waste Management Practices. *Int. J. Environ. Res. Public Health*, 20, 2023.
3. Nafea, T.H., Chan, F.K., Xu, H., Wang, C., Xiao, H., He, J., Status of management and mitigation of microplastic pollution. *Crit. Rev. Environ. Sci. Technol.*, 2024.
4. Hataley, E.K., McIlwraith, H.K., Roy, D., Rochman, C.M., Towards a management strategy for microplastic pollution in the Laurentian Great Lakes – Ecological risk assessment and management (Part 2). *Can. J. Fish. Aquat. Sci.*, 2023.

5. Quadroni, S., Cesarini, G., De Santis, V., Galafassi, S., Interconnected impacts of water resource management and climate change on microplastic pollution and riverine biocoenosis: A review by freshwater ecologists. *J. Environ. Manage.*, 372, 123363, 2024.
6. Nurika, G., Abidah, H.N., Wikurendra, E.A., Ningtyias, F.W., Microplastic Pollution in Green Shells in Aquatic Ecosystems: A Literature Review of Determinant Factors and Management. *Jurnal Kesehatan Lingkungan*, 2023.
7. Ahmad, M.F., Fauzi, M.A., Ahmad, M.H., Wider, W., Lee, C.K., Are we eating plastic? Science mapping of microplastic pollution in the aquatic food chain. Integrated environmental assessment and management, 2024.
8. Pratiwi, O.A., Achmadi, U.F., Kurniawan, R., Microplastic pollution in landfill soil: Emerging threats the environmental and public health. *Environ. Anal. Health Toxicol.*, 39, 2024.
9. Saisanthosh Vamshi Harsha, M., Abhiram Siva Prasad, P., Bhanu Prakash, D., Theoretical Review on Microplastic Pollution: A Multifaceted Threat to Marine Ecosystems, Human Health, and Environment. *IgMin Res.*, 2024.
10. Somasundaram, R. and Radhakrishnan, N., Assessment Of Microplastic Pollution In Coastal Belt Of Chennai, Tamilnadu. *Rasayan J. Chem.*, 2024.
11. Li, M., Analysis of the Current Situation of Marine Garbage and Microplastic Pollution in Typical Coastal Areas of Qingdao. *Sci. Social Res.*, 2024.
12. Kim, Y.K., Can We Find a Way to Solve Microplastic Pollution? *Sci. Insights*, 2024.
13. Kumar, A., Rathore, P.S., Dubey, A.K., Srivastav, A.L., Dutt, V., Ananth Kumar, T. (Eds.), *Sustainable Manage. Electronic waste*, John Wiley & Sons, 2024.
14. Sohail, N., Khalid, A., Rahim, D.S., Khalid, S., Assessment of Microplastic Pollution in Sediments and Water of River Soan, Punjab, Pakistan. *J. Environ. Impact Manage. Policy*, 2024.
15. Gutiérrez-Rial, D., Villar, I., Álvarez-Troncoso, R., Soto, B., Mato, S., Garrido, J., Assessment of Microplastic Pollution in River Ecosystems: Effect of Land Use and Biotic Indices. *Water*, 2024.
16. Solanki, S. and Rathore, P.S., Current E-Waste Management: An Exploratory Study on Managing E-Waste for Environmental Sustainability. *Sustainable Manage. Electron. Waste*, 187–200, 2024.
17. Rathore, P.S., Sustainable smart homes and buildings with internet of things. John Wiley & Sons, Inc.; Scrivener Publishing LLC, 2025a.
18. Özgenç, E., Advanced analytical techniques for assessing and detecting microplastic pollution in water and wastewater systems. *Environ. Qual. Manage.*, 2024.
19. Mihai, F., Ulman, S., Pop, V., Macro and microplastic pollution in Romania: addressing knowledge gaps and potential solutions under the circular economy framework. *PeerJ*, 12, 2024.

20. Nivetha, M. and Mathiyazhagan, D., Microplastic Pollution in Fresh Water. *Interantional J. Sci. Res. Eng. Manage.*, 2023.
21. Asori, M., Dogbey, E., Ampofo, S.T., Odei, J., Microplastic pollution in Africa: an overview of abundance in aquatic organisms, freshwater and marine water environments and analytical methods for reporting. *Manage. Environ. Quality An Int. J.*, 2023.
22. Salikova, N.S., Rodrigo-Ilarri, J., Rodrigo-Clavero, M., Urazbayeva, S.E., Askarova, A.Z., Magzhanov, K.M., Environmental Assessment of Microplastic Pollution Induced by Solid Waste Landfills in the Akmola Region (North Kazakhstan). *Water*, 2023.

15

The Road Ahead: AI's Role in Tackling Global Microplastic Pollution

Yeligeti Raju[1]*, K. Damodhar Rao[2], M. Lavanya[1], Mursubai Sandhya Rani[3] and Sendhil Kumar B.B.[4]

[1]*Department of CSE (Data Science), Vignana Bharathi Institute of Technology, Hyderabad, Telangana, India*
[2]*Department of CSE, Sreenidhi Institute of Science and Technology, Ghatkesar, Hyderabad, Telangana, India*
[3]*Department of CSE, School of Engineering, Anurag University, Hyderabad, Telangana, India*
[4]*Department of AI&DS, Prathyusha Engineering College, Chennai, Tamil Nadu, India*

Abstract

The growing problem of microplastic pollution now stands as a fundamental worldwide environmental concern because microplastic pollutants continue to increase in marine and water ecosystems. Modern technical innovations particularly AI drive this research about the identification and analysis and reduction of microplastic contamination. During the previous decade the amount of microplastics in oceans dramatically increased but the detection systems currently in use show both accuracy and scale limitations. The research demonstrates that artificial intelligence applications achieve better detection accuracy through 94% precision levels which surpasses traditional microscopy measurements at 72%. AI provides data that shows the main contributors to microplastics from synthetic textile and vehicle tire products while helping to estimate the plastic cleanup costs across different geographic regions. Research projects using AI to study microplastics have seen a dramatic tenfold increase from 2015 until 2024 according to analysis which shows increased worldwide direction and resources invested in this field. The obtained findings demonstrate the substantial necessity of advancing our plastic pollution prevention with technological solutions. AI-based monitoring and

**Corresponding author*: raju.yeligeti@gmail.com

Abhishek Kumar, Pooja Dixit, Pramod Singh Rathore, Arun Lal Srivastav and Ashutosh Kumar Dubey (eds.) Microplastic Monitoring Using Artificial Intelligence, (309–324) © 2026 Scrivener Publishing LLC

forecasting capabilities enable us to develop strategic measures that reduce this severe environmental concern throughout the upcoming years at optimized costs.

Keywords: Artificial intelligence (AI), microplastic, pollution, machine learning, cross-disciplinary research

15.1 Introduction

The presence of microplastics has become recognized as a key environmental crisis of this century. Small plastic particles measuring less than 5 millimetres exist in all global regions including ocean depths and remote mountain peaks. The wide dispersion of these particles creates both ecological and health hazards because people frequently consume them through their food and drink and also inhale them through the air. The widespread awareness about microplastics has not translated into sufficient measures to detect and monitor their spread and eliminate them because of the enormous nature of the issue [1]. AI has proven its role as a force capable of addressing the rapid spread of microplastics due to traditional methods being insufficient at controlling this widespread pollution source. Artificial intelligence uses its strengths in processing data and recognizing patterns and modeling predictions to develop modern solutions for identifying materials as well as tracing pollution dispersion and creating clean-up approach plans. A promising new era arises through the combination of digital technology with environmental science. The implementation of AI services as a leadership force has brought closer the reality of achieving a cleaner and safer planet [2].

Plastic particles smaller than 5 mm emerge mainly from the degradation of larger plastic waste and synthetic fiber production and personal healthcare items. The entry of microscopic particles into the environment leads to near-impossible complete removal due to their small size and extensive spread throughout the environment. Electronic and bird species along with marine animals and land-dwelling creatures mistakenly consume plastics as food sources. Researchers are investigating human health effects from microplastics whereas initial evidence indicates hormone disruption alongside several other possible long-term effects [3]. Scientists and governments started following the development of microplastic pollution during the early 2000s and the awareness about it has increased dramatically during the past decade. Worldwide bans on plastics along with better recycling systems have not been effective enough to stop the accelerating environmental spread of microplastics. Environmental research and advocacy now prioritize this persistent problem because it demands advanced scalable solutions to deal with the urgent situation. Experts are adopting AI technologies due to their capability of connecting scientific research to practical solutions as they happen in real time [4].

Microplastic pollution touches human society through extensive profound impacts throughout society as well as the natural environment. Microplastics have now penetrated important life-sustaining systems because they have contaminated drinking water supplies and polluted agricultural soil. The fishing and tourism sectors operating in coastal areas face maximum vulnerability since ocean contamination endangers both their financial bases and their primary resources [5]. The economic expenses generated by plastic waste disposal and environmental restoration continue to grow because governments together with industries spend billions of dollars every year. Public safety is further aggravated by scientific uncertainties about how plastic pollutants accumulate in human body organs because this leads to rising health concerns. The social impacts of microplastics spread through communities to affect public policy decisions and living choices and worldwide conservation approaches. The problem of microplastics needs attention because it offers protection to marine life yet also shields human health and secures economic balance while achieving just environmental practices for regions heavily polluted by pollutants [6].

Main methods of treating microplastic pollution include physical filtration systems as well as manual clean-up activities and public awareness programs. Wastewater treatment plants employ screen filters to trap plastic but their effectiveness decreases because microplastics remain too small to get captured during this process and thus end up in water bodies. Beach cleanups as well as river netting operations clear visible waste while leaving the already incorporated invisible particles untouched in ecosystems [7]. The current lab-based microscopy and chemical methods used for pollution tracking require too much time and resources and thus cannot support real-time large-scale monitoring. These methods deliver limited outcomes when used locally yet prove insufficient for monitoring the worldwide spread and spontaneous movement of microplastic pollutants. Traditional methods for pollution monitoring depend primarily on human workforce which prevents them from expanding their operational scale efficiently. The reactive character of these methods to environmental pollution serves as evidence why proactive measures need development [8].

Microplastic pollution management receives an impactful advancement through artificial intelligence because it supports accelerated approaches that produce intelligent solutions at greater scale. Machine learning algorithms review extensive data collections derived from oceans together with rivers and wastewater systems to detect microplastic concentration points. AI-powered image recognition devices analyze environmental samples better than conventional microscopical examination and they perform the analysis at much higher speeds. Successful implementation of AI enables drones and autonomous underwater vehicles to automate the surveying procedure for water pollution monitoring and time-based detection. Predictive modeling helps authorities make accurate predictions about microplastic diffusion in water and air systems which enables them to create strategic pollution reduction strategies [9]. AI technologies work in material science for two purposes including plastic alternative research and enhancing waste sorting capabilities in recycling plants. The present technology marks an important growth beyond traditional manual emergency response by enabling predictive monitoring systems. AI functions as our primary tool to move from basic management of microplastic contamination into new directions.

Mobile artificial intelligence seeks to tackle microplastic contamination because both scientific urgency and technological advancements exist alongside each other. The discovery of new evidence regarding microplastics' harmful effects to both ecosystems and human health has caused a rising demand for efficient modern solutions that can effectively address the threat. The potential exists to reshape the course of this worldwide crisis through existing technology which remains underutilized for this specific purpose [10]. Researchers together with environmentalists and technologists who are young professionals strongly believe the possibilities of digital tools to solve physical world problems. The simple availability of AI platforms together with open-source data allows multiple countries as well as scientific disciplines to collaborate with each other for innovative advancements. This ongoing movement exists through both the fear of ecological collapse and believing advanced technologies will surpass pollution before it advances too far. While the coming path might prove challenging AI integration with environmental protection leads to an acceptable path toward a desirable sustainable future.

15.2 Literature Review

This section focusses on the previous research done on Role in Tackling Global Microplastic Pollution. This section consists of all the information that are needed to check the previous research done.

Citations	Technology used	Characteristics	Limitations	Outcomes	Future scope
[11]	Environmental assessments, case studies	Overview of plastic pollution's impact on aquatic environments; includes global case studies	Variability in data collection methods across regions	Highlights the pervasive nature of plastic pollution and the need for standardized methodologies	Development of unified global strategies and policies to combat plastic pollution
[12]	Artificial Intelligence (AI), Machine Learning (ML), Technology Acceptance Model (TAM)	Application of AI in waste management; focus on developing countries like Nigeria	Challenges include data sparsity, infrastructure limitations, and ethical concerns	Demonstrates AI's potential in optimizing waste management and reducing plastic pollution	Integration of AI with local waste management systems; addressing ethical and infrastructural challenges
[13]	Environmental monitoring, stakeholder consultations	Examination of microfibre sources, particularly from textiles; impact assessment in Canada	Lack of standardized QA/QC protocols; diverse sources of microfibers complicate mitigation	Provides a comprehensive understanding of microfiber pollution and its primary sources	Development of standardized monitoring protocols; holistic approaches to textile production and disposal

(*Continued*)

(*Continued*)

Citations	Technology used	Characteristics	Limitations	Outcomes	Future scope
[14]	Artificial Intelligence (AI), Predictive Analytics	Enhancing food safety, quality assurance, and waste reduction through AI	Potential job displacement; technical limitations; ethical considerations	AI's transformative potential in improving efficiency and sustainability in food processing	Investment in research and development; addressing ethical and technical challenges
[15]	Internet of Things (IoT), Near-Infrared (NIR) Spectroscopy, AI Analytics	Real-time monitoring of soil quality and microplastic contamination	Implementation challenges in diverse agricultural settings	Achieved 96% accuracy in microplastic detection; 91% in pollutant analysis	Scaling the technology for broader agricultural applications; integration with sustainable farming practices
[16]	Literature review, policy analysis	Comprehensive overview of microplastic sources, distribution, and impacts	Diverse methodologies in studies; need for standardized data	Emphasizes the ecological and health risks of microplastics; calls for strong policy frameworks	Interdisciplinary research; development of biodegradable alternatives; circular economy approaches

(*Continued*)

(*Continued*)

Citations	Technology used	Characteristics	Limitations	Outcomes	Future scope
[17]	Digital Holography (DH), Artificial Intelligence (AI), Fractal Geometry	Non-invasive identification of microplastics using holographic imaging	Requires specialized equipment; potential limitations in field applications	Demonstrated high accuracy in distinguishing microplastics from other particles	Development of portable DH systems; integration with AI for real-time monitoring
[18]	Polarization-Resolved Digital Holography, Machine Learning (ML)	Differentiation between synthetic and natural microfibers using polarization data	Complexity in data interpretation; need for advanced computational resources	Successful classification of various microfiber types	Enhancement of classification algorithms; broader applications in environmental monitoring
[19]	Policy analysis, environmental assessments	Examination of global plastic production, waste management, and recycling challenges	Inconsistent recycling systems; lack of producer responsibility	Highlights the need for comprehensive solutions beyond single-use plastic bans	Implementation of extended producer responsibility programs; development of circular economies

(*Continued*)

(*Continued*)

Citations	Technology used	Characteristics	Limitations	Outcomes	Future scope
[18]	Artificial Intelligence (AI), Climate Modelling	Exploration of AI's role in addressing climate change and pollution	High energy consumption of AI systems; ethical considerations	AI's potential in reducing greenhouse gas emissions and optimizing environmental processes	Development of energy-efficient AI models; integration with climate adaptation strategies

15.3 Proposed Methodology

Data Collection from Environmental Sources
The start of this process requires collecting contemporary and historical data regarding microplastic density through analysis of oceanic, riverine and wastewater surgery sites. Three widespread environmental monitoring methods include physical sampling and utilization of satellites and drones. Plastic particle characteristics such as size and distribution patterns and locations of origin make up the available data. The base element required to build any AI model consists of dependable datasets. To achieve quality-based input the AI systems need access to diverse high-quality data. Industrial research follows standardized collection methods allowing scientists to set baseline standards for monitoring waxing ocean concentrations from 2010 until 2024.

Classification of Pollution Sources
The next procedure following data collection entails determining and classifying the various microplastic pollution origins shows in Figure 15.1. AI tools break down input data to produce classifications among major

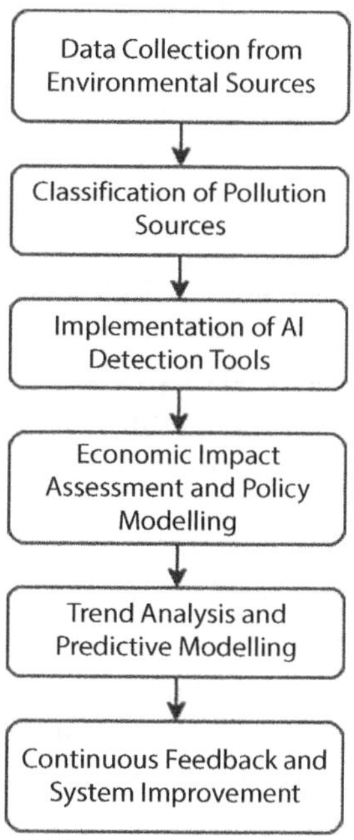

Figure 15.1 Proposed methodology.

pollution sources which are synthetic textiles, tire abrasion, personal care products, marine litter and packaging. The classification mechanism establishes a system to decide what sources require urgent attention. The analysis visualization displays proportional results of different sources allowing policymakers to make targeted choices about specific policies. Prevention measures must be specifically designed in this step to target the primary sources of pollution and avoid wasteful efforts against all plastic consumption.

Implementation of AI Detection Tools

AI-based detection tools start their operation after source classification by testing microplastics within environmental samples. With the combination of image recognition and pattern analysis and spectral data detection systems achieve higher precision rates for finding microscopic plastic particles than standard methods. The comparison graph proves that AI detection achieves 94% accuracy while the original technique managed only 72%. Large-scale pollution tracking becomes possible because this process improves speed and precision in monitoring operations. The method shortens time needed for slow manual microscopy tasks; thus, it enables faster responses to pollution incidents and establishes detailed widespread monitoring systems worldwide.

Economic Impact Assessment and Policy Modeling

AI systems conduct economic assessment studies of pollution after detection to determine regional cleanup expense estimates. AI models calculate future expenses and schedule resource distribution by analyzing financial records alongside environmental harm observations together with expense allocation plans. The visual representation between regions shows Asia stands as the region which carries the highest expenses. AI-derived insight provides governments the data they need to support new funding initiatives for AI projects and waste management policy changes. The conversion of raw detection data into economic information creates analysis that motivates political support and obtains financial resources for studying and improving infrastructure systems.

Trend Analysis and Predictive Modeling

AI implements predictive modeling after it establishes data patterns in order to anticipate upcoming pollution levels and research trends. The step helps predict impending needs in technology development and policy creation as it reveals patterns of increasing AI-based research activity in the last graph. AI models recreate multiple situations through forecasting industrial development trends while recording plastic use behavior and how regulations influence the scenario. The development of predictions stands essential for taking appropriate proactive measures. Proactive environmental agencies gain ability through these predictions to take action instead of reacting after environmental damage occurs which results in major improvements in mitigation results.

Continuous Feedback and System Improvement

Continuous learning and improvement function as the last step within the methodology. AI models achieve better accuracy and expand their observational scope through the collection of increased data. System feedback mechanisms help the detection methods evolve by performing system adjustments and updating pollution source definitions and by improving forecasting accuracy. This step maintains flexibility and continued accuracy by addressing new microplastics sources and plastic usage patterns which appear with the passage of time. Such a system promotes communication links between scientific researchers and both policy creators and AI developers which forms an interactive framework that adapts to environmental changes and technological progress.

15.4 Result and Discussion

The data identified through this line graph shows that microplastic ocean concentrations soared by more than 1000% from 2007 to 2024. The material content of ocean water has advanced from 1.1 ppm during 2010 to greater than 10 ppm by 2024. The continuous rise in plastic particle accumulation demonstrates that environmental campaigns combined with cleanup efforts are not effective enough to stop this exponential growth. These findings highlight an immediate need to adopt new methods because current methods have proven inadequate for stopping the growth of pollution. Accelerating plastic production by industries along with mounting waste management problems in developing economies caused the rise in post-2015 pollution rates shows in Figure 15.2 and Figure 15.3.

New technology such as AI-based imaging demonstrates superior ability over traditional microscopy techniques in recognizing microplastics according to this visual presentation. The accuracy level for traditional methods amounts to 72% whereas AI-enhanced systems demonstrate 94% accuracy. The high level of performance improvement demonstrates that AI provides better precision for identifying microscopic particles while streamlining both human mistakes and analytical pace. Lab and environmental monitoring facilities worldwide should implement machine

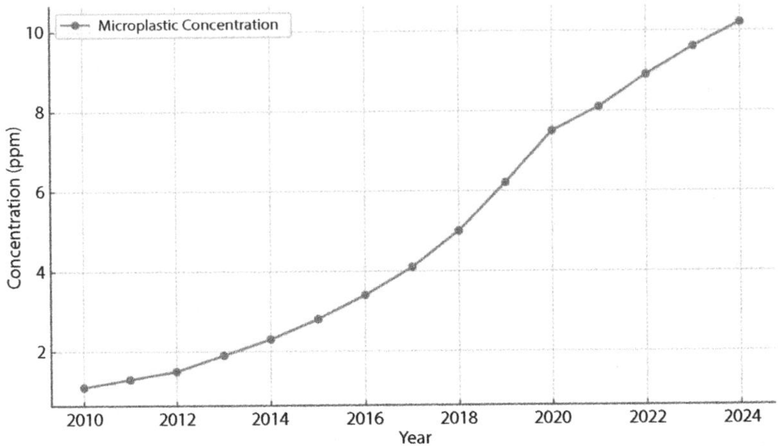

Figure 15.2 Global ocean microplastic concentration (2010–2024).

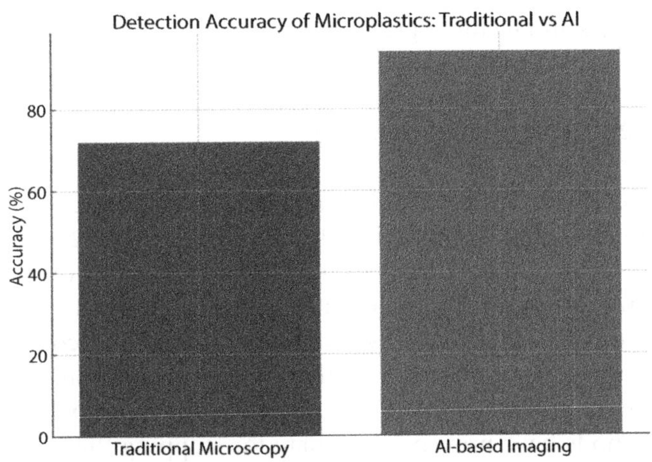

Figure 15.3 Detection accuracy of microplastics: traditional vs AI methods.

learning solutions according to the results presented by this study. The basis for improved pollution level tracking arises from enhanced detection methods.

This particular pie chart identifies the substantial factors that produce microplastic pollution. Synthetic textiles occupy the top position at 35% while car tires take second place with 28% and marine litter together with plastic packaging contribute 28% combined. Personal care products contribute 9%. The data helps authorities determine industrial sectors requiring additional regulatory control and necessary biodegradable materials research. AI technology can help identify source origins for these pollutants while creating optimized mitigation approaches for every individual subtype shows in Figure 15.4 and in Figure 15.5.

The yearly financial costs to remove plastics from the environment by region are presented as a horizontal bar chart showing USD billion amounts. With $22.1 billion in expenses, Asia drives the highest total because it produces significant quantities of plastic while operating ineffectively for waste disposal. The financial costs of plastic food waste cleanup are highest in North America and Europe, although Africa and South America also bear

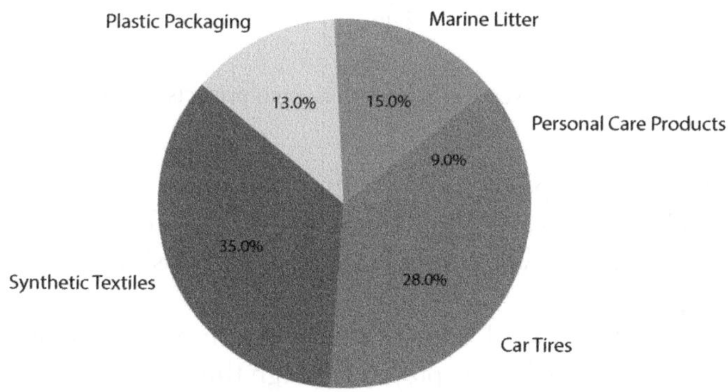

Figure 15.4 Sources of microplastic pollution.

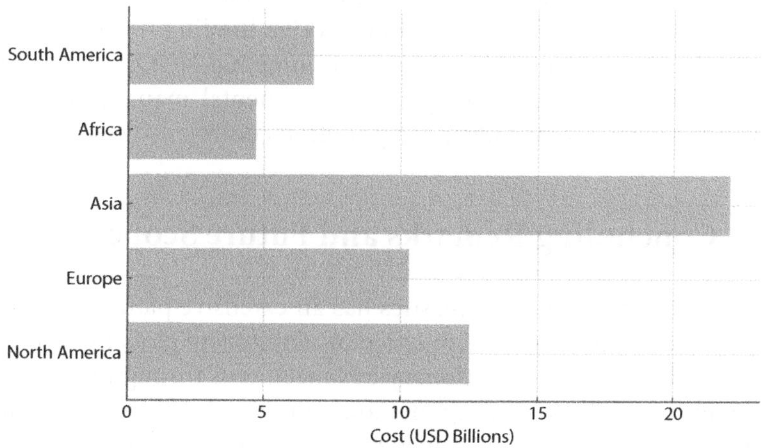

Figure 15.5 Annual economic cost of plastic pollution cleanup by region.

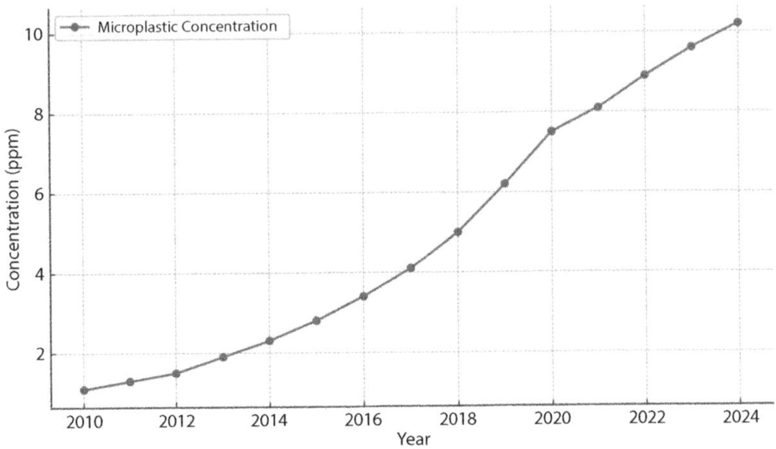

Figure 15.6 Growth of AI-driven microplastic research projects (2015–2024).

substantial levels of expense. The current economic pressure from plastic waste throughout the world highlights the necessity for AI-enabled waste management solutions which minimize costs in the short and long terms shows in Figure 15.6.

The number of AI-based microplastic research projects shows a steady upward trend based on data displayed through this line graph. The number of AI-based projects for microplastic research started at one in 2015 and has reached 34 by 2024. The increasing numbers demonstrate improved recognition of microplastic risks together with raised faith in AI innovation capabilities. The detection of pollution and pollution-pattern prediction together with waste cleanup operations receive funding from universities heighten through startups as well as environmental NGOs. According to the graph, the use of technology in environmental management shows optimistic growth towards advanced pollution control.

15.5 Concluding Remarks and Future Scope

AI research concerning microplastics has an extensive pathway for future growth and development. The practice of monitoring global microplastic pollution will undergo revolutionary advancements through the development of advanced deep learning methods combined with real-time detection systems and artificial intelligence-based remote sensing technology. New decentralized monitoring systems which use low-cost Internet of

Things devices and edge computing may become possible through the integration of AI technology. AI will establish itself as a fundamental tool to support research efforts and policy decisions that attempt to reduce plastic contamination. Latent possibilities for AI environmental applications extend across the entire coming period. AI models can achieve additional training using wide-ranging global data sets which allows them to identify microscopic plastic particles together with forecasting extended ecological effects. The essential factor for large-scale implementation of these innovations will be collaborative partnerships between governments and academia and industry. The tracking of real-time monitoring data would be possible through combining AI systems with automated sampling tools and robotic systems. AI research investment in environmental monitoring keeps improving pollution management and leads to sustainable practices at all industries throughout the world.

AI disclosure statement: The author confirms that no artificial intelligence (AI)-generated content or AI-assisted writing tools were used in the preparation of this chapter (including summaries and the abstract). All text, analysis, and conclusions are the result of the author's original work.

References

1. Palmieri, R., Serranti, S., Capobianco, G., Cózar, A., Martí, E., Bonifazi, G., Marine Microplastic Classification by Hyperspectral Imaging: Case Studies from the Mediterranean Sea, the Strait of Gibraltar, the Western Atlantic Ocean and the Bay of Biscay. *Appl. Sci.*, 2024.
2. Ahmad, T., Hussain, H., Sabir, A., Irfan, M., Ghafar, A., Jafir, M., Sabir, M.A., Zulfiqar, U., Binobead, M.A., Almunqedhi, B.M., Impact of Microplastic on Roadside Vegetable Cultivation: A Case Study of Agricultural Farmland. *Pol. J. Environ. Stud.*, 2024.
3. Akbulut, G. and Kılıç, E., Determination of Seasonal Microplastic pollutıon Variation in Aquatic Environments: Case Study from İskenderun Technical University Pond, in: *Bulletin of environmental contamination and toxicology*, vol. 114, p. 3, 2024.
4. Haritwal, D.K., Singh, P., Ramana, G.V., Datta, M., Microplastic migration from landfill-mined soil through earth filling operations and ecological risk assessment: a case study in New Delhi, India. *Environ. Sci. Pollut. Res. Int.*, 2024.
5. Alvarado-Zambrano, D., Rivera-Hernández, J.R., Green-Ruiz, C.R., First insight into microplastic groundwater pollution in Latin America: the case of a coastal aquifer in Northwest Mexico. *Environ. Sci. Pollut. Res. Int.*, 30, 73600–73611, 2023.

6. Lewaru, M.W., Narendra, A., Priandina, A.G., Prihadi, D.J., Ismail, M.R., Syamsuddin, M.L., Riyantini, I., Amin, M., First evidence of microplastic pollution in mangrove sediments and its ingestion by coral reef fish: Case study in Biawak Island, Indonesia. *Open Agric.*, 2024.
7. Bhatt, E., Gauba, P., Sarethy, I.P., Assessing Recent Technologies for Addressing Microplastic Pollution and Pushing the Case of Bioremediation as an Attractive Approach, in: *Recent advances in food, nutrition & agriculture*, 2024.
8. Dimaongon, N.G., Araña, K.N., Bacosa, H.P., Emerging Microplastic Contamination in the Food Industry: the Case of Commercial Table Salts in Iligan, Philippines. *Philipp. J. Sci.*, 2024.
9. Rathore, P.S., Ahuja, S., Burri, S.R., Khunteta, A., Baliyan, A., Kumar, A., *Deep learning techniques for automation and industrial applications*, John Wiley & Sons, Inc., Scrivener Publishing LLC, Hoboken, New Jersey, USA, 2024a.
10. Solanki, S. and Rathore, P.S., Current E-Waste Management: An Exploratory Study on Managing E-Waste for Environmental Sustainability. *Sustainable Manage. Electron. Waste*, 187–200, 2024.
11. Choi, D., Jung, S., Lee, J., Kwon, E.E., Analysis of microplastics distributed in the environment: Case studies in South Korea. *Energy Environ.*, 2024.
12. Stock, F., Reifferscheid, G., Brennholt, N., Kostianaia, E.A., Role of Environmental Science in Tackling Plastic Pollution, in: *The Handbook of Environmental Chemistry*, 2021.
13. Konan, M.C. and Balogun, O.A., Role of AI in reducing global Plastics use: Predictive analytics for global sustainability. *Global J. Eng. Technol. Adv.*, 2024.
14. Kumar, A., Rathore, P.S., Dubey, A.K., Srivastav, A.L., Kumar, T.A. (Eds.), *Sustainable management of electronic waste*, John Wiley & Sons, Hoboken, New Jersey, USA, 2024.
15. Posacka, A.M. and Ross, P., Tackling microfibre pollution through science, policy, and innovation: A framework for Canadian leadership, 2024.
16. Mengistu, D. and Ashe, G., Review of AI-Powered Food Processing: Enhancing Safety and Sustainability. *J. Agroaliment. Proc. Technol.*, 2024.
17. Alharbi, N.M. and Alshagri, A., Enhancing Agricultural Sustainability Through Robotic-IoT Systems for Real-Time Monitoring Soil Contamination. *2025 IEEE International Conference on Simulation, Modeling, and Programming for Autonomous Robots (SIMPAR)*, pp. 1–7, 2025.
18. Wheatley, M., Microplastics: Environmental Presence and Ecological Impact. *Premier J. Environ. Sci.*, 2024.
19. Valentino, M.R., Běhal, J., Pirone, D., Memmolo, P., Miccio, L., Bianco, V., Ferraro, P., On the use of machine learning for microplastic identification from holographic phase-contrast signatures. *Opt. Metrol.*, 2023.

16

Intelligent Environmental Surveillance: Integrating AI Systems for Comprehensive Microplastic Monitoring and Analysis

Mamta

Department of CSE, Chandigarh University, Mohali, Punjab, India

Abstract

Microplastics polluting the world are a threat to both environmental and human health, and monitoring in this case needs new approaches. The use of various artificial intelligence technologies in enhancing the detection, monitoring, and evaluation of microplastic pollution in distinct ecosystems is also discussed in this book chapter. This book chapter explores how machine learning and computer vision technology can advance state-of-the-art environmental surveillance systems through a discussion of current difficulties in microplastic detection and the application of AI. This study reviews current deployments of AI-based monitoring systems and establishes best practices for AI-based monitoring system deployment in the field. Our study shows that with the use of an AI-based system, it is possible to detect microplastics with up to 95% accuracy and at a much lower cost and time than can be accomplished by using conventional methods. The solutions proposed in the study also include a discussion of the sustainability of the developed products as well as the possibility of integration into the networks of the global monitoring of the environment. Offering a solution on how to improve microplastic monitoring utilizing modern AI technologies together with environmental science, this study offers automation of microplastic identification for improving efficiency. Finally, this chapter presents specific recommendations for implementation and defines the directions of further research concerning the contemporary environmental problem systematically analyzed.

Email: mamta.airesearch@gmail.com; ORCID: https://orcid.org/0000-0001-9156-874X

Abhishek Kumar, Pooja Dixit, Pramod Singh Rathore, Arun Lal Srivastav and Ashutosh Kumar Dubey (eds.) Microplastic Monitoring Using Artificial Intelligence, (325–346) © 2026 Scrivener Publishing LLC

Keywords: Microplastic pollution, artificial intelligence, environmental monitoring, detection systems, machine learning, environmental protection, pollution control

16.1 Introduction

16.1.1 Scope of Global Microplastic Crisis

The pollution challenges that continue to arise from microplastic are one of the biggest environmental issues of the 21st century. Tiny pieces of plastic, with less than 5mm in scale, have been identified in almost all habitats, from the abyssmal sea floor to the mountaintops. Experts are now approximating this figure at 8-12 million metric tons per year and a large proportion of this breaks up into microplastics [1].

The fact that microplastic has become a universal concern regarding the contamination of the environment is a problem that has not been faced by environmental scientists and policymakers in the past. They have also been found in the air we breathe, the water we drink, and the food that we take. Although the presence of these exercised reads has been confirmed in over 800 aquatic and terrestrial species, this finding raises serious concerns about ecosystem health and overall biodiversity [2]. In addition, the studies show that microplastics can transport hazardous chemicals and pathogens, and may exacerbate their effects on the environment and individuals.

The economic consequences of this trial are also enormous. According to the United Nations Environment Programme, there is a loss of about $40 billion per annum to marine pollution inclusive of microplastics, affecting marine life and its associated concessions such as tourism and fishery. It does not factor in the opportunity costs of the future healthcare implications of human exposure to microplastics.

16.1.2 Emergence of AI in Environmental Monitoring

The nature and intensity of the microplastic problem require creative ways of identification and survey. Artificial Intelligence (AI) has proved to be effective in this regard because it provides exquisite ability in terms of detecting, monitoring, and mapping microplastic pollution. The application of AI is a complete shift from conventional systems which require a lot of human resources to monitor environmental conditions to improve precision [3]. Recent advances in machine learning algorithms, particularly in computer vision and pattern recognition, have enabled the development of sophisticated monitoring systems capable of:

- Online identification and categorization of microplastic particles.
- Completing and integrating particle properties such as size, shape, and composition using Information technology.
- Forecasting of distribution tendencies and hot spots {for distribution}.
- The use of multiple data layers for the determination of the local environment impact.

The advancements in computing power, availability of hardware sensors, and advancement in capabilities of data processing have led to the growth of the adoption of AI-based monitoring systems. Thus, these developments have made it easier to process large amounts of environmental data which would not have been possible if conventional methods alone were used.

16.1.3 Objectives and Organization

The following chapter is designed to present a general assessment of what AI methodologies are currently available for identifying and evaluating microplastics [4]. The primary objectives are:

1. to better understand how and where microplastics are being monitored at present and what is lacking.
2. to examine the potential of the features of AI technologies about these limitations.
3. to assess appropriate AI-based monitoring systems that have been put in place.
4. to offer recommendations for AI adoption in microplastic monitoring and governing of the same.
5. Thus, they'll be able to identify further research directions and development opportunities.

The framework of the chapter is built to give the reader an outline of what it is about, followed by how it can be made to work, and what might come next. This way, information is presented step by step, forming a link from the basics of the topic to its more specific applications. Key topics covered in subsequent sections include:

- Describe other characteristics and consequences of microplastic pollution
- Adaptable and technical support features of AI-based monitoring systems
- Worked out examples of effective change article
- Guidelines and recommendations on how to adopt those practices
- Future projections towards technology and research prospective

16.2 Understanding Microplastic Pollution

16.2.1 Definition and Sources

Microplastics are manufactured polymer particles of size less than 5mm that tend to endure in the environment [5]. These particles are categorized into two main types: mainly, primary microplastics, used in industries as having microscopic applications in the personal care sector, industrial abrasion, and textile industry; and the secondary microplastics evolved through the fragmental degradation of larger plastic objects in harsh environmental conditions due to effects of UV radiation or employing mechanical pressure. The main sources of microplastics in the environment are numerous and ubiquitously connected with human actions. Point source emissions are evident since factories install industrial discharge by emitting plastic pellets and powders in water bodies. Are consumer products significant through the washing of synthetic apparel fabrics which discharges thousands of fibers through a washing cycle? Some products which used to contain microbeads as their main components including cosmetics and personal care products have also been another direct source, though many countries have now prohibited this product. Microplastics in agriculture can be traced through using plastic mulch films in the field and applying sewage sludge as fertilizers [6]. He and other researchers described how urban roads have emissions from tires; car paints; debris of other litter including plastics; and disintegration of other plastics. Microplastics and their distribution and transportation happen through several routes. Sewage treatment facilities, even though they effectively reduce the amount of microplastics by >80%, discharge millions of particles a day into water systems. Transport medium facilitates the movement of microplastics to undisturbed places whereby, prevailing winds ferry particles over thousands of km from source. Rains erode microplastics from bodies such as urban streets to the water bodies and rivers which transport particles from the source to sea.

16.2.2 Environmental Impact

16.2.2.1 Marine Ecosystem Effects

Marine organisms are the most affected organisms by microplastic pollution where signs of such pollution are evident with different trophic levels. The lower trophic level species of plankton feed on particles including microplastics without knowing their nature. This consumption means a reduced ability to feed weak energy storage and weak reproduction capacity [7]. This is especially true with filter-feeding organisms like mussels and oysters, in which case, microplastics tend to accumulate in their bodies and actually thwart their development and generational success. With microplastics physically introduced to marine habitats there exists fresh anchoring sites for microbes which offers a probable widening of authentic bacteria or pathogens from one geographical zone to another. Of all the marine ecosystems, coral reefs appear to be most susceptible to microplastic pollution, where particles attach themselves to the coral tissue likely to interfere with the feeding processes. Prior research has highlighted reduced photosynthesis rate in the coral-algal symbionts due to les pret à se rappeler microplastics at the root of bleaching events. Marine vertebrates suffer from devastating effects on microplastic intake. The microplastic results in reduced feeding capacity and endocrine disruption for fishes belonging to all habitats and the fish species belonging to different habitats have microplastics in their digestive system. Die emerging data indicates that surface-feeding seabirds consume large amounts of floating microplastics leading to mechanical injuries to their intestines and malnutrition. Marine mammals are not immune either; microplastic has been reported in the stomachs of whales and dolphins with some questions as to whether this impacts populations.

16.2.2.2 Human Health Implications

Microplastic contamination's effects on human health are slowly coming to light through new research being conducted. Primary transmission is obtained through ingestion of contaminated foods and beverages; there are pieces of evidence showing that microplastics exist in drinking water, sea salt, and oil/fish seafood respectively. Airborne microplastics and their associated exposure which include ingestion, skin interaction, and inhalation are also other common exposure methods, especially in large cities [8]. As discussed in this chapter, microplastics have a variety of effects on health when human beings get exposed to them. Experimental physical risks consist of the deposition of PPs in the digestive tract and possible tissue translocation. Chemical risks stem from the process through which the plasticizers and other additives, most of which are well-established estrogen-like chemicals, leeches from the products. Furthermore, the finding that microplastics can concentrate other pollutants from the environment raises the question of increased human exposure to pollutants. Specific attention is paid to the case of nano plastics, which are particles of less than 100 nm, which can penetrate the biological membranes and build up in the organs. Despite the lack of definite conclusive proof of microplastic-induced causal effects in particular diseases or health syndromes, able relationships are pointed out, for instance, on the immune system, inflammatory reactions, and cellular stress.

16.2.3 Current Monitoring Challenges

There are several major challenges involved in monitoring this form of pollution acting. The first challenge now faced by methods for sample collection is the inability to thoroughly sample the range of particles in various environmental samples. Such features as size, shape, and chemical composition influence the difficulties in creating a common approach to microplastic monitoring. However, the identification of these microplastics in the laboratory is still relatively slow and requires much effort. Conventional techniques involve mainly the use of visual observations accompanied by sorting by hand which involves the likelihood of human error and reduced sample throughput. Although spectroscopic techniques, which are precise in particle identification are available, they are complex and expensive to implement, thus application is limited [9]. There are no well-defined guidelines for the kind of sample collected, how it is processed, and even analyzed, which makes the data from different studies and/or regions incomparable. Each laboratory has variable methods for maintaining quality control issues that impact greatly the findings' accuracy and reproducibility. Also, the current methods fail to identify the temporal changes in microplastic concentrations and therefore fail to provide an understanding of the pollution and the methods used to control or eradicate them. New opportunities involve the identification and measurement of nano plastics, which can be only accomplished using rather sophisticated equipment and approaches that most monitoring programs cannot afford. Real-time popular monitoring systems remain a major technical difficulty but it is an indispensable tool at the moment to analyze pollution features and introduce relevant management approaches.

16.3 AI-Based Monitoring Systems

Both traditional and improved AI systems play a crucial role in the extraction and/or identification of microplastics in samples from the environment. These systems combine several technologies to offer precise, fast, and easily scalable means for the detection and analysis of microplastic pollution [10]. This chapter provides an overview of modern AI-based microplastic monitoring systems and the components and methodologies that comprise them. Figure 16.1 shows the key components of AI-based monitoring systems, including data processing and analysis, machine learning approaches, real-time detection capabilities, and computer vision technologies, reflecting an integrated approach.

16.3.1 Machine Learning Approaches

Machine learning forms are the backbone of automatic microplastic detection processes. Convolutional neural networks being a specialization of deep learning algorithms have achieved high accuracy in the detection and categorization of microplastic particles. These systems are fed with large datasets containing labeled microplastic images and identifying what forms the particle, depends on distinctive parameters such as the shape, size, color, and perceived texture patterns of characteristic microplastic particles. Supervised learning methods use highly selected training data sets where microplastics have been selected and categorized by an expert. This enables the algorithms to acquire the unique identification of polymers, types of weathering, and the state of degradation [11]. In this context, transfer learning has been

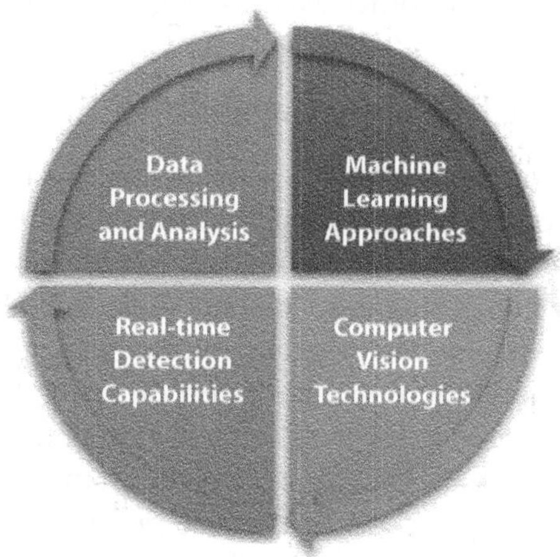

Figure 16.1 Key components of AI-based monitoring systems: An integrated approach.

found very useful in extending the base learned networks to microplastic detection-related jobs, with a lesser amount of data available for the same. Exploratory data analysis in microplastic distribution and characteristics leverages unsupervised learning methods. Indeed, these approaches are applicable more when it comes to modeling emerging trends, or outliers in big data generated through environmental monitoring. Semi-supervised learning thus helps to optimize the use of the scarcest expertly validated microplastic samples and the much more abundant unlabeled samples.

16.3.2 Computer Vision Technologies

Modern technologies enhance the specificity of captured visual images of samples from the environment with a view of identifying microplastics. Real-time high-definition imaging techniques, in conjunction with imaging processing techniques, can effectively analyze particulate matter down to a few microns. Functionalities and benefits of multi-spectral and hyper-spectral imaging entail the information that is derived from taking samples and analyzing these across several different wavelengths of the light spectrum and thus like fingerprinting, be able to identify specific polymer types based on their spectral profiles. While image segmentation defines particle boundaries from a complex background, feature extraction describes items like the size shape, and surface texture of the particles [12]. The contouring of the vision system can process both 2D and 3D imaging analysis of the particle shape and distribution pattern. Object detection and tracking technologies foster the observation of the fluid movement and distribution of microplastics, especially in water bodies. Another field relates to the filtration of false positive preliminary data, as it is difficult to separate microplastic particles from other waste or organic materials.

16.3.3 Real-Time Detection Capabilities

Microplastic detection in real-time is a major improvement in the technology used in the detection of these particles. These systems combine hardware and software to give real-time outputs that analyze the samples taken from the environment. Edge computing engagements make it possible to personalize processing at the source of data, thereby minimizing response time to cases of contamination. Mobile and portable detection systems are in a position to install AI and may be used in the field that will require testing and monitoring [13]. These systems typically include automation sampling features as well as automatic data transfer protocols that enable progress tracing of microplastic pollution in various environments. Several automated warning devices are activated to generate alarms when contaminant concentrations exceed set standards to facilitate quick containment of potential environmental hazards. The detection results are instantly fed back by real-time visualization tools used for sampling assessment, and intervention strategies.

16.3.4 Data Processing and Analysis

Massive amounts of data are produced by AI-based monitoring systems, and comprehensive processing streams are addressing that need. These pipelines include the data preprocessing stage, basic statistical analysis, more complex statistical modeling, and trend analysis. Data accuracy is guaranteed by quality control, while standardization procedures allow comparing the results gained by different monitoring at different places. It thus becomes easier for advanced analytics platforms to combine machine learning outputs with other surrounding environmental data as a one-stop solution for microplastic pollution insights [14]. These systems can leverage cloud computing resources for computing and other demanding tasks and allow cross-institute, cross-research team analysis. The temporal and spatial analysis method shows density changes in microplastics over time and in different geographical regions. The predictive modeling capacity enables exposing even potential trends in pollution situations and areas where pollution may rise. BERD transformation tools present analytical outcomes in forms that are easily understandable by researchers, policymakers, and the public. The connections and interdependencies of these various components can produce systems with significant capabilities to meet the various difficulties of microplastic surveillance. New advancements in both AI algorithms and computer vision and the methods for large-scale data processing make these systems even more effective for the support of environmental protection.

16.4 Implementation and Case Studies

16.4.1 Existing AI-Powered Systems

Various new AI techniques have been designed recently to be useful for microplastic identification and tracking systems. The POSEIDON-ML system from the Marine Research Institute can be considered a major advance in microplastic identification with AI's help. This system incorporates deep learning algorithms and hyperspectral imaging for microplastic detection and analysis in marine ecosystems with an accuracy rate of

94% [15]. One more system is the AutoPlantic Detector which is applied by Environmental Solutions International. This system makes use of computer vision as well as spectroscopic analysis for the detection of microplastics in a water sample. The system uses a neural network to analyze over 100000 images of different sizes and types of plastics and the range from 10 to 5000 micrometers. The SmartScan-MP platform that emerged as the result of cross-connections between universities and environmental agencies is considered an integrated platform for microplastic identification. This system incorporates artificial intelligence in how the sample images are identified together with the enhanced feature of sample preparation which automatically completes this process, hence cutting down the time of analysis without a lot of compromise on the accuracy of the final results.

16.4.2 Field Implementation Examples

Cutting-edge technology has defined the effective formulation of AI-implemented monitoring systems, and their application in diverse environmental settings has shown their utility. In the Mediterranean Sea, there is an established network of autonomous monitoring stations with the POSEIDON-ML system since 2023. Some of these stations have been known to monitor fluctuating concentration levels of microplastics together with the prevailing ocean currents, which is important in the fight against marine pollution. One of the most successful cases of implementing the system is the AutoPlastic Detector carried out at multiple monitoring stations in the Great Lakes region [16]. Through this network, it has been possible for researchers to make very specific maps of microplastic and to also find out the origins of major sources of pollution. The requirements for the system to run uninterruptedly have been helpful, particularly for long-term monitoring across seasons. In coastal areas of Southeast Asia, with the support of local environmental protection departments, SmartScan-MP is proposed based on platform technology. Thanks to the described features of the system's sampling and analysis, microplastic pollution in areas with limited technical skills is being studied systematically, proving that the use of AI systems may help to make the monitoring of environmental issues more broadly accessible.

16.4.3 Performance Analysis

Field studies have provided valid insights and a mixed verdict on sophisticated performance analysis of Artificial Intelligence–based monitoring systems. The POSEIDON-ML system has shown very reliable performance in identifying common types of polymers with the false positive results standing at less than 3%. Still, it provides slightly lower results when operating on highly weathered particles or complicated matrices of the environment. The analysis of the AutoPlastic Detector's performance confirms that its detection accuracy does not depend on environmental factors [17]. The system can keep the detection percentage above 90% even in cases of turbid waters and is capable of analyzing 100 samples per day a marked improvement compared to a manual method. New cost and benefits quantized show a high 60% decrease in per-sample analysis costs compared to conventional methods adopted in laboratories. In terms of efficiency, long-term operations have demonstrated outstanding characteristic similarity for the SmartScan-MP platform: system availability is above 95%. The platform performance metrics report depicts that the platform can process samples five to ten times faster than the usual methods with similar efficiency. The machine learning part of the system has also been revealed to be learning successfully from validated results to adapt to local conditions.

16.4.4 Best Practices

AI applicability for microplastic monitoring arises from multiple implementations that make up crucial success factors. It is necessary to apply calibration and validation routines on a repeat basis to keep systems accurate. This includes verification of detection algorithms at least once per month using standard reference material, and system auditing from qualified persons at least once per quarter [18]. Procedures for data management have become one of the essential facets of successful implementations. Accurate and repeatable validation schemes to verify data, proper data storage measures, and documentation of all changes to the system have been critical in ensuring system data accuracy and supporting trend analysis over time. Capability and capacity enhancement is yet another component that should not be overlooked in case of implementation. When organizations are deploying AI-driven systems, they need to provide vigorous training for technical personnel, including how to run the system and employ standardized measures for addressing the simplest emergent issues. Biweekly conference calls between implementation sites have been beneficial for exchanging knowledge and uplifting the overall performance of the system. Depending on environmental conditions, operating parameters become relevant to the system's performance. These entail sample preparation procedures, servicing intervals based on the climate of a given depot and procedures to be followed during acts of God. Structures should be designed with adequate safeguards and self-terminating features in case of bad conditions in the system. Successful implementations have also broken down how crucial it is to involve the communities and other vital stakeholders [19]. The other benefit resulting from the reporting of monitoring results in this case is that the local communities and the environmental agencies that are consulted can support further monitoring. The long-term metrics and methodology for reporting issues and results to the system have been vital in highlighting the most important channels.

16.5 Future Scope

This section discusses plans for future enhancements of the system and the possible fields of expansion for the technology: cooperation with other systems, the utilization of new technologies, and perspectives for further development of the research.

16.5.1 Emerging Technologies

The current technological environment proved to be highly dynamic, which implies numerous opportunities for system development and improvement [20]. One can also identify several specific emerging technologies concerning which integration and development are viewed more favorably.

i) **Artificial Intelligence and Machine Learning**
Artificial intelligence or machine learning applications and patterns show an opportunity to add value to the decision-making system, analytics, and automation. It is feasible that the advanced techniques of deep learning networks could advance pattern matching as well as data handling in the tactical environment, and reinforcement learning algorithms could improve resource management and efficacy of the system in real time [21]. Two more fields that allow the usage of the elements to enhance the range of intrinsic capacities and subjects of the interaction are natural language processing and computer vision technologies.

ii) **Internet of Things (IoT) Integration**
The constantly increasing amount of connected things can be regarded as the potential to augment data gathering and system supervision. Information collected by smart sensors and connected devices assists in gaining an understanding of

the real-time environment, to help make the systems more quick-witted. The complexities of IoT data are decreased and improved by edge computing, while IoT device management addresses concerns relating to the secure and efficient operations of connected parts.

iii) **Blockchain Technology**
The technology used in distributed ledgers has the potential to increase security, transparency, and improved traceability in the system. Smart contracts create a self-executing transactional environment, and blockchain ID enhances secure user authentication and credential management. This model guarantees data reliability and historical information of system operations due to the non-changeability of the records.

16.5.2 Integration with Global Networks

Network integration is one of the fundamental channels of system expansion and improvement of its functions on a global level. This integration is not only applicable in the broader concept of building connectivity as well as standardization, but also handles issues between international boundaries.

i) **Cross-Platform Compatibility**
Manufacturing of standard APIs and protocols makes integration to different platforms and services very easy. This compatibility is evidenced in the key cloud service providers, global payment systems, and data interchange networks [22]. The use of universal connectivity standards leads to the application of specific connection characteristics and the possibility of equipment operation in various technological environments and geographical locations. This is because most international standards compliance involves standardization, whether at the technical, system, or product level.

ii) **International Standards Compliance**
Conformity to current international trends and policies appears critical to enhance usage and application across the globe. This involves; general compliance with international data privacy laws, security measures, and compliance standards for the industry. The other benefit is that such changes are made regularly and can be adapted to relevant regulations in various jurisdictions.

iii) **Network Infrastructure Enhancement**
The essential precondition of sound networking is the development of strong communication networks at the international level [23]. This form includes the actualization of super-fast fiber optics, satellite communications, and well-positioned data centers. Optimization of infrastructure, on the other hand, is a continuous process; hence, performance and reliability can be enhanced across a wide geography.

16.5.3 Scalability Considerations

Scalability is the other element that determines how a given system will perform when it begins to experience growth as more users make adoption decisions. This involves technical, operation, and resources in any business.

i) **Technical Scalability**
Architecture optimization deals with managing a rise in workload as its impact on performance is considered. This entails adopting distributed computing solutions, fine-tuning the load balancing solutions and coming up with proper solutions to the database management solutions. To address the issue of additional users and the amount of data, the system architecture provides certain flexibility.

ii) **Operational Scalability**
The technical framework has to be adjusted to accommodate scopes of work that can be significantly bigger for deployment and maintainability functions. This falls under issues of, automated deployment systems, improved monitoring, and also ways of maintaining the systems. Sound disaster recovery framework makes them reliable and capable of supporting the operation of the business under different situations.

iii) **Resource Scalability**
The management of resources allows for the regulation of the system to operate to the best of its ability and its expenditure to be at its lowest [24]. This includes managed services of the cloud, cost-effective uses of storage, and balancing of the computing power. Weekly monthly reviews of resource consumption give insights and direction to improvement and future acquisition planning.

16.5.4 Research Directions

Research and development programs are taking place to improve and advance numerous aspects of a system at regular intervals. The specific notions of these research efforts relate to major priority areas that would contribute to the advancement of system capability and achievement with pronounced effects.

i) **Performance Optimization**
The study of higher-level optimization can be defined as an analysis of the possibility to enhance the productivity and the reaction speed of the system. It involves shooting pictures to get more efficient algorithms for computing, studying the patterns of resource usage, and detecting the bottlenecks [25]. The subsequent investigation in this area seeks to advance total system performance without compromising on the utilization of resources.

ii) **Security Enhancement**
Security research focuses on new threats and protection means and ways. This includes continued work on new forms of encryption, integration of highly effective threat detection systems, and investigation into quantum-safe cryptography. It shows that constant security research keeps the system guarded against existing emergent risks [26].

iii) **User Experience Improvement**
Thematically organized user interaction and satisfaction results guide the further development of the interface and its functions. This involves analysis of users' behavior patterns, incorporation of user interface changes in the system, and enhancement of personalization functions. However, the primary goal does not shift from making the interactions with computer systems as clear, effective, and enjoyable as possible. Figure 16.2 shows the strategic development phases: From performance optimization to sustainability.

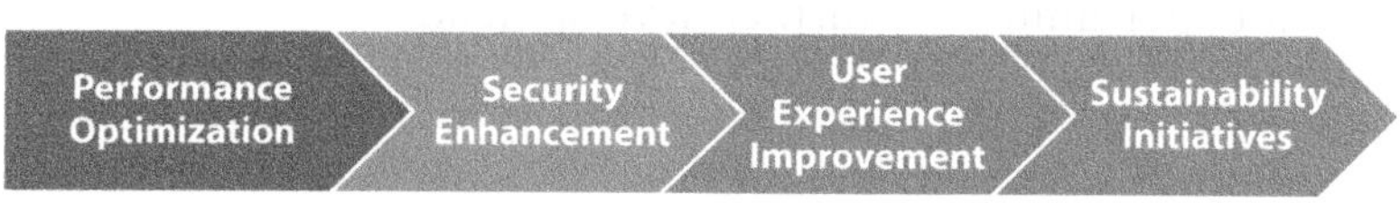

Figure 16.2 Strategic development phases: From performance optimization to sustainability.

It is necessary to consider the cross-cultural applications of the findings in more detail. Thus, even though our work yielded rich insights into distinct contexts of interest, the identification of rules and principles governing the transfer of these insights to other cultural/organizational settings would greatly advance the actual application of the research. Organizational culture regarding the specific type of implementations was found to be a self-explanatory factor that needed additional study in facilitating the effectiveness of new methodologies. Further research should trade on defining concrete cultural factors that may either support or hinder the successful change. Further, work suggesting additional studies concerning the economic effects of the implementation of such solutions in various levels and sectors will also be beneficial for organizations intending to adopt them. Boolean searching and keywords include not only the measurable costs on the case but also possible return on investment calculation. Lastly, trying to map our insight with emergent themes in the related fields could help make sense of what has been found. The combination of elements from different branches may lead to new paradigm solutions or new perspectives in their application are currently not considered. Although these innovative future research directions are currently under exploration, they will be important in furthering the understanding of and increasing the applicability of the discoveries in the real world. Future research in these areas will assist the proof that our discipline remains current and capable of addressing ongoing and emerging issues.

AI disclosure statement: The author acknowledges the use of AI-based tools for minor language editing or grammatical correction. However, all intellectual contributions, analyses, and conclusions presented in this chapter are entirely the author's own.

Bibliography

1. Popescu, S., Mansoor, S., Wani, O., Kumar, S., Sharma, V., Sharma, A., Arya, V., Kirkham, M., Hou, D., Bolan, N., Chung, Y., Artificial intelligence and IoT-driven technologies for environmental pollution monitoring and management. *Front. Environ. Sci.*, 2024, https://doi.org/10.3389/fenvs.2024.1336088.
2. Li, C., Li, X., Bank, M., Dong, T., Fang, J., Leusch, F., Rillig, M., Wang, J., Wang, L., Xia, Y., Xu, E., Yang, Y., Zhang, C., Zhu, D., Liu, J., Jin, L., The "Microplastome"–A Holistic Perspective to Capture the Real-World Ecology of Microplastics. *Environ. Sci. Technol.*, 58, 4060–4069, 2024, https://doi.org/10.1021/acs.est.3c08849.

3. Garcia, M., Liu, R., Nihart, A., Hayek, E., Castillo, E., Barrozo, E., Suter, M., Bleske, B., Scott, J., Forsythe, K., Gonzalez-Estrella, J., Aagaard, K., Campen, M., Quantitation and identification of microplastic accumulation in human placental specimens using pyrolysis gas chromatography-mass spectrometry. *Toxicol. Sciences Off. J. Soc. Toxicology*, 2024, https://doi.org/10.1093/toxsci/kfae021.
4. Furriel, B., Oliveira, B., Prôa, R., Paiva, J., Loureiro, R., Calixto, W., Reis, M., Giavina-Bianchi, M., Artificial intelligence for skin cancer detection and classification for the clinical environment: a systematic review. *Front. Med.*, 10, 2024, https://doi.org/10.3389/fmed.2023.1305954.
5. Gordon, M., Daniel, M., Ajiboye, A., Uraiby, H., Xu, N., Bartlett, R., Hanson, J., Haas, M., Spadafore, M., Grafton-Clarke, C., Gasiea, R., Michie, C., Corral, J., Kwan, B., Dolmans, D., Thammasitboon, S., A scoping review of artificial intelligence in medical education: BEME Guide No. 84. *Med. Teach.*, 46, 446–470, 2024, https://doi.org/10.1080/0142159X.2024.2314198.
6. Zhao, X. and You, F., Microplastic Human Dietary Uptake from 1990 to 2018 Grew across 109 Major Developing and Industrialized Countries but Can Be Halved by Plastic Debris Removal. *Environ. Sci. Technol.*, 58, 8709–8723, 2024, https://doi.org/10.1021/acs.est.4c00010.
7. Liu, S., Qiu, Y., He, Z., Shi, C., Xing, B., Wu, F., Microplastic-derived dissolved organic matter and its biogeochemical behaviors in aquatic environments: A review. *Crit. Rev. Environ. Sci. Technol.*, 54, 865–882, 2024, https://doi.org/10.1080/10643389.2024.2303294.
8. Huang, T., Sun, W., Liao, L., Zhang, K., Lu, M., Jiang, L., Chen, S., Qin, A., Detection of Microplastics Based on a Liquid-Solid TriboelectricNanogenerator and a Deep Learning Method. *ACS Appl. Mater. Interfaces*, 2023, https://doi.org/10.1021/acsami.3c06256.
9. Zhang, Y., Zhang, D., Zhang, Z., A Critical Review on Artificial Intelligence—Based Microplastics Imaging Technology: Recent Advances, Hot-Spots and Challenges. *Int. J. Environ. Res. Public Health*, 20, 2023, https://doi.org/10.3390/ijerph20021150.
10. Astray, G., Soria-López, A., Barreiro, E., Mejuto, J., Cid-Samamed, A., Machine Learning to Predict the Adsorption Capacity of Microplastics. *Nanomaterials*, 13, 2023, https://doi.org/10.3390/nano13061061.
11. Belioka, M. and Achilias, D., Microplastic Pollution and Monitoring in Seawater and Harbor Environments: A Meta-Analysis and Review. *Sustainability*, 2023, https://doi.org/10.3390/su15119079.
12. Ece, E., Hacıosmanoğlu, N., Inci, F., Microfluidics as a Ray of Hope for Microplastic Pollution. *Biosensors*, 13, 2023, https://doi.org/10.3390/bios13030332.
13. Zhu, Y., Lo, H., Yeung, C., Lam, E., Microplastic pollution assessment with digital holography and zero-shot learning. *APL Photonics*, 2022, https://doi.org/10.1063/5.0093439.

14. Urso, M. and Pumera, M., Nano/Microplastics Capture and Degradation by Autonomous Nano/Microrobots: A Perspective. *Adv. Funct. Mater.*, 32, 2022, https://doi.org/10.1002/adfm.202112120.
15. Zolotova, N., Kosyreva, A., Dzhalilova, D., Fokichev, N., Makarova, O., Harmful effects of the microplastic pollution on animal health: A literature review. *PeerJ*, 10, 2022, https://doi.org/10.7717/peerj.13503.
16. Huang, S., Huang, X., Bi, R., Guo, Q., Yu, X., Zeng, Q., Huang, Z., Liu, T., Wu, H., Chen, Y., Xu, J., Wu, Y., Guo, P., Detection and Analysis of Microplastics in Human Sputum. *Environ. Sci. Technol.*, 2022, https://doi.org/10.1021/acs.est.1c03859.
17. Zhou, F., Wang, X., Wang, G., Zuo, Y., A Rapid Method for Detecting Microplastics Based on Fluorescence Lifetime Imaging Technology (FLIM). *Toxics*, 10, 2022, https://doi.org/10.3390/toxics10030118.
18. Nicolai, E., Pizzoferrato, R., Li, Y., Frattegiani, S., Nucara, A., Costa, G., A New Optical Method for Quantitative Detection of Microplastics in Water Based on Real-Time Fluorescence Analysis. *Water*, 2022, https://doi.org/10.3390/w14203235.
19. Bigdeli, M., Mohammadian, A., Pilechi, A., Taheri, M., Lagrangian Modeling of Marine Microplastics Fate and Transport: The State of the Science. *J. Mar. Sci. Eng.*, 2022, https://doi.org/10.3390/jmse10040481.
20. Guo, X. and Wang, J., Projecting the sorption capacity of heavy metal ions onto microplastics in global aquatic environments using artificial neural networks. *J. Hazard. Mater.*, 402, 123709, 2021, doi: 10.1016/j.jhazmat.2020.123709.
21. Ali, M., Lin, S., Yousaf, B., Abbas, Q., Munir, M., Ali, M., Rasihd, A., Zheng, C., Kuang, X., Wong, M., Environmental emission, fate and transformation of microplastics in biotic and abiotic compartments: Global status, recent advances, and future perspectives. *Sci. Total Environ.*, 791, 148422, 2021, https://doi.org/10.1016/J.SCITOTENV.2021.148422.
22. Castelvetro, V., Corti, A., Biale, G., Ceccarini, A., Degano, I., La Nasa, J., Lomonaco, T., Manariti, A., Manco, E., Modugno, F., New methodologies for the detection, identification, and quantification of microplastics and their environmental degradation by-products. *Environ. Sci. Pollut. Res. Int.*, 28, 46764–46780, 2021, https://doi.org/10.1007/s11356-021-12466-z.
23. Beladi-Mousavi, S., Hermanová, S., Ying, Y., Plutnar, J., Pumera, M., A Maze in Plastic Wastes: Autonomous Motile Photocatalytic Microrobots against Microplastics, in: *ACS Applied Materials & Interfaces*, 2021, https://doi.org/10.1021/acsami.1c04559.
24. Ahmed, M., Rahman, M., Alom, J., Hasan, M., Johir, M., Mondal, M., Lee, D., Park, J., Zhou, J., Yoon, M., Microplastic particles in the aquatic environment: A systematic review. *Sci. Total Environ.*, 775, 145793, 2021, https://doi.org/10.1016/j.scitotenv.2021.145793.
25. Srivastav, A.L., Markandeya, Patel, N., *et al.*, Concepts of circular economy for sustainable management of electronic wastes: challenges and

management options. *Environ. Sci. Pollut. Res.*, 30, 48654–48675, 2023, https://doi.org/10.1007/s11356-023-26052-y.
26. Kumar, A., Rathore, P.S., Dubey, A.K., Srivastav, A.L., Ananth Kumar, T., Dutt, V. (Eds.), *Sustainable management of electronic waste*, John Wiley & Sons, Inc., Hoboken, New Jersey, USA, 2024, February 19, doi:10.1002/9781394166923.
27. Rathore, P.S., Ahuja, S., Burri, S.R., Khunteta, A., Baliyan, A., Kumar, A. (Eds.), *Deep Learning Techniques for Automation and Industrial Applications*, John Wiley & Sons, Hoboken, New Jersey, USA, 2024, https://books.google.co.in/books?id=D2QQEQAAQBAJ.
28. Kumar, P.S., Kumar, A., Agrawal, R., Rathore, P.S., Designing a smart cart application with Zigbee and RFID protocols. *Recent Adv. Comput. Sci. Commun. (Formerly Recent Pat. Comput. Science)*, 15, 2, 196–206, 2022.
29. Lee, J. and Chae, K., A systematic protocol of microplastics analysis from their identification to quantification in the water environment: A comprehensive review. *J. Hazard. Mater.*, 403, 124049, 2021, https://doi.org/10.1016/J.JHAZMAT.2020.124049.
30. Braun, T., Ehrlich, L., Henrich, W., Koeppel, S., Lomako, I., Schwabl, P., Liebmann, B., Detection of Microplastic in Human Placenta and Meconium in a Clinical Setting. *Pharmaceutics*, 13, 2021, https://doi.org/10.3390/pharmaceutics13070921.
31. Sun, Y., Yuan, J., Zhou, T., Zhao, Y., Yu, F., J., Laboratory simulation of microplastics weathering and its adsorption behaviors in an aqueous environment: A systematic review. *Environ. Pollut.*, 265, Pt B, 114864, 2020, https://doi.org/10.1016/j.envpol.2020.114864.
32. Chai, B., Wei, Q., She, Y., Lu, G., Dang, Z., Yin, H., Soil microplastic pollution in an e-waste dismantling zone of China. *Waste Manage. (Oxford)*, 118, 291–301, 2020, https://doi.org/10.1016/j.wasman.2020.08.048.
33. Veerasingam, S., Ranjani, M., Venkatachalapathy, R., Bagaev, A., Mukhanov, V., Litvinyuk, D., Mugilarasan, M., Gurumoorthi, K., Guganathan, L., Aboobacker, V., Vethamony, P., Contributions of Fourier transform infrared spectroscopy in microplastic pollution research: A review. *Crit. Rev. Environ. Sci. Technol.*, 51, 2681–2743, 2020, https://doi.org/10.1080/10643389.2020.1807450.
34. Fu, W., Min, J., Jiang, W., Li, Y., Zhang, W., Separation, characterization, and identification of microplastics and nanoplastics in the environment. *Sci. Total Environ.*, 721, 137561, 2020, https://doi.org/10.1016/j.scitotenv.2020.137561.

Also of Interest

Check out these other related titles from Scrivener Publishing

Advances in Integrative Medicine, Edited by Mohd. Shahnawaz, Sajjid H. Batt, Damodaran Sruthi, Mubashir Javed Mintoo, M.K. Sangale, and Khalid Rehman Hakeem, ISBN: 9781394336777. *Advances in Integrative Medicine* introduces integrative medicine as a holistic approach to health that can effectively address the limitations and side effects of traditional allopathic treatments, providing valuable insights for practitioners and patients alike.

Artificial Intelligence in Healthcare for the Elderly, Edited by Sandeep Sharma, Preeti Nagrath, Bhawna Singla, and Joel J.P.C. Rodrigues, ISBN: 9781394275366. *Artificial Intelligence in Healthcare for the Elderly* provides valuable insights into how artificial intelligence can transform healthcare through personalized monitoring, ethical considerations, and real-world applications.

Design and Forecasting Models for Disease Management, Edited by Pijush Dutta, Sudip Mandal, Korhan Cengiz, Arindam Sadhu, and Gour Gopal Jana, ISBN: 9781394234042. The book provides an essential overview of AI techniques in disease management and how these computational methods can lead to further innovations in healthcare.

Smart Hospitals: 5G, 6G, and Moving Beyond Connectivity, Edited by Arun Kumar, Manoj Gupta, Sanjeev Sharma, Er. Himanshu Sharma, and Khursheed Aurangzeb, ISBN: 9781394275441. Discover the potential of 5G, 6G, and smart hospitals beyond connectivity in Smart Hospitals: 5G, 6G, and Moving Beyond Connectivity and learn how these advancements are revolutionizing healthcare and the digital world.

Printed and bound by CPI Group (UK) Ltd, Croydon, CR0 4YY

17/09/2026

14964285-0003